T5-CQF-323

MTP International Review of Science

# Biochemistry
# Series One

Consultant Editors
**H. L. Kornberg, F.R.S. and**
**D. C. Phillips, F.R.S.**

# Publisher's Note

The MTP International Review of Science is an important venture in scientific publishing, which is presented by Butterworths in association with MTP Medical and Technical Publishing Co. Ltd. and University Park Press, Baltimore. The basic concept of the Review is to provide regular authoritative reviews of entire disciplines. Chemistry was taken first as the problems of literature survey are probably more acute in this subject than in any other. Biochemistry and Physiology followed naturally. As a matter of policy, the authorship of the Review of Science is international and distinguished, the subject coverage is extensive, systematic and critical.

The Review has been conceived within a carefully organised editorial framework. The overall plan was drawn up and the volume editors appointed by seven consultant editors. In turn, each volume editor planned the coverage of his field and appointed authors to write on subjects which were within the area of their own research experience. No geographical restriction was imposed. Hence the 500 or so contributions to the Review of Science come from many countries of the world and provide an authoritative account of progress.

Biochemistry Series One (12 volumes) and Physiology Series One (8 volumes) are being published in the period 1973–1975. The 33 text volumes and 3 index volumes comprising Series One of Inorganic Chemistry, Physical Chemistry and Organic Chemistry were published in 1972–1973. In accordance with the stated policy of issuing regular reviews to keep the series up to date, volumes of Chemistry Series Two are being published in the period 1975–1976. In Biochemistry Series One, Physiology Series One and Chemistry Series Two a subject index is incorporated in each volume and there is no separate index volume.

Butterworth & Co. (Publishers) Ltd.

## ORGANIC CHEMISTRY SERIES TWO

Consultant Editor
D. H. Hey, F.R.S., *formerly of the Department of Chemistry, King's College, University of London*

*Volume titles and Editors*

1 **STRUCTURE DETERMINATION IN ORGANIC CHEMISTRY**
Professor L. M. Jackman, *Pennsylvania State University*

2 **ALIPHATIC COMPOUNDS**
Professor N. B. Chapman, *University of Hull*

3 **AROMATIC COMPOUNDS**
Professor H. Zollinger, *Eidgenossische Technische Hochschule, Zurich*

4 **HETEROCYCLIC COMPOUNDS**
Professor K. Schofield, *University of Exeter*

5 **ALICYCLIC COMPOUNDS**
Professor D. Ginsburg, *Technion-Israel Institute of Technology, Haifa*

6 **AMINO ACIDS, PEPTIDES AND RELATED COMPOUNDS**
Professor H. N. Rydon, *University of Exeter*

7 **CARBOHYDRATES**
Professor G. O. Aspinall, *York University, Ontario*

8 **STEROIDS**
Dr. W. F. Johns, *G. D. Searle & Co., Chicago*

9 **ALKALOIDS**
Professor K. Wiesner, F.R.S., *University of New Brunswick*

10 **FREE RADICAL REACTIONS**
Professor W. A. Waters, F.R.S., *University of Oxford*

---

## INORGANIC CHEMISTRY SERIES TWO

Consultant Editor
H. J. Emeléus, C.B.E., F.R.S. *Department of Chemistry University of Cambridge*

*Volume titles and Editors*

1 **MAIN GROUP ELEMENTS—HYDROGEN AND GROUPS I–III**
Professor M. F. Lappert, *University of Sussex*

2 **MAIN GROUP ELEMENTS—GROUPS IV AND V**
Dr. D. B. Sowerby, *University of Nottingham*

3 **MAIN GROUP ELEMENTS—GROUPS VI AND VII**
Professor V. Gutmann, *Technical University of Vienna*

4 **ORGANOMETALLIC DERIVATIVES OF THE MAIN GROUP ELEMENTS**
Professor B. J. Aylett, *Westfield College, University of London*

5 **TRANSITION METALS—PART 1**
Professor D. W. A. Sharp, *University of Glasgow*

6 **TRANSITION METALS—PART 2**
Dr. M. J. Mays, *University of Cambridge*

7 **LANTHANIDES AND ACTINIDES**
Professor K. W. Bagnall, *University of Manchester*

8 **RADIOCHEMISTRY**
Dr. A. G. Maddock, *University of Cambridge*

9 **REACTION MECHANISMS IN INORGANIC CHEMISTRY**
Professor M. L. Tobe, *University College, University of London*

10 **SOLID STATE CHEMISTRY**
Dr. L. E. J. Roberts, *Atomic Energy Research Establishment, Harwell*

---

## PHYSICAL CHEMISTRY SERIES TWO

Consultant Editor
A. D. Buckingham, F.R.S., *Department of Chemistry University of Cambridge*

*Volume titles and Editors*

1 **THEORETICAL CHEMISTRY**
Professor A. D. Buckingham, F.R.S., *University of Cambridge* and Professor C. A. Coulson, F.R.S., *University of Oxford*

2 **MOLECULAR STRUCTURE AND PROPERTIES**
Professor A. D. Buckingham, F.R.S., *University of Cambridge*

3 **SPECTROSCOPY**
Dr. D. A. Ramsay, F.R.S.C., *National Research Council of Canada*

4 **MAGNETIC RESONANCE**
Professor C. A. McDowell, F.R.S.C., *University of British Columbia*

5 **MASS SPECTROMETRY**
Professor A. Maccoll, *University College, University of London*

6 **ELECTROCHEMISTRY**
Professor J. O'M Bockris, *The Flinders University of S. Australia*

7 **SURFACE CHEMISTRY AND COLLOIDS**
Professor M. Kerker, *Clarkson College of Technology, New York*

8 **MACROMOLECULAR SCIENCE**
Professor C. E. H. Bawn, C.B.E., F.R.S., *formerly of the University of Liverpool*

9 **CHEMICAL KINETICS**
Professor D. R. Herschbach *Harvard University*

10 **THERMOCHEMISTRY AND THERMO-DYNAMICS**
Dr. H. A. Skinner, *University of Manchester*

11 **CHEMICAL CRYSTALLOGRAPHY**
Professor J. M. Robertson, C.B.E., F.R.S., *formerly of the University of Glasgow*

12 **ANALYTICAL CHEMISTRY—PART 1**
Professor T. S. West, *Imperial College, University of London*

13 **ANALYTICAL CHEMISTRY—PART 2**
Professor T. S. West, *Imperial College, University of London*

**BIOCHEMISTRY SERIES ONE**

Consultant Editors
H. L. Kornberg, F.R.S.
*Department of Biochemistry University of Leicester* and
D. C. Phillips, F.R.S., *Department of Zoology, University of Oxford*

*Volume titles and Editors*

1 **CHEMISTRY OF MACRO-MOLECULES**
Professor H. Gutfreund, *University of Bristol*

2 **BIOCHEMISTRY OF CELL WALLS AND MEMBRANES**
Dr. C. F. Fox, *University of California, Los Angeles*

3 **ENERGY TRANSDUCING MECHANISMS**
Professor E. Racker, *Cornell University, New York*

4 **BIOCHEMISTRY OF LIPIDS**
Professor T. W. Goodwin, F.R.S., *University of Liverpool*

5 **BIOCHEMISTRY OF CARBO-HYDRATES**
Professor W. J. Whelan, *University of Miami*

6 **BIOCHEMISTRY OF NUCLEIC ACIDS**
Professor K. Burton, F.R.S., *University of Newcastle upon Tyne*

7 **SYNTHESIS OF AMINO ACIDS AND PROTEINS**
Professor H. R. V. Arnstein, *King's College, University of London*

8 **BIOCHEMISTRY OF HORMONES**
Professor H. V. Rickenberg, *National Jewish Hospital & Research Center, Colorado*

9 **BIOCHEMISTRY OF CELL DIFFER-ENTIATION**
Professor J. Paul, *The Beatson Institute for Cancer Research, Glasgow*

10 **DEFENCE AND RECOGNITION**
Professor R. R. Porter, F.R.S., *University of Oxford*

11 **PLANT BIOCHEMISTRY**
Professor D. H. Northcote, F.R.S., *University of Cambridge*

12 **PHYSIOLOGICAL AND PHARMACO-LOGICAL BIOCHEMISTRY**
Dr. H. K. F. Blaschko, F.R.S., *University of Oxford*

---

**PHYSIOLOGY SERIES ONE**

Consultant Editors
A. C. Guyton,
*Department of Physiology and Biophysics, University of Mississippi Medical Center* and
D. F. Horrobin,
*Department of Physiology, University of Newcastle upon Tyne*

*Volume titles and Editors*

1 **CARDIOVASCULAR PHYSIOLOGY**
Professor A. C. Guyton and Dr. C. E. Jones, *University of Mississippi Medical Center*

2 **RESPIRATORY PHYSIOLOGY**
Professor J. G. Widdicombe, *St. George's Hospital, London*

3 **NEUROPHYSIOLOGY**
Professor C. C. Hunt, *Washington University School of Medicine, St. Louis*

4 **GASTROINTESTINAL PHYSIOLOGY**
Professor E. D. Jacobson and Dr. L. L. Shanbour, *University of Texas Medical School*

5 **ENDOCRINE PHYSIOLOGY**
Professor S. M. McCann, *University of Texas*

6 **KIDNEY AND URINARY TRACT PHYSIOLOGY**
Professor K. Thurau, *University of Munich*

7 **ENVIRONMENTAL PHYSIOLOGY**
Professor D. Robertshaw, *University of Nairobi*

8 **REPRODUCTIVE PHYSIOLOGY**
Professor R. O. Greep, *Harvard Medical School*

MTP International Review of Science

# Biochemistry
# Series One

## Volume 5
## Biochemistry of Carbohydrates

Edited by **W. J. Whelan**
University of Miami

**Butterworths** · London
**University Park Press** · Baltimore

THE BUTTERWORTH GROUP

ENGLAND
Butterworth & Co (Publishers) Ltd
London: 88 Kingsway, WC2B 6AB

AUSTRALIA
Butterworths Pty Ltd
Sydney: 586 Pacific Highway, NSW 2067
Also at Melbourne, Brisbane, Adelaide and Perth

NEW ZEALAND
Butterworths of New Zealand Ltd
Wellington: 26–28 Waring Taylor Street, 1

SOUTH AFRICA
Butterworth & Co (South Africa) (Pty) Ltd
Durban: 152–154 Gale Street

ISBN 0 408 70499 3

UNIVERSITY PARK PRESS
U.S.A. and CANADA
University Park Press
Chamber of Commerce Building
Baltimore, Maryland, 21202



**Library of Congress Cataloging in Publication Data**

Main entry under title:
Biochemistry of carbohydrates.

(Biochemistry, series one; v. 5) (International review of science)
Includes index.
1. Carbohydrates. I. Whelan, William Joseph. II. Series. III. Series: International review of science. [DNLM: 1. Carbohydrates. 2. Carbohydrates—Metabolism. W1 BI633 ser. 1 v. 5/QU75 B615]
QP501.B527 vol. 5 [QP701] 574.1′92′08s
[574.1′9248]
ISBN 0–8391–1044–8 75–8754

First Published 1975 and © 1975



Typeset, printed and bound in Great Britain by
REDWOOD BURN LIMITED
Trowbridge & Esher

# Consultant Editors' Note

The MTP International Review of Science is designed to provide a comprehensive, critical and continuing survey of progress in research. Nowhere is such a survey needed as urgently as in those areas of knowledge that deal with the molecular aspects of biology. Both the volume of new information, and the pace at which it accrues, threaten to overwhelm the reader: it is becoming increasingly difficult for a practitioner of one branch of biochemistry to understand even the language used by specialists in another.

The present series of 12 volumes is intended to counteract this situation. It has been the aim of each Editor and the contributors to each volume not only to provide authoritative and up-to-date reviews but carefully to place these reviews into the context of existing knowledge, so that their significance to the overall advances in biochemical understanding can be understood also by advanced students and by non-specialist biochemists. It is particularly hoped that this series will benefit those colleagues to whom the whole range of scientific journals is not readily available. Inevitably, some of the information in these articles will already be out of date by the time these volumes appear: it is for that reason that further or revised volumes will be published as and when this is felt to be appropriate.

In order to give some kind of coherence to this series, we have viewed the description of biological processes in molecular terms as a progression from the properties of macromolecular cell components, through the functional interrelations of those components, to the manner in which cells, tissues and organisms respond biochemically to external changes. Although it is clear that many important topics have been ignored in a collection of articles chosen in this manner, we hope that the authority and distinction of the contributions will compensate for our shortcomings of thematic selection. We certainly welcome criticisms, and solicit suggestions for future reviews, from interested readers.

It is our pleasure to thank all who have collaborated to make this venture possible—the volume editors, the chapter authors, and the publishers.

Leicester H. L. Kornberg

Oxford D. C. Phillips

# Preface

The task that I undertook in securing a compilation of reviews on the biochemistry of carbohydrates was both easy and difficult. Easy because of the wide-ranging terrain under survey, presenting numerous significant topics for selection. Difficult because of the constraints of space and the need to assign priorities to the possible topics of review. The selection reflects the editor's bias from within those topics, plus his unbiased opinion that carbohydrates are quite the most important of biological molecules, beginning from standpoint that the nucleic acids are substituted polysaccharides.

Biochemical studies of carbohydrates, and particularly glycogen and starch, have brought to light many findings that have had repercussions through the whole of biology, for example the discovery of cyclic AMP and regulation of metabolism by protein phosphorylation/dephosphorylation. Given that continuing seminal findings of this nature continue to emanate from studies of glycogen and starch metabolism, it was essential that reviews of both polysaccharides be included (Chapters 6 and 7).

There is one aspect of polysaccharides where an early opportunity to conduct pioneering research was not taken up. This concerns three-dimensional structure and the influence of the higher orders of structure on the physical and biological properties of polysaccharides. Physical studies of starch and cellulose, made 40 years ago, had revealed their helical and linear structures, respectively, and C.S. Hanes in 1937 made a far-sighted correlation between the helical structure of the starch chain and the action pattern of α-amylase. However, the study of secondary, tertiary and quaternary structure in relation to polysaccharide properties dropped from sight until relatively recently. It was more tempting to turn to nucleic acids and proteins, which looked more amenable to structural description, than did the apparently amorphous polysaccharides, seemingly lacking in anything of a fixed nature other than primary structure, with doubt even whether there was anything of a regular repeating nature in the primary structures. The editor considers that it has been a disadvantage that natural product chemists, himself included, working with polysaccharides, have been preconditioned to believe that if a substance cannot be crystallised, then the next best state in which to isolate it is as a fine, white, free-flowing powder. In order to achieve this end, they will subject the polysaccharide to brutalities besides which the extremes of the Spanish Inquisition pale into insignificance. One need only think of the Pflüger method of obtaining glycogen. Isolated polysaccharides therefore are almost always not only chemically degraded, but thoroughly denatured, liable to display the same non-physiological behaviour as does a denatured protein. Fortunately for the polysaccharides, and carbohydrate biochemistry in general, a new look at the subject of three-dimensional structure was pioneered by Dr. D. A. Rees, the author of the first chapter.

Glycolysis, the oldest known metabolic pathway, also continues to throw up new facets of metabolic control, and when reviewed (Chapter 5), as it must be, in conjunction with gluconeogenesis, we see examples of virtually every known form of metabolic regulation. Generalisations are also in order for inborn errors of carbohydrate metabolism (Chapter 8), where the pin-pointing of the lesion in terms of an absent enzyme has proved possible in numerous cases and where, for example, the first case of a lysosomal storage disease was described.

Returning to structural considerations, the vastly documented connection between phenomena occurring at the cell surface, and that surface's complement of specific carbohydrate structures linked to protein or lipid, has been revealing of the fundamental processes involved in specific adhesion, contact phenomena, morphogenesis and carcinogenesis (Chapter 2). Structures of basically similar chemical composition, linked to protein, i.e. the proteoglycans, primarily molecules of the extracellular space, are extremely important in vertebrates for their contribution to the overall structure of the organism, and the authors of Chapters 3 and 4 cover the subject in an integrated manner by dealing with structure and biosynthesis. These three chapters (2–4), where hetero-oligosaccharides and heteropolysaccharides feature prominently, provide examples of an area of carbohydrate research where the complexity of structure was once a deterrent to the potential investigator. New mèthodology, and persistence by pioneering investigators, among them our reviewers, have opened this exciting field, which includes situations where carbohydrates are seen to carry and transmit information in as clear a fashion as the genetic code.

My warmest thanks go out to Drs. D. A. Rees, M. M. Burger, K. W. Talmadge, L. Rodèn, N. B. Schwartz, H. Muir, T. E. Hardingham, H. A. Lardy, M. G. Clark, D. French, E. G. Krebs, J. Preiss, D. H. Brown and B. I. Brown, for providing this compellingly fascinating overview of carbohydrate biochemistry, and to Drs. H. L. Kornberg and D. C. Phillips for their invitation to me to act as editor.

Miami W. J. Whelan

# Contents

# 1
# Stereochemistry and Binding Behaviour of Carbohydrate Chains

**D. A. REES**
Unilever Research, Bedford



## 1.1 GENERAL INTRODUCTION

This chapter was first planned to cover the subject of 'Polysaccharide Structure', but perhaps more interesting than structure itself, and how it is determined, is a subject that is wider and more difficult to write about, namely the relation between structure and function. Recent interest has shifted from all-carbohydrate macromolecules to the carbohydrate parts of complex biopolymers which contain protein or lipid residues in addition. 'Clean polysaccharides' are now important, not only for traditional reasons, but also as models with which ideas and techniques can be developed for application to more elaborate substances whose chemistry and biology are, for the moment, the objects of considerable biological speculation and fascination.

For these reasons we shall attempt to review the relation between the structure of carbohydrate chains, whether or not they are bonded to non-carbohydrate material, and properties that might point to biological function. This subject is still only vaguely understood and any general conclusions must be very preliminary. The most we can hope for is that this review will serve as a staging post to the satisfying picture that must surely emerge within the next 5–10 years. In any case, so many excellent and up-to-date reviews[1-18] cover between them the aspects of 'polysaccharide structure' that it would be pointless to cover this ground again.

## 1.2 CONFORMATION AND FUNCTION—THE GENERAL ISSUES

### 1.2.1 The state of present understanding

For chains of certain types, our knowledge of conformation is considerable and has indeed been reviewed in a companion volume[18]. This chapter will therefore include selected conformations only, to illustrate particular arguments.

In contrast, very little is known about detailed biological function and its basis in conformation. Some loose correlations can be perceived: cellulose, for example, is the major component of the skeleton of land plants and the strength and rigidity for this function can be traced to the ribbon-like and eminently packable conformation which has hydroxy groups accessible for efficient hydrogen bonding. The chains therefore exist in thread-like crystallites which are dense and strong[19]. In the vertebrate body the film-forming and viscous characteristics of fluid barriers that protect and lubricate inner surfaces which communicate with the environment can be correlated with the expanded conformations of branched, polyanionic glycoproteins that are characteristically present[2]. The properties of connective tissues, including mechanical properties, derive to a significant extent from a high water content which in turn depends on the osmotic behaviour of proteoglycan molecules trapped within the meshwork of collagen. This osmotic behaviour is a function of the charge density and structure of the carbohydrate chains[20]. Conceivably the conformational properties of these chains are involved in binding or trapping the proteoglycan within the meshwork.

Many other examples could be given of carbohydrate chains having functions in structure and support, lubrication and rheological properties, water binding, and in holding ions and other species in the neighbourhood of cells.

It is generally believed that these functions are unlikely to employ such intricate interactions as do the functions of those other biopolymers, the proteins and nucleic acids. Nevertheless, the argument of this chapter will be that carbohydrate structures probably contain more fine details having biological significance than we have so far been able to appreciate. A statement such as 'this polysaccharide forms a matrix to hold the extracellular, aqueous phase' may be as true and yet as superficial as a statement like 'this protein is a catalyst', which credits the protein with a catalytic role that is no more subtle than that of, say, a hydrogen ion. Enzyme catalysis is actually more specific than proton catalysis and its control by interaction with other molecules is essential for life. The corresponding subtleties in polysaccharide function, if they exist, are hardly visible in 1975.

### 1.2.2 Why are polysaccharide structures so elaborate?

Our understanding of function does not yet suggest any selection pressures to explain the evolution of the complexity and wide variety of known polysaccharides. Perhaps these pressures have never existed and all the embellishments of structure are by-products of the mainstream of evolution. This question becomes more challenging as more knowledge is acquired of the conformations of heteropolysaccharides, as we shall illustrate by the relationships in the series of polysaccharides derived from covalent structure (1).

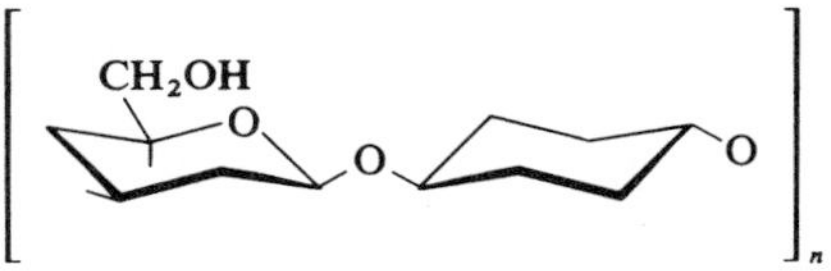

$CH_2OH$

(1)

These include the glycosaminoglycans, whose chemistry[21] and biosynthesis[22] is reviewed elsewhere in this volume, and certain marine algal polysaccharides[23-25] such as agarose and carrageenans.

The algal polysaccharides form double helices (Section 1.2.3)[26-28] which may be either right or left handed, whereas the glycosaminoglycan conformations are single helices which are related in that they are left handed and differ like states of extension and twist of a wire spring (Figure 1.1 and Table 1.1). Some polysaccharides such as hyaluronic acid can be persuaded to adopt as many as four out of five of the known types of conformation (Table 1.1)[30]. Conversely, some conformation types such as the two-fold or three-fold single helices are accessible to all or almost all of the glycosaminoglycans (Table 1.1).

When polysaccharides adopt conformations of the same type, their shapes need not become absolutely identical. For example, conformation (f) is longer in pitch for chondroitin 6-sulphate than it is for dermatan sulphate (Table 1.1). Even when glycans are alike in their ranges of accessible conformations, their preferences for particular states would of course differ under a given set of conditions.

**Table 1.1 Conformations of alternating polysaccharides**

| *Polysaccharide* | *Regular part of structure* | *Form used for investigation* | *Conformations (for key, see Figure 1.1) and the values for the length (in ångstrom units) of each disaccharide residue projected on to the helix axis* (a) | (b) | (c) | (d) | (e) | (f) | (g) | *Refs.* |
|---|---|---|---|---|---|---|---|---|---|---|
| *ι*-Carrageenan | 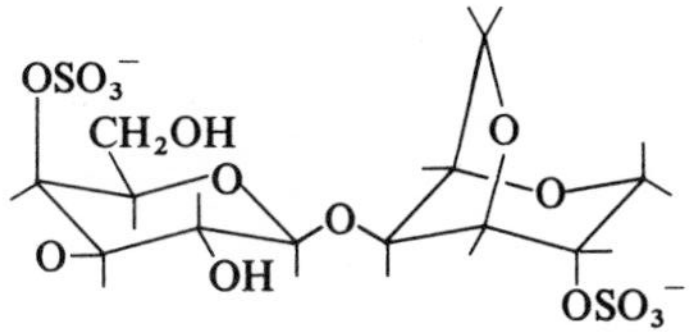 $OSO_3^-$, $CH_2OH$, O, O, OH, O, O, O, $OSO_3^-$ | Various salts | 8.7 | | | | | | | 26, 27 |
| *κ*-Carrageenan | 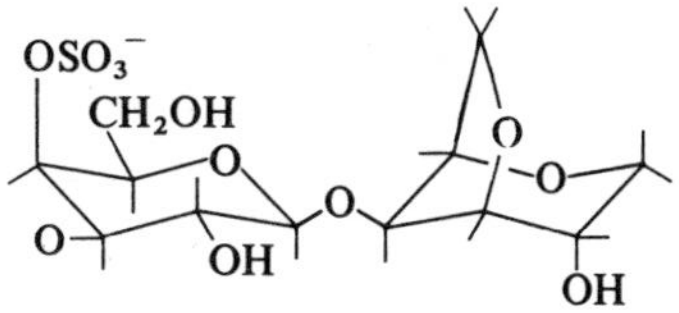 $OSO_3^-$, $CH_2OH$, O, O, OH, O, O, O, OH | Various salts | 8.2 | | | | | | | 26 |
| Agarose | 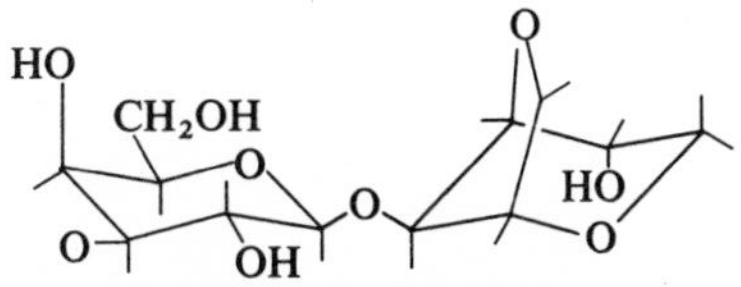 HO, $CH_2OH$, O, O, OH, O, O, HO, O | Various derivatives | | 6.3 | | | | | | 28 |

| Polysaccharide | Structure | Form | Value 1 | Value 2 | Value 3 | Value 4 | Value 5 | Ref. |
|---|---|---|---|---|---|---|---|---|
| Hyaluronic acid | 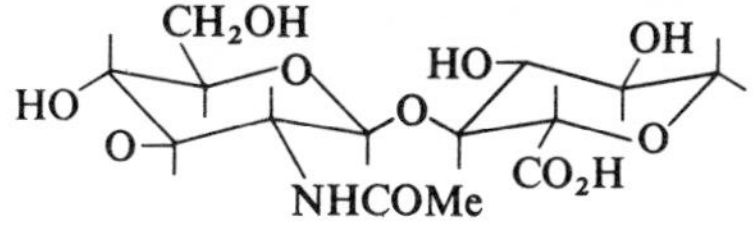 $CH_2OH$, HO, O, O, NHCOMe, HO, OH, O, $CO_2H$ | Univalent salt | 8.4 | 9.3 | 9.5 | | | 30a, 48 |
| | | Free acid | | | | | 9.8 | 48 |
| Chondroitin 6-sulphate | 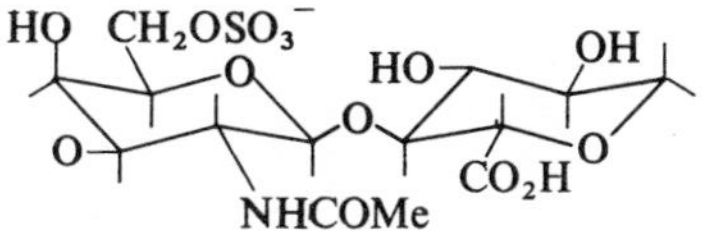 HO, $CH_2OSO_3^-$, O, O, NHCOMe, HO, OH, O, $CO_2H$ | Univalent salt | | | 9.5 | 9.8 | | 49, 50 |
| | | Free acid | | | | | 9.3 | 50 |
| Chondroitin 4-sulphate | 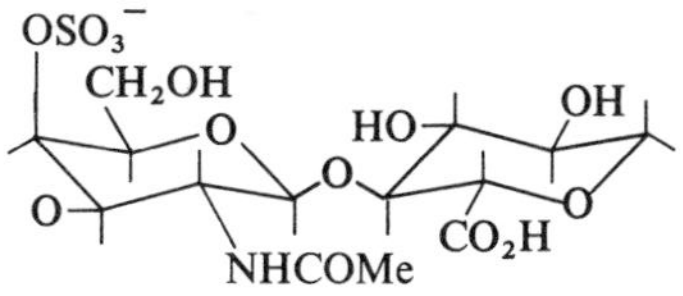 $OSO_3^-$, $CH_2OH$, O, O, NHCOMe, HO, OH, O, $CO_2H$ | Univalent salt | | | 9.6 | | | 51 |
| | | Free acid | | | | | 9.8 | 51 |
| Dermatan sulphate | 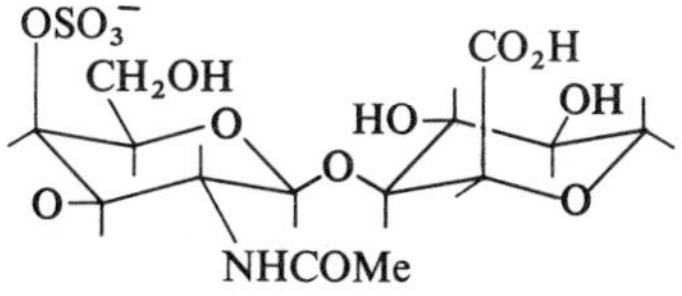 $OSO_3^-$, $CH_2OH$, O, O, NHCOMe, HO, $CO_2H$, OH, O | Univalent salt | | | 9.5 | 9.3 | 9.6 | 52, 53 |
| | | Free acid | | | 9.5 | | 9.7 | 52 |
| Keratan sulphate | 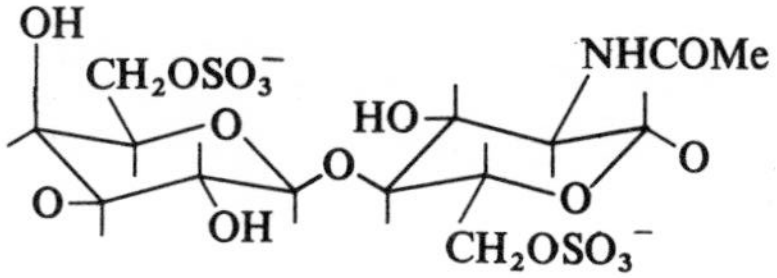 OH, $CH_2OSO_3^-$, O, O, OH, HO, NHCOMe, O, $CH_2OSO_3^-$ | Univalent salt | | | | | 9.5 | 54 |

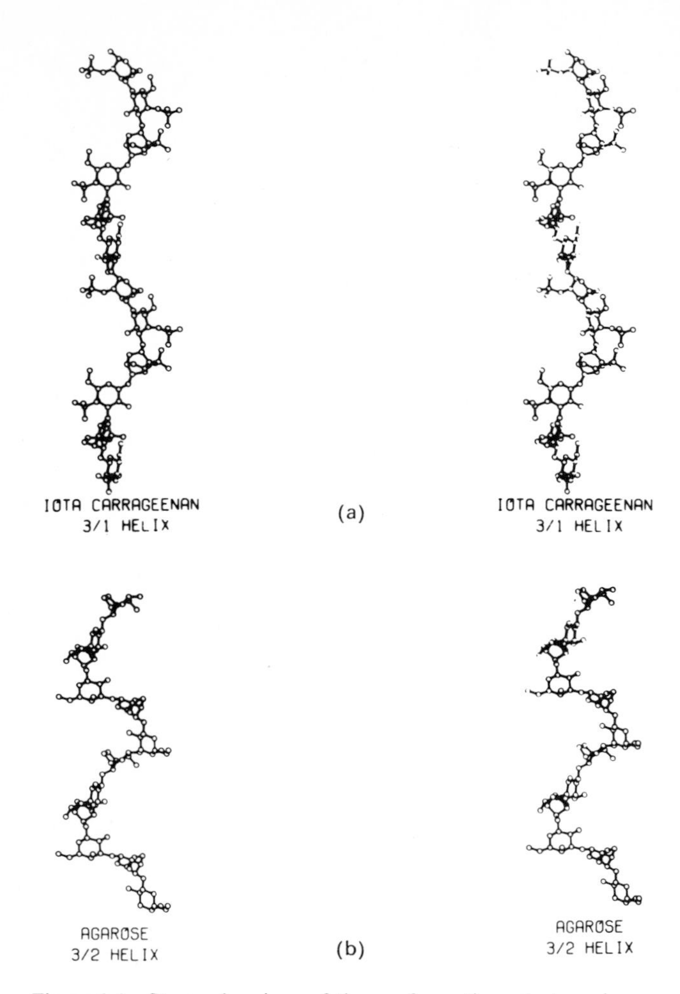


**Figure 1.1** Stereo drawings of the conformations that are known for alternating copolysaccharides having the general formula (1). (a) One strand of the right-handed, three-fold double helix in which the length of each disaccharide residue, projected on to the helix axis, $h$, is 8.7 Å. (b) One strand of the left-handed three-fold double helix in which $h = 6.3$ Å. (c) Left handed four-fold double helix in which $h = 8.5$ Å. (d) Left-handed four-fold single helix in which $h = 9.3$ Å. (e) Left-handed three-fold single helix in which $h = 9.3$ Å. (f) Left-handed helix having eight residues in three turns and $h = 9.8$ Å. (g) Two-fold helix in which $h = 9.6$ Å. For details of the occurrence of these and related conformations, see Table 1.1. These drawings were kindly made available by Prof. S. Arnott and Dr. J. M. Guss and were derived by computer model-building to achieve the most reasonable fit to the diffraction evidence given in the references listed in Table 1.1, except (c) which was derived through a detailed refinement[30a] and was plotted by Dr. P. J. C. Smith. Stereoviewers may be obtained, for example, from Ward's Natural Science Establishment, Inc., Rochester, N.Y., U.S.A., Model 25, W2951

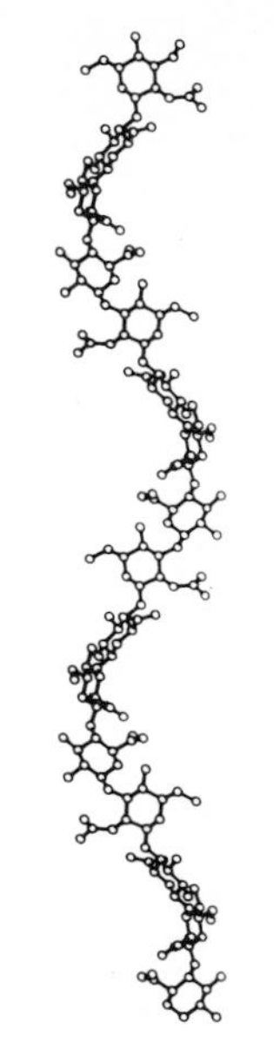
HYALURONIC ACID
4/3 HELIX(C=33.9)

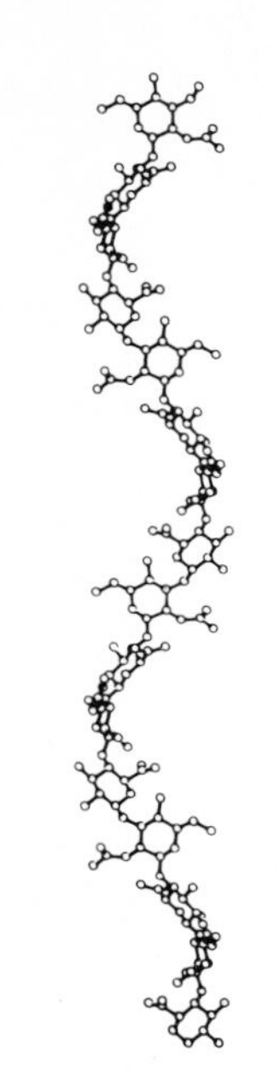
HYALURONIC ACID
4/3 HELIX(C=33.9)

(c)

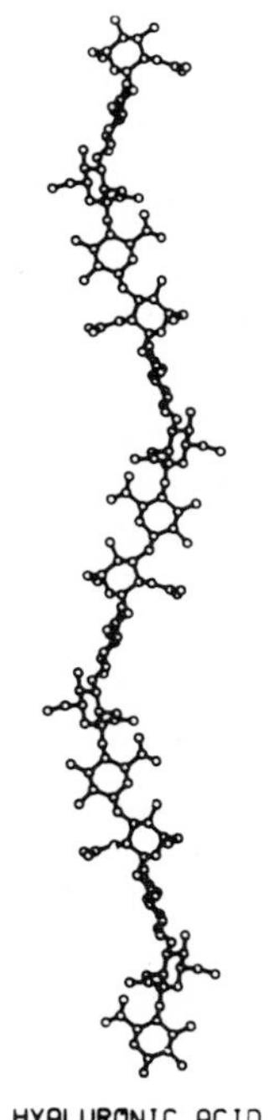
HYALURONIC ACID
4/3 HELIX(C=37.1)

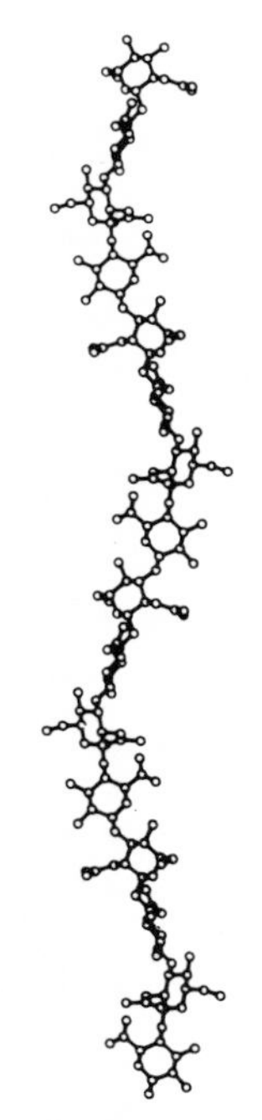
HYALURONIC ACID
4/3 HELIX(C=37.1)

(d)

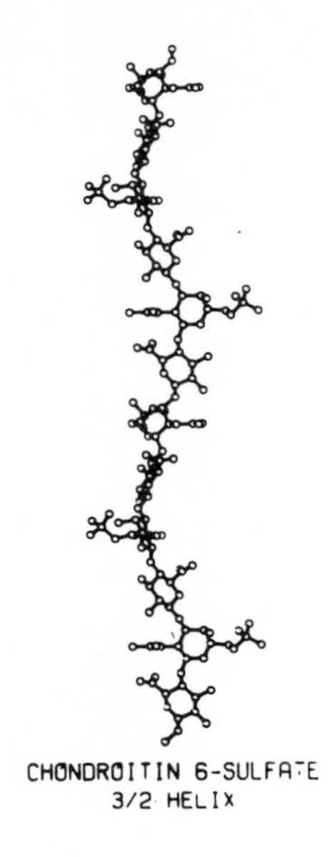


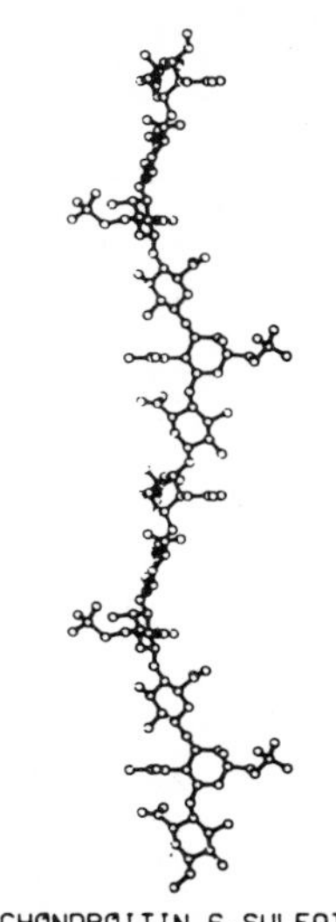


(e)

CHONDROITIN 6-SULFATE
8/5 HELIX

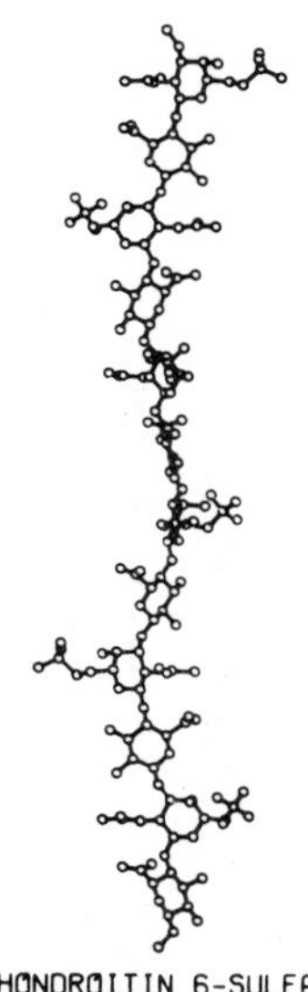


(f)

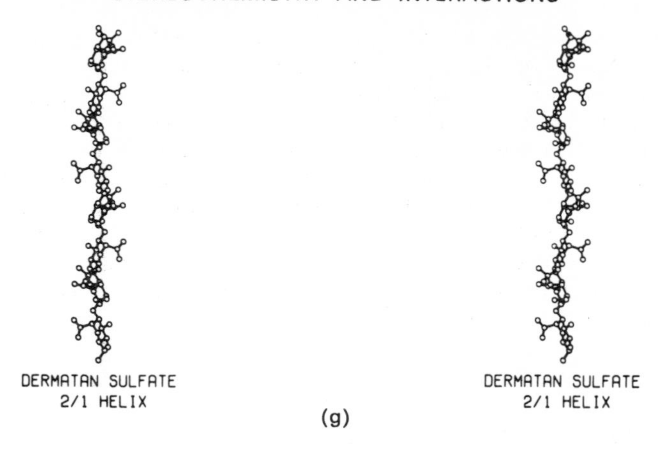


(g)

All these analogies and differences, in covalent structure as well as in conformation, would appear to suggest similarities in biological function. This is supported by close similarities in their occurrence in connective tissues[21]—sometimes they are even combined in the same proteoglycan macromolecule[21]. Their proportions change with development of the tissue in a way[21] that suggests that, for biological purposes, they are variations on a common theme. The similarities are also echoed in the routes to biosynthesis[22], which are indeed so closely related that polysaccharide products can be structural hybrids of the idealised formulae shown in Table 1.1.

The knowledge of these conformations (Figure 1.1 and Table 1.1) has not yet illuminated the biological functions, except perhaps for the polysaccharides discussed in the following section. Even here, however, the understanding is only in general terms. It is not even known whether the single helical forms ever exist under biological conditions.

### 1.2.3 Further discussion of some polysaccharide helices

The function of each of these polysaccharides is to form a suitable aqueous phase around or between cells. Such matrices, or gels, are also described in Section 1.3.4. Here we discuss three examples—agarose, ι-carrageenan and hyaluronic acid—to illustrate the extent to which biological function can be seen in terms of structure and conformation.

Agarose and ι-carrageenan occur in different species of marine red algae in which they form thick gels[31] to maintain the osmotic environment, give physical protection, and perhaps permit transport of metabolites to and from the cells[32]. Hyaluronic acid is extracellular in animals and certain bacteria[33, 34]. In animals it is a component of connective tissues[34], and is often in large amounts in connective tissues having a high water content, such as synovial fluid, umbilical cord and certain types of skin such as rooster comb and sexual skin of apes[35].

The biological condition and apparent function of each of these polysaccharides is matched by an ability when isolated to support large amounts of water in an open, porous gel or gel-like structure[15,36,37]. For ι-carrageenan the evidence is now very good indeed that the double helix exists in solutions and gels as well as in the condensed state. A cooperative transition can be demonstrated by optical rotation[38], $^{13}$C n.m.r.[39] and $^{1}$H n.m.r.[40] which clearly corresponds to the formation of double helix because it is bimolecular[40], is accompanied by an exact doubling of number-average and weight-average molecular weights[41], shows a shift in optical rotation which agrees in sign and magnitude with that expected for the double helix derived by diffraction analysis[42], and because the van't Hoff and calorimetric heats of formation are consistent for dimerisation by a two-state all-or-none mechanism[43]. For agarose the evidence is also good because a disorder-to-order transition is seen by optical rotation when the gel forms, and this has properties that are consistent with the coil to double helix conversion in the way it is influenced by introduction of substituent groups and in the sign and magnitude of the shift[28]. For hyaluronate the chain conformation (c) is so sinuous and compressed that it seemed likely at first[29,30] that it existed as one strand of a double helix. A thorough x-ray refinement[30a] has now shown that this is not so, however. Ordered forms of hyaluronate do seem to exist in solution[44,45,45a] and from first principles[18] they might be expected to be involved in the network formation that evidently occurs.

In spite of the analogies in the covalent structure and 'water-binding' properties of agarose, ι-carrageenan and hyaluronic acid, there are crucial differences between the three polysaccharides. In physical properties, ι-carrageenan gels are optically clear, elastic and can be rehydrated after the water

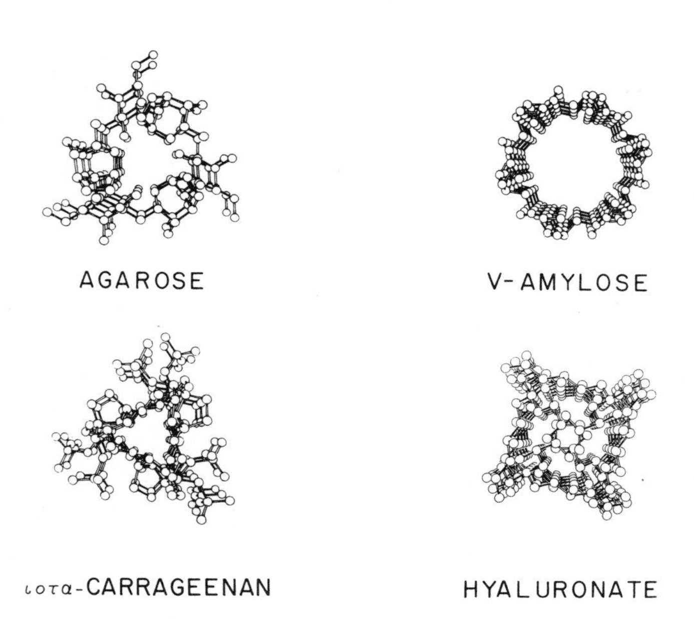


**Figure 1.2** Top view of four polysaccharide helices[28]. For further details, see text. Computer drawings kindly supplied by Drs. S. Arnott, J. M. Guss and P. J. C. Smith

has been removed by drying. Typical agarose gels are turbid or even opaque, and although they do show an elastic response to deformation, they are much more rigid and brittle in texture than the carrageenan gels. They dehydrate to a horny solid which will not rehydrate readily. Hyaluronic acid shows yet further variation in being viscoelastic[36, 37] rather than purely elastic (i.e. it flows slowly under stress). Although a common covalent skeleton (1) exists in all three[46], it is varied from one to another by changes in the nature and stereochemistry of substituents (Table 1.1). These molecular differences must in some way account for the differences in physical properties. In conformation there are the fundamental differences in shape, type and dimensions, shown in Figures 1.1 and 1.2, and Table 1.2.

**Table 1.2 Features of polysaccharide helices**

| *Feature* | *Agarose** | *ι-Carrageenan* | *Hyaluronic acid** |
|---|---|---|---|
| Type | Double | Double | Single |
| Relative chain sense | Parallel | Parallel | Antiparallel |
| Chain handedness | Left | Right | Left |
| Length of disaccharide residue, projected on to the helix axis/Å | 6.3 | 8.7 | 8.4 |
| Number of disaccharide residues per turn of helical chain | 3 | 3 | 4 |

* This refers to one of several hyaluronate conformations (see Table 1.1)

It does seem possible to trace a few connections between covalent structure, conformation, bulk properties and biological function. The double helix of ι-carrageenan has a polyelectrolyte surface which gives water solubility, whereas the double helix of agarose is electrically neutral and is insoluble. Consequently the helices differ in their states of aggregation with effects for the gel framework that are illustrated schematically in Figure 1.3. This in turn implies that the distributions of pore sizes, as well as the ion-exchange characteristics, differ between the two gels. The viscoelasticity of hyaluronic acid systems is presumed to arise from a dynamic 'making and breaking' character in the associations between helical strands which can separate more easily since they are not combined in a double helix and therefore do not need to untwist. In the purely elastic gels of agarose and ι-carrageenan the double helical associations are stable and long-lived so that moderate stress causes the network to deform rather than to rearrange. The biological value of the viscoelastic properties of hyaluronic acid could be that the viscous character permits, for example, the surfaces of bone joints to move while the elastic component would support and protect against shock[47]. Because plants are not capable of locomotion, however, it is presumably more appropriate that they should have matrix material which, like agarose or carrageenan, is purely elastic in its properties.

In summary, we can perceive in these systems some connections between

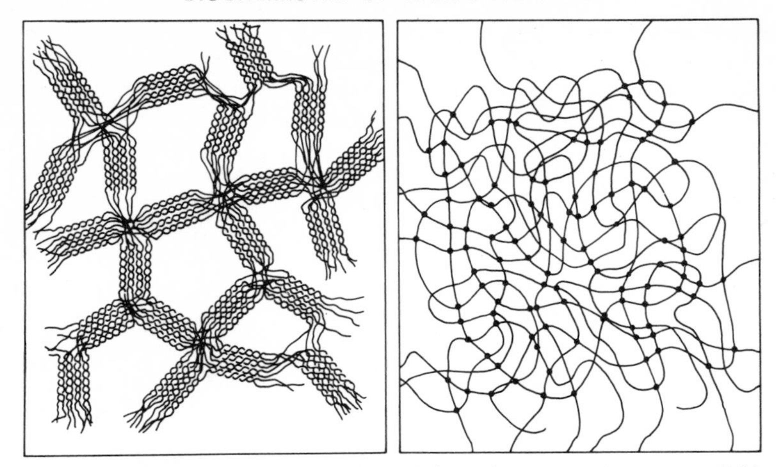

**Figure 1.3** Schematic representation of the gel structure in agarose (left) compared with Sephadex (right). Note that the aggregation of helices in agarose gels leads to an increase in the average pore size and strengthens the framework to make it less easily deformed and more brittle. In ι-carrageenan (not shown) the helices are not aggregated and the properties would be expected to be intermediate

covalent structure, conformation, physical properties and biological function, at a very general level. We may suspect that the helices have been optimised, through evolution for interactions with other extracellular components and with the cell surface itself, to mediate in the interactions and communications between cells. However, we cannot even begin to speculate about significance at that level of detail.

### 1.2.4 Conclusions: the importance of binding behaviour

This 'case study' of alternating copolysaccharides of certain algae and animal connective tissues has shown a polysaccharide group for which it remains difficult to identify structure–function relationships in detail, despite a wealth of information about covalent structure and about conformations that have been characterised by diffraction analysis. A similar conclusion would have been reached had we chosen to consider any other group of carbohydrate chains instead.

Such clues as we do have about biological function would suggest that binding behaviour is likely to be involved. It would seem that the polysaccharide helices function as somewhat elaborate mechanisms for controlled binding between strands. The regulation of this binding can lead to all kinds of extracellular texture, from a viscoelastic fluid through an elastic gel to a strong and porous framework. Each of these forms may have ion-exchange and other interaction properties as a result of appropriate substitution. Binding that involves carbohydrate chains is certainly involved in a number of important cell recognition events (see Section 1.3.6), although little is known yet about these phenomena at the molecular level. Analogy with non-carbohydrate systems would also point to the importance of binding behaviour. For example, enzyme function involves binding to substrate and often to allosteric effectors and other species. Nucleic acid functions in replication, transcription and protein synthesis likewise involve specific non-covalent binding.

In the hope that a knowledge of the binding behaviour of carbohydrate chains will help us to work towards a better understanding of biological function, most of the remainder of this chapter is devoted to this topic.

## 1.3 BINDING BEHAVIOUR OF CARBOHYDRATE CHAINS

### 1.3.1 Lipid binding*

Some principles that can be involved in the interactions between carbohydrate chains and lipids will be illustrated in this section by reference to the synthesis of fatty acids by a multi-enzyme complex from *Mycobacterium phlei* which is considerably affected by the addition of certain polysaccharides from the same micro-organism[55].

The polysaccharides concerned are a glucan[56] and a mannan[55,57], respectively, both of which carry further substituents. The glucan is the better characterised of the two. It is a chain of 17 α-D-glucose residues which carries a single-residue side chain and is terminated at the reducing end by D-glyceric acid, and which also carries *O*-methyl, *O*-succinyl and other *O*-acyl substituents, as shown in (2)[58]. Several different forms exist, but these appear to

-Me-4,6-Acyl$_2$-α-D-G*p*(1 [→ 4)(6-Acyl-α-D-G*p*)(1 ]$_3$ [→ 4)(6-Me-α-D-G*p*)(1 ]$_9$ → 4)(6-Me-α-D-G*p*)(1
3
↑
1
6-Suc-α-D-G*p*

→ 4)(6-Suc-α-D-G*p*)(1 → 3)(6-Suc-α-D-G*p*)(1 → 6)α-D-G*p*(1 → 2) (3-Oct-D-GlyA)

(2)

differ only in their levels of *O*-succinylation. The evidence for this structure is derived from full use[56-61] of almost all the established methods for polysaccharide structure determination, including methylation analysis, partial enzymic hydrolysis, replacement of *O*-acyl substituents by *O*-methyl[62] and Smith degradation. Several unconventional methods were also used very effectively, such as *O*-propylation to locate *O*-methyl substituents and a variant of the Lossen degradation[63] to remove the glyceric acid residue. In the mannan structure also, the main glycosidic linkage is α-1,4 and most residues occur as adjacent monomethyl ethers, see (3). No other substituents have

α-D-Man*p*(1 → 2)α-D-Man*p*(1 [→ 4)(3-Me-α-D-Man*p*)(1 ]$_9$ → 4) (3-Me-α-D-Man*p*)

(3) Partial structure only – see text

* I am grateful to Dr. K. Bloch for helpful correspondence on the subject matter of this section, and especially in comparing possible models for the binding behaviour.

yet been identified in this structure[57]. However, the low reducing power compared with the chain length estimate from sedimentation equilibrium, and certain circumstantial evidence, would suggest[57] that the reducing terminal is blocked by some group which remains to be identified, or that the structure is cyclic (compare the cycloamyloses[64], but see, however, the objections recently raised against cyclic models[70]).

Synthesis of fatty acids by the enzyme systems of *Mycobacterium phlei* proceeds from the usual substrates, namely acetyl-CoA and malonyl-CoA. The first stage is growth of the acyl chain to a length of 16 or 18 carbon atoms, after which the chain is either discharged from the enzyme by the action of palmityl transacylase to form palmityl-CoA, or as enzyme-bound palmitate it may be elongated to as many as 24 carbon atoms. The course of this synthesis is influenced by the presence of heat-stable substances from extracts of *M. phlei* that were eventually shown[55] to correspond to the two polysaccharides. It was found that: (a) synthesis of palmitate is stimulated by addition of either polysaccharide, especially at lower concentrations of acetyl-CoA[65]; (b) this stimulation is accompanied by a dramatic lowering of the apparent $K_m$ for acetyl-CoA and a similar but less striking effect on the apparent $K_m$ for malonyl-CoA[65]; (c) amongst the requirements for fast elongation of palmitate are a high concentration of palmityl-CoA and the addition of polysaccharide[66]; (d) the mixture of fatty acid products is shifted in composition by addition of polysaccharides towards products of shorter chain length[67].

All these effects were eventually traced[68] to an ability of each of the polysaccharides to form a 1:1 complex with palmityl and other longer-chain acyl-CoA derivatives. The two polysaccharides differed little in their influence on the biosynthetic system or in the properties of their complexes, except that: (a) the mannan was significantly more effective than the glucan in stimulating fatty acid synthesis at low levels of acetyl-CoA[65], and (b) the ability of mannan to stimulate fatty acid synthesis and to complex with long-chain acyl-CoA derivatives is not affected by further *O*-methylation, whereas both properties of the glucan are abolished by such treatment[68].

Analogies between the two polysaccharides are reinforced by the observation that the glucan (2) retains[68] its biological and complexing activities after de-*O*-acylation followed by digestion with α-amylase to remove[59] a trisaccharide or (less frequently) a tetrasaccharide from the 'non-reducing' terminal. Thus it is reasonable to propose[68] that the important properties are to be traced to the sequence which exists in both polysaccharides, of 10 consecutive, α-1,4-linked mono-*O*-methylhexose residues. A mechanism by which this sequence could bind palmityl-CoA and its analogues is suggested by consideration of the conformation and binding behaviour of amylose—another polysaccharide in which the residues are linked α-1,4, and which exists in the variety of conformational states that has been described and classified in a companion volume[69]. Briefly, the only conformations possible for amylose are members of a family of helices which correspond to different states of extension of a wire spring. This was shown by computer-aided conformational analysis and supported by analogy with crystal structures of related di- and oligo-saccharides, as well as by direct characterisation of a number of known forms of the polysaccharide itself in the solid state. As with other polysaccharide helices, the more compressed forms tend to have cavities which run down the helix axis and which may be filled with some other molecular species. Figure 1.2 shows this effect for four such helices, three of which have already been discussed in Sections 1.2.2 and 1.2.3. The most extended of these chains is hyaluronate and, in the crystal structure, the cavity is filled by interdigitating, side-by-side packing. Next is ι-carrageenan for which the cavity is snugly filled by a second coaxial chain. Because the agarose helix is still extended, a second chain can only fill the cavity incompletely; water molecules are therefore included as well. The most compressed of all four helices is the form shown of amylose; this has an even larger cavity which cannot accommodate a second polysaccharide chain, but it can accommodate a variety of small molecules and, indeed, this conformation is induced by such molecules which happen to be mainly apolar. Strict compatibility is unnecessary because the helix geometry can to some extent adapt to the size of the guest

**Figure 1.4** Helical conformation of a sequence of residues in the *O*-methylglucose polymer (2), and its role in binding fatty acids and their CoA derivatives. Top right: End view of the proposed helical sequence—note that the cavity is lined with $CH_3$ groups. Left: Side view showing the approximate match of the lengths of the polysaccharide and the palmitic acid residue. Note also that the helix exterior carries hydroxy groups for hydrogen bonding to water and that they also stabilise the helix by hydrogen bonding to each other. Bottom right: Proposed complex, showing steric 'fit' and hydrophobic contacts between the palmitic acid residue and the methyl ether groups. These models were built by E. J. Handson and D. A. Rees. Analogous models are possible for the *O*-methylmannose polymer (3)[70]

molecule[69]. When the chain of ten 6-*O*-methyl-D-glucose or 3-*O*-methyl-D-mannose residues is arranged in the best characterised conformation of this type, which is a left-handed helix having six hexose residues per turn and (for the glucose polymer only) a hydrogen bond between O-2 and O-6 of residues on neighbouring turns, it is seen that the methyl groups project inwards to enhance the hydrophobic character of the cavity (Figure 1.4). This helix matches the palmitate chain closely in length and diameter, and a very plausible inclusion complex therefore seems possible (Figure 1.4), stabilised by hydrophobic interactions. Optical rotation changes are observed when fatty acids or their CoA derivatives are complexed by the *O*-methylmannose polymer[70]; calculations show that the sign and magnitude of the shift is consistent with a transition from random coil to V-amylose-like helix, induced by binding[70a].

Small differences in properties that exist between the two helices can probably be traced to consequences of the different orientations of O-2 in each series. To explain the complex pattern of specificity differences in the binding of fatty acids of different chain lengths, and of free fatty acids compared with their CoA thioesters[70], we would probably need to assess the relative free energies of competing states for each component, including fatty acid micelles and folded chains for the longer acids. This would lie outside the scope of this chapter.

Although these polysaccharides do seem to be designed for specific, conformation-dependent binding behaviour towards fatty acid chains, it is unlikely that we have yet grasped their full biological significance. Possible functions have been suggested[65-68] in the regulation of feedback inhibition of fatty acid synthesis and in the control of the chain-length distribution of fatty acid products. The fact that the polysaccharides provide a 'hydrophilic overcoat' for the fatty acid chain would also suggest a role in transport through aqueous environments which is similar to, but the inverse of, the function of 'carrier' lipids which carry carbohydrate chains in the environment of the membrane during biosynthesis[71].

### 1.3.2 Water binding

A good example of water binding by carbohydrate chains is in connective tissues in which the interstices of a meshwork of collagen fibres are occupied by proteoglycan molecules which[31], being flexible and polyanionic, give rise to an internal osmotic pressure. Water is imbibed to create a turgor which is contained by tension in the meshwork and which supports and gives flexibility to an otherwise brittle structure.

Joint cartilage contains about 25 parts of water for each part of polysaccharide by weight. The relative importance of various contributions to this 'osmotic binding' is shown by investigation of model gel systems which can imbibe 5–50 parts by weight of water. The interpretation derives from the classical theory of rubber elasticity[72, 73] which, of course, is equally applicable to biopolymer systems[74]. The network is considered to imbibe water because the chains are 'soluble', even though they cannot pass into true solution because they are tied together in a network. These two tendencies reach a compromise in a swollen gel when the drive to chain separation (i.e. the osmotic pressure) is balanced by the elastic resistance to further swelling.

In non-ionic systems the 'solubility' arises because of favourable solvation and/or because the chain separation gives access to more configurational states. This is equivalent to saying that the driving force may contain a heat term and an entropy of polymer dilution term. That a practical, quantitative treatment can be developed on this basis is shown by the very satisfactory and useful applications to Sephadex[75, 76] (a neutral polysaccharide gel that is made by cross-linking dextran with epichlorohydrin) and polyacrylamide gel[77]. We start[75] by recalling that the osmotic pressure, $\pi$, of a polymer solution is given by:

$$\frac{\pi}{RT} = \frac{c}{M} + A_2c^2 + A_3c^3 \tag{1.1}$$

where $c$ is the concentration, $M$ is the number-average molecular weight, $R$ and $T$ have their usual meanings, and $A_2$ and $A_3$ are the so-called virial coefficients which, being correction terms for departure from ideal solution behaviour, are a mathematical device for including the entropy and heat terms already mentioned. If this equation is applied to the network which of course has infinite molecular weight, it becomes:

$$\left(\frac{\pi}{RT}\right)_{\text{network}} = A_2c^2 + A_3c^3$$

Since the elastic resistance to swelling is a property of the network structure, any attempt to estimate its magnitude must require a model for that structure. For an idealised, randomly cross-linked network, the theory of rubber elasticity[72, 73] suggests that this term has an entropic origin and is related to the partial specific volume ($v$) and the concentration ($c$) of the network polymer by:

$$\text{Elastic contribution, } E = \text{constant} \times \left[(v\,c)^{\frac{1}{3}} - \frac{v\,c}{2}\right] \tag{1.2}$$

Thus, equilibrium is reached for the gel in contact with water when:

$$\left(\frac{\pi}{RT}\right)_{\text{network}} = E \tag{1.3}$$

or, if the gel is in contact with a solution of a polymer which does not penetrate the network, then:

$$\left(\frac{\pi}{RT}\right)_{\text{diff}} = \left(\frac{\pi}{RT}\right)_{\text{inner phase}} - \left(\frac{\pi}{RT}\right)_{\text{outer phase}} = E \tag{1.4}$$

By assuming that $v$ for Sephadex is the same as for the parent dextran, and by the determination of $c$ and $(\pi/RT)_{\text{outer phase}}$ for a series of gel–polymer systems, these equation could be used to find the best values for the virial coefficients for Sephadex and to show that these are not very different from the values for dextran itself[75]. This provides an internal consistency which is good support for our ideas about the origin of 'water binding' by the gels. Alternatively, the swelling of Sephadex[76] or polyacrylamide[77] may be calibrated for the determination of the molecular weights of polymers in solutions with which the gel is equilibrated. Excellent results are obtained by such 'gel osmometry', especially with special techniques for the measurement of gel swelling and for the elimination of complications caused by penetration of polymer into the gel[77].

An even better model for real connective tissues would, of course, be a polysaccharide gel in which the chains carry ionic substituents. The introduction of such groups—as, for example, when Sephadex is converted into sulphoethyl-Sephadex (i.e. to $^{-}O_3SCH_2CH_2O$—Sephadex)—usually causes an increase in the 'water binding'. This increase can be traced to the influence of the counter ions (e.g. $Na^+$), which are free and mobile but are confined to the gel domain by the need to maintain overall electrical neutrality. The state of these ions is discussed more fully in Section 1.3.3.1, and here we need only note that, to the extent that they do have free and independent motions within the gel phase, they exert an osmotic pressure which is additional to that of the polysaccharide network itself. Actually the motions of the cations are somewhat constrained by proximity to the polyanionic chains, so that their activity and therefore the osmotic pressure is lower than in a simple salt solution of the same concentration. This effect can, however, be estimated[78-81] (see Section 1.3.3.1) so that the 'water-binding' can once again be interpreted[82] in terms of equation (1.4). This shows that the swelling pressure of the gel—which is the term $(\pi/RT)_{diff}$ in equation (1.4) and is equivalent to the pressure that must be applied to squeeze out the 'bound' water—can be of the order of several atmospheres. This pressure, as well as the uptake capacity of the gel, is depressed by simple electrolytes such as occur in physiological systems, but even under these conditions the dominant contribution is from the counter ions rather than from the network itself.

The constant which determines the magnitude of elastic resistance to swelling [see equation (1.2)] is found to increase with the extent of swelling, showing that more and more distortion of the gel structure occurs in approaching the full capacity for water binding.

To construct systems that might provide even better models for connective tissues, gelatin gels have been formed in the presence of glycosaminoglycans and proteoglycans so that polysaccharide is actually trapped in a matrix of collagen-derived protein[83]. (Gelatin is manufactured from the collagen of animal skin, bone and white connective tissue by an extraction which causes extensive denaturation and some hydrolysis. It owes its gel structure to cross-links that form by partial renaturation to the collagen triple helix[84].) The behaviour of the gelatin gel itself is readily accounted for in terms of the treatment that is outlined above[85], but when a sulphated proteoglycan or a hyaluronate–protein complex is trapped within it, the swelling pressures are higher than expected from the osmotic activities of the individual components by a factor of two or more. The most likely explanation is that the gelatin and the polysaccharide components tend to avoid overlapping their hydrodynamic domains for reasons[86, 87] discussed in Section 1.3.5.3. This has the effect (Section 1.3.5.3) of increasing the thermodynamic activity of each component and hence increasing the overall osmotic pressure. Whether or not this interpretation is correct, however, it is a matter of empirical fact that mixtures of proteoglycan and denatured collagen show a synergistic increase in the osmotic binding of water when they are mixed. It is very likely that similar effects operate in the extracellular spaces of native tissues.

### 1.3.3 Cation binding

Three types of cation binding can be distinguished. Combinations of these can occur together in real systems, but we shall discuss the idealised extremes because this is the best way to illustrate the principles. In 'statistical binding' the cations merely have a statistical tendency to occur in the neighbourhood of the negative charges on the polysaccharide chain, and any complexes are ion pairs which form by chance collision and do not necessarily have a given geometry. The ions can be described as forming an atmosphere around the polyanion chain, like the atmosphere of gases that may surround a planet. 'Site binding' involves specific complexes in which hydroxy groups of the sugar residues usually enter the coordination sphere of the cation in a way that is geometrically defined. The known examples of 'cooperative binding' involve site binding of several cations in such a way that each binding event facilitates the next.

#### *1.3.3.1 Statistical binding*

The driving force for this binding is the electrostatic repulsion between negative charges that are held together by the covalent structure of the polymer. These repulsions could conceivably be screened completely by clustering all the cations around the polymer chain, but the loss of repulsive energy would then be offset by the loss of degrees of freedom of the cations, i.e. a loss of entropy. What actually happens is that these tendencies compromise in a state in which the cations remain mobile but have a higher probability of being near the negatively charged chain than elsewhere in solution. The probability is biased to the extent that will minimise the free energy of the system, which depends on the structure of the polysaccharide, the nature of the cation, the nature and concentration of any other solutes and the temperature. It is possible to estimate this result from the polysaccharide structure using equations that have been derived from a statistical mechanical model for a generalised polyelectrolyte[78, 80]. As would be expected intuitively, these equations show that the fraction of cations bound, $F$, increases as the average distance between charges, $D$, diminishes. For univalent cations associated with a polysaccharide which carries univalent charges, the result at low concentration in the absence of added salt is:

$$F = 1 - \frac{\varepsilon k T D}{e^2} \tag{1.5}$$

where $\varepsilon$ is the bulk dielectric constant and $e$, $k$ and $T$ have their usual meanings. The behaviour with added salt can also be estimated: the difference, $\phi$, between the observed osmotic pressure and that expected in the absence of binding, is:

$$\phi = \left(\frac{XY}{2} + 2\right) \Big/ \left(X + 2\right)$$

where $Y$ is the second term on the right-hand side of equation (1.5), and $X$ is the number of polymer charges (summed over all molecules), as a ratio relative to the concentration of added uni-univalent salt. This expression does not take account of deviations from ideal behaviour that are properties of the salt solution itself, but these effects are well known and tabulated, and may be added in empirically[81].

These equations have been derived from a generalised model for any polyelectrolyte without the use of adjustable parameters. As an illustration of their applicability to polysaccharides, the binding of cations by hyaluronate, chondroitin sulphate, two carboxymethylcelluloses and carboxymethylcellulose sulphate were measured by equilibrium dialysis and interpreted in terms of the equations to estimate that the average distances between charges along the polymer backbone are 15.5, 6.4, 11.9, 8.4 and 9.4 Å, respectively[88]. The corresponding distances for molecular models of the fully extended chains are 12.0, 5.1, 8.2, 5.2 and 6.7 Å, showing that the order of magnitude and the directions of trends are successfully predicted. Detailed measurements of sodium chloride activity and of osmotic pressure for the system sodium dextran sulphate + sodium chloride + water likewise agree well with calculation[20], and this agreement carries to higher concentrations of sodium chloride when the extended version[81] of the theory[78] is used.

### *1.3.3.2 Binding at isolated sites*

In contrast to the statistical binding of cations, the driving force for specific binding is not necessarily dominated by formal negative charges on the carbohydrate ligands. Indeed, 1:1 complexes of significant stability are known for certain neutral sugars (Table 1.3)[89]. Such stability is possible when sets of two or three hydroxy groups are stereochemically arranged to fit into

**Table 1.3 Stability constants* for metal complexes with sugars**

| *Complex* | *Cation radius*/Å † | $K_s$/ l mol$^{-1}$ | *Ref.* |
|---|---|---|---|
| α-D-Allopyranose–NaCl | 0.95 | 0.12 | 89 |
| α-D-Allopyranose–$MgCl_2$ | 0.65 | 0.19 | 89 |
| α-D-Allopyranose–$CaCl_2$ | 0.99 | 6.2 | 89 |
| α-D-Allopyranose–$Y(NO_3)_3$ | 0.93 | 1.9 | 89 |
| α-D-Allopyranose–$LaCl_3$ | 1.15 | 10.4 | 89 |
| *cis*-Inositol–$CaCl_2$ | 0.99 | 21 | 89 |
| Epi-inositol–$CaCl_2$ | 0.99 | 3.2 | 89 |
| Calcium αβ-glucuronate | 0.99 | 32 | 96 |
| Calcium α-glucuronate | 0.99 | 28 | 96 |
| Calcium β-glucuronate | 0.99 | 37 | 96 |
| Calcium αβ-galacturonate | 0.99 | 68 | 96 |
| Calcium α-galacturonate | 0.99 | 101 | 96 |
| Calcium β-galacturonate | 0.99 | 61 | 96 |
| Calcium methyl α-D-glucosiduronate | 0.99 | 40 | 96 |
| Calcium methyl β-D-galactosiduronate | 0.99 | 56 | 96 |

* The stability constant, $K_s$, is for a 1:1 complex; $K_s$ = $[M^{n+}]$[sugar]/[complex]
† These values are from L. Pauling[95]

the coordination sphere of the cation with the displacement of water molecules. The interaction therefore depends on the size and preferred coordination geometry of the cation. This discussion will be centred around the carbohydrate complexes of $Ca^{2+}$ because these would seem to be important in biological systems.

Evidence has existed for many years that neutral sugars can complex with alkali and alkaline earth cations, both in the solid state and in solution[90], but only recently have major advances been made in defining the geometry involved. The first clue is that *cis*-inositol (4) has particularly high electrophoretic mobility towards the cathode in solutions of calcium and certain other salts[91], and therefore that it has an arrangement of hydroxy groups that

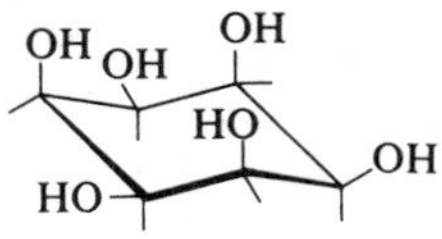

(4)

is especially favourable for complexation. When the behaviour of many carbohydrate structures is compared in this way, it turns out[89] that their mobility is significantly affected when they contain an *ax-eq-ax* sequence of oxygen atoms on the six-membered ring, i.e. the partial structure (5). If such an arrangement represents a preferred site for cation binding, the properties of *cis*-inositol are explained because it occurs three times in this molecule with the three axial oxygens as a possible extra site, because they form a triangle of similar dimensions to that of the *ax-eq-ax* sequence. These ideas are confirmed by $^{1}H$ n.m.r. spectroscopy[92]; for example, the proton attached to the central carbon of the supposed site in epi-inositol (6) shows the biggest shift and this could be caused by proximity to the cation. The ring conformation (6) is retained because there is no change in coupling constants. The spectra

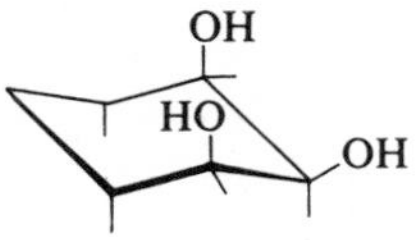

(5)

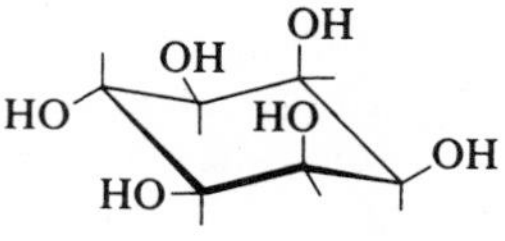

(6)

show that rapid equilibrium exists between complexed and non-complexed forms.

The mutarotation equilibrium of α-D-allose (7) and β-D-allose (8) can be displaced by the addition of cations because the α-anomer has the *ax-eq-ax* binding site whereas the β-anomer has not. On the assumption that binding to other sites is unimportant, the change in anomeric equilibrium may be used to estimate the stability constant. This shows that, for cations of the same charge, strongest complexes are formed when the radius is close to 1.0 Å. The most strongly complexing cations identified so far in the alkali metal, alkaline earth and lanthanide series are $Na^{+}$, $Ca^{2+}$ and $La^{3+}$. At similar radius the stability constant increases with charge, as shown by the values for these cations (Table 1.3). Such alteration of chemical equilibria by complex formation can be useful by carbohydrate synthesis[94] and could conceivably occur in biological processes.

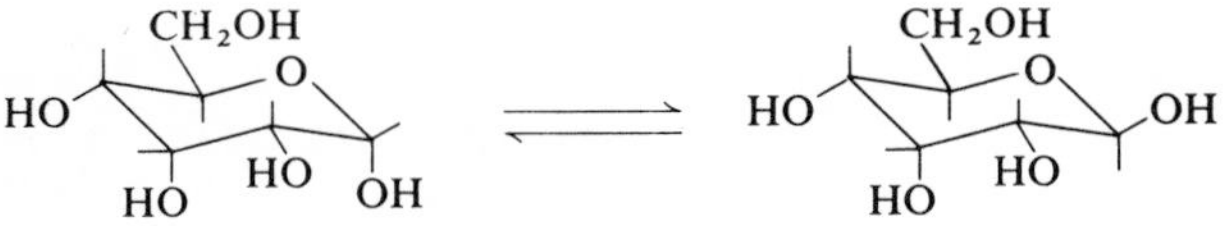

$CH_2OH$ HO HO HO OH (7) ⇌ $CH_2OH$ HO HO HO OH (8)

The calcium complexes of sugar anions are more stable than those of neutral sugars, presumably because of electrostatic contributions, but stereochemical influences exist for them also (Table 1.3)[96]. The complexes of galacturonates are stronger than of glucuronates, suggesting that O-4 in galacturonate is coordinated to the cation[96]. This order of complex stability is confirmed[97,98] by use of $Eu^{3+}$ as an $^1H$ n.m.r. probe for $Ca^{2+}$. Because of its paramagnetic characteristics[99], $Eu^{3+}$ causes large changes in the chemical shifts of the ligand which can be related to the geometry of interaction. A model for calcium galacturonate has thus been proposed[98] in which O-4, O-5 and a carboxylate oxygen above the plane of the sugar ring [see (9)] form an equilateral triangle having a side of *ca.* 2.8 Å; this site is analogous to that formed by the *ax-eq-ax* arrangement in calcium complexation by neutral sugars.

Further evidence for important interactions between sugar hydroxy groups and calcium ions is provided by crystal structures of adducts formed by co-crystallization of neutral sugars with calcium halides, and also by the crystal structures of calcium salts of sugar acids. Such structures include β-D-mannofuranose calcium chloride tetrahydrate[100], α-lactose calcium bromide heptahydrate[101,102], α-D-galactose calcium bromide trihydrate[101,105], *myo*-inositol calcium bromide tetrahydrate[101], α-D-xylose calcium chloride trihydrate[103], α,α-trehalose calcium bromide monohydrate[104], calcium arabonate pentahydrate[106], calcium D-*xylo*-5-hexulosonate dihydrate[107], calcium 'α'-D-glucoisosaccharinate[108], the mixed salt calcium bromide lactobionate tetrahydrate[109], sodium calcium galacturonate hexahydrate[110] and strontium 4-*O*-(4-deoxy-α-L-*threo*-hex-4-enopyranosyluronate)-D-galacturonate in a form with four and a half water molecules for each molecule of disaccharide[111]. In the adducts with neutral sugars, interactions between calcium and oxygen are evidently so favourable that they exclude halide from the coordination sphere against coulombic attaction. Calcium is surrounded either by seven oxygen atoms in a distorted pentagonal bipyramid[103,104] or by eight oxygen atoms in a distorted square antiprism[101,102,105] (Figure 1.5). In the calcium salts of sugar acids the coordination is usually eight-fold to form a distorted square antiprism, but there is one example[110] of nine-coordination to form a trigonal prism (Figure 1.5). Although the idealised polyhedra (Figure 1.5) are always distorted in the carbohydrate structures, they can be used to visualise and rationalise the experimental results. Within each structure the calcium–oxygen distances are rather variable and they are of course somewhat shorter on average when fewer atoms are in the coordination sphere; for example, they are shorter by about 0.1 Å for seven-coordination compared with eight-coordination. The oxygen–oxygen separations on the edges of each polyhedron (Figure 1.5) are correspondingly variable but are usually in the range 2.7–3.0 Å. This distance corresponds very closely to the separation of 2.7–2.9 Å between adjacent *eq-eq* or *eq-ax* oxygen atoms[112] or between 1,3-diaxial oxygen atoms[113] in crystal structures of pyranose sugars in the absence of cations. This would suggest a fundamental compatibility between the geometry of pyranose sugars and the coordination geometry of calcium.

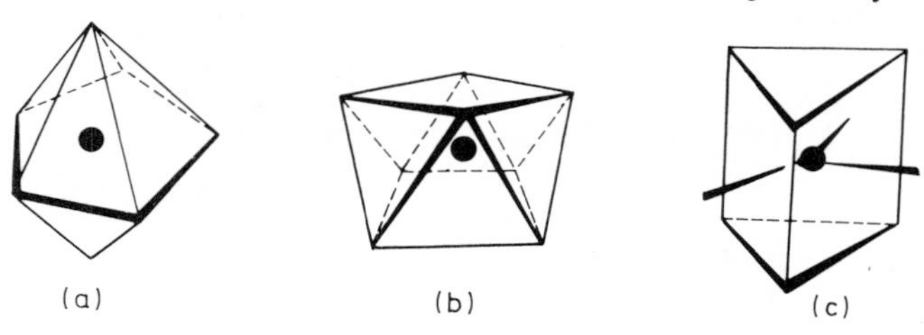


**Figure 1.5** Idealised drawings of the coordination geometries found in calcium complexes of carbohydrates: The oxygen atoms form either (a) a pentagonal bipyramid (seven-coordination), (b) a square antiprism (eight-coordination) or (c) a trigonal prism with three additional oxygens outside the centres of the three vertical faces (nine-coordination). The calcium ion is shown as a filled circle. The oxygen atoms are at each corner of each geometrical body, except for the three additional oxygens in (c). Note that in all carbohydrate crystals so far studied the oxygen atoms are significantly distorted from the arrangements shown here

Each coordination polyhedron (Figure 1.5) could be assembled by putting together two triangular arrangements of oxygen atoms of the types (5) or (9) that have been proposed for complexation in solution, with addition of single oxygens to fill the other sites. Such a mechanism would be especially favourable in solution because it enables each set of three coordination positions to be filled for the loss of only three translational degrees of freedom rather than the nine that would be lost if each site were occupied by an independent oxygen function. In the solid state, however, since all translational movement is lost, we do not necessarily expect this mechanism to be favoured. Indeed, only two crystal structures are known in which a triangular face is formed from the oxygens of the same ligand molecule. These involve O-1, O-2 and O-3 of β-D-mannofuranose (10)[100] and O-1, O-2 and O-6 of

(9) (10) (11)

D-*xylo*-5-hexulosonate (11)[107]. It is, however, very usual in complexes of neutral sugars to find that pairs of hydroxy groups in adjacent *eq-eq* or *eq-ax* positions on the sugar ring are used together to form polyhedron edges. These oxygen atoms are squeezed closer by about 0.2 Å than they are in the absence of cations[101, 102, 104, 105] and are closer than other pairs of oxygens in the coordination shell. Presumably this aids packing around the cation and would suggest that the non-bonded interactions in this slight distortion of the sugar ring are less severe than in compression of oxygens of separate ligand molecules. The sugar anions also distort such pairs of oxygens in forming complexes. The carboxylate oxygen usually enters the polyhedron and in cyclic sugars it forms an edge with the ring oxygen. The galacturonate residues in known crystal structures[110, 111] differ from the structure suggested for the solution complex (see above)[98] in that O-4 is not involved in coordination and in that the conformation about C-5—C-6 has the carboxylate group almost coplanar with C-5—O-5 rather than perpendicular to it.

### *1.3.3.3 Cooperative binding*

Certain carbohydrate systems show ion-binding that cannot entirely be explained in terms of statistical binding, or specific binding at individual sites, or any combination of these. A striking example is in a comparison[114] of the calcium-binding behaviour of the 1,4-linked oligosaccharides of α-L-guluronic acid and β-D-mannuronic acid, the structures and conformations[115-117] of which are as shown [(12) and (13)]. The calcium ion activity, $\gamma_{Ca^{2+}}$, was

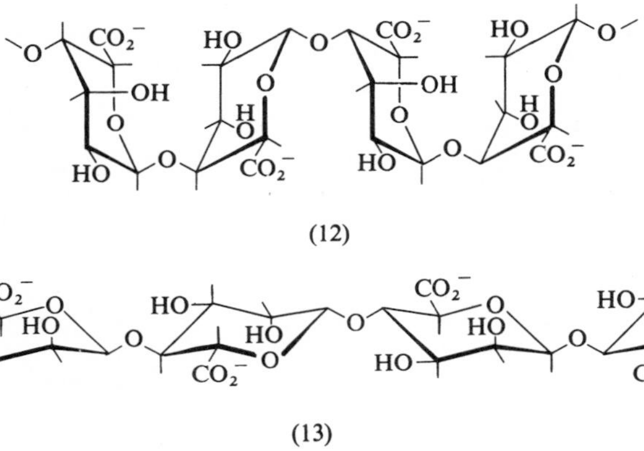

(12)

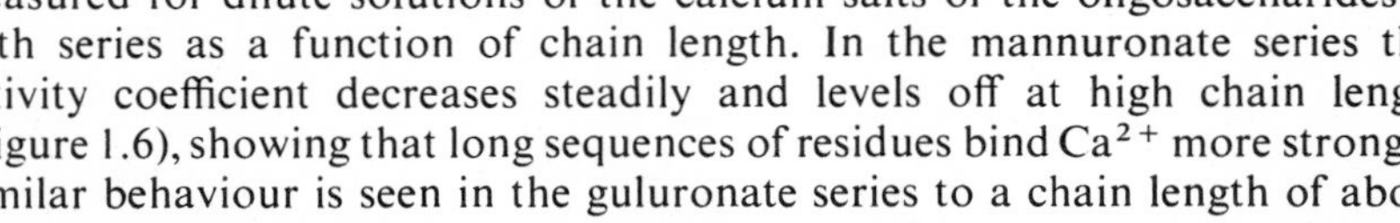

(13)

measured for dilute solutions of the calcium salts of the oligosaccharides in both series as a function of chain length. In the mannuronate series this activity coefficient decreases steadily and levels off at high chain length (Figure 1.6), showing that long sequences of residues bind $Ca^{2+}$ more strongly. Similar behaviour is seen in the guluronate series to a chain length of about 20 (Figure 1.6), although with a sharper increase with chain length. All this is readily explained as a combination of specific binding by individual residues (Section 1.3.3.2), augmented by polyelectrolyte effects (Section 1.3.3.1) to an extent that depends on chain length and average charge separation [compare (12) and (13)]. For the guluronate series it is suggested[89, 97, 118], from model building and analogy with the monosaccharide systems that are discussed above, that a cavity exists between each pair of neighbouring residues to form a very favourable coordination site [see (14)]. Quite inexplicable in such terms alone, however, is the sudden increase in binding which occurs at about chain

(14)

length 20 in the guluronate series, and shows as a 'second step' in the curve (Figure 1.6). For reasons outlined elsewhere[119], such a sharp change over a critical range of chain length shows that a cooperative mechanism has intervened. Another pointer to this conclusion is the ability to bind $Ca^{2+}$ in competition with $Na^{+}$ over a wide range of relative concentrations of the two ions[120]. Yet further evidence for cooperativity is to be seen in various properties of the $Ca^{2+}$-induced gelation of these polymers[15, 121].

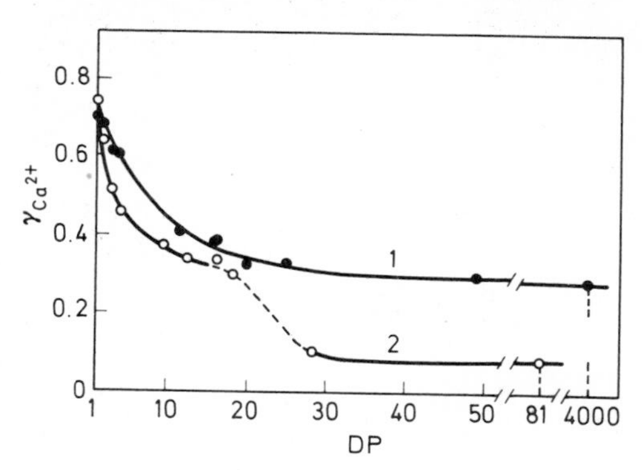


**Figure 1.6** Activity coefficient $\gamma_{Ca^{2+}}$ in solutions of calcium oligo- and poly-mannuronates and calcium oligo- and poly-guluronates dependent on the degree of polymerisation (DP). 1. Calcium oligo- and poly-mannuronates. 2. Calcium oligo- and poly-guluronates. (From Kohn and Larsen[114], by courtesy of Dr. R. Kohn.)

This means that, for oligoguluronate chains of sufficient length, each binding event somehow facilitates the next so that a total population of bound $Ca^{2+}$ is held much more strongly than the same number of ions bound individually and independently. Such cooperative behaviour is possible only for suitable geometrical relationships between individual binding sites[119] and a general examination of cooperativity in polysaccharide systems soon shows that polyguluronate (together with other polysaccharides of the so-called Type A) has little if any scope for cooperativity within an individual chain[122].

Indeed, all the experimental evidence would indicate that cooperative binding in this system involves chain association[114,120,121,125]. Computer model-building was therefore used to generate packing arrangements that could give an ordered array of sites to explain the cooperative behaviour[123]. This turns out to be possible with chains in a conformation similar to (14) and also similar to the conformation that exists in the solid state for the free acid, poly(guluronic acid)[116]. The packing leaves oxygen-lined cavities reminiscent of the cyclic ethers which are powerful and selective binding agents for alkali metal and alkaline earth cations[124]. Each cavity provides a good fit for a calcium ion and may place as many as ten oxygens within an appropriate distance for coordination, as shown by space filling models in Figure 1.7.

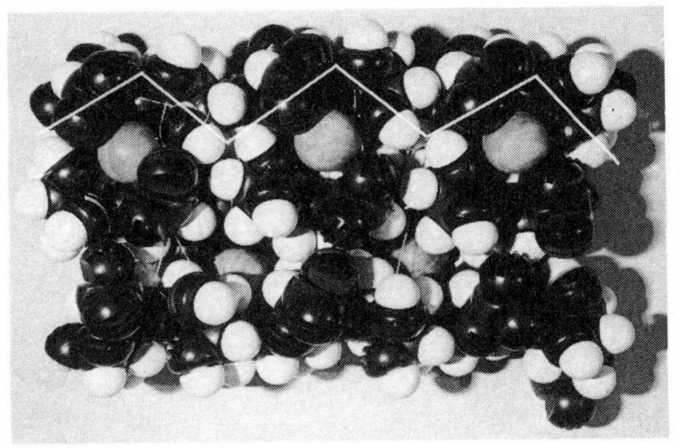

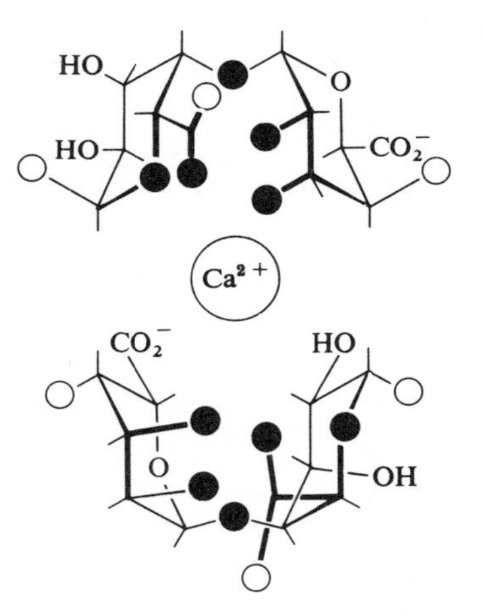


**Figure 1.7** (Top) Photograph of an array of calcium binding sites in space-filling models, as proposed in the 'egg-box model'[123]. The contour of one chain is traced in white to show its buckled character. This model was made by Dr. D. Thorn. (Bottom) Sketch of the disposition of groups around the calcium ion in the egg-box model; oxygen atoms coordinated to calcium are shown as filled circles. For clarity the diagram has been exploded in the vertical direction, i.e. the chains have been moved apart

The proximity of the cation to the n-orbitals of the carboxylate group causes a profound perturbation (indeed, a reversal) of the optical activity of the $n \rightarrow \pi^*$ transition[123, 125]. The packing is schematically shown in (15), with

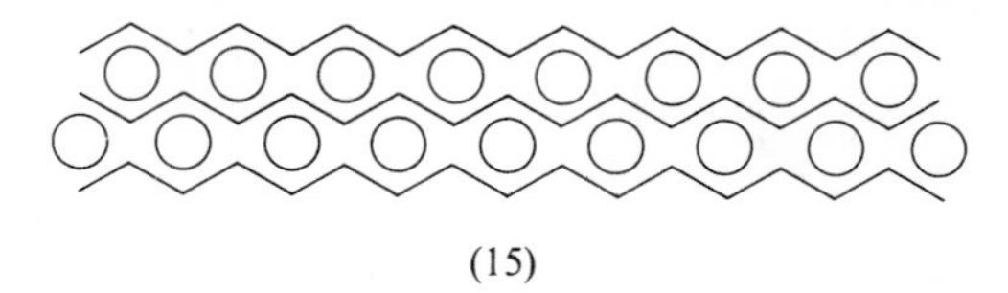

(15)

each polysaccharide chain as a zig-zag line and each cation by an open circle. This is like an edge-on drawing of a pile of corrugated egg-trays with calcium ions corresponding to eggs in the interstices. The analogy is that the strength and selectivity of cooperative binding is determined by the comfort with which 'eggs' may pack in the cavities, and with which the 'trays' pack with each other around the 'eggs'. This type of binding is therefore known as the 'egg-box mechanism'[123]. Polygalacturonate also binds alkaline earth cations in this way. It shows a different relative selectivity for $Ca^{2+}$ and $Sr^{2+}$ which is readily predicted from the computer models and which therefore provides some confirmation of the models.

Few other examples have been reported of the cooperative binding of cations by carbohydrate chains. Instances yet to be discovered need not necessarily involve chains with large numbers of charges and which bind large numbers of cations in the cooperative event. The binding of a single ion could perhaps be cooperative, if it were linked to some other source of free energy such as a conformation change.

#### *1.3.3.4 Conclusions*

Of the three types of binding of cations by carbohydrate systems, cooperative binding is the strongest and most specific. It could have biological functions related to the change of conformation and state of association of the carbohydrate chains that accompany the binding.

In many systems, statistical binding and specific binding probably contribute together to the same process. Even when binding is very largely determined by charge density, ligand oxygens are always likely to contribute by competitive displacement of water molecules from the hydration shell. These sorts of binding are important in biological systems chiefly through their contribution to osmotic effects which govern the hydration and elasticity of tissues, although more subtle effects may yet become apparent.

### 1.3.4 Carbohydrate binding

Two of the mechanisms have already been discussed by which carbohydrate chains may bind each other, namely in double helices and aggregated double helices (Section 1.2.3) and in egg-box structures with simultaneous binding of cations (Section 1.3.3.3). The sequences in carbohydrate chains which can bind in these or other ways are often separated by other sequences which cannot bind. The result is then a three-dimensional network that can be expanded to a gel by osmotic effects (Section 1.3.2). This might function to hold water and ions in the environment of cells, and to mediate in the passage of molecules in and out of cells. The properties of such gels and the way they are determined by polysaccharide structure and conformation have been reviewed several times recently[15-17]. Here we need only add a brief account of very recent developments.

Several of the polysaccharide types that occur in plant cell walls, including galactomannans, glucomannans, arabinoxylans and derivatised celluloses, do not readily gel from aqueous solution but may do so when mixed with a double-helical polysaccharide such as agarose or κ-carrageenan. Polysaccharide chains of the two different types are combined in the network by the binding of unsubstituted sequences of β-1,4-linked mannose, glucose or xylose residues, which are ribbon-like in shape, to double helices or aggregates of them[126]. Such mixed networks can also form by the binding of the β-1,4-linked sequences to polysaccharides from the outer layers of certain bacteria. The most thoroughly studied systems of this sort are mixtures of galactomannans and extracellular polysaccharides from *Xanthomonas* species (xanthans), particularly from *Xanthomonas campestris*[127]. The sequence[128, 129] of the polysaccharide from *X. campestris* has recently been characterised as (16) and the conformation in the solid state has been shown to be helical[130].

[→4)-β-D-G*p*-(1→4)-β-D-G*p*-(1→

3
↑
1

α-D-Man*p*-6-OAc

2
↑
1

β-D-GA*p*

4
↑
1

β-D-Man*p*

4 6

C

Me $CO_2H$ ]*n*

(16)

This polysaccharide has also been shown to have an ordered conformation in solution at ambient temperature which 'melts out' with heating[132]. It is not known whether this conformation is the same as the helix which exists in the solid state. Many of the characteristics of the binding of galactomannan to xanthan, such as the specificity and the need for the ordered rather than

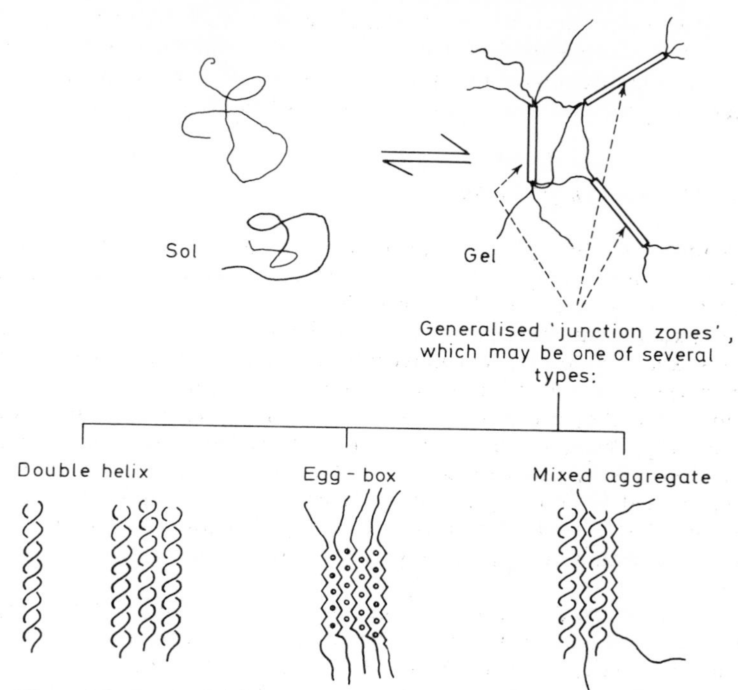


**Figure 1.8** Some mechanisms by which a carbohydrate chain may bind another carbohydrate chain of the same or different structure, with the formation of a gel network. From left to right, the drawings correspond to: isolated double helix (e.g. ι-carrageenan), aggregated double helices (e.g. agarose), egg-box (e.g. calcium polyguluronate), mixed aggregate (e.g. agarose–galactomannan).

disordered form of xanthan, are similar to the binding to agarose and carrageenan helices. This would suggest a similar mechanism, i.e. that binding proceeds by formation of mixed aggregates of xanthan helices and mannan ribbons. This process could provide the means by which the bacterial cell attaches to plant cell surfaces that are identified by characteristic hemicellulose compositions. Since *Xanthomonas* species are plant pathogens (responsible for blights of cabbage, beans, cotton and other crop plants), this recognition would be an important step in the invasion and colonisation of plant tissues[132].

We have discussed only three of the mechanisms by which a carbohydrate chain may bind another carbohydrate chain of the same kind or a different kind. They happen to be the most biologically significant of the mechanisms that are presently understood and are alike in that each leads to a gel network. A schematic summary is shown in Figure 1.8.

### 1.3.5 Protein binding

Proteins are so important and well known for their wide range of binding behaviour that it may seem eccentric to imply as we do here that the protein can be seen as a substrate for binding by carbohydrate, rather than the converse. We shall merely list and classify the types of carbohydrate–protein interactions that are important in biology, so that we can see them in relation to other phenomena that are discussed in this Chapter.

### *1.3.5.1 'Lock and key' binding*

This binding involves globular proteins such as enzymes, antibodies and lectins. Of course, many enzymes exist to act on carbohydrates and each is specific for a particular carbohydrate or a small group of carbohydrates. Likewise, a large number of antibody specificities exist against carbohydrates, including such important carbohydrate antigens as blood-group antigens and bacterial antigens. Lectins are proteins which in many ways resemble antibodies, but they are isolated from plants. They have been much investigated recently, because their specific binding to carbohydrate chains in glycoproteins and glycolipids can be a valuable tool in cell biology[133, 134]. They have been used for the agglutination of cells, the induction of mitosis, the alteration of cell-surface structures and the purification of polysaccharides, glycoproteins and glycopeptides by precipitation and by affinity chromatography.

The most complete picture of the geometry and energetics of carbohydrate binding to a globular protein is still provided by the crystal structure of the lysozyme complexes with tri-*N*-acetyl-β-chitotriose[135], and with a tetrasaccharide analogue of the transition state[131]. The trisaccharide fits snugly into a cleft in the enzyme structure and makes many favourable molecular contacts. These include six hydrogen bonds and over 40 van der Waals' and hydrophobic contacts. Some idea of the relative importance of the various contributions to binding emerges by comparison of the binding of a range of related saccharides[136]. One of the acetamido groups of the trisaccharide appears to be particularly important. This is on the reducing sugar residue and seems to contribute as much as a third of the total free energy of binding or 3 kcal mol$^{-1}$ by hydrogen bonds involving the amide nitrogen and carbonyl and by hydrophobic bonding of the methyl group (Figure 1.9).

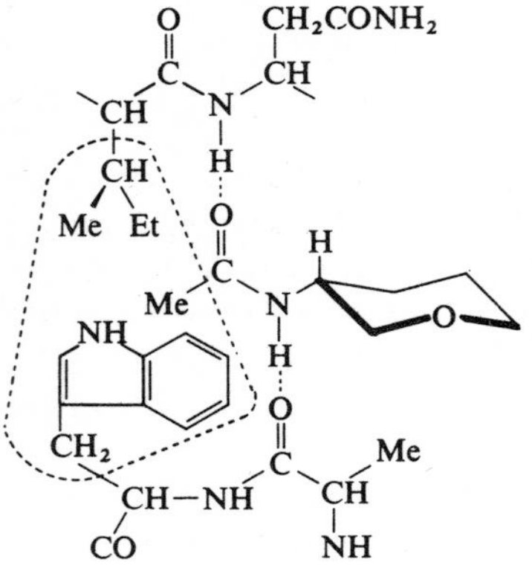

**Figure 1.9** Very schematic representation of the binding of lysozyme of the acetamido group at the reducing terminal of tri-*N*-acetylchitotriose. The 'hydrophobic pocket' is enclosed by a broken boundary and the hydrogen bonds are shown by dotted lines. To a good approximation, the three-atom $CH_3$-CONH sequence is sandwiched between two antiparallel sequences RCHNHCO and RCHCONH, where R is a hydrophobic group in each case. The sandwiching sequences are formed from atoms in the dipeptide segments Tyr 108–Ala 107 and Ile 58–Asn 59, which are below and above the sugar, respectively. For a more complete and accurate representation, see the stereoview published in the lecture by Lipscomb[135]

This binding site was first characterised by diffraction analysis of lysozyme crystals into which the saccharide had been diffused, but much evidence now confirms that the mechanism is the same in solution[136]. This includes an accurate confirmation by $^{19}F$ n.m.r. methods using 2-(*N*-trifluoroacetamido)-2-deoxy-D-glucose that the internal coordinates of the trifluoromethyl group in the solution complex are the same as those of the analogous methyl in the crystalline complex[137].

The primary sequence[138] and the electron density map to high resolution[138,139] have now been determined for the lectin concanavalin A. The binding sites for *myo*-inositol[139] and for *o*-iodophenyl β-D-glucoside[138] in the crystalline complexes have also been established and found to occur in the same region of the molecule. However, these conclusions are in fundamental disagreement with those derived for the solution state by comparing the $^{13}C$ n.m.r. relaxation times of labelled methyl glucosides bound to the $Mn^{2+}$ and $Zn^{2+}$ derivatives of concanavalin A. The paramagnetic contribution to the relaxation was calculated and used to compute the distance and orientation of the sugar with respect to the metal cation[140,141]. It seems likely that the conformation of concanavalin A differs between the crystal and solution and therefore that more work will need to be done to characterise the binding site in solution.

#### *1.3.5.2 Electrostatic binding and related systems*

The polysaccharide chains that are trapped in the meshwork of collagen fibres in connective tissues, and whose water-binding functions (Section 1.3.2) and possible conformations (Sections 1.2.2 and 1.2.3) have already been discussed, are often suggested to interact specifically with collagen and other protein components of the same tissues. A possible model for this biological system is the interaction of glycosaminoglycans with cationic homopolypeptides which drive the polypeptide from the random coil to the α-helix[142-146]. The change is demonstrated by circular dichroism and would seem from the sharp reversal on heating and from the stoichiometry to be caused by stabilisation of the cationic α-helix by regular ionic bonding to the anionic polysaccharide. The relative affinities for binding within the polypeptide series, and within the polysaccharide series, are assessed in the various mixed systems from the tendency of the α-helix to form and from its melting temperature. In general, the interaction is facilitated by increased exposure of the charges (poly-L-arginine > poly-L-lysine > poly-L-ornithine), by increased charge density (e.g. heparin > chondroitin sulphates > hyaluronic acid) and by stereochemical effects which are not understood but which arise from the presence of iduronic acid residues (dermatan sulphate > chondroitin 6-sulphate > chondroitin 4-sulphate).

At least two other systems are known in which proteins are bound to glycosaminoglycan chains by cooperative electrostatic effects that are enhanced by high charge density and the presence of iduronic acid residues in the polysaccharide. Certain serum lipoproteins known to be involved in lipid deposition in the development of atherosclerosis, are bound to polysaccharide chains in the order[147] heparin > dermatan sulphate > heparan sulphate ≈ chondroitin sulphate > hyaluronic acid. Of the glycosaminoglycans that occur naturally in the arterial wall, only dermatan sulphate binds the lipoprotein at physiological salt concentrations, suggesting that this polysaccharide could be involved in atherosclerosis. Dermatan sulphate is also stronger than chondroitin sulphate in binding to soluble collagen[148]. The complex is stable at physiological ionic strength and probably involves clusters of arginine and lysine residues on the collagen. Not all charged groups on the polysaccharide are available for binding[149], which would suggest that long lengths of suitable structures are required—as expected for a cooperative mechanism.

Other binding processes in connective tissues involve hyaluronic acid and proteoglycan[150, 151]. This can be demonstrated by an increase in apparent molecular weight by viscometry or gel filtration. Evidence that this binding exists in native tissue is that proteoglycan is obtained as hyaluronate-linked aggregates when mild conditions are used for isolation. Only one binding site is present on each proteoglycan molecule and this is thought to be situated on the protein part. Inhibition experiments with oligosaccharides would suggest that the binding involves a segment of hyaluronic acid which is about five or six disaccharide residues in length.

### *1.3.5.3 Non-cooperative interactions*

There remains a third, distinct type of polysaccharide–protein interaction which is likely to occur in biological systems and to contribute to their properties. It is actually a general property of polymer mixtures and has been discussed more fully than is possible in the space available here, sometimes with a slant towards particular systems or applications in several books[73, 152, 153] and reviews[87, 154]. Experimentally it may be observed in solution as a visible separation into two or more liquid phases or, if the interaction is weaker, by thermodynamic measurements described below. These phenomena depend on the resultant of all contributions to the free energy of mixing the components and need not include specific binding effects. Such contributions are in essence the same as determine, for example, whether any substance will dissolve in water. The entropy of mixing will amost always favour solution—that is, a single phase system—but two phases will result if this is opposed and outweighed by energy terms. An example is the system barium sulphate + water, in which the lattice energy is so much lower than the energies of the hydrated ions.

The entropy of mixing depends to a first approximation on the number of molecules involved in the mixing and therefore it is independent of molecular weight. In contrast, the interaction energy between two mixed polymers depends on the number of segment–segment contacts and must, for a given number of molecules, increase with molecular weight. Polymers in expanded, disordered conformations will make statistically more segment–segment contacts than those with compact, ordered conformations and, indeed, they are found to show the effects described here to greater degree. The entropy term becomes insignificant at high molecular weight so that the behaviour of mixed polymers is determined by the interaction energies even when the segment–segment contacts are fluctuating and the energies are small. If the interaction between like polymer segments is more favourable than between unlike segments, the two aqueous solutions may separate into two distinct phases which behave like immiscible liquids. This behaviour is often described as 'polymer incompatibility'. If, however, the interaction between unlike segments is more favourable than between like segments, it can happen that the two polymers collect together and separate into a common phase that is either liquid- or solid-like. This is often called 'complex coascervation'.

Polymer incompatibility may be used to create two immiscible aqueous phases for the separation of particles by partition procedures, often using a polysaccharide such as dextran to form one of the phases[87]. Gelatin, being a random-coil protein, shows phase separation with many neutral polysaccharides[155]. Complex coascervates of polysaccharides and proteins which have opposite charges—especially gum arabic and gelatin—form the basis of modern microencapsulation technology[156].

It is found[86] that the osmotic pressure of a solution of incompatible polymers is in excess of the value expected from the sums of the osmotic pressures of the components. (Similarly, the osmotic pressure will be lower than this value if complex coascervation occurs.) This is interpreted as an exclusion of each polymer from the domain of the other by mutual repulsion, which increases the effective concentration, i.e. the activity coefficient of each. Even though actual phase separation has not been seen in mixtures such as hyaluronic acid and bovine serum albumin that are thought to be good models for natural polysaccharide–protein systems in the extracellular parts of tissues, they do show excess osmotic pressures[157, 158]. Another consequence of the changed activity coefficients is that the binding of each polymer to other tissue components may be changed. For example, the apparent equilibrium constants for enzyme–substrate and enzyme–inhibitor binding may be shifted in the presence of an incompatible polysaccharide[159].

When a polysaccharide is in sufficient concentration and has a form that is sufficiently expanded that chain segments can distribute uniformly over solution space, the steric and repulsive interactions may hinder the translational movement of globular proteins while having little effect on their rotational movement[160-162]. A rod-like macromolecule can only travel freely through such a 'network' in the direction of its long axis[163]. These effects can occur at and below physiological polysaccharide concentrations and could therefore be important in modifying transport processes.

### 1.3.6 Functional binding by unknown mechanisms

In earlier sections we have discussed interactions of carbohydrate chains by mechanisms that are well understood, even though the biological functions could not often be seen in detail. By contrast we shall now point to some clear biological functions which involve carbohydrate interactions, but for which molecular explanations cannot yet be given.

Attached to the walls and membranes of Gram-positive bacteria are polymers of glycitol phosphates or sugar 1-phosphates, known as the teichoic acids. The activity of relevant biosynthetic enzymes in the presence and absence of these and other wall and membrane components would suggest that they promote metabolic activity in the outer surfaces of the cell by their accumulation of, and interaction with, $Mg^{2+}$ ions[164]. It has been proposed that teichoic acids bound to the wall may scavenge $Mg^{2+}$ from the medium, while the membrane polymer is more intimately associated with various enzyme complexes and mediates in transfer of $Mg^{2+}$ to them. Perhaps this transfer occurs in such a way as to avoid formation of hydrated magnesium cation as an intermediate, to prevent rate-limitation of the overall enzyme-catalysed reactions by the slow process of displacement of water from the hydration shell of this cation. The membrane teichoic acid is a poly(glycerol phosphate) derivative attached covalently to glycolipid or phosphatidyl-glycolipid, and its interactions with $Mg^{2+}$ seem to be controlled by the extent to which the glycerol residues are esterified with D-alanine[165].

It is becoming increasingly evident that carbohydrate chains are often involved in recognition events at the surfaces of animal cells. Removal of serum glycoproteins from circulation, for example, may occur by an absorption by the liver that is triggered by loss of terminal sialic acid residues[166]. If this exposes an appropriate interior sequence, the carbohydrate structure can bind to the plasma membranes of hepatocytes and uptake follows[167]. This sequence of residues is terminated by β-D-galactose, but inhibition experiments with oligosaccharides of known structure would suggest that the specificity extends further, possibly to the trisaccharide structure *O*-β-D-galactopyranosyl-(1→6)-*O*-(2-acetamido-2-deoxy-β-D-glucopyranosyl)-(1→2)-D-mannose, or even beyond[168]. Because binding is abolished by removal of sialic acid residues from the membrane and restored if they are subsequently replaced[167], it follows that appropriate carbohydrate structures are required on the cell membrane as well as on the glycoprotein substrate. The fact that $Ca^{2+}$ ions must also be present[167] might suggest a conformation change with the binding of $Ca^{2+}$ by sialic acid (*cf.* Section 1.3.3.3). A glycoprotein has now been isolated which seems to be responsible for the binding activity of the membrane[169]; this has been suggested to be a galactosyltransferase[170].

Sialic acid residues have, of course, long been known to be associated with the receptor sites for viruses at cell surfaces[13]. Evidently they also participate in the aggregation of blood platelets by 5-hydroxytryptamine, because the rate and extent of this aggregation are substantially increased when extra sialic acid residues are added to the cell surface by use of a suitable transferase[171].

Many changes occur in the carbohydrate structures of animal cells with different stages of the cell cycle, with malignant transformation and with changes in growth density. These effects have been reviewed several times already[13, 14, 172, 173]. Two further examples will be mentioned here, to suggest that carbohydrate interactions are involved in the recognition of one cell by another that leads to tissue formation.

Certain marine sponges may be dissociated into individual cells which can then recombine, in some cases to form complete functional organisms. Cells from sponges of different pigmentations can, when mixed, sort into species-specific aggregates. The recombination is known to be brought about by soluble 'aggregation factors' which cross-link the cells by specific non-covalent binding. The aggregation factor from *Microciona parthena* has a molecular weight of several million and contains roughly equal amounts by weight of residues of sugars and amino acids[174]. Proteolytic digestion gives a product of high molecular weight that consists mainly of carbohydrate, suggesting that the sugars, which are mainly galactose, mannose, uronic acid and glucosamine, are present in chains of about 100 residues[175]. Electron microscopy suggests an extended, branched macromolecule made up from rod-like segments *ca.* 45 Å in diameter. The substance binds $Ca^{2+}$ strongly and forms a gel in so doing, but loses this property when denatured by heat or EDTA [174]. The factor from *Microciona prolifera* likewise requires $Ca^{2+}$ to cause cell aggregation, but this activity is inhibited by glucuronic acid or by pretreatment with a mixture of glycosidases[176]. A receptor glycoprotein is released from the cell surface by hypotonic treatment[177]. All these facts together indicate that dissociated sponge cells can be recombined by the interaction of two components, namely (i) a multivalent, intercellular aggregation factor which is some kind of polysaccharide–protein complex with a specificity that depends on terminal carbohydrate residues, and on a conformation that is stabilised by the binding of $Ca^{2+}$, and (ii) a receptor protein or glycoprotein that is bound to the cell surface.

It is generally believed that the coordinated cell interactions during development of higher animal tissues are due to specific adhesion effects between cells. The sponge system could therefore be regarded as a model for these processes. A direct indication that carbohydrate groups can be involved in the higher organisms has been provided[178] by incubating cultured fibroblast cells with Sephadex beads of similar size to which various monosaccharide residues had been covalently coupled. Mixed aggregates were formed with beads that had been derivatised with D-galactose, but not with D-glucose or *N*-acetyl-D-glucosamine.

## 1.4 CONCLUSION: SEQUENCE, STEREOCHEMISTRY AND BINDING

For consideration of the stereochemistry and interactions of carbohydrate chains, it is convenient to group the covalent structures into three classes. As discussed below and shown in the diagrammatic summary (Figure 1.10), each class has its own characteristic properties as well as shared properties with the other types.

(i) *Periodic type.* In these structures, regular sequences of sugar residues have the same glycosidic configurations and are linked through the same positions in each 'repeating unit'. Examples are amylose, cellulose and hyaluronic acid. Quite complicated repeating sequences of sugar residues might be present as, for example, in the O-antigen chains of the lipopolysaccharides of Gram-negative bacteria in which up to five residues may be present in each repeating unit[6]. The chains may exist either in ordered or disordered conformations, depending on the balance between conformational entropy and any possibilities for cooperative stabilisation under the particular conditions[119]. If the conformation is ordered, the regular periodic sequence will generate a regular periodic conformation such as a helix or a ribbon[18]. The binding and interaction possibilities for these various stereochemical

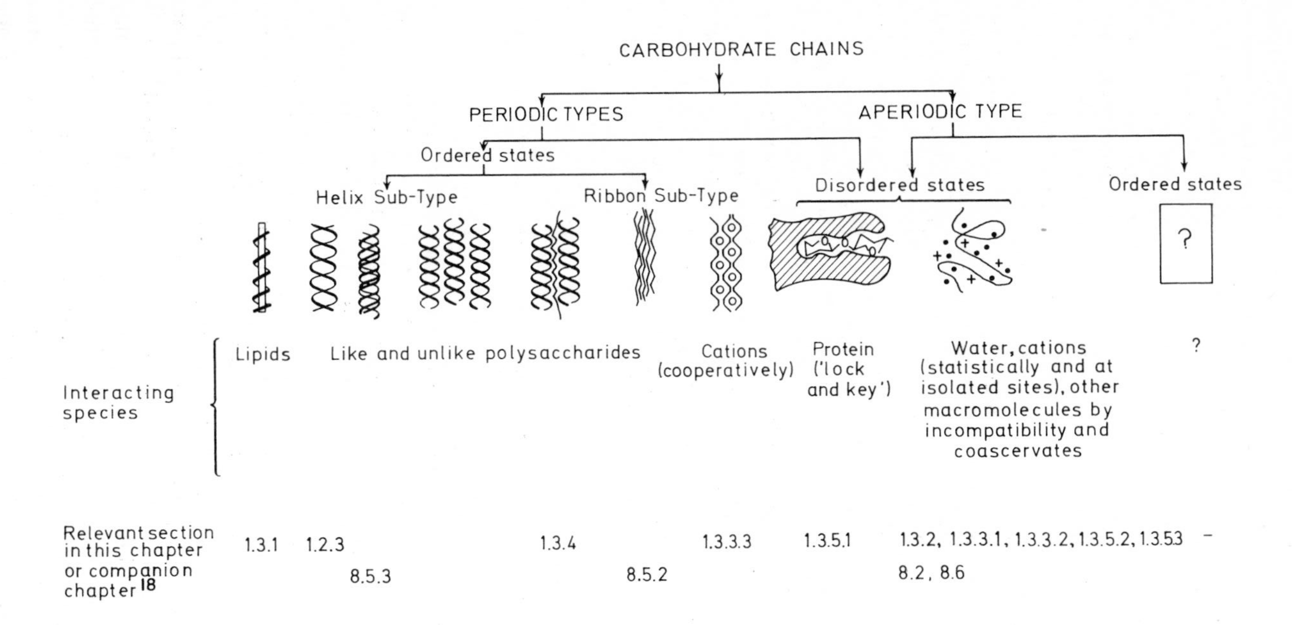


**Figure 1.10** Schematic summary of the stereochemistry and interactions of carbohydrate chains. For the commentary, see Section 1.4 of text

states have been discussed earlier in this chapter and are summarised in Figure 1.10.

(ii) *Periodic type with interruptions*. These polysaccharide chains also have sequences of repeating units, but these are separated, or interrupted, by departures from the regularity. Examples are in alginates, carrageenans and certain glycosaminoglycans. Because each regular region will follow the conformational rules described above, the heterogeneity within each chain can lead to overall conformations that are partly ordered and partly disordered with the corresponding variety of interaction behaviour. This is observed when gel formation results from interrupted periodic structures (Figure 1.8) and the cooperative binding of carbohydrate chains and of cations, the statistical binding of cations and the osmotic binding of water can all occur together in the same system.

(iii) *Aperiodic type*. These carbohydrate chains are very commonly found in covalent attachment to polypeptide and to lipid in components of animal cell surfaces and secretions. There is a growing appreciation that they are likely to have important biological functions that depend on their shape and interactions with proteins, and perhaps with cations, small molecules and other carbohydrate chains (*cf.* Section 1.3.6). In disordered conformations their interaction behaviour is of course likely to be qualitatively similar to that of the other types of carbohydrate chain (Figure 1.10). If, however, they can exist in ordered conformations (and this is not yet known), they must be quite

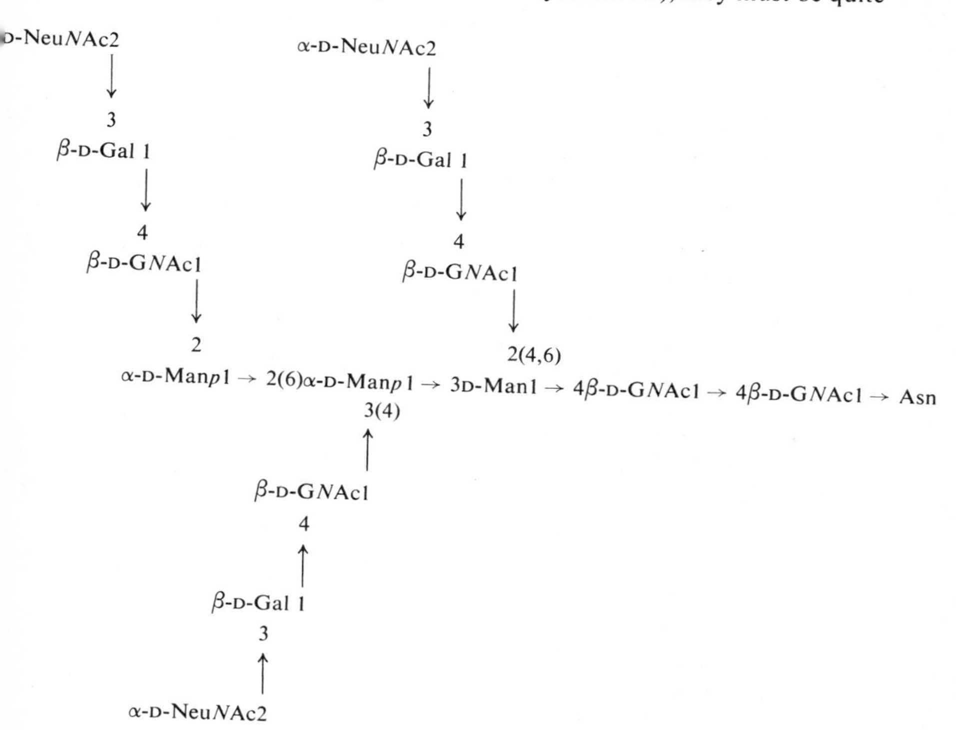


(17)

different in this state from the other chain types. These molecules are characterised by short chain lengths and irregular sequences of sugar units, linkage positions and sometimes configurations[2, 12-14]. As a result, it is geometrically and/or energetically impossible for these molecules to form extended helices or ribbons [see, for example, the structure proposed for a carbohydrate unit in fetuin (17)[12]]. It is tempting to speculate that the dense branching that often exists can serve to gather the chains together into stable 'globular' structures. If this does indeed occur in natural systems we predict that it will have immense implications for biological mechanisms.

## References


1. Aspinall, G. O. (1970). *Polysaccharides* (Oxford: Pergamon)
2. Gottschalk, A. (1972). *Glycoproteins* (A. Gottschalk, editor) (Amsterdam: Elsevier Publishing Co.)
3. (1967–1972). *Specialist Periodical Reports* (*Carbohydrate Chemistry*), Vols. 1–6 (London: The Chemical Society)
4. Montgomery, R. (1973). In *MTP International Review of Science, Organic Chemistry Series One*, Vol. 7, *Carbohydrates* (G. O. Aspinall, editor), 213 (London: Butterworths)
5. Aspinall, G. O. and Stephen, A. M. (1973). Ref. 4, p. 285
6. Lindberg, B. and Svensson, S. (1973). Ref. 4, p. 319
7. Pazur, J. H. and Aronson, N. N. (1972). *Advan. Carbohyd. Chem. Biochem.*, **27,** 301
8. Marshall, R. D. and Neuberger, A. (1970). *Advan. Carbohyd. Chem. Biochem.*, **25,** 407
9. Aspinall, G. O. (1969). *Advan. Carbohyd. Chem. Biochem.*, **24,** 333
10. Kiss, J. (1969). *Advan. Carbohyd. Chem. Biochem.*, **24,** 382
11. Heath, E. C. (1971). *Ann. Rev. Biochem.*, **40,** 29
12. Spiro, R. G. (1973). *Advan. Protein Chem.*, **27,** 349
13. Hughes, R. C. (1973). *Prog. Biophys. Mol. Biol.*, **26,** 189
14. Kraemer, P. M. (1971). *Biomembranes*, **1,** 67
15. Rees, D. A. (1969). *Advan. Carbohyd. Chem. Biochem.*, **24,** 267
16. Rees, D. A. (1972). *Biochem. J.*, **126,** 257
17. Rees, D. A. (1972). *Chem. Ind.* (*London*), 630
18. Rees, D. A. (1973). Ref. 4, p. 251
19. Rees, D. A. (1967). *The Shapes of Molecules: Carbohydrate Polymers* (Edinburgh: Oliver and Boyd)
20. Wells, J. D. (1973). *Proc. Roy. Soc.* **B183,** 399
21. Hardingham, T. E. and Muir, H. This Volume, chapter 4
22. Rodén, L. and Schwartz, N. B. This volume, chapter 3
23. Anderson, N. S., Dolan, T. C. S. and Rees, D. A. (1973). *J. Chem. Soc. Perkin Trans. I*, 2173
24. Lawson, C. J., Rees, D. A., Stancioff, D. J. and Stanley, N. F. (1973). *J. Chem. Soc. Perkin Trans. I*, 2177
25. Penman, A. and Rees, D. A. (1973). *J. Chem. Soc. Perkin Trans. I*, 2182
26. Anderson, N. S., Campbell, J. W., Harding, M. M., Rees, D. A. and Samuel, J. W. B. (1969). *J. Mol. Biol.*, **45,** 85
27. Arnott, S., Scott, W. E., McNab, C. G. M. and Rees, D. A. (1974). *J. Mol. Biol.*, **90,** 253
28. Arnott, S., Fulmer, A., Scott, W. E., Dea, I. C. M., Moorhouse, R. and Rees, D. A. (1974). *J. Mol. Biol.*, **90,** 253
29. Dea, I. C. M., Moorhouse, R., Rees, D. A., Arnott, S., Guss, J. M. and Balazs, E. A. (1973). *Science*, **179,** 560
30. Atkins, E. D. T. and Sheehan, J. K. (1973). *Science*, **179,** 562

30a. Guss, J. M., Hukins, D. W. L., Smith, P. J. C., Winter, W. T., Arnott, S., Moorhouse, R. and Rees, D. A. (1975). *J. Mol. Biol.*, **95,** 359

31. Gordon, E. M. and McCandless, E. L. (1973). *Proc. Nova Scot. Inst. Sci.*, **27,** Supp. 111
32. Rees, D. A. (1962). *Brit. Phycol. Bull.*, **2,** 180

33. Brimacombe, J. S. and Webber, J. M. (1964). *Mucopolysaccharides* (Amsterdam: Elsevier)
34. Laurent, T. C. (1970). In *Chemistry and Molecular Biology of the Intercellular Matrix*, Vol. 2, 703 (E. A. Balazs, editor) (London: Academic Press)
35. Szirmai, J. A. (1966). In *The Amino Sugars*, Vol. IIB, 129 (E. A. Balazs and R. W. Jeanloz, editors) (New York: Academic Press)
36. Balazs, E. A. (1966). *Fed. Proc.*, **25,** 1817
37. Gibbs, D. A., Merrill, E. W., Smith, K. A. and Balazs, E. A. (1968). *Biopolymers*, **6,** 777
38. McKinnon, A. A., Rees, D. A. and Williamson, F. B. (1969). *Chem. Commun.*, 701
39. Bryce, T. A., McKinnon, A. A., Morris, E. R., Rees, D. A. and Thom, D. (1974). *Faraday Discuss. Chem. Soc.*, **57,** 221
40. Bryce, T. A., Rees, D. A., Reid, D. S. and Williamson, F. B. (1975). *Biopolymers*, submitted
41. Jones, R. A., Staples, E. J. and Penman, A. (1973). *J. Chem. Soc. Perkin Trans. II*, 1608
42. Rees, D. A., Scott, W. E. and Williamson, F. B. (1970). *Nature (London)*, **227,** 390
43. Reid, D. S., Bryce, T. A., Clark, A. H. and Rees, D. A. (1974). *Faraday Discuss. Chem. Soc.*, **57,** 230
44. Chakrabarti, B. and Balazs, E. A. (1973). *J. Mol. Biol.*, **78,** 135
45. Chakrabarti, B. and Balazs, E. A. (1973). *Biochem. Biophys. Res. Commun.*, **52,** 1170
45a. Darke, A., Finer, E. G., Moorhouse, R. and Rees, D. A. (1975). *J. Mol. Biol.*, in press
46. Rees, D. A. (1969). *J. Chem. Soc. B*, 217
47. Balazs, E. A. and Gibbs, D. A. (1970). In *Chemistry and Molecular Biology of the Intercellular Matrix*, Vol. 3, 1241 (E. A. Balazs, editor) (London: Academic Press)
48. Atkins, E. D. T., Phelps, C. F. and Sheehan, J. K. (1972). *Biochem. J.*, **128,** 1255
49. Arnott, S., Guss, J. M., Hukins, D. W. L. and Matthews, M. B. (1973). *Science*, **180,** 743
50. Atkins, E. D. T., Gaussen, R., Isaac, D. H., Nandanwar, V. and Sheehan, J. K. (1972). *J. Polymer Sci. B*, **10,** 863
51. Isaac, D. H. and Atkins, E. D. T. (1973). *Nature New Biol.*, **244,** 252
52. Atkins, E. D. T. and Isaac, D. H. (1973). *J. Mol. Biol.*, **80,** 773
53. Arnott, S., Guss, J. M., Hukins, D. W. L. and Matthews, M. B. (1973). *Biochem. Biophys. Res. Commun.*, **54,** 1377
54. Arnott, S., Guss, J. M., Hukins, D. W. L., Dea, I. C. M. and Rees, D. A. (1974). *J. Mol. Biol.*, **88,** 175
55. Ilton, M., Jevans, M. A. W., McCarthy, E. D., Vance, D., White, H. B. and Bloch, K. (1971). *Proc. Nat. Acad. Sci. USA*, **68,** 87
56. Ballou, C. E. (1968). *Accounts Chem. Res.*, **1,** 366
57. Gray, G. R. and Ballou, C. E. (1971). *J. Biol. Chem.*, **246,** 6835
58. Smith, W. L. and Ballou, C. E. (1973). *J. Biol. Chem.*, **248,** 7118
59. Saier, M. H. and Ballou, C. E. (1968). *J. Biol. Chem.*, **243,** 4332
60. Saier, M. H. and Ballou, C. E. (1968). *J. Biol. Chem.*, **243,** 992
61. Keller, J. M. and Ballou, C. E. (1968). *J. Biol. Chem.*, **243,** 2905
62. De Belder, A. N. and Norrman, B. (1968). *Carbohyd. Res.*, **8,** 1
63. Hoare, D. G., Olson, A. and Koshland, D. E. (1968). *J. Amer. Chem. Soc.*, **90,** 1638
64. Thoma, J. A. and Stewart, L. (1965). In *Starch: Chemistry and Technology*, Vol. 1, 209 (R. L. Whistler and E. Paschall, editors) (New York: Academic Press)
65. Vance, D. E., Mitsuhashi, O. and Bloch, K. (1973). *J. Biol. Chem.*, **248,** 2303
66. Vance, D. E., Esders, T. W. and Bloch, K. (1973). *J. Biol. Chem.*, **248,** 2310
67. Knoche, H., Esders, T. W., Koths, K. and Bloch, K. (1973). *J. Biol. Chem.*, **248,** 2317
68. Machida, Y. and Bloch, K. (1973). *Proc. Nat. Acad. Sci. USA*, **70,** 1146
69. Rees, D. A. (1973). Ref. 4, p. 268
70. Bergeron, A., Machida, Y. and Bloch, K. (1975). *J. Biol. Chem.*, **250,** 1223
70a. Rees, D. A. (1975). Unpublished work
71. E.g., Osborn, M. J. (1969). *Ann. Rev. Biochem.*, **38,** 501
72. Flory, P. J. and Rehner, J. J. (1943). *J. Chem. Phys.*, **11,** 512
73. Flory, P. J. (1953). *Principles of Polymer Chemistry* (New York: Cornell University Press)
74. Katchalsky, A. (1954). *Progr. Biophys.*, **4,** 1
75. Edmond, E., Farquhar, S., Dunstone, J. R. and Ogston, A. G. (1968). *Biochem. J.*, **108,** 755

76. Ogston, A. G. and Wells, J. D. (1970). *Biochem. J.*, **119,** 67
77. Ogston, A. G. and Preston, B. N. (1973). *Biochem. J.*, **131,** 843
78. Manning, G. S. (1969). *J. Chem. Phys.*, **51,** 924
79. Manning, G. S. (1969). *J. Chem. Phys.*, **51,** 3249
80. Bailey, J. M. (1973). *Biopolymers*, **12,** 1705
81. Wells, J. D. (1973). *Biopolymers*, **12,** 223
82. Ogston, A. G. and Wells, J. D. (1972). *Biochem. J.*, **128,** 685
83. Meyer, F. A., Comper, W. D. and Preston, B. N. (1971). *Biopolymers*, **10,** 1351
84. Traub, W. and Piez, K. A. (1971). *Advan. Protein Chem.*, **25,** 243
85. Preston, B. N. and Meyer, F. A. (1971). *Biopolymers*, **10,** 35
86. Edmond, E. and Ogston, A. G. (1968). *Biochem. J.*, **109,** 569 and references given there
87. Albertsson, P.-Å. (1970). *Advan. Protein Chem.*, **24,** 309
88. Preston, B. N., Snowden, J. McK. and Houghton, K. T. (1972). *Biopolymers*, **11,** 1645
89. Angyal, S. J. (1973). *Pure Appl. Chem.*, **35,** 131
90. Rendleman, J. A. (1966). *Advan. Carbohyd. Chem. Biochem.*, **21,** 209
91. Mills, J. A. (1961). *Biochem. Biophys. Res. Commun.*, **6,** 418
92. Angyal, S. J. and Davies, K. P. (1971). *Chem. Commun.*, 500
93. Angyal, S. J. (1972). *Aust. J. Chem.*, **25,** 1957
94. Evans, M. E. and Angyal, S. J. (1972). *Carbohyd. Res.*, **23,** 43
95. Pauling, L. (1960). *The Nature of the Chemical Bond*, 3rd Ed., 514 (New York: Cornell University Press)
96. Gould, R. O. and Rankin, A. F. (1970). *Chem. Commun.*, 489
97. Anthonsen, T., Larsen, B. and Smidsrød, O. (1972). *Acta Chem. Scand.*, **26,** 2988
98. Anthonsen, T., Larsen, B. and Smidsrød, O. (1973). *Acta Chem. Scand.*, **27,** 2671
99. Bleaney, B., Dobson, C. M., Levine, B. A., Martin, R. B., Williams, R. J. P. and Xavier, A. V. (1972). *J. Chem. Soc. Chem. Commun.*, 791
100. Craig, D. C., Stephenson, N. C. and Stevens, J. D. (1972). *Carbohyd. Res.*, **22,** 494
101. Bugg, C. E. and Cook, W. J. (1972). *J. Chem. Soc. Chem. Commun.*, 727
102. Bugg, C. E. (1973). *J. Amer. Chem. Soc.*, **95,** 908
103. Richards, G. F. (1973). *Carbohyd. Res.*, **26,** 448
104. Cook, W. J. and Bugg, C. E. (1973). *Carbohyd. Res.*, **31,** 265
105. Cook, W. J. and Bugg, C. E. (1973). *J. Amer. Chem. Soc.*, **95,** 6442
106. Furberg, S. and Helland, S. (1962). *Acta Chem. Scand.*, **16,** 2373
107. Balchin, A. A. and Carlisle, C. H. (1965). *Acta Crystallogr.*, **19,** 103
108. Norrestam, R., Werner, P. E., and von Glehn, M. (1968). *Acta Chem. Scand.*, **22,** 1395
109. Cook, W. J. and Bugg, C. E. (1973). *Acta Crystallogr.*, **B29,** 215
110. Gould, S. E. B., Gould, R. O., Rees, D. A., and Scott, W. E. (1975). *J. Chem. Soc. Perkin Trans. II*, 237
111. Gould, S. E. B., Gould, R. O., Rees, D. A. and Wight, A. W. (1975). *J. Chem. Soc. Perkin Trans. II*, in press
112. Arnott, S. and Scott, W. E. (1972). *J. Chem. Soc. Perkin Trans. II*, 324
113. Girling, R. L. and Jeffrey, G. A. (1971). *Carbohyd. Res.* **18,** 339
114. Kohn, R. and Larsen, B. (1972). *Acta Chem. Scand.*, **26,** 2455
115. Penman, A. and Sanderson, G. R. (1972). *Carbohyd. Res.*, **25,** 273
116. Atkins, E. D. T., Nieduszynski, I. A., Mackie, W., Parker, K. D. and Smolko, E. E. (1973). *Biopolymers*, **12,** 1879
117. Atkins, E. D. T., Nieduszynski, I. A., Mackie, W., Parker, K. D. and Smolko, E. E. (1973). *Biopolymers*, **12,** 1865
118. Smidsrød, O., Haug, A. and Whittington, S. G. (1972). *Acta Chem. Scand.*, **26,** 2563
119. Rees, D. A. (1973). Ref. 4, p. 252
120. Boyd, J., Morris, E. R., Rees, D. A., Thom. D. and Turvey, J. R. (1974). In preparation
121. Smidsrød, O. and Haug, A. (1972). *Acta Chem. Scand.*, **26,** 2063
122. Rees, D. A. (1973). Ref. 4, pp. 263, 279
123. Grant, G. T., Morris, E. R., Rees, D. A., Smith, P. J. C. and Thom, D. (1973). *FEBS Lett.*, **32,** 195
124. e.g. Helgeson, R. C., Timko, J. M. and Cram, D. J. (1973). *J. Amer. Chem. Soc.*, **95,** 3023
125. Morris, E. R., Rees, D. A. and Thom, D. (1973). *J. Chem. Soc. Chem. Commun.*, 245
126. Dea, I. C. M., McKinnon, A. A. and Rees, D. A. (1972). *J. Mol. Biol.*, **68,** 153
127. Dea, I. C. M., and Morrison, A. (1974). *Advan. Carbohyd. Chem. Biochem.*, **29,** in the press

128. Jansson, P.-E., Kenne, L. and Lindberg, B. (1975). *Carbohyd. Res.*, in the press
129. Melton, L. D., Mindt, L., Rees, D. A. and Sanderson, G. R. (1975). *Carbohyd. Res.*, in the press
130. Arnott, S., Moorhouse, R. and Rees, D. A. (1975). In preparation
131. Ford, L. O., Johnson, L. N., Machin, P. A., Phillips, D. C. and Tjian, R. (1974). *J. Mol. Biol.*, **88**, 349
132. Morris, E. R., Rees, D. A. and Walkingshaw, M. (1974). *J. Mol. Biol.*, submitted
133. Sharon, N. and Lis, H. (1972). *Science*, **177**, 949
134. Burger, M. M. (1973). *Fed. Proc.*, **32**, 91
135. Blake, C. C. F., Mair, G. A., North, A. C. T., Phillips, D. C. and Sarma, V. R. (1967) *Proc. Roy. Soc.*, **B167**, 365; for stereo illustrations of the enzyme–oligosaccharide complex, see Lipscomb, W. N. (1972). *Quart. Rev. Chem. Soc.*, **20**, 1 and Harte, R. A. and Rupley, J. A. (1968). *J. Biol. Chem.*, **243**, 1663; see also Ref. 131
136. Chipman, D. M. and Sharon, N. (1969). *Science*, **165**, 454
137. Butchard, C. G., Dwek, R. A., Kent, P. W., Williams, R. J. P. and Xavier, A. V. (1972). *Eur. J. Biochem.*, **27**, 548
138. Edelman, G. M., Cunningham, B. A., Reeke, G. N., Becker, J. W., Waxdal, M. J. and Wang, J. L. (1972). *Proc. Nat. Acad. Sci. USA*, **69**, 2580
139. Hardman, K. D. and Ainsworth, C. F. (1972). *Biochemistry*, **11**, 4910
140. Brewer, C. F., Sternlicht, H., Marcus, D. H. and Grollman, A. P. (1973). *Proc. Nat. Acad. Sci. USA*, **70**, 1007
141. Brewer, C. F., Sternlicht, H., Marcus, D. H. and Grollman, A. P. (1973). *Biochemistry.*, **12**, 4448
142. Gelman, R. A., Rippen, W. B. and Blackwell, J. (1973). *Biopolymers*, **12**, 541
143. Gelman, R. A., Glaser, D. N. and Blackwell, J. (1973). *Biopolymers*, **12**, 1223
144. Gelman, R. A. and Blackwell, J. (1973). *Biopolymers*, **12**, 1959
145. Gelman, R. A. and Blackwell, J. (1973). *Biochim. Biophys. Acta*, **297**, 452
146. Gelman, R. A. and Blackwell, J. (1973). *Arch. Biochem. Biophys.*, **159**, 427
147. Iverius, P. H. (1972). *J. Biol. Chem.*, **247**, 2607
148. Öbrink, B. (1973). *Eur. J. Biochem.*, **33**, 387
149. Öbrink, B. and Sundelöf, L.-O. (1973). *Eur. J. Biochem.*, **37**, 226
150. Hardingham, T. E. and Muir, H. (1972). *Biochim. Biophys. Acta*, **279**, 401
151. Hardingham, T. E. and Muir, H. (1973). *Biochem. J.* **135**, 905
152. Morawetz, H. (1965). *Macromolecules in Solution* (New York: Wiley-Interscience)
153. Albertsson, P.-Å. (1960). *Partition of Cell Particles and Macromolecules* (New York: Wiley)
154. Koningsveld, R. (1968). *Advan. Colloid Interface Sci.*, **2**, 2
155. Grinberg, V. Y. and Tolstoguzov, V. B. (1972). *Carbohyd. Res.*, **25**, 313
156. Ranney, M. W. (1969). *Microencapsulation Technology*, New Jersey: Noyes Development Corporation)
157. Laurent, T. C. and Ogston, A. G. (1963). *Biochem. J.*, **89**, 249
158. Preston, B. N., Davies, M. and Ogston, A. G. (1965). *Biochem. J.*, **96**, 449
159. Laurent, T. C. (1971). *Eur. J. Biochem.*, **21**, 498
160. Laurent, T. C. and Öbrink, B. (1972). *Eur. J. Biochem.*, **28**, 94
161. Preston, B. N., Öbrink, B. and Laurent, T. C. (1973). *Eur. J. Biochem.*, **33**, 401
162. Öbrink, B. and Laurent, T. C. (1974). *Eur. J. Biochem.*, **41**, 83
163. Laurent, T. C., (1974). Personal communication
164. Hughes, A. H., Hancock, I. C. and Baddiley, J. (1973). *Biochem. J.*, **132**, 83
165. Archibald, A. R., Baddiley, J. and Heptinstall, S. (1973). *Biochim. Biophys. Acta*, **291**, 629
166. Morell, A. G., Gregoriadis, G., Scheinberg, I. H., Hickman, J. and Ashwell, G. (1971). *J. Biol. Chem.*, **246**, 1461
167. Price, W. E. and Ashwell, G. (1971). *J. Biol. Chem.*, **246**, 4825
168. Rogers, J. C. and Kornfeld, S. (1971). *Biochem. Biophys. Res. Commun.*, **45**, 622
169. Morell, A. G. and Scheinberg, I. H. (1972). *Biochem. Biophys. Res. Commun.*, **48**, 808
170. Aronson, N. N., Tan, L. Y. and Peters, B. P. (1973). *Biochem. Biophys. Res. Commun.*, **53**, 112
171. Mester, L., Szabados, L., Born, G. V. R. and Michal, F. (1972). *Nature New Biol.*, **236**, 213
172. Oseroff, A. R., Robbins, P. W. and Burger, M. M. (1973). *Ann. Rev. Biochem.*, **42**, 647

173. Kemp, R. B., Lloyd, C. W. and Cook, G. M. W. (1973). *Progr. Surface Membrane Sci.*, **7,** 271
174. Henkart, P., Humphreys, S. and Humphreys, T. (1973). *Biochemistry*, **12,** 3045
175. Cauldwell, C. B., Henkart, P. and Humphreys, T. (1973). *Biochemistry*, **12,** 3051
176. Turner, R. S. and Burger, M. M. (1973). *Nature*, **244,** 509
177. Weinbaum, G. and Burger, M. M. (1973). *Nature*, **244,** 510
178. Chipowsky, S., Lee, Y. C. and Roseman, S. (1973). *Proc. Nat. Acad. Sci. USA*, **70,** 2309

# 2
# Carbohydrates and Cell-surface Phenomena

**K. W. TALMADGE and M. M. BURGER**
University of Basel

## 2.1 INTRODUCTION

Bacteria, yeast, plants, i.e. most lower forms of life, contain a cell surface which is composed primarily of polysaccharide. It is believed that one important function of these cell surface polysaccharides is structural—to maintain a rigid cell wall. In contrast, mammalian cells have no rigid cell surface and contain only small amounts of polysaccharide, mostly in the form of mucopolysaccharides (proteoglycans); in addition, a large portion of their carbohydrate is covalently linked to protein in the form of glycoprotein or to lipid in the form of glycolipid. These cell-surface carbohydrates are of particular interest in mammalian systems because of their possible role as cell recognition sites complementary to receptor sites on other cells or on environmental effectors. These cell recognition sites are involved in many fundamental processes governing various functions present in multicellular organisms, such as specific adhesion, contact phenomena and morphogenesis. Histogenesis during development, and loss of organised structure during carcinogenesis, are manifestations of the acquisition and loss of the ability of cells to recognise each other as a part of an organised multicellular unit[1].

In this review we will discuss the current knowledge concerning the chemical nature of the carbohydrate-containing components present on mammalian cell surfaces, and we will try to indicate some of the functions in which these may be involved.

In Figure 2.1 is a diagrammatic representation of a mammalian cell surface. The 'cell surface' as used in this review, unless otherwise stated, includes cell components found outside the unit membrane structure, particularly the carbohydrate-rich exterior region of the cell which electron microscopists have referred to as the 'fuzzy coat'[2]. Structures are assumed to be in this location if they can be detected with specific reagents (such as

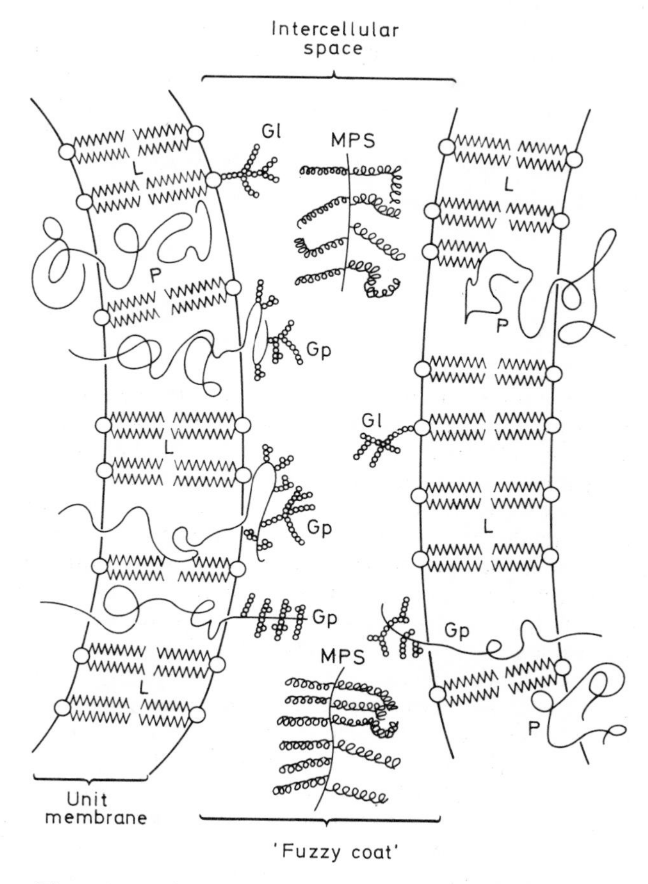


**Figure 2.1** Diagrammatic representation of the cell surface, including the basic components of the unit membrane and intercellular space. Abbreviations are as follows: L = lipid, P = protein, Gl = glycolipid, Gp = glycoprotein, MPS = mucopolysaccharide

antibodies and lectins), or if they can be liberated from intact cells by treatment with various enzymes which do not destroy the integrity of the membrane. This definition is somewhat arbitrary because it is impossible to draw a sharp line between surface and environment, i.e. between components intrinsic to the membrane and extrinsic factors such as secreted products and serum molecules which may bind specifically or adventitiously to the membrane surface. However, there is increasing evidence, which will be discussed later, that both types of material may be structurally and functionally important in cell-surface phenomena.

## 2.2 GLYCOPROTEINS, GLYCOLIPIDS AND MUCOPOLYSACCHARIDES IN MAMMALIAN CELL SURFACES

Before discussing in detail the role of cell surface carbohydrates in biological systems, it would be helpful to discuss cursorily at least the structure and metabolism of the various sugar-containing cell surface components. In this section we will review general studies on glycoproteins, glycolipids and mucopolysaccharides and will also indicate, within each class, specific cell surface components which have been examined and mention their possible functions.

### 2.2.1 Metabolism

#### *2.2.1.1 Biosynthesis*

Complex carbohydrate-containing materials are formed in a completely different manner than polymers such as proteins and nucleic acids, where the genetic information is determined by a template and the polymers are then formed by relatively non-specific enzyme systems. In the synthesis of heterosaccharide chains of glycoproteins, mucopolysaccharides or glycolipids, the sequence of sugars in the heterosaccharide is under more indirect genetic control; it resides in the specificities of different glycosyltransferases which add the monosaccharides to appropriate acceptor molecules[3]. The monosaccharides are added via activated intermediates which are the appropriate nucleotide derivatives. The general reaction for this sugar transfer is given in Figure 2.2. All of the glycosyltransferases within one family, such as sialyltransferases or galactosyltransferases, utilise the same sugar nucleotide as the glycosyl donor, e.g. cytidine monophosphate (CMP)–sialic acid, uridine diphosphate (UDP)–galactose. Thus, the biosynthesis of carbohydrate units with defined structures is determined by the specificities of these enzymes, which are directed toward both the sugar nucleotide and the acceptor molecule. The glycosyltransferases appear to work in concert to assemble a unit, with the product of one enzymatic reaction becoming the substrate for the next reaction. It is currently believed that specific complexes

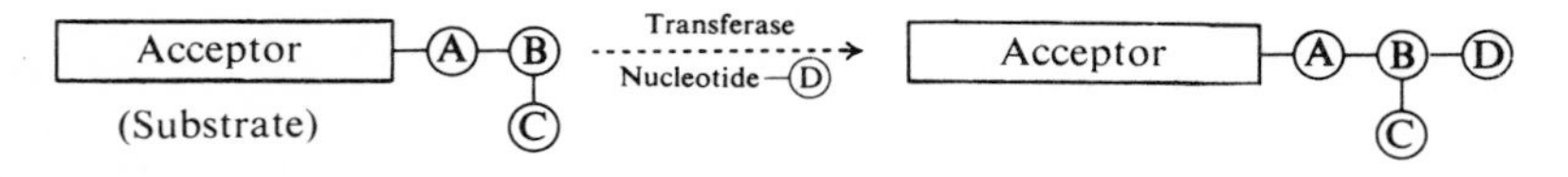

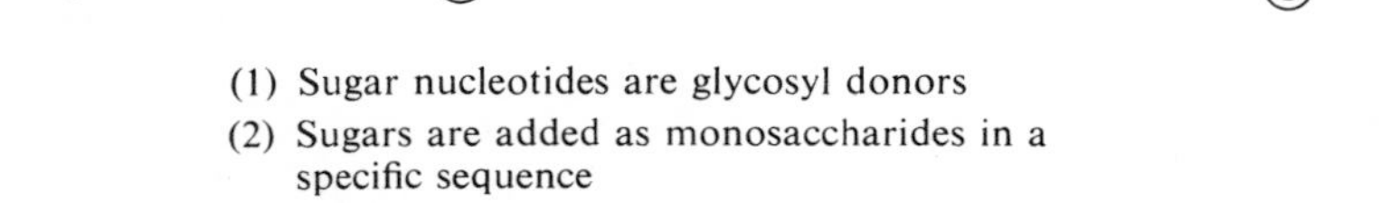

(1) Sugar nucleotides are glycosyl donors
(2) Sugars are added as monosaccharides in a specific sequence
(3) Chain elongation occurs at non-reducing ends or branch points

**Figure 2.2** Synthesis of oligosaccharide chains in glycoproteins, mucopolysaccharides and glycolipids. Acceptor is protein or lipid. The letters signify carbohydrate residues

of glycosyltransferases, called multiglycosyltransferase systems[4], catalyse the synthesis of the oligosaccharide chains, and that different transferase systems are required for synthesis of glycoproteins, mucopolysaccharides and glycolipids.

An example of one such system is the biosynthesis of ganglioside $GM_1$, which is depicted in Figure 2.3. This is an important glycolipid of brain cells and is formed by the stepwise addition of sugars to the terminal hydroxyl of ceramide. The two galactose molecules in $Gm_1$ are added by two different galactosyltransferases. The final structure is determined then by the specificities of the five glycosyltransferases involved.

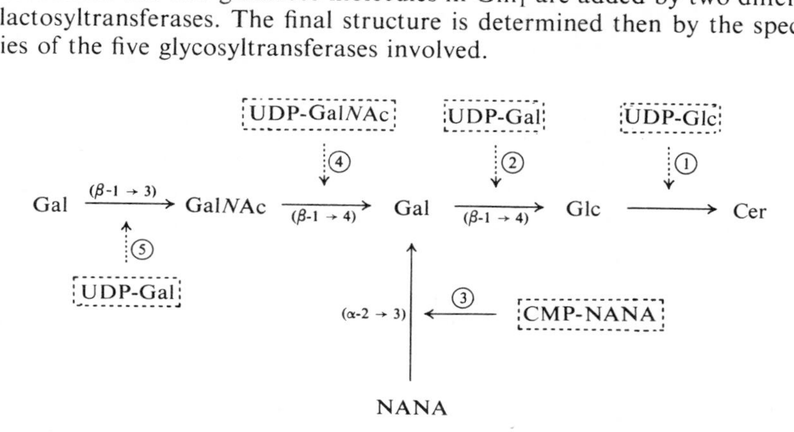


**Figure 2.3** Biosynthesis of ganglioside $GM_1$. The monosaccharides are transferred to the indicated acceptors from the sugar nucleotides by the five specific glycosyltransferases: ①-glucosyltransferase, ②-galactosyltransferase, ③-sialyltransferase, ④-*N*-acetylgalactosaminyltransferase, ⑤-galactosyltransferase. The following abbreviations are used: Cer = ceramide, Glc = glucose, Gal = galactose, Gal*N*Ac = *N*-acetylgalactosamine, Glc*N*Ac = *N*-acetylglucosamine, Fuc = fucose, NANA = *N*-acetylneuraminic acid (sialic acid). (From Roseman[3], by courtesy of *Chem. Phys. Lipids*)

In the case of glycoprotein synthesis, present evidence[5-7] indicates that the polypeptide moiety is first assembled on membrane-bound ribosomes of the rough endoplasmic reticulum and that the majority of the sugars are added in smooth membrane fractions and the Golgi apparatus during the later stages of protein maturation. The role of the Golgi apparatus in this process has been recently reviewed by Whaley *et al.*[8].

Regulation of carbohydrate synthesis in glycoproteins, mucopolysaccharides and glycolipids is poorly understood at the present time, but must certainly involve availability of the appropriate acceptors, transferases and nucleotide sugars. Recent data from mammalian systems[9,10] suggest the occurrence of polyisoprenoid lipid intermediates similar to those which have been shown to be important in the synthesis of certain bacterial carbohydrate components[11].

In bacteria these lipid intermediates function as carriers to convert water-soluble activated sugars into hydrophobic molecules that can then react with macromolecules within the hydrophobic environment of the membrane. If further work does establish such a role for lipid carriers in mammalian heterosaccharide synthesis, then these lipid intermediates could fulfill an important role in regulatory mechanisms.

### *2.2.1.2 Degradation*

Experimental evidence has accumulated in recent years indicating that lysosomes play an important role in the degradation of cellular materials[12]. A large number of hydrolases, most of which have acid pH optima, are contained within the lysosomal vesicles and it has been demonstrated that these can function to degrade a wide variety of cellular components[12]. The largest group of enzymes in lysosomes are glycosidases, many of which have now been characterised[13]. These lysosomal glycosidases are very specific and apparently act sequentially to degrade in a stepwise manner the oligosaccharide units of glycoproteins, mucopolysaccharides and glycolipids in a pathway that is in many instances the reverse of biosynthesis.

The Golgi apparatus, in addition to being important in biosynthesis within the cell, may also play a role in the catabolic functioning of the cell through the formation of the membrane-bound vesicles containing the lysosomal enzymes[8]. The membranes of these lysosomal vesicles have the capacity to fuse in certain cases with the plasma membrane or with plasma membrane-derived endocytic vesicles, which may bring surface material back into the cell for degradation[14]. The specificity of this fusion may be related to membrane characteristics developed in the Golgi apparatus, since both the plasma membrane and the lysosomal vesicles have their origin in this organelle.

In recent years, considerable attention has been given to a number of pathological conditions in which there appears to be a disorder in the catabolism of carbohydrate containing compounds, including mucopolysaccharides, glycolipids and glycoproteins[15,16]. These storage diseases are characterised by the toxic accumulation of specific compounds which are not cleaved

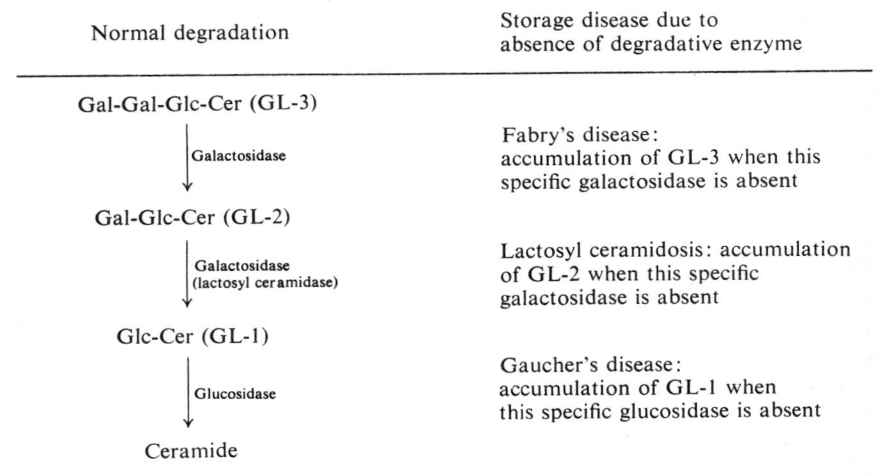

Normal degradation
Storage disease due to absence of degradative enzyme
Gal-Gal-Glc-Cer (GL-3)
Galactosidase
Fabry's disease: accumulation of GL-3 when this specific galactosidase is absent
Gal-Glc-Cer (GL-2)
Galactosidase (lactosyl ceramidase)
Lactosyl ceramidosis: accumulation of GL-2 when this specific galactosidase is absent
Glc-Cer (GL-1)
Glucosidase
Gaucher's disease: accumulation of GL-1 when this specific glucosidase is absent
Ceramide


**Figure 2.4** Biochemical derangement of glycolipid metabolism. Abbreviations are given in Figure 2.3

owing to the lack of a specific glycosidase. In Figure 2.4 several examples are given of diseases which are associated with malfunctions in the degradation of glycolipids.

### 2.2.2 Glycoproteins

The general structural features of glycoproteins have been covered in a number of reviews[17, 18] and will not be dealt with in detail here. Glycoproteins comprise a diverse class of compounds ranging in molecular weight from 15 000 to over one million and include such molecules as immunoglobulins, hormones and enzymes. Some glycoproteins, as the serum proteins, contain a few, branched, oligosaccharide chains of 20–30 sugar residues, while other glycoproteins, e.g. mucins, may contain up to 800 disaccharide units per molecule. Schematic structures of serum type and mucin type glycoproteins are illustrated in Figure 2.5. Seven sugars account for nearly all types of carbohydrates found in mammalian glycoproteins: L-fucose, D-mannose, D-galactose, D-glucose, *N*-acetylglucosamine, *N*-acetylgalactosamine and sialic acid (*N*-acetylneuraminic acid). In general, fucose and sialic acid are found only at the non-reducing ends of chains. The oligosaccharide units are glycosidically linked through the C-1 position of the reducing terminal sugar residues to a functional group on an amino acid in the peptide chain

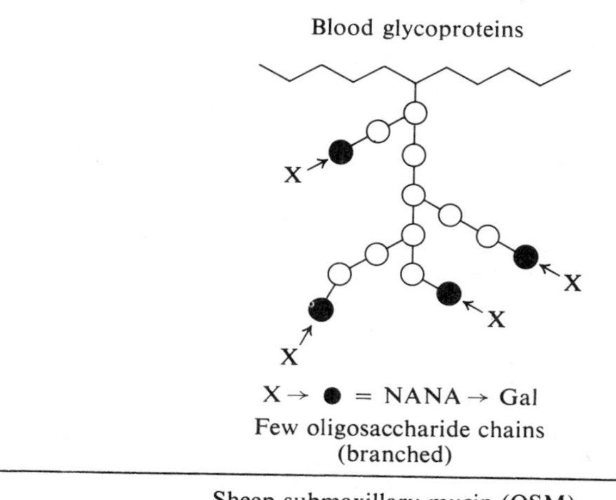


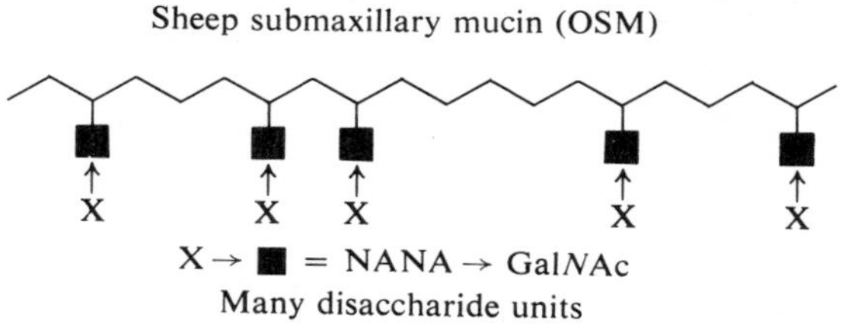


**Figure 2.5** Schematic structures of serum type glycoproteins and mucins. Abbreviations are the same as in Figure 2.3. (Taken in part from Roseman[3], by courtesy of *Chem. Phys. Lipids*)

of the protein. The three types of linkages most commonly found between the carbohydrate chains and amino acid residues are summarised below:
(a) The amide nitrogen of asparagine in a *N*-glycosidic bond with *N*-acetyl-glucosamine. This is common in serum type glycoproteins (Figure 2.6a).
(b) The hydroxy group of serine or threonine occuring in an *O*-glycosidic linkage with *N*-acetylgalactosamine. This type of linkage is present in mucins (Figure 2.6b).

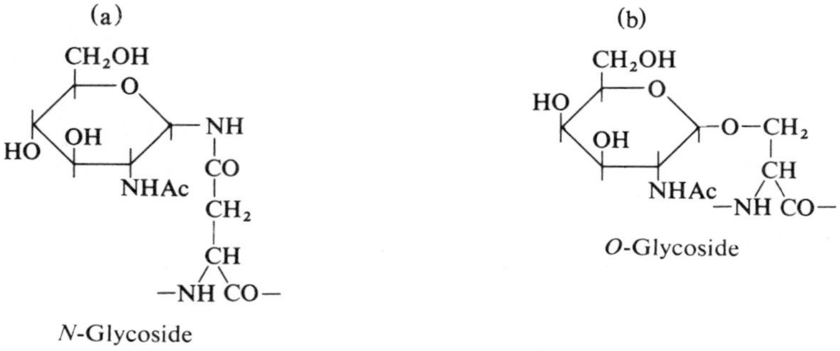

(a)

$CH_2OH$, O, HO, OH, NHAc, NH, CO, $CH_2$, CH, —NH CO—

*N*-Glycoside

(b)

$CH_2OH$, O, HO, OH, NHAc, —O—$CH_2$, CH, —NH CO—

*O*-Glycoside

**Figure 2.6** Basic types of linkage between oligosaccharides and peptide chains in glycoproteins

(c) The hydroxy group of hydroxylysine involved in an *O*-glycosidic linkage with a galactose residue. This is found only in structural proteins such as collagen, and basement membranes which have the carbohydrate units as either galactose or the disaccharide α-glucosyl-(1→2)-galactose[19]. The occurence of glucose in glycoproteins is restricted to this type of structural glycoprotein.

Thioglycosidic linkages have recently been reported to occur in glycopeptides isolated from urine[20] and from erythrocyte membranes[21].

Most glycoproteins contain oligosaccharides attached by only one type of linkage. There are, however, examples of glycoproteins such as immuno-globulin G, which contain two types of oligosaccharide chain that are attached by two types of linkage[22].

The similar carbohydrate units in glycoproteins may not be precisely identical, either within the same molecules or within different molecules from the same tissue. This is a common situation and has been termed 'peripheral' heterogeneity or 'microheterogeneity'[23, 24] and is believed to be due to variability during synthesis or to variable degrees of degradation *in vivo*. This microheterogeneity has complicated and protracted the fine structural analysis of the carbohydrate unit in many glycoproteins.

### *2.2.2.1 Cell-surface glycoproteins*

This subject has been covered very extensively in a number of recent reviews[25-29] and only the recent material will be emphasised here. There is a great deal of microscopic evidence for the presence of carbohydrate material at cell surfaces. Likewise it has been well documented that glycoproteins are present as integral components of the plasma membrane of mammalian cells. The evidence for this comes from studies of the chemical composition of isolated plasma membranes and also from the liberation of glycopeptides from intact cells by the use of proteolytic enzymes. One of the most significant facts that has come out of these early studies is that the carbohydrate associated with the plasma membrane is asymmetrically distributed, being primarily on the external surface of the cell.

It is also worthwhile noting here that many structural features present in the carbohydrate chains of the soluble glycoproteins, discussed in the previous section, have also been found to occur in the limited number of glycoproteins that have been isolated from membranes. These membrane glycoproteins, however, do show some properties which differ from the soluble glycoproteins, but these appear to arise primarily from the protein portion and are those that would be expected from the different solubilities of the two types of molecules. The isolated membrane glycoproteins are sparingly soluble in water and appear to be somewhat enriched in hydrophobic amino acids, although not as much as was originally suspected.

(a) *Erythrocyte*—Concerning studies of membrane glycoproteins, certainly the largest amount of work has been done on red blood cells. The isolated human erythrocyte membrane typically contains *ca.* 8% carbohydrate material, 42% lipid and 50% protein, of which about 10% is glycoprotein[30]. Polyacrylamide gel electrophoresis of total human erythrocyte ghosts solubilised in sodium dodecyl sulphate (SDS) indicates[30] that the membrane contains 7–9 main polypeptide chains, ranging in molecular size from 23 000 to 240 000. Two of these peptide chains contain carbohydrate material, as shown by reaction with periodic acid–Schiff stain. These two glycoproteins[31] are the only polypeptide components that have clearly been shown to be located with some parts on the exterior surface of the erythrocyte membrane[32, 33]. One of these surface proteins has a molecular weight of approximately 100 000, contains 5–8% carbohydrate material and represents *ca.* 30% of the total membrane protein[31]. The surface location is indicated by the fact that the protein reacts in intact cells with labelling reagents impermeable through the membrane[33-36] and is hydrolysed in intact cells by pronase[37-39]. Nothing has been reported concerning the structure or function of the carbohydrate residues in this glycoprotein component.

The other surface component is the major sialoglycoprotein of the human erythrocyte membrane. This molecule consists of 60% carbohydrate and contains most of the sialic acid as well as over half of the hexose and hexosamine of the human erythrocyte[40]. It has been estimated that there are *ca.* $1 \times 10^6$ molecules of this glycoprotein per human erythrocyte[40]. The carbohydrate moieties of this molecule contain the AB and MN blood group activities and also carry the receptor sites for influenza virus and several plant lectins[40, 41]. Sialic acid has been shown to be an important determinant group in influenza virus binding[42] as well as in the M and N activities[43, 44]. The large net negative charge of human erythrocytes has been attributed to the sialyl residues of this molecule[45, 46], and this negative charge may be important on the cell surface to allow the cells to repel each other and not to clump. In this regard it is interesting that if most of the sialic acid is removed from the erythrocytes and the cells returned to the circulation, there is an accumulation of the de-sialysed cells in the spleen. This phenomenon also appears to occur as erythrocytes 'age'; there is a decrease in sialic acid and subsequently a removal of these cells from circulation.

Chemical characterisation of the major sialoglycoprotein of human erythrocytes has been greatly facilitated by the fact that it can be extracted preferentially from erythrocyte ghosts[41,47-49]. Marchesi and co-workers, using a lithium di-iodosalicylate (LIS) extraction[41], have purified this glycoprotein to homogeneity, as shown by acrylamide gel electrophoresis and chromatography. This molecule is estimated[50] to have a molecular size of 50 000–55 000. (However, these authors caution that the exact size is not known with certainty, since the presence of large amounts of carbohydrate complicates analytical studies). Some of the most extensive work on the carbohydrate structure of this molecule has been carried out by Winzler and his colleagues[40,51], who have examined the glycopeptides obtained from trypsin treatment of intact erythrocytes and also similar glycopeptides that have been obtained from trypsin digestion of the purified glycoprotein. These isolated glycopeptides[40] appear to consist of a family of closely related molecules with molecular weights of *ca.* 10 000. They have the M and N serological activity and the carbohydrate accounts for about 80% of their weight. There are at least two types of carbohydrate unit attached to the polypeptide chain in these glycopeptides. One is a tetrasaccharide containing galactose, *N*-acetylgalactosamine and sialic acid linked through an *O*-glycosidic linkage to serine or threonine[52], while the other carbohydrate unit is more complex and is most probably attached to asparagine residues[53]. Proposed structures of these two chains are given in Figure 2.7. Jackson *et al.*[54] have estimated that the intact molecule contains *ca.* 18 of the tetrasaccharide units and 2 or 3 of the large carbohydrate units.

The structure of the asparagine-linked carbohydrate unit given in Figure 2.7 bears a marked resemblance to the structure proposed by Kornfeld and Kornfeld[55,56] for the erythrocyte receptor for phytohaemagglutinin. The phytohaemagglutinin receptor appears to be a singly branched molecule apparently lacking the side chain sequence α-fucosyl-(1→2)-β-galactosyl-(1→6)-*N*-acetylglucosaminylmannose, and thus may be an incomplete version of the molecule presented in Figure 2.7.

The molecular structure and membrane orientation of the human erythrocyte major glycoprotein has been partly characterised by analysis of the products from both cyanogen bromide cleavage[57] and from trypsin digestion[40,54] of the isolated glycoprotein. One product of the proteolysis was an insoluble peptide which contained less than 4% carbohydrate and approximately one-third of the total peptide content of the intact glycoprotein[54]. The partial amino acid sequence analysis of this insoluble tryptic peptide and also the analysis of a cyanogen bromide fragment, which represents the carboxyl terminus of the molecule, indicate that there is a unique sequence of approximately 23 residues which is extremely hydrophobic and which contains no charged residues[57]. There appears to be a clustering of charged residues at both ends of this 23-residue stretch. These results suggest the presence of a special hydrophobic region that is associated with the lipids

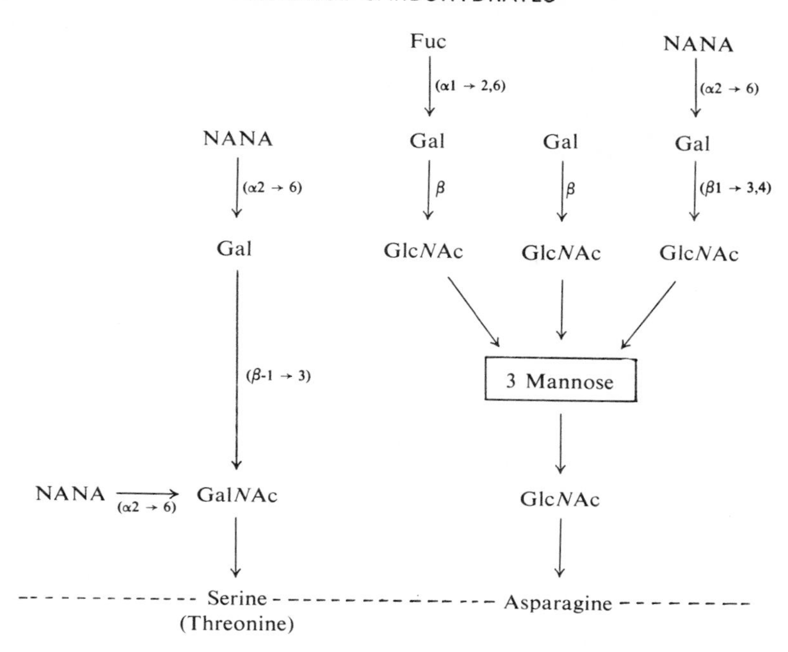


**Figure 2.7** Tentative partial structures for the two basic types of carbohydrate chains present in the major glycoprotein of human erythrocytes. Smaller versions of both of these carbohydrate units lacking some of the peripheral sugars are also present in the membrane. The abbreviations are given in Figure 2.3. (Taken in part from Thomas and Winzler[52, 53], by courtesy of the American Society of Biological Chemists and the Biochemical Society

of the membrane and may possibly serve to hold the protein in the membrane.

The model that has been suggested[40, 58, 59] for the major erythrocyte glycoprotein is shown in Figure 2.8. The molecule can be divided into three chemically distinct regions: the terminal portion exposed to the external environment of the cell and containing all the carbohydrate and receptor sites, a hydrophobic region which passes through the centre of the membrane, and the carboxyl-terminal portion which is hydrophilic but which contains no detectable carbohydrates and may be exposed in the interior of the lipid bilayer.

Bretscher[34, 37] has suggested that the main erythrocyte glycoprotein and also the 100 000 molecular weight glycoprotein traverse the lipid bilayer so that parts of these proteins are exposed on both the outer and inner surfaces of the membrane. The evidence for this conclusion comes from labelling experiments and enzymatic digestion of membranes[60] in which part of the above molecules react in intact cells (when only the outer surface is exposed) and other parts of the molecules react in ghosts (when both sides of the membrane are exposed to the reagent). However, this interpretation is correct only if the arrangement of protein and lipids and the permeability properties

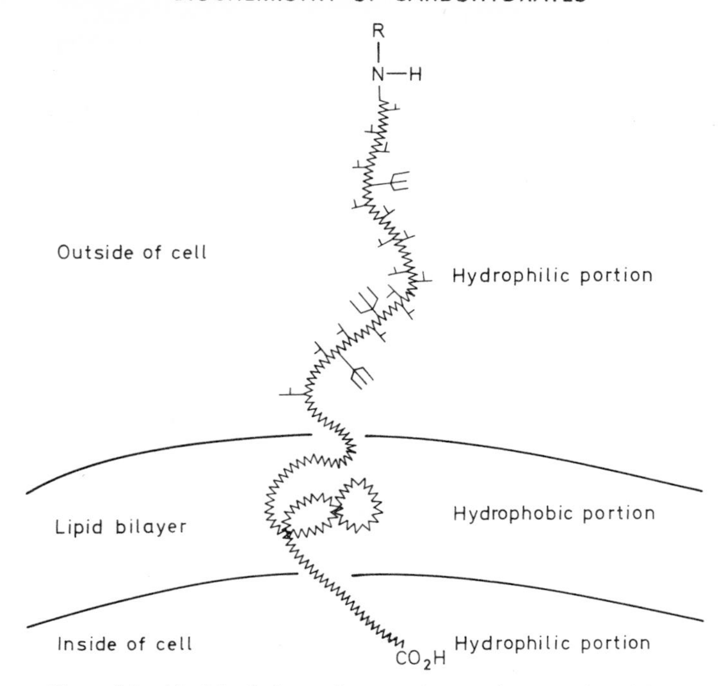


**Figure 2.8** Model of the major membrane glycoprotein of human erythrocytes. The glycoprotein contains a single polypeptide with a blocked *N*-terminal amino acid residue and substituted with numerous side chain oligosaccharides. Two basic types of carbohydrate chain are present, as shown in the diagram: ⊢ = tetrasaccharide attached to serine or threonine; –⋲ = branched-chain oligosaccharides attached to asparagine. The three main parts are indicated: a hydrophilic portion containing oligosaccharides, a hydrophobic portion containing only uncharged amino acids, which is thought to be embedded in the lipid matrix of the membrane, and a hydrophilic carboxyl terminal portion containing no sugars but numerous charged amino acids. (Taken in part from Marchesi *et al.*[58], by courtesy of the National Academy of Sciences)

are the same in intact cells and in isolated ghost membranes, an assumption which is still unproven.

Additional evidence that the major glycoprotein traverses the lipid bilayer comes from the observation that the membrane of the erythrocyte, when cleaved by freeze-fracture to expose the inner hydrophobic surface, presents smooth areas and particles[61, 62] (see Figure 2.9). The particles protrude through to the outer surface and seem to carry the heterosaccharide ABO[63] determinants and receptors for phytohaemagglutinin and influenza virus[64], activities associated with the major glycoprotein. Unambiguous evidence that the freeze-fracture particles are identical with the major erythrocyte glycoprotein is not available as yet.

(b) *Platelets*—The cell surface components on the small blood platelets have been studied using similar techniques as described above for erythrocyte membranes. The isolated plasma membrane of human platelets has a similar overall composition as the erythrocyte membrane (7% carbohydrate, 56% lipid and 32% protein); however, the carbohydrate and lipid compositions are clearly different in the two cell types[65].

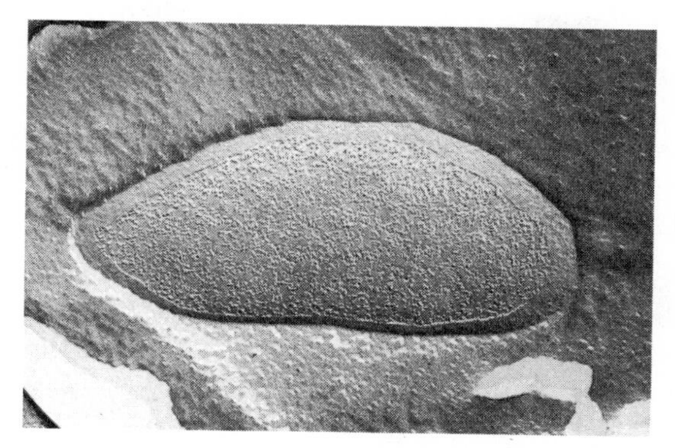

**Figure 2.9** Electron micrograph of the human red-cell membrane produced by freeze-etching ghost membranes. The cleavage process splits the membrane roughly in half and thereby exposes internal regions of the membrane where numerous globular particles can be seen. (By courtesy of Dr. D. Branton)

Pepper and Jamieson[66, 67] have treated isolated platelet membranes with proteases and obtained a mixture of glycopeptides which were subsequently fractionated into three classes of compounds differing in molecular weight and carbohydrate composition. The largest glycopeptide class has a molecular weight of *ca.* 120 000 (which is considerably larger than the erythrocyte tryptic glycopeptides discussed earlier which have molecular weights of *ca.* 10 000). The large glycopeptides from platelets contain two types of carbohydrate unit attached to both serine and threonine residues and to asparagine residues. Acrylamide gel electrophoresis of purified platelet membranes indicated 2-3 major glycoprotein bands[68, 69]. Iodination of intact platelets led to intense labelling of one of these major glycoproteins, a polypeptide of molecular weight *ca.* 100 000. This surface glycoprotein also has the concanavalin A (Con A) receptor activity and has been partially purified from isolated cell membranes by lithium di-iodosalicylate extraction and affinity chromatography using bound Con A[69].

(c) *Histocompatibility antigens*—Histocompatibility antigens are the cellular products of the histocompatibility genes, and they carry the immunological determinants involved in tissue transplant rejection[70]. In the mouse about 30 genetic loci controlling histocompatibility have been described. However, of primary importance are the H-2 gene products which elicit the strongest graft rejection. In man the HL-A antigen is especially important.

Nathenson and co-workers[71] have solubilised H-2 antigens from mouse-cell membranes using the non-ionic detergent NP-40 and have then further isolated the antigens by immune precipitation. These antigens have been further purified and characterised as having a subunit molecular weight of *ca.* 45 000. The isolated molecules are rather water insoluble and tend to form aggregates in the absence of detergent.

Solubilisation of the H-2 antigen has also been accomplished by papain treatment[72] of membrane fractions which results in a water-soluble fragment of subunit molecular weight of *ca.* 40 000. This proteolytic fragment is active and contains the same glycopeptides as the native molecule as well as the major portion (75–85%) of the native H-2 glycoprotein subunit. From these results it would appear that a large portion of the native molecule is exposed on the outer surface, and that a small hydrophobic portion of the molecule may be responsible for the membrane integration. These results appear to be in accord with the general model proposed for the major glycoprotein of the erythrocyte membrane in Figure 2.8, although more extensive work is necessary to substantiate this fully.

The solubilised H-2 antigens contain 10–15% carbohydrate which appears to be linked only in the *N*-glycosidic type of linkage between asparagine and *N*-acetylglucosamine[73]. It has been estimated that there are two hetero-saccharide chains per H-2 molecule, each chain containing 10–15 sugar units. It is not yet clear whether the carbohydrate moieties are related to the immunological specificities of the antigen[73], but there are arguments against this idea[74].

(d) *Rhodopsin*—The outer membranes of adult retinal rods contain several major proteins, of which at least one is a glycoprotein. This is the apoprotein of rhodopsin[75, 76] of molecular weight 28 000. This glycoprotein contains a single hexasaccharide unit of *N*-acetylglucosamine, galactose and mannose which is linked to the polypeptide chain through an *N*-aspartylglucosamine linkage[77]. About 50% of the amino acids in rhodopsin are hydrophobic[77]. The surface location of the carbohydrate units in the intact disc membrane was determined using fluorescein-labelled Con A, which binds specifically to the membrane in a stoichiometric manner (one mole Con A per mole retinal)[78]. Rhodopsin could be separated from other components of the rod outer segment by affinity chromatography using Con A bound to agarose.

It has been hypothesised that during the growth and assembly in the membrane of the retinal rods the single oligosaccharide group of each glycoprotein molecule is oriented to the aqueous phase, thus enabling the hydrophobic protein part of the molecule to associate properly with neighbouring molecules[77]. Indeed, many membrane proteins are known to have carbohydrate attached and it is attractive to hypothesise that one general function of these carbohydrate residues is to act as a surface marker to ensure the proper orientation in the growing membrane mosaic.

(e) *Nervous tissue*—The synaptic junctions appear to be particularly rich in carbohydrate material, as shown by various staining techniques. Many glycoproteins appear to be associated with the membraneous elements of synaptosomes, axons and microsomes and a number of these have been isolated and partially characterised[79]. Also a number of different glycosyl-transferases have been found to be located primarily at the ends of nerves in embryonic brain[80]. Concerning glycoprotein synthesis, Barondes[81] has suggested that the polypeptides are synthesised in the cell body and then transported through neurofilaments to the synaptosomal junction where the sugars are added; the glycoproteins thus formed presumably pass through the synaptic membrane and become part of the surface of the presynaptic membrane or are contained in the intercellular cleft material[82]. These glycoproteins of nervous tissue are of special interest in view of the speculations[3, 83-85] concerning their involvement in storage information and in the transmission capacity of the synaptic region. The evidence at present to support these speculations is still very questionable as there has been little structural characterisation of any of these nervous system glycoproteins.

(f) *Glycoprotein differences in virus-transformed cells*—Transformed cells are characterised by a loss of growth control present in normal cells. This loss of control leads *in vivo* to invasiveness, metastasis and unlimited growth. *In vitro* transformed cells grow to higher cell densities under identical conditions. These characteristics of transformed cells may well depend on alterations of cell surface components in malignant cell membranes, and such changes could represent key events in transformation. The burgeoning literature for the involvement of an altered cell surface in neoplasma will not be discussed here, as this subject has been very extensively reviewed[26, 86-89]. We will restrict our discussion to glycosyltransferase and glycoprotein differences that have been found between virus-transformed cells and their untransformed parental cells.

As can be seen in Table 2.1, numerous comparative studies on the chemical differences have been carried out. However, the earlier expectation of finding a clear chemical difference between normal and transformed cells which could be correlated with transformation has not been achieved. Indeed, the

**Table 2.1 Summary of glycoprotein and glycosyltransferase alterations following viral transformation of a number of tissue culture cell lines***

| *Alterations in transformed cells* | *Cell lines* | *Refs.* |
|---|---|---|
| *Less* glycosylation of membrane glycoproteins | SV3T3 | 90 |
| | PyNil | 91, 94 |
| | PyBHK | 92, 93, 95, 98, 99 |
| *More* glycosylation of a unique membrane glycoprotein | RSV-CEF | 96, 105 |
| | RSV-, Py-BHK | 100, 101 |
| | MSV-, Balb-3T3 | |
| | RSV-, Balb/3T3 | |
| of secreted glycoproteins | PyBHK | 95 |
| Similar patterns of glycoprotein (fewer molecules in transformed cells) | SV3T3 | 97 |
| Decrease in glycosyltransferases | PyBHK | 103 |
| | SV3T3 | 98, 99 |
| Increase in glycosyltransferases | RSV-, SV-, Py-3T3 | |
| | RSV-CEF | 104, 106 |
| | PyBHK | 101, 102 |

* The cell lines are as follows: 3T3 and Balb/3T3, mouse embryo fibroblasts; BHK, baby hamster kidney fibroblasts; Nil, hamster fibroblasts; CEF, chick embryo fibroblasts. The viruses are: SV, simian virus 40; Py, Polyoma virus; RSV, Rous sarcoma virus; MSV, murine sarcoma virus

present results still leave many controversies unresolved and there is a growing awareness that many of the discrepancies may be attributed to differences in the growth phase, clonal variation or physical conditions of the cells. In Table 2.1 it can be seen that a number of cases have been reported where membrane glycoproteins of transformed cells were less glycosylated than those of normal cells, and as a corollary to this there are reports of decreased levels of glycosyltransferases in these transformed cells. Grimes[98, 99] has found decreased levels of sialic acid and sialyltransferase in a number of transformed cell lines, and has very nicely demonstrated a correlation between these values and cell saturation density. Opposite results have been found by Warren and co-workers[96, 105]. However, they are examining glycopeptides obtained from Sephadex chromatography and these are specific components which account for only a small part of the total carbohydrate of the membrane. These workers have shown that several different types of cells transformed by both RNA and DNA viruses contain an enrichment of fucose–sialic acid-containing glycopeptides of apparently higher molecular weight than the major group of glycopeptides from untransformed cells. This higher molecular weight material obtained from transformed cells appears to have extra residues of sialic acid, and may represent a unique component which is different from the glycopeptides of normal non-growing cells[298]. They have also found in transformed cells an increase in the activity of a specific sialyltransferase which transfers sialic acid to an early-eluting, asialo-acceptor material derived from the surface of growing transformed cells[102]. It is difficult to determine whether these differences are a result of transformation of merely of growth dependence, as the early-eluting material also appears in rapidly dividing normal cells[299]. In this regard Muramatsu *et al.*[300] have found growth-dependent alterations in the carbohydrate–peptide linkage region of glycopeptides isolated from the surface of human diploid cells (KL2).

It is also worthwhile noting here that most of the differences described above have been detected in trypsin-released material. Similar comparisons made between entire cell-surface membranes have not shown such great differences between normal and transformed cells.

(g) *Lectin receptors*—Lectins have recently become very popular tools for studying the architecture of cell surfaces as these compounds bind specifically to saccharide moieties on the surface membrane. This binding is reversible as the interaction of lectins with cells can in many instances be inhibited specifically by simple sugars. For more information on the various lectins and their sugar specificities the reader is referred to several recent reviews[107, 108].

Certain lectins have been demonstrated to be able to agglutinate red blood cells and also stimulate 'resting' lymphocytes to undergo mitotic division. However, the greatest interest in this field has come from the observation that some lectins can preferentially agglutinate mammalian tissue culture cells that have been transformed by oncogenic viruses[109, 110]. Thus these lectins are capable of differentiating the surfaces of normal and malignant cells and a great deal of work has gone into attempting to define this surface architectural difference. This subject will be discussed in more detail in a later section (2.4.7), as will the mechanism of the lectin agglutination reaction.

The cellular receptors on the surface membrane for lectins are assumed to be glycoproteins and several laboratories are presently isolating and characterising a number of these lectin receptor sites[111-114]. At present the detailed chemical characterisation of only a few of these receptor molecules has been achieved. The most thoroughly studied is the phytohaemagglutinin receptor on the human erythrocyte membrane[55,56], which is present on the major membrane glycoprotein that was discussed in detail previously (see Section 2.2.2.1a). It is worthwhile mentioning that the recently introduced technique of affinity chromatography using matrix-bound lectins will undoubtedly play an increasingly important role in the isolation and purification of these lectin receptor sites[108].

### 2.2.3 Glycolipids

Glycolipids have been shown to be concentrated primarily in the plasma membrane[115], with some localisation also occuring in the endoplasmic reticulum[116]. These compounds are generally minor lipid components of the membrane, accounting for from 0.5–5% of the total membrane lipid. Glycolipids are responsible for most of the serological specificity exhibited by mammalian cells, and distribution studies of these compounds show that many organs have their specific type of glycolipid[117]. Glycolipids consist

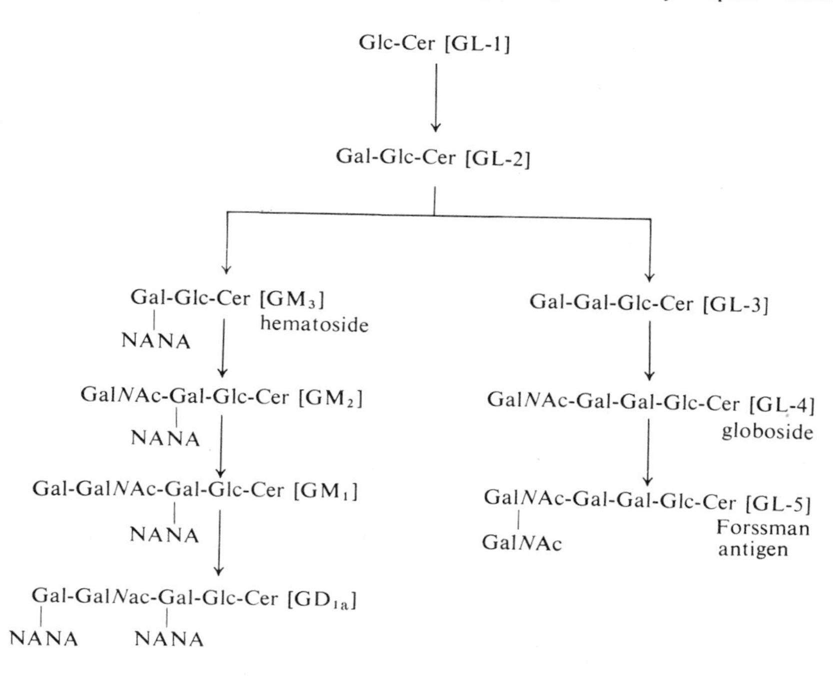


**Figure 2.10** Proposed mechanism for the biosynthesis of the major glycolipids found in humans. (Taken in part from Sweeley and Dawson[118], by courtesy of J. B. Lippincott)

of a carbohydrate unit composed of one or more monosaccharides glycosidically linked to the terminal hydroxyl of ceramide (*N*-acylsphingosine). The amphoteric nature of these molecules, with their polar sugar residues and non-polar acyl chains, makes them ideal as possible mediators of interaction at the aqueous–lipid interface of the surface membrane. Glycolipids contain all of the sugars commonly found in glycoproteins, with the exception of mannose. It is noteworthy that similar sugar sequences occur in the carbohydrate units of some glycolipids and glycoproteins.

In Figure 2.10 is a simplified diagram of the proposed pathway for the biosynthesis of the major glycolipids found in humans. As indicated previously in Section 2.2.1.1, the complex oligosaccharide chains are made by the stepwise addition of sugar units from appropriate sugar nucleotides using specific glycosyltransferases. As can be seen from Figure 2.10, lactosyl ceramide (GL-2) is a key intermediate in glycolipid synthesis; it can be converted into hematoside ($GM_3$) and initiate the biosynthesis of gangliosides (glycolipids containing sialic acid) or it can be converted into GL-3.

### *2.2.3.1 Erythrocyte and blood-group substances*

In man, glycolipids constitute less than 5% of the total lipids in the isolated erythrocyte membrane[118]. The four neutral glycolipids, GL-1, GL-2, GL-3 and GL-4 (see Figure 2.10), are the major components that account for more than 95% of the total glycolipid, the remainder consisting of gangliosides and blood-group active glycolipids. There appear to be relatively wide variations in the composition and concentrations of the glycolipids present in erythrocytes of various mammalian species[118]. A very important group of carbohydrate containing components on the surface of erythrocytes are the blood-group active substances, which are responsible for the strong antigenic activity of the erythrocytes and of such importance in connection with the compatibility of blood (for recent reviews see Refs. 25, 119–122). Although there is still some controversy as to whether these blood-group active substances on erythrocytes are glycoproteins or glycolipids, the current evidence suggests that the major antigenic activities are associated with glycolipids.

The individual members of a species may be divided into blood groups according to their complement of cell-surface antigens. The first blood-group system to be discovered was the ABO system and this is the most important one in connection with the compatibility of transfused blood. These blood-group antigens, however, are also present on other cell surfaces in the body and are also found on water-soluble glycoproteins in tissue fluids and secretions. This is an example of the same sugar sequences being present in both glycoproteins and glycolipids.

The largest amount of information on the chemical basis of blood-group specificity has come from chemical and immunological examination of these soluble glycoproteins. The conclusion that has emerged so clearly from a large number of studies is that the antigenic specificity of these blood-group substances, whether they be glycoprotein or glycolipid, resides in the three-dimensional pattern of the three or four sugar residues at the non-reducing ends of the carbohydrate chains. In Figure 2.11 are given the sugar sequences and biosynthesis of some of the structures responsible for the ABO and Lewis blood types in man[123]. The formation of these unique carbohydrate structures is genetically determined by the presence or absence of four glycosyltransferases which are responsible for the attachment of the outer or immunodominant sugars to a central chain common to all phenotypes. Thus the A gene is responsible for the formation of the *N*-acetylgalactosaminyltransferase, the B gene a galactosyltransferase, while the H gene and Le$^{a}$ genes produce two different fucosyltransferases. Thus the addition of a new sugar residue to the end of an existing carbohydrate chain produces a new serological specificity primarily determined by the added sugar, but also dependent on those sugars already present in the chain. This specificity of the blood-group substances points out an apparent anomaly in the current dogma about the biosynthesis of the carbohydrate units in glycoproteins. On the one hand there is the appreciable microheterogeneity mentioned previously that exists in the sugar chains of most glycoproteins. However, in blood group substances this heterogeneity does not exist.

A final point to be made here, one that has been proposed by Albersheim *et al.*[124], is that sugars and sugar-containing molecules are uniquely sensitive to specific recognition by protein, and there is probably in biology no molecular interaction more specific, more sensitive to structural alteration than this recognition of carbohydrate by protein. This is probably best exemplified in the A and B blood-group specificities, where a protein is able to distinguish, ultimately, the small structural variation at C-2 in a sugar with a galactose configuration, namely whether there is a hydroxy group or an *N*-acetylamino group.

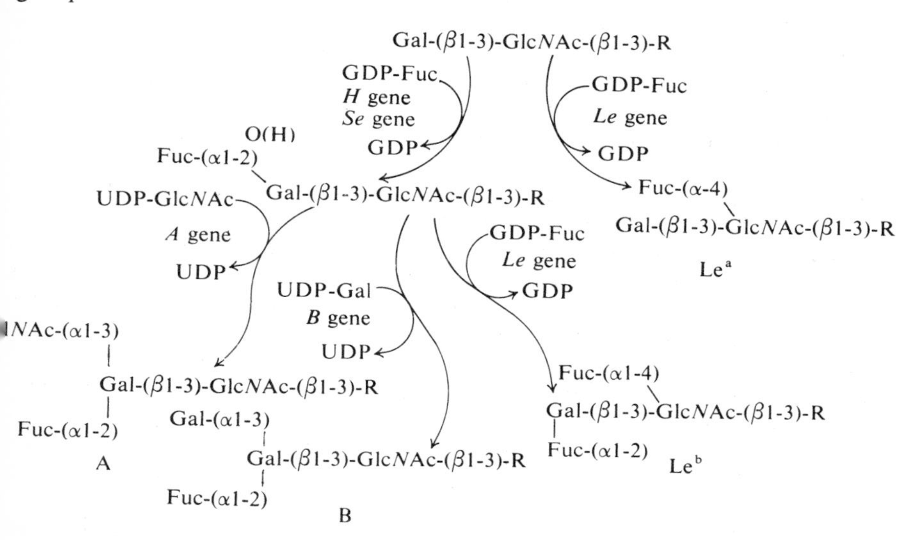


**Figure 2.11** Biosynthesis of the structures responsible for ABO and Lewis blood types in man. (From Kokata *et al.*[123], by courtesy of Academic Press)

The concept that thus emerges is that the informational capacities present in proteins and lipids can be greatly enhanced by the addition of carbohydrate chains that may confer different types of specificity or allow for a greater range of variability in specificity. Carbohydrates are ideally suited because of the enormous number of specific structures that can be formed from relatively few monomeric units. The monosaccharides can be linked to each other through any of several hydroxy groups, the linkages can be either α or β, and, in addition, extensive branching can occur. Per carbohydrate unit we have therefore many more possibilities to create diversity than are available for amino acids, where only peptide linkages of one type are seen in mammalian cells. The role of carbohydrates in blood-group specificity offers a classic example of the fact that unique patterns of sugars present on the surface membrane can serve as recognition sites.

#### *2.2.3.2 Glycolipid differences in virus-transformed cells*

A number of laboratories have studied the changes in glycolipid patterns associated with *in vitro* transformation by tumor viruses, and a great deal of data is now available. However, a major question still unanswered is the relation of these biochemical alterations to malignant transformation. Biochemical studies on glycolipids have been carried out on cell lines from several different sources; the most thoroughly studied are the hamster and mouse lines. In both of these cell lines it was found that the more complex glycolipids disappeared or decreased following viral transformation. However, the specific glycolipid changes were not the same in the hamster and mouse cells. The results of the two systems are summarised in Table 2.2.

Hamster cell lines contain mainly the neutral glycolipids (GL-5, GL-4, GL-3, GL-2) and hematoside ($GM_3$) with only small amounts of other gangliosides[127-129]. One of the most interesting observations concerning these components in normal cells is the finding that these glycolipids change with cell density; with increasing culture density there is a specific increase in the large neutral glycolipids (GL-3, GL-4, GL-5)[132] and parallel to this increase Kijimoto and Hakamori[130, 131] have observed higher levels of the enzyme which transfers galactose to GL-2 to form GL-3, the first component of this series (see Figure 2.11 for biosynthetic scheme). In hamster cells transformed by a number of tumor viruses, these larger glycolipids or 'density-dependent' glycolipids were much reduced or undetectable and they no longer change with cell culture density. Transformed cells also have reduced levels of the α-galactosyltransferase mentioned above, which catalyses the synthesis of GL-3. It is to be noted that in the hamster cell lines no general correlation could be found between tumorigenicity and presence or absence of complex glycolipids[128].

The mouse cell lines contain appreciable amounts of the gangliosides ($GD_{1a}$, $GM_1$, $GM_2$, $GM_3$)[133] and, in contrast to the hamster cells, the mouse cells do not show density-dependent changes in any of the glycolipids[134]. However, Brady *et al.*[133] have shown that the mouse cell lines transformed by tumor viruses contain decreased amounts of gangliosides with an oligosaccharide chain length larger than hematoside[135]. The block has been

**Table 2.2 Summary of glycolipid and glycosyltransferase alteration during cell growth and following viral transformation***

| *Major glycolipid components* | | *Density dependent changes in glycolipids* | | *Glycolipid changes following viral transformation* | |
|---|---|---|---|---|---|
| | | HAMSTER CELL LINES | | | |
| *Acidic* | *Neutral* | | *Refs.* | | *Refs.* |
| | GL-2 | Increase with cell density (a) GL-3, GL-4 GL-5 | 125, 127, 128, 131, 132 | Increase in GL-2 Decrease in $GM_3$ GL-3, GL-4, GL-5 | 126–128, 130 |
| | ↓ Gal'ase ↓ | | | | |
| $GM_3$ | GL-3 | | | No density dependent changes | 131 |
| | | (b) Gal'ase | 130 | | |
| | GL-4 | | | Decreased levels of Gal'ase | 130 |
| | GL-5 | | | | |
| | | MOUSE CELL LINES | | | |
| $GM_3$ | | No changes in glycolipids | 134 | Increase in $GM_3$ Decrease in $GM_2$, $GM_1$, $GD_{1a}$ | 133, 135, 136 |
| ↓ GalNAc'ase ↓ | | | | | |
| $GM_2$ | | | | Decrease in GalNAc'ase | 136 |
| $GM_1$ | | | | | |
| $GD_{1a}$ | | | | | |

* The glycolipids are listed as in the biosynthetic diagram given in Figure 2.10. The structures of the various glycolipids are also given in Figure 2.10. Gal'ase = galactosyltransferase which catalyses the formation of GL-3; GalNAc'ase = *N*-acetylgalactosaminyltransferase which catalyses the formation of $GM_2$

characterised as being the loss of the enzyme that catalyses the transfer of *N*-acetylgalactosamine to hematoside[136]. However, it should be mentioned that spontaneously transformed, highly tumorigenic cells do not show this decrease in the *N*-acetylgalactosaminyltransferase activity[133], so this type of biochemical alteration does not appear to be a general function of malignant transformation. Recently, Laine and Hakomori[137] have tried another approach to this problem. They found that when globoside (GL-4) was added exogenously to the growth medium of hamster cells it was taken up by these cells and incorporated in plasma the membrane. Concomitantly there was a reduction in growth rate and saturation density.

### 2.2.4 Mucopolysaccharides (proteoglycans or glycosaminoglycans)

The glycoproteins we have discussed in Section 2.2.2 contain one or more heterosaccharide units made up of a relatively low number of sugar residues covalently linked to a polypeptide chain. The sugar units of these glycoproteins do not contain any kind of serially repeating units. Animal tissues also contain a group of carbohydrate–protein compounds in which the carbohydrate moiety is a polysaccharide, composed of characteristic repeating disaccharide

**Table 2.3 Major constituents of mucopolysaccharides***

| | *Classification* | *Polysaccharides* | *Hexosamine* | *Other major monomer* |
|---|---|---|---|---|
| Group I | Hexuronic acid | Hyaluronic acid | D-Glucosamine | D-Glucuronic acid |
| | | Chondroitin | D-Galactosamine | D-Glucuronic acid |
| Group II | Hexuronic acid and *O*-sulphate groups | Chondroitin 4-sulphate | D-Galactosamine | D-Glucuronic acid |
| | | Chondroitin 6-sulphate | D-Galactosamine | D-Glucuronic acid |
| | | Dermatan sulphate | D-Galactosamine | L-Iduronic acid or D-Glucuronic acid |
| Group III | *O*-Sulphate groups | Keratan sulphate | D-Glucosamine | D-Galactose |
| Group IV | Hexuronic acid *O*-sulphate and *N*-sulphate (sulphamate) groups | Heparin | D-Glucosamine | D-Glucuronic acid |
| | | Heparan sulphates | D-Glucosamine | D-Glucuronic acid or L-Iduronic acid |

* Excluding polysaccharide–protein linkage sequences and other minor sugar components

units, usually consisting of an amino sugar and a hexuronic acid. These mucopolysaccharides or proteoglycans are of particular interest in that they are most often found as constituents of the amorphous extracellular matrix[138-140].

Mucopolysaccharides are highly charged polyelectrolytes of very large molecular weight. Their polyanionic character results from the presence of carboxy groups of constituent hexuronic acids or of ester-bound sulphate groups, or both. The mucopolysaccharides at present consist of only eight different polysaccharide components. These have been classified according to the chemical nature of their anionic groups and are listed in Table 2.3 along with the characteristic repeating disaccharide units.

The most common type of carbohydrate–protein linkage found in the proteoglycans is an *O*-glycosidic linkage between galactosyl-galactosyl-xylose residues and serine, i.e. xylose is glycosidically attached to the hydroxy group of the amino acid. In fact, this linkage type is specific for the connective tissue mucopolysaccharides and has not been demonstrated in other classes of compounds. This linkage is found in chondroitin 4- and 6-sulphate, dermatan sulphate, heparin and heparan sulphate. Only in keratan sulphate are found the other types of linkages which we discussed in the previous section on glycoproteins. The most thoroughly studied of the above mucopolysaccharides is probably chondroitin sulphate, which is mainly carbohydrate and contains only 10–15% protein. A schematic structure that

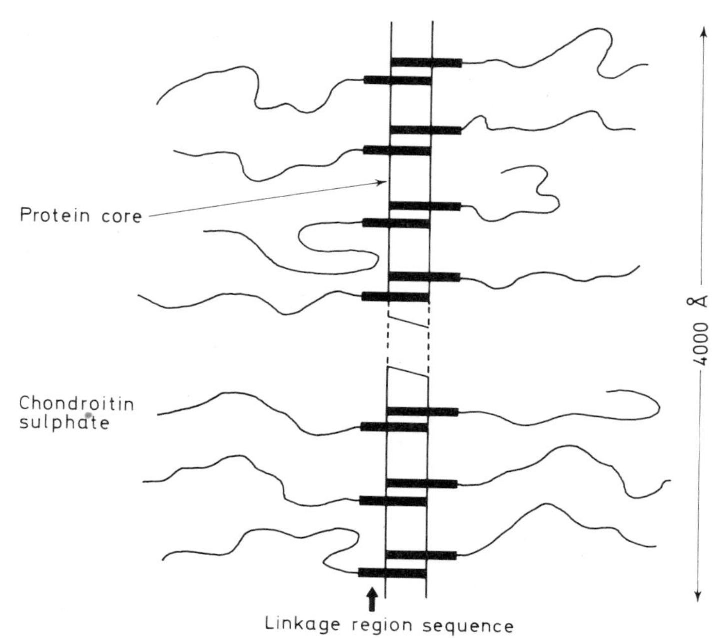


**Figure 2.12** Schematic representation of the major chondroitin sulphate–protein macromolecule component of vertebrate hyaline cartilage. (Taken in part from Mathews[141], by courtesy of the Biochemical Society)

has been proposed for chondroitin sulphate is presented in Figure 2.12[141] (see also Chapter 4). It consists of a protein core to which about 100 polysaccharide chains are attached. The size of this macromolecule may vary considerably, depending upon the tissue or species of origin, with molecular weights ranging from $3 \times 10^5$ to $4 \times 10^6$. Recent work by Hascall and Sajdera[142] indicates that these large molecular weight complexes are made up of a 30S mucopolysaccharide subunit and of a small glycoprotein subunit. This latter component is thought to be responsible for the aggregation of the 30S mucopolysaccharide subunits. Rosenberg *et al.*[143] observed in the electron microscope that some of these large aggregates appear in a sunburst pattern with smaller subunits radiating outward from a central core.

For recent studies on the biosynthesis of chondroitin sulphate in cartilage, see Richmond *et al.*[144, 145] and DeLuca *et al.*[146], and also Chapter 3.

#### 2.2.4.1 *Mucopolysaccharides in transformed cells*

Although this area has been very poorly investigated so far, there are several recent reports indicating differences in mucopolysaccharide components following viral transformation. Concerning the occurrence of mucopolysaccharides at cell surfaces, Kraemer[147, 148] has examined six different cell lines of both connective and epithelial origin and found heparan sulphate to be a surface component in all of these cell lines. Also, numerous microscopic studies have demonstrated the presence of mucopolysaccharides at the cell membrane and there have been reports from electron microscopic studies that there are increases in these surface components in cells transformed by viruses[149, 150]. Chiarugi and Urbano[151] have found that material released from BHK hamster cell by EDTA was made up of glycoprotein and mucopolysaccharide; the latter component was identified as heparan sulphate. They noted little difference in the compositions of the EDTA-released material from BHK cell and polyoma- or RSV-transformed cells. Also, a similar composition was found in the material removed from normal cells by trypsin[152]. However, of interest was the finding that the trypsinate from transformed cells was different in that it contained significantly smaller amounts of heparan sulphate. Similar results have also been obtained by Minnikin and Allen[153].

### 2.2.5 Turnover of membrane components

The current concept visualises the plasma membrane as a very active organelle which is continuously being broken down and replaced. Receptor glycoproteins, especially transplantation antigens[154, 155], viral receptors[156] and hormone receptors[157], have been shown to undergo rapid turnover as reflected by their rapid regeneration following treatment of intact cells with specific hydrolytic enzymes. Warren and co-workers have published a number of papers dealing with the problem of plasma membrane synthesis and turnover. They have demonstrated that growing and non-growing L cells synthesise approximately the same amount of plasma membrane[158]. In growing cells the material is incorporated with net increase in membrane mass, while in non-growing cells the incorporation results in turnover without net synthesis. In this study no gross differential turnover was noted in the protein, lipid and carbohydrate of surface membranes. Gerner *et al.*[159], in a study of synthesis in mammalian KB cells during different stages of the cell cycle, found there was a marked increase in incorporation of various labelled precursors into the surface membrane just after cell division.

*In vivo* studies on membrane turnover by Schimke and co-workers indicate that the membrane proteins of liver turn over rapidly compared with the soluble proteins[160, 161]. Also, specific membrane proteins have very different rates of turnover[162], with the plasma membrane fractions enriched in glycoprotein showing higher degradation rates than those deficient in glycoprotein[163]. Dehlinger and Schimke[161] also noted that there is an interesting correlation between relative degradation rates of membrane proteins and the molecular size of the protein subunits, with higher molecular weight polypeptides being degraded more rapidly than those of low molecular weight.

## 2.3 MOLECULAR ORGANISATION OF GLYCOPROTEINS AND GLYCOLIPIDS IN CELLULAR MEMBRANES

A review of cell-surface components is not complete unless some discussion is given to the molecular arrangement of these molecules in the unit membrane. This area has proven to be very exciting in recent years, as experimental evidence from a variety of different sources has indicated that most biological membranes are not rigid, but exist in a fluid state. We will only very briefly discuss this topic, as full accounts have been presented in other articles in this volume (Chapters 3 and 4) and in Volume 2 of this series[302].

A discussion of the integration of glycoproteins and glycolipids into biological membranes is closely tied with the current view of the ultrastructure of the unit membrane[164, 165]. In most models that have been proposed, the integrity of the membrane structure resides in interactions between lipid and protein, involving lipid–lipid, lipid–protein and protein–protein couples. The carbohydrate is probably not directly involved in maintaining the integrity of the membrane. The major portion of the lipids exist in the form of a discontinuous bilayer with their ionic and polar heads in contact with water. The proteins in the membrane are predominantly globular and appear to have their hydrophilic portions protruding from the membrane, and their hydrophobic sections embedded in the lipid bilayer. The proteins make up the membrane 'active sites'. Some are embedded on one side or the other, while others may pass through the bilayer. The glycoprotein and glycolipid components are asymmetrically distributed across the membrane, with the hydrophilic carbohydrate chains being external and interacting with the aqueous environment, while the hydrophobic regions of the peptide and the hydrocarbon chains of the glycolipid interact with the internal hydrophobic matrix.

One of the most important recent concepts to have appeared, and for which considerable evidence has now accumulated, is the fluid mosaic model[166]: the membrane components are not rigidly fixed but exist in a fluid and dynamic state in which the protein and glycoprotein components embedded in the lipid bilayer are able to undergo lateral diffusion within the membrane. Some of the most elegant experiments in favour of and even fundamental to the development of this membrane model were the results of Frye and Edidin[167], who showed that surface antigens are mobile, and also the results of Taylor *et al.*[168], who demonstrated that immunoglobulin molecules on the surface of lymphocytes can be cross-linked by specific divalent antibodies and redistributed to form patches, and ultimately caps, at one pole of the cell.

Thus the picture we now have is that large membrane molecules (in most cases these have been demonstrated to be glycoproteins) are able to migrate in the plane of the surface membrane and the membrane is sufficiently fluid in some circumstances to allow components to move relative to one another. Several important consequences of such a fluid membrane structure are that the membrane will form a heterogeneous mosaic and that the subunits of this mosaic will be liable to redistribution, depending upon the physiological conditions operating either on the whole membrane or locally on certain portions of it. The biological activity of cell surface receptors resulting from interactions with various ligands may depend upon the relative membrane mobility of these receptors as well as their distribution[169,170].

## 2.4 SURFACE CARBOHYDRATES IN CELLULAR RECOGNITION PROCESSES

The development of a characteristic morphology of cells and tissues remains one of the most intriguing and yet poorly understood biological phenomena. In many organisms morphogenesis requires adhesion between cells destined to assume identical or associated functions. It has long been known that dissociated embryonic cells when cultured in appropriate medium reaggregate to form organotype structures that are defined by the tissue and not by species of origin[171,172]. Such recognition of one cell by another has been assumed to involve recognition sites on the cell membrane[173]. The chemical basis for these recognition processes, which have been repeatedly demonstrated in all forms of life from viruses to mammals, remains to be elucidated. In the following sections a number of different biological systems will be discussed in which there is biochemical evidence that carbohydrate surface components may be involved in some basic manner in these recognition processes. The major emphasis will be given to mammalian systems. However, in order to be complete and as a model for study in mammalian systems, a discussion is included of the role of carbohydrates in cell surface interactions involving bacteria, as well as a discussion of the membranes of certain animal envelope viruses.

### 2.4.1 Lipid-bound carbohydrates of bacterial O-antigens

The O-antigen system of the *Salmonella* bacteria offers an excellent example of specific polysaccharide–protein interaction[174]. The genus *Salmonella* encompasses hundreds of serotypes which are believed to differ phenotypically only in the structure of their O-antigens on the cell surface. The O-antigenic behaviour of the *Salmonella* serotypes is determined by polysaccharides which are composed of repeating units of up to four different sugars. There are a number of *Salmonella* serotypes in which the O-antigens are composed of repeating units containing identical sequences of sugars, but which differ only in the nature of the linkages between sugar molecules. Structural differences between O-antigens are sufficient to define hundreds of specific reactions between these molecules and antibodies.

Bacterial viruses recognise their hosts through an interaction between molecules on the tail fibres of the bacteriophage, and the O-antigenic polysaccharides of the bacterial cell surface. Robbins and co-workers[175-177] have contributed significantly in elucidating the role of the O-antigens in this cell surface recognition process. They have observed that the structure of the normal O-antigenic polysaccharide of *Salmonella anatum* is modified when the bacterium is lysogenic for the bacteriophage $\varepsilon^{15}$. Such bacteria contain the viral genome in a non-replicating state. The O-antigen of the uninfected *S. anatum* cells contains the repeating unit (1→4)-α-galactosyl-(1→6)-α-mannosyl-(1→4)-rhamnose. The repeating unit of the O-antigenic polysaccharide of the bacteria containing the $\varepsilon^{15}$ is the same, except that the galactosyl-1(→6)-mannose linkage is altered from the α- to the β-configuration, and the *O*-acetyl substituent, which in the repeating unit of the normal antigen is attached to the C-6 position of the galactose residue, is absent in the lysogenic strain. This conversion is brought about by phage-specific proteins which (*a*) inhibit the synthesis or activity of host enzymes catalysing the specific reactions in the biosynthetic sequence, and (*b*) substitute for the host enzyme a phage-specific enzyme which generates β-galactosyl linkages. It is of particular interest that the strain of *S. anatum* which is lysogenic for $\varepsilon^{15}$ exhibits an altered susceptibility to further bacteriophage infection. Viruses which are normally able to infect *S. anatum*, including $\varepsilon^{15}$ itself, cannot attach themselves to the surface of the lysogenic bacterium containing the $\varepsilon^{15}$. Thus, $\varepsilon^{15}$ infection alters the resistance of the host bacterium to attack by other viruses through a minor structural modification of a bacterial cell wall polysaccharide. The O-antigenic polysaccharide, then, is a critical factor in determining susceptibility to bacteriophage infection.

In our opinion these results reflect, in an elegant manner, the high specificity that can be exhibited by surface carbohydrates in cell surface phenomena. They are classic examples of the involvement of surface carbohydrates and perhaps can be used as examples for similar specificities occuring in mammalian systems.

### 2.4.2 Virus envelopes and membrane glycoproteins

Purified virus particles consist of a nucleocapsid core surrounded by a membrane which the virus acquires when it leaves the host cell by budding through the plasma membrane[178]. The lipids of the viral envelope resemble those of the host cell plasma membrane[179], while the proteins appear to be coded for by the virus. The envelope of virus particles appears to be one of the simplest membrane systems yet studied. It consists of lipid, probably arranged in a bilayer structure, and from one to four proteins. Some of these membrane proteins are glycoproteins and, in all cases studied to date, the glycoproteins occur in part on the exterior surface of the membrane (influenza[180, 181], Sindbis[182], SV5[183], vesicular stomatits[184], Rous sarcoma[185] and Semliki forest[186]). In a number of these viruses the glycoprotein molecules are responsible for (*a*) the haemagglutinin that contains the specific sites for attachment of the virus to receptors on the plasma membrane, and (*b*) the viral neuraminidase which plays an essential role in the life cycle of the virus. These neuraminidase and haemagglutinin glycoproteins are seen in electron micrographs as projections or spikes that protrude from the envelope of the intact virus particles. The projections can be selectively removed by treatment with proteases without affecting the remainder of the surface structure. The infectivity of the protease-treated viruses is markedly reduced, as are the hemagglutinin and neuraminidase activities. After limited proteolysis these two activities have been recovered in the supernatant fractions after removal of the spikeless viruses[187]. In Semliki forest virus[186] part of the envelope glycoprotein was found to be left in the spikeless particles after protease treatment, and this fraction was enriched in hydrophobic amino acids, indicating that these membrane glycoproteins may be intercalated in the lipid bilayer with a hydrophobic peptide sequence.

It is not clear at present to what extent the carbohydrate units of the surface glycoproteins are involved in the haemagglutinin and neuraminidase activities. However, the ubiquitous presence of the carbohydrate in the viral envelope suggests that it may have a biological function. One such function may be to provide a hydrophilic surface which allows the virus to be freely dispersed in an aqueous medium[181].

An important function of the viral surface that may well involve the membrane carbohydrate chains is in the resistance of the envelope viruses to the immune response of the host animal. In this regard it is interesting that there are antigenic similarities between the membranes of many viruses and the host cells in which they grow. Since the antigenic determinates probably residue in the carbohydrates on the external surface of the cells[188, 189], it would appear that there are similar carbohydrate structures present in virus glycoproteins and in the heterosaccharides normally present on the cell surface of the host cell. It may be that the carbohydrate units of the virus membrane are largely specified by host cell glycosyltransferases and can thus escape the immunological response in the host animal[36]. The incorporation of a host-cell heterosaccharide structure, using host transferases, into its own unique polypeptide chain would also be important to the virus, as the array of glycosyltransferases needed could not be coded for by the small genome content of most of these viruses.

### 2.4.3 Cell-surface carbohydrates in simple eukaryotes

#### *2.4.3.1 Green algae and yeast*

For *Chlamydomonas*, Wiese[190] has shown that flagella tips or isolated isoagglutinins from one mating type can agglutinate cells from the opposite mating type. These isoagglutinins appear to be high molecular weight glycoproteins. It is interesting that only the female gamete appears to be sensitive to low doses of trypsin, and only the male gamete can be inactivated after binding to low concentrations of the agglutinin Con A[191]. Thus it appears that the mating adhesion is brought about by an interaction of a carbohydrate component on the surface of the male gamete and a protein component of the female gamete. Such conclusions are tentative and require considerably more biochemical evidence. Thus, the inhibition of the mating agglutination by Con A may simply be due to steric affects by the bulky Con A.

A similar system is now available in yeast mating interaction. Crandall and Brock[192] have partially isolated some glycoproteins from the surfaces of opposite yeast mating types which agglutinated or neutralised the opposite mating type. These surface components, as well as those in the *Chlamydomonas* system mentioned above, require better chemical characterisation to establish clearly their direct involvement in the mating interaction.

### *2.4.3.2 Cellular slime mold*

The developmental patterns seen in the slime molds make these organisms ideal for studying the changes in the cell surface during morphogenesis. The life cycle of the cellular slime mold *Dictyostelium discoideum* has two distinct phases: a non-social vegetative state in which separate amoebae feed on bacteria and divide by fission, and a social phase initiated by a period of starvation, in which the free-living amoebae aggregate to form a multicellular structure before fruiting body formation[193]. The initial aggregation after termination of the growth phase is species specific, and thus may involve specific cell-surface molecules[194].

Beug *et al.*[195] have shown that univalent antibodies directed against discrete cell-surface antigens are able to block the formation of intercellular contacts, while having no influence on cell mobility or chemotactic reactivity. Using a different approach, Rosen *et al.*[196] have examined the synthesis of a specific carbohydrate-binding protein during the life cycle of *Dictyostelium discoideum* and found that the synthesis of this component is correlated with the development of cohesiveness of these cells. This protein, which is assayed by its ability to agglutinate sheep erythrocytes, has a strong affinity for sugars with a galactose configuration. This protein component appears to be present on the surface of cohesive but not vegetative slime mold cells and the developmental appearance of this factor closely parallels the appearance of the antigens reported above by Beug *et al.*[195]. Further characterisation of this protein is necessary, as is direct evidence for its involvement in cell adhesion.

A further indication that cell-surface carbohydrates may be involved in slime mold development is the preliminary report that there is an alteration in the cell surface during the growth cycle that is detected by agglutination with Con A. Cells harvested from exponential growth phase are agglutinated by lower concentrations of Con A than are cells harvested during stationary growth phase or during differentiation[197]. It should be pointed out that the above evidence is only circumstantial and that at present there is no direct evidence for the involvement of surface sugar in this developmental system.

### *2.4.3.3 Sponges*

One area in which cell-surface carbohydrate molecules may well be involved in cellular recognition is in the species-specific reaggregation of dissociated marine sponge cells (for reviews see Refs. 198, 199). In fact this simple system may serve as a useful model in studying cell interactions. Marine sponges can be dissociated into single cells by squeezing through a fine mesh cloth. The resulting single cells will reaggregate, and in some cases form a functional sponge. In a number of cases the aggregation process has been found to be species specific, e.g. if the dissociated sponge cells from two different coloured sponges were mixed, the respective coloured cells of each species aggregate only with their own kind[200]. It has been demonstrated that for reaggregation to occur, divalent cations are required[201]. In an extensive series of studies, Humphreys[202-204] has isolated a soluble factor from single cell suspensions which enhances the ability of the cells to reaggregate. This cell-surface factor is necessary for aggregation of sponge cells which have been dissociated in $Ca^{2+}$- and $Mg^{2+}$-free sea water and then returned to normal

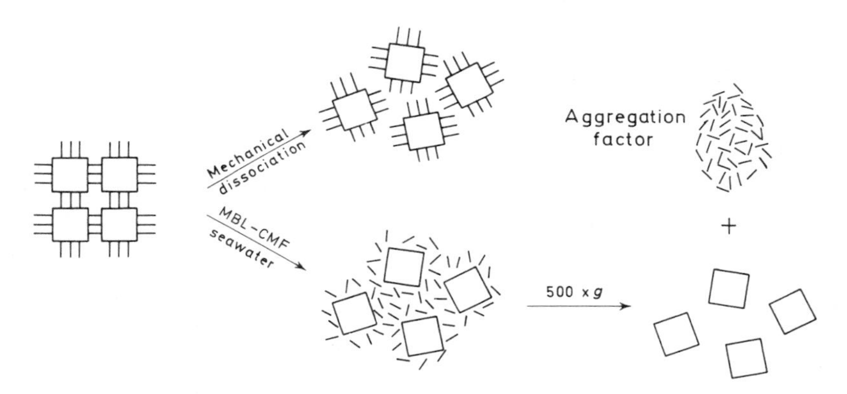


**Figure 2.13** Schematic representation of the results of mechanical (MD) and chemical (CMF-SW) dissociation of sponge cells. ⌗ = sponge cell with factor still attached; □ = sponge cell stripped of aggregation factor; ⁂ = aggregation factor. (From Turner and Burger[199], by courtesy of *Rev. Physiol.*)

sea water (see Figure 2.13). The aggregation factor is presumably removed from the cells while they are being dissociated in the divalent cation-free sea water. These surface components have been demonstrated to be species specific, with the aggregation factor from one species only enhancing re-aggregation of the homologous species[198, 204].

This simple system has the advantages for biochemical study that both the dissociation of the sponges into single cells and the release of aggregation factor from the cell surface can be accomplished by very gentle means. Also, a relatively simple *in vitro* assay system is available to test the surface factor for cell aggregation and specific cell sorting. In aggregation systems in higher animals the cells cannot be easily dissociated without the use of enzymes such as trypsin, which may destroy intercellular molecules as well as any specific receptors on the cell surface.

Using the sponge aggregation system described above, a number of studies have been done on the biochemical properties of the aggregation factor from several different sponge species[198, 202, 206-212]. The most extensively studied of these aggregation components is that from the marine sponge *Microciona parthena*, which has been purified and characterised by Humphreys and co-workers[196, 207]. They found this surface component to have a number of very unusual properties. The aggregation factor appears to be homogeneous by differential centrifugation and gel filtration on Sepharose 4B. It is a large, acidic, protein–polysaccharide complex of several million dalton in size; it contains about equal amounts of protein and carbohydrate. The major sugars are galactose and mannose (26%), uronic acid (10%), glucosamine and galactosamine (11%). It contains only minor amounts of sulphates. On the basis of high uronic acid content and large size of the molecule, the aggregation factor would appear to be of the mucopolysaccharide class (see Section 2.2.4). In the electron microscope the factor appears as a large, fibrous molecule with the fibres arranged in a type of sunburst pattern with radiations emanating from an inner circle. This inner circle has a diameter of *ca.* 800 Å and there are about a dozen arms some 1000 Å in length coming from the circle. All of these arms are composed of fibres that are *ca.* 45 Å in diameter. EDTA causes a loss in biological activity of the aggregation component and at the same time causes the native factor to dissociate into 9 or 10 subunits, which come from the arms, and a core which is the inner circle that does not break down in EDTA. The subunit structures are mucopolysaccharide-type molecules of about $2 \times 10^5$ dalton with large polysaccharide side chains of an estimated molecular weight of 20 000–40 000.

The aggregation factor has the interesting property of being able to form a gel in sea water. There appears to be a correlation between this gel formation and activity of the factor, as the calcium dependence of factor activity roughly follows the calcium dependence of gel formation. (The possibility that aggregation factor activity is due to physical entrapment of cells is rigorously excluded by its species specificity.) The correlation of gel formation and activity may be a reflection of the need for two or more of these surface molecules to interact in order for them to bring cells together into an aggregate. The aggregation factor appears to be a unique molecule, quite unlike any others which have been described. It seems to be a mucopolysaccharide, at least from the uronic acid content and the large size of the carbohydrate chains, although most mucopolysaccharides which have been studied contain more sulphate and less protein. Further analysis of the carbohydrate portion, and chemical analysis of the subunits, should resolve this problem. It is noteworthy that the unique shape of the aggregation factor, with its sunburst pattern, has also been observed in electron micrographs of certain fractions of cartilage[143]. See Section 2.2.4.

The aggregation factor from *Microciona parthena* is one of the first mucopolysaccharides for which a biological function can be assayed. It will be interesting to see how universal this type of molecule is in the sponge aggregation phenomenon, and also to see if these molecules may have similar functions in the more complicated organogenic processes in vertebrates.

Concerning the aggregation factors obtained from other species, Humphreys has indicated that the factor from *Microciona prolifera* is similar to the *Microciona parthena* factor described above with regard to many physical and chemical properties[208]. The aggregation factor from *Geodia cydonium*[210] appears in the electron microscope to have a similar sunburst pattern as seen above. However, preliminary evidence indicates that this factor consists solely of protein, in contrast to most other aggregation factors which have been studied.

Recent investigations into the importance of the carbohydrate portion of the aggregation factor have demonstrated glucuronic acid to be a very effective and specific inhibitor of *Microciona prolifera* aggregation[205]. Glucuronic acid did not inhibit the aggregation of *Microciona cliona* and *Microciona parthena*, implicating the carbohydrate portion of aggregation factors as a possible determinant of species specificity. Furthermore, preliminary experiments indicate that glucuronic acid cleaving enzymes, as well as sodium metaperiodate, were able to destroy the activity of the aggregation factor[205].

The aggregation factors described are able to come off the cells and can be put back on the cells. It is presumed that they are the primary molecules involved in the linkage between the cells. This factor must interact with a second macromolecular component which remains on the cell surface after removal of the aggregation factor. Weinbaum and Burger[213] have demonstrated that this second component can be released by hypotonic shock from the surface of *Microciona prolifera* cells, previously stripped of aggregation factor in divalent cation free-sea water (see Figure 2.14). This released surface component, termed 'baseplate', can inactivate the aggregation factor if the two components are pre-incubated and then added to chemically dissociated cells. Furthermore, cells stripped of the surface component by hypotonic shock can no longer be aggregated by the aggregation factor unless the 'shockate' is returned to the aggregation assay. The cell-surface component which reacts with the aggregation factor has also been obtained by trypsinisation of the sponge cells[208]. The material obtained by this treatment is able to inhibit the activity of the aggregation factor; however, it does not stop the gelling of the factor. No further biochemical data are available on this second cell-surface component.

Weinbaum and Burger[213] have covalently attached the crude cell-surface component, as well as the aggregation factor, to Sepharose beads. Preliminary results suggest that the beads might be used as a cellular aggregation system; i.e. when aggregation factor coated beads were added to a suspension of cells dissociated in divalent cation-free sea water, the cells bound the beads; if aggregation factor coated beads were added to hypotonically shocked cells the cells had no affinity for the beads. This use of beads covered with specific cell-surface components is certainly very promising as a model system for studying cell–cell interactions.

Research is still required with unequivocal biochemical evidence to demonstrate that carbohydrates are involved in sponge-cell aggregation. Further purification and characterisation of the aggregation factors from different species, as well as the chemistry of the interaction between the surface components, are needed. As a population of dissociated sponge cells consists of several different cell types, it will also be necessary to investigate

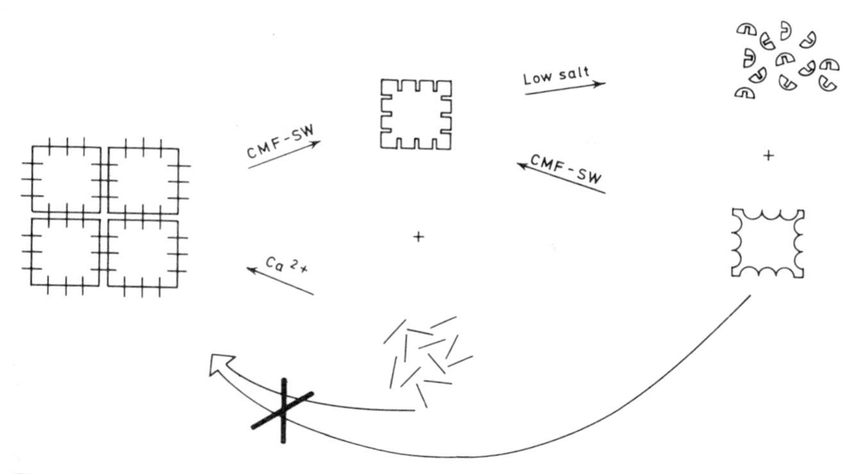


**Figure 2.14** Requirements of a membrane-bound baseplate for the specific sorting-out process. Sponge cells dissociated in calcium–magnesium-free seawater (CMF-SW) will release aggregation factor (long black bars) that is required in the presence of $Ca^{2+}$ for reaggregation of the cells lacking the aggregation factor. Low salt treatment (hypotonic swelling) of such aggregation factorless cells will release a baseplate or receptor for the aggregation factor (kidney shaped receptors in right upper corner) that is necessary for the cell to respond to the aggregation factor. (Taken in part from Turner and Burger[199], by courtesy of *Rev. Physiol.*)

the role of these different cell types in the cell aggregation process[214]. Preliminary evidence indicates that different types may have dissimilar adhesive properties and that there may be sorting-out between different types of cells within one single species of sponge[215].

## 2.4.4 Mammalian cell surfaces in development

### *2.4.4.1L ectin studies in embryonic differentiation*

It is generally believed that specific changes in the cell surface are involved in embryonic differentiation. Studies on the adhesion-dependent processes of reaggregation of dissociated embryonic cells to form multi-cellular aggregates and the sorting out and rearrangement of histotypically dissimilar cells within such aggregates have contributed supporting evidence for the involvement of specific cell-surface determinants in developmental processes[171, 216]. Another technique that has been used to study mammalian cell surfaces is the utilisation of lectins that are able to bind to specific carbohydrate groups on the cell surface and result in cell agglutination.

Kleinschuster and Moscona[217], using lectins with different sugar specificities, have studied changes at the surface of chick embryonic neural retina cells during their development and differentiation. Crude castor-bean lectin agglutinated chick and foetal retina cells from all developmental ages tested, while wheat germ agglutinin (WGA) did not agglutinate any of these cells, unless they had first been treated with trypsin. In contrast, Con A appeared to be able to discern differences during development as this lectin agglutinated young (up to 12 days old) retina cells while it did not agglutinate older ones. The latter cells, however, were agglutinable after protease treatment. Similarly, liver cells taken from 10-day-old chick embryos were agglutinated by Con A, while agglutination with WGA occurred only after the cells had been treated with trypsin[218]. The studies on the agglutination of embryonic cells by Con A have been extended to human tissue[219], where intestinal cells from the human foetus were markedly agglutinated by Con A whereas cells from adult intestine were not. In yet another aspect of the use of lectins, Lallier[220] has demonstrated that Con A is able to interfere with various aspects of development in the sea urchin.

Whether these membrane changes detected by lectins are incidental expressions of the cell surface or are actual cell behaviour determinants is not known. In an attempt to answer this question, Steinberg and Gepner[221] have examined several adhesion-dependent processes in the presence of univalent Con A (Con A capable of specific binding but no longer able to cause agglutination). They found that covering of the Con A receptor sites with this univalent lectin had little affect on cell aggregation and cell sorting in chick embryo cells, thus indicating that these particular sugar sites on these cells seem to play no major role as sites during the formation of intercellular adhesions *in vitro*. It should be added that this general area is only in its infancy and will certainly yield more concrete results in the next few years.

### *2.4.4.2 Embryonic antigens or oncofoetal antigens*

Tumor tissues often express embryonic antigens which are absent from normal adult tissues[222, 223]. Recently, Ting *et al.*[224] have found several foetal antigenic specificities in cells transformed *in vitro* by SV40 or polyoma virus. Embryonic antigen expression has also been found in chemically induced rat tumors[225, 226]. A number of embryo-specific murine antigens have also been reported to reappear in adult cells following malignant conversion[227]. These antigens were not detected in normal cells with one exception, but were demonstrated to be present in a large variety of murine sarcomas in which malignant transformation had been induced by virus, chemicals or radiation. Also, primative terato-carcinoma cells in culture have been shown to possess cell-surface antigens, in common with normal mouse cleavage-stage embryos[228]. These surface antigens were absent from 1-cell eggs, and increased to a maximum at the 8-cell stage. They were not present in any other mouse cells tested.

The presence of oncofoetal antigens on tumor cells has been considered to be an example of their dedifferentiated character. It is assumed that synthesis of these antigens is repressed in adult tissues, but it is not known whether normal adult cells might possess certain cryptic embryonic antigens. The similarity of the surface properties of tumor cells to those of embryonic cells may be related to the increased cell mobility of early embryonic cells that is expressed similarly in tumor cells, although to a much higher degree, as these are capable of indiscriminantly infiltrating neighbouring tissues.

#### *2.4.4.3 Mucopolysaccharides in morphogenesis*

Toole and co-workers[229-231], in studying the involvement of hyaluronate in embryonic cartilage induction, have found that hyaluronate synthesis is generally associated with the morphogenetic phase of cell migration and proliferation, while its removal by the action of a hyaluronidase accompanies subsequent differentiation. They propose that hyaluronate may act as an inhibitor of precartilage cell aggregation in embryogenesis, and prevent differentiation; its synthesis and removal are part of the mechanism of timing of migration, aggregation and subsequent differentiation.

Bernfield and associates[232-234] have found a possible regulatory role for extracellular mucopolysaccharides in morphogenesis in the mouse embryo salivary gland. They have shown a selective activation of mucopolysaccharide synthesis localised at positions of future cleft formation and it is suggested that the newly deposited mucopolysaccharides induce collagen polymerisation[232]. The surface-associated mucopolysaccharide is also believed to be a stimulus for contraction of intracellular microfilaments which would stabilise the newly formed clefts.

### 2.4.5 Cell-surface carbohydrates in intercellular adhesion

Intercellular adhesion is a fundamental biological property of cells in multicellular organisms and is certainly important in a wide variety of physiological processes, including growth, morphological differentiation and metastasis[235]. There has been much published on the biological descriptions of these processes. However, we will restrict our attention here to those studies aimed at elucidating the underlying biochemical mechanisms.

Oppenheimer *et al.*[236] have obtained single cell suspensions of mouse teratoma tumors using trypsin treatment, a procedure which dissociates the cells and decreases adhesiveness by partially degrading cell-surface receptors. When the single cell suspensions from the teratoma tumors were placed in a synthetic tissue culture medium, the cells were observed to reassociate rapidly. However, this recovery of adhesiveness required the presence of glutamine in the growth medium. Glutamine could be specifically replaced by glucosamine or mannosamine, suggesting that it is required because of its participation in the formation of glucosamine 6-phosphate, which is an important precursor in the synthesis of several carbohydrate molecules. These results suggest that a crucial event in the conversion of non-adhesive into adhesive cells is the synthesis of complex carbohydrates. Similar studies have now shown that glutamine promotes the adhesion of several type of ascites tumor cells and may also be involved in cellular adhesion of chick-embryo neural retina cells[237].

The involvement of cell-surface carbohydrates in cellular adhesion has also been studied by treating cells with specific carbohydrate-releasing enzymes which remove terminal sugar residues from the cell surface. Roth *et al.*[238] have shown that treatment of dissociated embryonic neural retina cells with β-galactosidase alters the adhesive properties of these cells, and suggest that terminal β-galactosyl residues on the surface may be partly responsible for the adhesive selectivity in these cells. Vickers and Edwards[239] have reported that neuraminidase treatment increases the adhesiveness of trypsinised hamster fibroblasts (BHK-21 cells), while having little affect on the BHK-21 cells transformed by polyoma virus. However, Kemp[240] has found that neuraminidase treatment of freshly dissociated primary embryonic chick muscle cells resulted in a decrease in cell aggregation.

When liver cells dissociated from 10-day-old chick embryos were cultured as monolayers, the re-aggregability of the harvested cells declined steeply with the length of time in monolayer culture[241]. The loss of re-aggregability of these cells was correlated with an increase in some carbohydrate cell-surface determinants. An early rapid increase in Forssman antigen (GL-5) was demonstrated, as well as increase in the receptor activities of these cultured cells for Japanese encephalitis virus and influenza virus. It is interesting that re-aggregability in 4-day cultured cells could be recovered by removal of sialic acid using neuraminidase. Matsuzaua[241] also found, in agreement with Kemp[240], that neuraminidase treatment of freshly dissociated primary cells resulted in a decline in cell aggregation.

Sialic acid appears to contribute partially to the negative surface charge of mammalian cells[242]. Removal or masking of cell-surface sialyl residues could result in lowering of the surface charge and could increase cell adhesiveness by lowering electrostatic repulsion between cells[243]. Sialic acid is also believed to confer structural rigidity to glycoprotein molecules[244]. Weiss[245] has shown that removal of sialic acid from the surface of Sarcoma 37 cells increased the overall cellular deformability, which he pointed out was an important parameter in cell adhesion.

Recently, Chipowsky *et al.*[246] have attempted another approach to the examination of the role of cell-surface carbohydrates in adhesion processes. Insoluble carbohydrate analogues were prepared by coupling monosaccharides to Sephadex beads, and the interaction of these derivatives and various cells was studied. It was found that SV40-transformed 3T3 fibroblasts adhered to the beads derivatised with galactose, but did not adhere to the corresponding beads derivatised with glucose or *N*-acetylglucosamine. Cells that adhered to the galactose-beads appeared to initiate a nucleation process; they became more adhesive toward the cells in suspension, leading to the formation of large aggregates containing both cells and galactose beads. These insoluble analogues of cell-surface components may prove very useful in our further understanding of cellular recognition and adhesion.

### *2.4.5.1 Glycosyltransferases*

Much of the recent speculations concerning the role of cell-surface carbohydrates in cell–cell adhesion have centred around Roseman's model (Figure 2.15) that cell recognition and aggregation result from enzyme–substrate interaction between glycosyltransferases on the surface of one cell, which bind to and may glycosylate an oligosaccharide acceptor present on another cell surface[3]. The cells would then be held together by the affinity of the enzymes for the substrates. An extension of this theory is that since glycoproteins and glycolipids are synthesised by a series of glycosyltransferases, the adhesive properties of cells could be changed or modified at various stages of development. The adhesive properties and specificity of this adhesion would be determined by the nature and number of (*a*) incomplete carbohydrate chains in the receptor, (*b*) glycosyltransferases, (*c*) sugar nucleotides and (*d*) the product glycoprotein or glycolipid. Experimental support for this hypothesis has come from the demonstration that whole chick-embryo neural retina cells are able to incorporate galactose from UDP-galactose into their surface components[247]. The nucleotide sugar was not taken up into the cells, and radioactive galactose became attached to components on

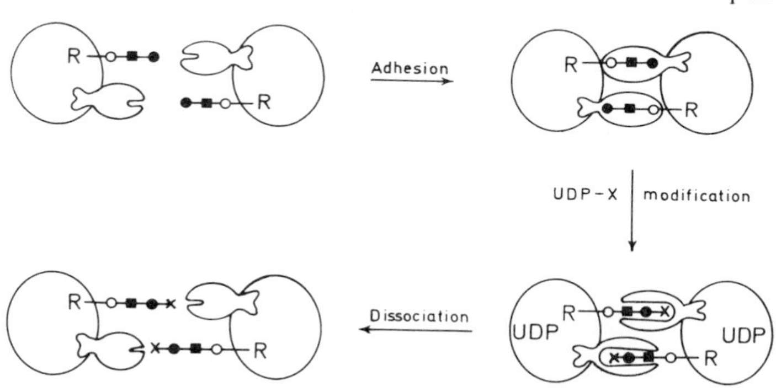


**Figure 2.15** Roseman's model[3] for the role of glycosyltransferases in intercellular adhesion. The fish-shaped figures represent membrane-bound glycosyltransferases

the cell periphery, implying that both enzyme and substrate were present on the cell surface. Similar studies have demonstrated the presence of cell-surface glycosyltransferases on normal and virus-transformed 3T3 cells[106], blood platelets[248, 249] and intestinal cells[250]. Evidence for intercellular glycosylation has been deduced from the fact that the rate and extent of sugar incorporation were greatest when the reaction took place in a cell suspension under conditions allowing for maximum contact between cells[251]

According to Roseman's theory, transformed cells do not bind strongly to each other because of deficiencies in certain transferases or acceptors which are present on the surface of normal cells. The results presented earlier [see Sections 2.2.2.1(f) and 2.2.3.2] on glycosyltransferase deficiencies in virus-transformed cell lines would be in accord with this. In addition, Roth and White[251] have indicated that, in contrast to normal cells, where glycosylations are essentially intercellular, glycosyltransferases on tumour cells would also be able to glycosylate substrates on the same cell, presumably because enzyme and substrate are closer to each other than on normal cells. The ability not to depend on intercellular glycosylations would be of obvious advantage to transformed cells that do not have normal intercellular contacts.

Further evidence for the involvement of glycosyltransferases comes from the work of Weiser[250, 252], who has studied surface membrane glycoproteins during differentiation of rat intestinal epithelial cells. The epithelial cells begin as undifferentiated mitotically active cells at the base of the crypt (see Figure 2.16), and differentiate as they move up the villus to reach full maturity at the upper third of the villus in 36–72 hours. The ability to separate villus cells from crypt cells has made it possible to study surface membrane changes during this differentiation process. Labelling of these cells with sugar

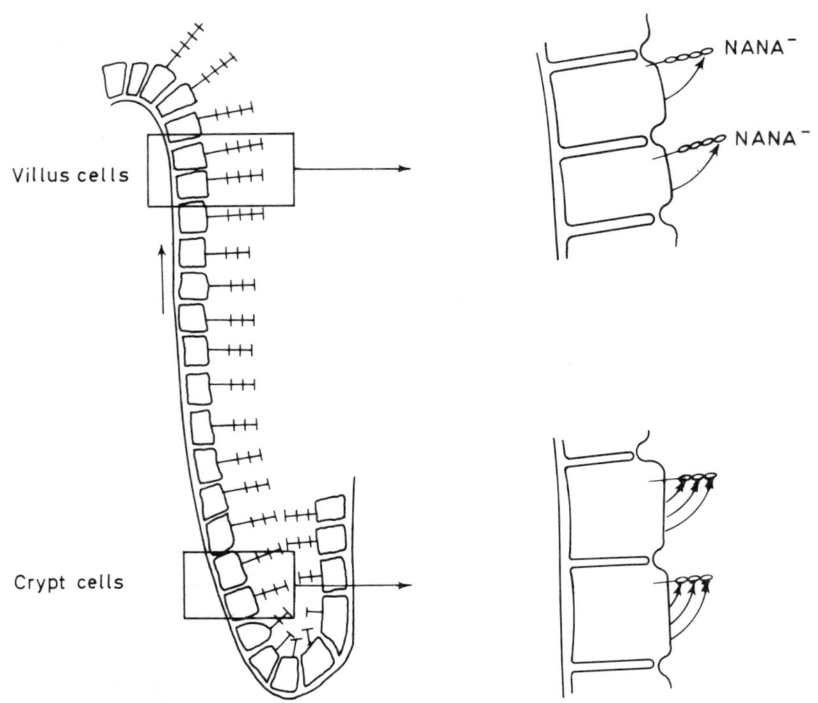


**Figure 2.16** Schematic representation of the intestinal epithelium with the attachment of membrane glycoproteins as the cells move up the villus

precursors, either *in vivo* in rats or *in vitro*, resulted in a high incorporation into membrane glycoproteins of the more differentiated cells (the villus cells), with little incorporation occurring in the undifferentiated crypt cells.

Surface-membrane glycosyltransferase activity and endogenous acceptors on these epithelial cells were examined by incubating intact cells with labelled sugar nucleotides[250]. The glycosyltransferase:endogenous acceptor activity found in the differentiated villus cell fractions was only 5–10% of the activity present in the undifferentiated crypt cells. The only exception was the sialyltransferase, which was higher in the villus cells. Cell-surface differences between the crypt and villus cells have also been demonstrated using Con A derivatised to nylon fibres. This insolubilised Con A selectively bound the undifferentiated crypt cells[253], which would agree with the results presented earlier (Section 2.4.4.1) on the Con A agglutination of undifferentiated embryonic cells.

These studies on epithelial cell differentiation would suggest that the undifferentiated state is characterised by the presence of glycosyltransferases and incompletely glycosylated glycoproteins on the surface membrane. The evidence presented earlier on the incompletely glycosylated sugar chains in transformed cells [Sections 2.2.2.1(f) and 2.2.3.2] and the dedifferentiated character of transformed cells (Section 2.4.4.2) would be in accord with this hypothesis.

Probably the best evidence in support of Roseman's glycosyltransferase theory comes from studies on platelet:collagen adhesion. The primary step in haemostasis involves the adhesion of circulating blood platelets to collagen that is exposed by damage to the vascular endothelium[254]. This is followed by the aggregation of increasing numbers of platelets to themselves to form a plug at the site of injury. Collagen is known to contain small and variable quantities of carbohydrate linked to the hydroxy group of δ-hydroxylysine[255]. These sugar residues can consist of a complete α-glucosyl-(1→2)-β-galactosyl unit, or of only a singly β-linked galactosyl residue. The proportion of incomplete chains of single galactosyl residues on collagen varies from 3% to 60%[256].

Isolated platelet plasma membranes contain the two glycosyltransferases capable of adding the corresponding monosaccharides to the appropriate acceptors on collagen[248, 249]. However, the presence of these two very specific collagen transferases is unexplained, as platelets apparently neither contain nor synthesise collagen-like molecules[257, 258]. The interesting supposition has been proposed that these enzymes may be involved in platelet:collagen adhesion[248, 259]. In particular, Jamieson *et al.*[259] have proposed that the collagen:glucosyltransferase on the surface of the platelet membrane mediates platelet:collagen adhesion by formation of a complex with the incomplete carbohydrate chains of collagen. As evidence for the involvement of this enzyme, these workers found that (*a*) glucose was incorporated into collagen from UDP-glucose during the process of collagen adhesion, and that (*b*) several inhibitors gave a parallel inhibition of the membrane glucosyltransferase activity and the platelet:collagen adhesion reaction[259].

Evidence for the involvement of the carbohydrate residues of collagen in this adhesion process has come from Katzman *et al.*[260], who have examined the structural features of collagen that might be involved in collagen-induced platelet aggregation. They prepared cyanogen bromide peptides of the $\alpha_1$ chain of chicken-skin collagen and found that only one fragment, representing less than 4% of the $\alpha_1$ chain, was active in inducing platelet aggregation. This is a glycopeptide that contains one residue of glucosylgalactose linked to hydroxylysine. Subsequently it was found that the platelet aggregation induced by collagen could be inhibited specifically by pure α-glucosyl-(1→2)-β-galactosyl-(1→5)-lysine. This evidence strongly suggests the involvement of these carbohydrates as determinants necessary for inducing aggregation.

In summary, the present evidence for Roseman's theory of the involvement of cell-surface glycosyltransferases in cell-surface recognition is only indirect. The discovery of cell-surface glycosyltransferases is a first prerequisite.

However, the involvement of these enzymes as specific recognition sites is still in doubt. One question that has been raised concerns the existence of extracellular nucleotide sugars, which are required if these transferases are indeed involved in transferring sugars to receptors on a different cell.

Recently, Grimes and Patt[301] have made some very interesting observations which may prove to be of fundamental importance in connection with the concept of cell-surface glycosyltransferases. They have detected a number of different glycolipid and glycoprotein glycosyltransferase activities which are associated with intact mouse fibroblast cell lines. They conclude that these glycosyltransferase activities observed in intact cells probably measure normal intracellular synthetic systems. Homogenisation of cells either decreased or did not affect the activities observed when whole cells were incubated with nucleotide sugars as precursors. The most conclusive evidence came from the observation that the glycolipid products formed when intact cells were incubated with UDP-[$^{14}$C]galactose (which cannot enter the cell) were the same as those being synthesised when the cells were incubated with [$^{3}$H]-galactose.

A major goal in this field is to determine the chemical structures of the surface receptors that are involved in cell recognition. However, this will be difficult to determine, utilising the present techniques available for measuring cell recognition, which measure the affinity of cells for each other. Merrell and Glaser[261] have recently demonstrated specific recognition using plasma membranes, and this may prove to be an important step towards the evaluation of the molecular parameter. The use of insoluble analogues of cell-surface components may also prove to be important, both in studying cellular recognition and in isolating complementary cell-surface components.

### 2.4.6 Homing phenomena

Recognition phenomena can also be manifested in the interaction between cells and soluble molecules. The role of glycoproteins in this type of interaction has been clearly demonstrated in the highly interesting studies of Morell, Ashwell and their associates[262-266], who have investigated the role of the terminal sugar residues on the fate of a number of intravenously administered glycoproteins. They have shown that many serum glycoproteins (including orosomucoid, fetuin, ceruloplasmin, haptoglobin, $\alpha_2$-macroglobulin, thyroglobulin) and two pituitary hormones (human chorionic gonadotrophin and follicle-stimulating hormone) are rapidly cleared from the circulation upon removal of sialic acid from these molecules. Thus, while the normal half-life of unmodified ceruloplasmin in the circulation is 54 hours, the removal of *any* two of the possible 10 terminal sialyl residues produces a material which is removed from the circulation within minutes[262]. The prompt disappearance of asialo-ceruloplasmin from the serum was found to be accompanied by an equally rapid accumulation of the labelled material within the liver. At 24 minutes after injection, most of the modified ceruloplasmin was recoverable from this organ and was located exclusively in the parenchymal cells; no activity was found in the reticulo-endothelial system at any time. These molecules were found to be subsequently degraded within the lysosomes of the liver.

Sialic acid-free glycopeptides, prepared from the glycoproteins, inhibited uptake of the asialo-glycoproteins, indicating that the recognition sites responsible for the uptake of glycoproteins probably involve the carbohydrate moieties rather than the overall stereochemistry of the macromolecules. Removal of sialic acid from these glycoproteins leaves galactose in non-reducing terminal positions; in Figure 2.17 is illustrated the unique

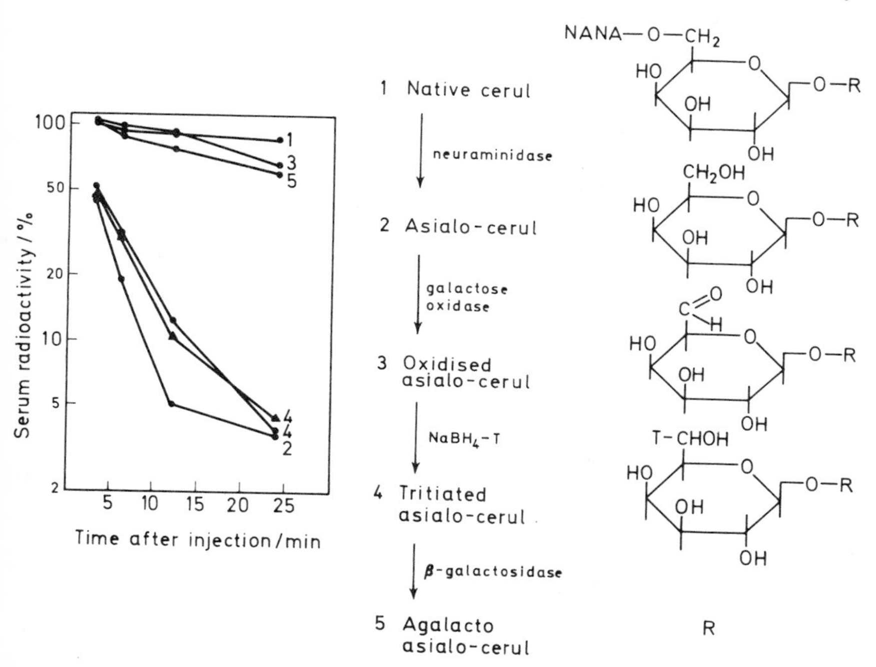


**Figure 2.17** Disappearance from serum of $^{64}$Cu-labelled native ceruloplasmin and modified rabbit ceruloplasmins (Cerul). On the right are given the reaction sequences used in preparing the various modified derivatives along with the structures of the terminal residues. (Taken from Ashwell and Morell[262], by courtesy of J. B. Lippincott)

role of the terminal galactosyl residues in affecting the rapid clearance of asialo-ceruloplasmin. Treatment of asialo-ceruloplasmin with galactose oxidase, whereby 94% of the primary alcohol groups of galactose are converted into aldehyde functions, results in a striking reversal in the rate of disappearance (curve 3) compared with that of the untreated asialo-ceruloplasmin (curve 2). Reduction of the aldehyde group back to the original chemical group in asialo-ceruloplasmin is accompanied by a rapid clearance from the serum (curve 4). A prolonged survival time is again observed following removal of the galactosyl residues with β-galactosidase (curve 5). It should be added that these modified derivatives of ceruloplasmin were carefully characterised both chemically and biochemically. These *in vivo* experiments clearly point to the role of the carbohydrates and demonstrate that it is specifically the galactosyl residues which are intimately involved in the recognition process which leads to the removal of these glycoprotein molecules from the circulation.

The isolated liver plasma membrane has been identified as the primary binding site of the asialo-glycoproteins *in vitro*, and has provided a biological model for detailed examination of the physiological mechanisms involved in the *in vivo* binding phenomenon[264]. There is a close parallelism between the two systems that is manifested by the similar organ specificity and analogous susceptibility to competitive inhibition. Effective binding on the liver plasma membrane requires the absence of sialyl residues on the glycoprotein. However, binding is dependent on the presence of sialic acid on the plasma membranes, as neuraminidase treatment of the isolated plasma membrane results in loss of binding. This effect could be reversed in the presence of CMP–*N*-acetylneuraminic acid. A membrane sialyltransferase catalyses the transfer of sialic acid from the nucleotide to the partially desialyated membrane and this is accompanied by a significant restoration of the binding activity. The binding of the asialo-glycoproteins by plasma membranes also exhibits a specific requirement for calcium ion, and changes in the concentration of this ion can result in a rapid reversal of the binding process.

Future research will certainly be centred around determining the molecules present on the liver plasma membrane responsible for this binding phenomenon. The results of Morell and Scheinberg[266], indicating that a liver membrane glycoprotein binding complex can be solubilised with detergent and still retain the specific binding activity in solution, should aid in the isolation of these receptors. It will certainly be interesting to see what role the membrane sialyltransferase or other membrane glycosyltransferases may have in this binding phenomenon. A preliminary report by Aronson *et al.*[267] suggests that a liver membrane galactosyltransferase may be involved in the binding.

These excellent studies give support to the postulate that a predominant role of glycoprotein prosthetic groups is to serve as specific probes for translocation and adsorption at some specific site. This type of interaction may also be involved in the adsorption of cells. It has been mentioned previously that erythrocytes which have been treated with neuraminidase are removed from the circulation. Another well known example of this is the homing of lymphocytes[268, 269]. Sialic acid appears to be necessary on the lymphocytes for these cells to traverse their unique route through the body and these molecules may act by recognising complementary structures on the surface of endothelial cells in lymphoid tissues. Woodruff and Gesner[269] have found that lymphocytes which have been treated with neuraminidase and transfused become trapped in the liver and there is a reduction in the usual accumulation of these cells in the lymph nodes and spleen. A similar mechanism may be operative here, as in the case of the asialo-serum glycoprotein discussed above.

### 2.4.7 Lectin agglutination studies

As we have mentioned previously, lectins have become very popular tool for studying cell surfaces. They are of particular interest to us here becaus they bind to specific carbohydrate groups on the cell surface. Lectins are able to distinguish between some transformed cells and their untransformed parental cells[109, 110], embryonal and adult cells[218] as well as mitotic and interphase cells[270, 271].

In Table 2.4 are summarised some positive correlations between increased lectin agglutinability and growth capacity of cells. It has been found that a large number of lines of transformed cells agglutinate well with various lectins[110, 272, 273], while the normal cells are not agglutinable at the same low concentrations of agglutinin unless they have been treated with very low

**Table 2.4 Lectin agglutination**

| | |
|---|---|
| Transformed cell agglutination | Some 40–50 growing tissue culture lines of transformed cells have been found to agglutinate well with various lectins. (For a complete discussion of this topic see review articles by Rapin and Burger[88] and Lis and Sharon[108]) |
| Correlation of growth control and agglutinability | Various cell lines derived from a single stem line display increases in agglutinability in direct proportion to which the cells have lost density-dependent growth control[274] |
| Revertants studies | (a) Revertant clones have been isolated from transformed cells which have regained some density-dependent inhibition of growth and these cells do not further agglutinate[274, 275] |
| | (b) Transformed cells selected for loss of agglutinability were found to have regained some of their growth control[276, 277] |
| Mutants studies | Conditionally non-transformed viral mutants[278] and temperature-sensitive viral mutants[279-281] cannot achieve the agglutinable state under non-permissive conditions. Temperature-sensitive host mutants lose some transformed properties at the non-permissive temperature and lose their agglutinability simultaneously[282, 283] |

concentrations of protease[109, 110]. Like all generalisations, however, there are some exceptions to the above. A good correlation has been found between the agglutination of mouse and hamster fibroblasts and the degree to which they have lost density-dependent growth control under similar conditions regarding serum concentration and pH[274]. That these surface changes are indeed part of the transformation process has been demonstrated with the aid of conditional (host range, temperature-sensitive) mutants. Cells infected with certain viruses under permissive conditions (usually the lower temperature) were neoplastically transformed and concomitantly they lost their density-dependent inhibition of growth and became agglutinable. Under the non-permissive conditions, however, the viruses did not transform the cells; these did not grow beyond a low saturation density and were only poorly agglutinated by lectins[279-283].

At the present time there is no clear answer to the question to what extent agglutinability of *in vitro* tissue culture cells is a good index of tumorigenicity *in vivo*. The malignant state is almost certainly a multiparameter phenomenon.

The surface change observed in transformed cells with a given lectin may detect only a single function of these transformed cells, like increased growth, loss of density-dependent inhibition of growth, metastasis, adhesion and the like.

### 2.4.7.1 *Possible mechanisms of agglutination*

The exact mode of interaction of a given lectin with its particular cell-surface receptor, and the mechanism of the subsequent agglutination, is not known, but in some regards it may be analogous to an antigen–antibody reaction. A decrease in valency by chemical alteration or a partial degradation[284, 285] leads to a lectin which still can bind to the receptor site, but can no longer agglutinate, a situation comparable to that with monovalent antibodies. Below we will consider several possible mechanisms of agglutination. It should be made clear that all of these mechanisms are hypothetical and are presented only as possibilities to be tested.

(a) *Possible increase in number of available lectin receptor sites on transformed cells*—The most obvious explanation for the increased agglutinability would be an increase in the number of lectin binding sites. Transformation would lead to an exposure of saccharide receptors present in cryptic form on the surface of the normal cell. However, a variety of laboratories[286-289] have suggested that there is little or no difference in agglutinin binding between normal and transformed cells. Noonan and Burger[290] have recently demonstrated that if care is taken to avoid endocytosis and non-specific binding of the agglutinin molecules to the cell surface (binding studies carried out at 0 °C), virially transformed fibroblasts bind 3–5 times more Con A per unit surface area than do the untransformed parental cells. At room temperature, where pinocytosis and other phenomena increase the 'binding', this difference was not present.

(b) *Receptors for lectin have different chemical structures on normal and transformed cells*—The model for this mechanism comes from the changes observed in lipopolysaccharide structures during lysogenic conversion of *Salmonella anatum*[175-177] and discussed briefly in a previous Section (2.4.1). Several suggestions have appeared that chemical changes might occur in the surface structures of cells transformed by viruses, and these data have been presented earlier (Sections 2.2.2.4 and 2.2.3.2). Transforming viruses could cause a modification of one or more glycosyltransferases and thus modify lectin receptors. If this were the case, it would imply that all malignant transformations, including chemical or spontaneous transformations or those induced by radiation, must ultimately depend on viruses, a possibility which as yet has neither been proven nor ruled out. An argument against the virus involvement and the formation of new receptor sites during transformation is the fact that normal cells also contain lectin-receptor sites, and that protease treatment of these cells renders them as fully agglutinable as are transformed cells. However, the ultimate answer to this possible difference must await the fine structural analysis of the carbohydrate-bearing lectin receptors obtained from both transformed and untransformed parental cells.

(c) *Transformed cells may have a more fluid membrane*—Membranes of agglutinable cells (i.e. transformed cells, or normal cells during mitosis or after protease treatment) may be somehow more fluid, so that receptor sites can more easily be gathered together into a configuration favourable for agglutination. This conclusion has been reached by several workers from electron microscopic observations on the interaction of derivatised lectins with normal and transformed cells[291, 292]. It has also been found that cells treated with glutaraldehyde are no longer agglutinable by Con A, and since binding of the lectin was found not to be decreased by this treatment, the effect may be on immobilisation of membrane proteins by cross-linking[293, 294]. The temperature dependence of agglutination may also be an indication of a need for membrane fluidity[290]. It may be that movement of receptor sites is induced by the cross-linking lectin and this movement only becomes significant at temperatures above the freezing point of the membrane lipids[295, 296].

(d) *Other possible mechanisms*—The above are only a few of the possibilities to explain the increased agglutinability of transformed cells. Here are briefly listed some other possibilities which may play a role in binding and agglutination:

(*a*) Concentration of sites in transformed cells owing to surface shrinkage[288].

(*b*) Differences in surface charge between normal and transformed cells.

(*c*) Increased flexibility of the whole membrane.

(*d*) Lipophilic interactions.

(*e*) Clustering of sites on microvilli[297].

There is still no unambiguous evidence that lectin molecules are spanning between the cell surfaces of agglutinated cells, and that this bridge is the cause of agglutination. One therefore has to still consider the possibility that lectins alter the surface structure and thereby allow interaction between parts of the membrane which do not directly carry the lectin molecules.

## References


1. Kalckar, H. M. (1968). *Science*, **150**, 305
2. Revel, J.-P. and Ito, S. (1967). *The Specificity of Cell Surfaces*, 211 (B. D. Davis and L. Warren, editors) (Englewood Cliffs, New Jersey: Prentice-Hall)
3. Roseman, S. (1970). *Chem. Phys. Lipids*, **5**, 270
4. Roseman, S. (1968). *Biochemistry of Glycoproteins and Related Substances*, 244 (E. Rossi and E. Stoll, editors) (Basel: S. Karger)
5. Schachter, H., Jabbal, I., Hudgin, R. L., Pinteric, L., McGuire, E. J. and Roseman, S. (1970). *J. Biol. Chem.*, **245**, 1090
6. Melchers, F. (1971). *Biochemistry*, **10**, 653
7. Choi, Y. S., Knopf, P. M. and Lennox, E. S. (1971). *Biochemistry*, **10**, 668
8. Whaley, W. G., Dauwalder, M. and Kephart, J. E. (1972). *Science*, **175**, 596
9. Evans, P. J. and Hemming, F. W. (1973). *FEBS Lett.*, **31**, 335
10. Lennarz, W. J. and Scher, M. G. (1972). *Biochim. Biophys. Acta*, **265**, 417
11. Osborn, M. J. (1969). *Ann. Rev. Biochem.*, **38**, 301
12. Aronson, N. N. and de Duve, C. (1968). *J. Biol. Chem.*, **243**, 4564
13. Patel, V. and Tappel, A. L. (1971). *Glycoproteins of Blood Cells and Plasma*, 133 (G. A. Jamieson and T. J. Greenwalt, editors) (Philadelphia: J. B. Lippincott)
14. Dingle, J. T. (1969). *Lysosomes in Biology and Pathology*, Vol. 2, 421 (J. T. Dingle and H. B. Fell, editors) (Amsterdam: North-Holland)
15. Raivio, K. O. and Seegmiller, J. E. (1972). *Ann. Rev. Biochem.*, **41**, 543

16. O'Brien, J. S., Okada, S., Ho, M. W., Fillerup, D. L., Veath, M. L. and Adams, K. (1971). *Fed. Proc.*, **30,** 956
17. Spiro, R. G. (1969). *New Engl. J. Med.*, **281,** 991
18. Marshall, R. D. (1972). *Ann. Rev. Biochem.*, **41,** 673
19. Spiro, R. G. (1967). *J. Biol. Chem.*, **242,** 1923
20. Weiss, J. B., Lote, C. J. and Bobinski, H. (1971). *Nature (London) New Biol.*, **234,** 25
21. Lote, C. J. and Weiss, J. B. (1971). *FEBS Lett.*, **16,** 81
22. Smyth, D. S. and Utsumi, S. (1967). *Nature (London)*, **216,** 332
23. Gottschalk, A. (1969). *Nature (London)*, **222,** 452
24. Cunningham, L. W. (1971). *Glycoproteins of Blood Cells and Plasma*, 16 (G. A. Jamieson and T. J. Greenwalt, editors) (Philadelphia: J. B. Lippincott)
25. Ginsburg, V. and Kobata, A. (1971). *Structure and Function of Biological Membranes*, 439 (L. I. Rothfield, editor) (New York: Academic Press)
26. Hughes, R. C. (1973). *Progress in Biophysics and Molecular Biology*, Vol. 26, 189 (J. A. V. Butler and D. Noble, editors) (Oxford: Pergamon Press)
27. Winzler, R. J. (1972). *Glycoproteins: Their Composition, Structure, and Function*, B.B.A. Library Vol. 5B, 1268 (A. Gottschalk, editor) (Amsterdam: Elsevier)
28. Oseroff, A. R., Robbins, P. W. and Burger, M. M. (1973). *Ann. Rev. Biochem.*, **42,** 647
29. Guidotti, G. (1972). *Ann. Rev. Biochem.*, **41,** 731
30. Rosenberg, S. A. and Guidotti, G. (1969). *J. Biol. Chem.*, **244,** 5118
31. Guidotti, G. (1972). *Ann. N.Y. Acad. Sci.*, **195,** 139
32. Bretscher, M. S. (1971). *J. Mol. Biol.*, **58,** 775
33. Phillips, D. R. and Morrison, M. (1971). *FEBS Lett.*, **18,** 95
34. Bretscher, M. S. (1971). *Nature (London), New Biol.*, **231,** 229
35. Berg, H. C. (1969). *Biochim. Biophys. Acta*, **183,** 65
36. Gahmberg, C. G. and Hakomori, S. (1973). *J. Biol. Chem.*, **248,** 4311
37. Bretscher, M. S. (1971). *J. Mol. Biol.*, **59,** 351
38. Bender, W. W., Garan, H. and Berg, H. C. (1971). *J. Mol. Biol.*, **58,** 783
39. Phillips, D. R. and Morrison, M. (1971). *Biochem. Biophys. Res. Commun.*, **45,** 1103
40. Winzler, R. J (1969) *Red Cell Membranes: Structure and Function*, 157 (G A. Jamieson and T. J. Greenwalt, editors) (Philadelphia: J. B. Lippincott)
41. Marchesi, V. T. and Andrews, E. P. (1971). *Science*, **174,** 1247
42. Suttajit, M. and Winzler, R. J. (1971). *J. Biol. Chem.*, **246,** 3398
43. Springer, G. F. and Ansell, N. J. (1958). *Proc. Nat. Acad. Sci. USA*, **44,** 182
44. Klenk, K. E. and Uhlenbruck, G. (1960). *Z. Physiol. Chem.*, **319,** 151
45. Eylar, E. H., Madoff, M. A., Brody, O. V. and Oncley, J. L. (1962). *J. Biol. Chem.*, **237,** 1992
46. Seaman, G. V. F. and Uhlenbruck, G. (1963). *Arch. Biochem. Biophys.*, **100,** 493
47. Kathan, R. H., Winzler, R. J. and Johnson, C. A. (1961). *J. Exp. Med.*, **113,** 37
48. Blumenfeld, O. O., Callop, P. M., Howe, C. and Lee, L. T. (1970). *Biochim. Biophys. Acta*, **211,** 109
49. Lenard, J. (1970). *Biochemistry*, **9,** 1129
50. Segrest, J. P., Jackson, R. L., Andrews, E. P. and Marchesi, V. T. (1971). *Biochem. Biophys. Res. Commun.*, **44,** 390
51. Winzler, R. J., Harris, E. D., Pekas, D. J., Johnson, C. A. and Weber, P. (1967). *Biochemistry*, **6,** 2195
52. Thomas, D. B. and Winzler, R. J. (1969). *J. Biol. Chem.*, **244,** 5943
53. Thomas, D. B. and Winzler, R. J. (1971). *Biochem. J.*, **124,** 55
54. Jackson, R. L., Segrest, J. P., Kahane, I. and Marchesi, V. T. (1973). *Biochemistry* **12,** 3131
55. Kornfeld, S. and Kornfeld, R. (1969). *Proc. Nat. Acad. Sci. USA*, **63,** 1439
56. Kornfeld, R. and Kornfeld, S. (1970). *J. Biol. Chem.*, **245,** 2536
57. Segrest, J. P., Jackson, R. L. and Marchesi, V. T. (1972). *Biochem. Biophys. Res. Commun.*, **49,** 964
58. Marchesi, V. T., Tillack, T. W., Jackson, R. L., Segrest, J. P. and Scott, R. E. (1972). *Proc. Nat. Acad. Sci. USA*, **69,** 1445
59. Morawiecki, A. (1964). *Biochim. Biophys. Acta*, **83,** 339
60. Steck, T. L., Fairbanks, G. and Wallach, D. F. H. (1971). *Biochemistry*, **10,** 2617
61. Pinto da Silva, P. and Branton, D. (1970). *J. Cell. Biol.*, **45,** 598
62. Tillack, T. W. and Marchesi, V. T. (1970). *J. Cell Biol.*, **45,** 649

63. Pinto da Silva, P., Douglas, S. D. and Branton, D. (1971). *Nature (London)*, **232**, 194
64. Tillack, T. W., Scott, R. E. and Marchesi, V. T. (1972). *J. Exp. Med.*, **135**, 1209
65. Barber, A. J. and Jamieson, G. A. (1970). *J. Biol. Chem.*, **245**, 6357
66. Pepper, D. S. and Jamieson, G. A. (1969). *Biochemistry*, **8**, 3362
67. Pepper, D. S. and Jamieson, G. A. (1970). *Biochemistry*, **9**, 3706
68. Jamieson, G. A., Fuller, N. A., Barber, A. J. and Lombart, C. (1971). *Ser. Haematol.*, **4**, 125
69. Nachman, R. L., Hubbard, A. and Ferris, B. (1973). *J. Biol. Chem.*, **248**, 2928
70. Kahan, B. D. and Reisfeld, R. A. (1969). *Science*, **164**, 514
71. Schwartz, B. D., Kato, K., Cullen, S. E. and Nathenson, S. G. (1973). *Biochemistry*, **12**, 2157
72. Shimada, A. and Nathenson, S. G. (1969). *Biochemistry*, **8**, 4048
73. Nathenson, S. G. and Muramatsu, T. (1971). *Glycoproteins of Blood Cells and Plasma*, 245 (G. A. Jamieson and T. J. Greenwalt, editors) (Philadelphia: J. B. Lippincot)
74. Nathenson, S. G. and Cullen, S. E. (1974). *Biochim. Biophys. Acta*, **344**, 1
75. Heller, J. (1968). *Biochemistry*, **7**, 2906
76. Shichi, H., Lewis, M. S., Irreverre, F. and Stone, A. L. (1969). *J. Biol. Chem.*, **244**, 529
77. Heller, J. and Lawrence, M. A. (1970). *Biochemistry*, **9**, 864
78. Steinemann, A. and Stryer, L. (1973). *Biochemistry*, **12**, 1499
79. Brunngraber, E. G., Brown, B. D. and Aguilar, V. (1969). *J. Neurochem.*, **16**, 1059
80. Den, H., Kaufman, B. and Roseman, S. (1970). *J. Biol. Chem.*, **245**, 6607
81. Barondes, S. H. (1968). *J. Neurochem.*, **15**, 699
82. Zatz, M. and Barondes, S. H. (1971). *J. Neurochem.*, **18**, 1125
83. Bogoch, S. (1968). *The Biochemistry of Memory* (New York: Oxford University Press)
84. Brunngraber, E. G. (1969). *Perspect. Biol. Med.*, **12**, 467
85. Barondes, S. H. (1970). *The Neurosciences: Second Study Program*, 747 (F. O. Schmitt, editor) (New York: Rockefeller University Press)
86. Burger, M. M. (1971). *Current Topics in Cellular Regulation*, Vol. 3, 135 (B. L. Horecker and E. R. Stadtman, editors) (New York: Academic Press)
87. Emmelot, P. (1973). *Eur. J. Cancer*, **9**, 319
88. Rapin, A. and Burger, M. M. (1974). *Advan. Cancer Res.*, 191
89. Herschman, H. R. (1972). *Membrane Molecular Biology*, 471 (C. F. Fox and A. D. Keith, editors) (Stamford, Connecticut: Sinaver Associates)
90. Onodera, K. and Sheinin, R. (1970). *J. Cell Sci.*, **7**, 337
91. Ohta, N., Pardee, A. B., McAuslan, B. R. and Burger, M. M. (1968). *Biochim. Biophys. Acta*, **158**, 98
92. Meezan, E., Wu, H. C., Black, P. H. and Robbins, P. W. (1969). *Biochemistry*, **8**, 2518
93. Wu, H. C., Meezan, E., Black, P. H. and Robbins, P. W. (1969). *Biochemistry*, **8**, 2509
94. Makita, A. and Seyama, Y. (1971). *Biochim. Biophys. Acta*, **241**, 403
95. Chiarugi, V. P. and Urbano, P. (1972). *J. Gen. Virol.*, **14**, 133
96. Buck, C. A., Glick, M. C. and Warren, L. (1970). *Biochemistry*, **9**, 4567
97. Sakiyama, H. and Burge, B. W. (1972). *Biochemistry*, **11**, 1366
98. Grimes, W. J. (1970). *Biochemistry*, **9**, 5083
99. Grimes, W. J. (1973). *Biochemistry*, **12**, 990
100. Warren, L., Chritchley, D. and Macpherson, I. (1972). *Nature (London)*, **235**, 275
101. Warren, L., Fuhrer, J. P. and Buck, C. A. (1972). *Proc. Nat. Acad. Sci. USA*, **69**, 1838
102. Warren, L., Fuhrer, J. P. and Buck, C. A. (1973). *Fed. Proc.*, **32**, 80
103. Den, H., Schultz, A. M., Basu, M. and Roseman, S. (1971). *J. Biol. Chem.*, **246**, 2721
104. Bosmann, H. B., Hagopian, A. and Eylar, E. H. (1968). *J. Cell. Physiol.*, **72**, 81
105. Buck, C. A., Glick, M. C. and Warren, L. (1971). *Science*, **172**, 169
106. Bosmann, H. B. (1972). *Biochem. Biophys. Res. Commun.*, **48**, 523
107. Sharon, N. and Lis, H. (1972). *Science*, **177**, 949
108. Lis, H. and Sharon, N. (1973). *Ann. Rev. Biochem.*, **42**, 541
109. Burger, M. M. (1969). *Proc. Nat. Acad. Sci. USA*, **62**, 994
110. Inbar, M. and Sachs, L. (1969). *Proc. Nat. Acad. Sci. USA*, **63**, 1418
111. Wray, W. P. and Walborg, E. F. Jr. (1971). *Cancer Res.*, **31**, 2027
112. Allan, D., Auger, J. and Crumpton, M. J. (1972). *Nature (London) New Biol.*, **236**, 23
113. Jansons, V. K. and Burger, M. M. (1973). *Biochim. Biophys. Acta*, **291**, 127
114. Hayman, M. J. and Crumpton, M. J. (1972). *Biochem. Biophys. Res. Commun.*, **47**, 923

115. Weinstein, D. B., Marsh, J. B., Glick, M. C. and Warren, L. (1970). *J. Biol. Chem.*, **245**, 3928
116. Critchley, D. R., Graham, J. M. and Macpherson, I. (1973). *FEBS Lett.*, **32**, 37
117. Carter, H. E., Johnson, P. and Weber, E. J. (1965). *Ann. Rev. Biochem.*, **34**, 109
118. Sweeley, C. C. and Dawson, G. (1969). *Red Cell Membrane: Structure and Function*, 172 (G. A. Jamieson and T. J. Greenwalt, editors) (Philadelphia: J. B. Lippincott)
119. Marcus, D. M. (1969). *New Engl. J. Med.*, **280**, 994
120. Watkins, W. M. (1966). *Science*, **152**, 172
121. Ginsburg, V., Kobata, A., Hickey, C. and Sawicka, T. (1971). *Glycoproteins of Blood Cells and Plasma*, 114 (G. A. Jamieson and T. J. Greenwalt, editors) (Philadelphia: J. B. Lippincott)
122. Watkins, W. M. (1972). *Glycoproteins: Their Composition, Structure and Function*, B.B.A. Library Vol. 5B, 830. (A. Gottschalk, editor) (Amsterdam: Elsevier)
123. Kobata, A., Grollman, E. F. and Ginsburg, V. (1968). *Biochem. Biophys. Res. Commun.*, **32**, 272
124. Albersheim, P., Jones, T. M. and English, P. D. (1969). *Ann. Rev. Phytopathol.*, **7**, 171
125. Hakomori, S. (1970). *Proc. Nat. Acad. Sci. USA*, **67**, 1741
126. Hakomori, S. and Murakami, W. T. (1968). *Proc. Nat. Acad. Sci. USA*, **59**, 254
127. Critchley, D. R. and Macpherson, I. (1973). *Biochim. Biophys. Acta*, **296**, 145
128. Sakiyama, H. and Robbins, P. W. (1973). *Fed. Proc.*, **32**, 86
129. Sakiyama, H., Gross, S. K. and Robbins, P. W. (1972). *Proc. Nat. Acad. Sci. USA*, **69**, 872
130. Kijimoto, S. and Hakomori, S. (1971). *Biochem. Biophys. Res. Commun.*, **44**, 557
131. Kijimoto, S. and Hakomori, S. (1972). *FEBS Lett.*, **25**, 38
132. Hakomori, S. and Kijimoto, S. (1972). *Nature (London) New. Biol.*, **239**, 87
133. Brady, R. O., Fishman, P. H. and Mora, P. T. (1973). *Fed. Proc.*, **32**, 102
134. Mora, P. T., Cumar, F. A. and Brady, R. O. (1971). *Virology*, **46**, 60
135. Mora, P. T., Fishman, P. H., Bassin, R. H., Brady, R. O. and McFarland, V. W (1973). *Nature (London) New. Biol.*, **245**, 226
136. Cumar, F. A., Brady, R. O., Kolodny, E. H., McFarland, V. W. and Mora, P. T. (1970). *Proc. Nat. Acad. Sci. USA*, **67**, 757
137. Laine, R. A. and Hakomori, S. (1973). *Biochem. Biophys. Res. Commun.*, **54**, 1039
138. Dodgson, K. S. and Lloyd, A. G. (1968). *Carbohydrate Metabolism and Its Disorders*, Vol. 1, 169 (F. Dickens, P. J. Randle and W. J. Whelan, editors) (London: Academic Press)
139. Lindahl, U. and Rodén, L. (1972). *Glycoproteins: Their Composition, Structure and Function*, B.B.A. Library Vol. 5A, 491 (A. Gottschalk, editor) (Amsterdam: Elsevier)
140. Marshall, R. D. and Neuberger, A. (1970). *Advan. Carbohyd. Chem. Biochem.*, **25**, 407
141. Mathews, M. B. (1971). *Biochem. J.*, **125**, 37
142. Hascall, V. C. and Sajdera, S. W. (1969). *J. Biol. Chem.*, **244**, 2384
143. Rosenberg, L., Hellmann, W. and Kleinschmidt, A. K. (1970). *J. Biol. Chem.*, **245**, 4123
144. Richmond, M. E., DeLuca, S. and Silbert, J. E. (1973). *Biochemistry*, **12**, 3898
145. Richmond, M. E., DeLuca, S. and Silbert, J. E. (1973). *Biochemistry*, **12**, 3904
146. DeLuca, S., Richmond, M. E. and Silbert, J. E. (1973). *Biochemistry*, **12**, 3911
147. Kraemer, P. M. (1971). *Biochemistry*, **10**, 1437
148. Kraemer, P. M. (1971). *Biochemistry*, **10**, 1445
149. Martinez-Palomo, A., Braislovsky, C. and Bernhard, W. (1969). *Cancer Res.*, **29**, 925
150. Smith, H. S., Hiller, A. J., Kingsbury, E. W. and Roberts-Dory, C. (1973). *Nature (London) New. Biol.*, **245**, 67
151. Chiarugi, V. P. and Urbano, P. (1973). *Biochim. Biophys. Acta*, **298**, 195
152. Chiarugi, V. P., Vannucchi, S. and Urbano, P. (1974). *Biochim. Biophys. Acta*, **345**, 283
153. Minnikin, S. M. and Allen, A. (1973). *Biochem. J.*, **134**, 1123
154. Hughes, R. C., Sanford, B. and Jeanloz, R. W. (1972). *Proc. Nat. Acad. Sci. USA*, **69**, 942
155. Turner, M. J., Strominger, J. L. and Sanderson, A. R. (1972). *Proc. Nat. Acad. Sci. USA*, **69**, 200
156. Philipson, L., Lonberg-Holm, K. and Pettersson, U. (1968). *J. Virol.*, **2**, 1064
157. El-allawy, R. M. M. and Gliemann, J. (1972). *Biochim. Biophys. Acta*, **273**, 97

158. Warren, L. and Glick, M. C. (1968). *J. Cell Biol.*, **37**, 729
159. Gerner, E. W., Glick, M. C. and Warren, L. (1970). *J. Cell Physiol.*, **75**, 275
160. Arias, I. M., Doyle, D. and Schimke, R. T. (1969). *J. Biol. Chem.*, **244**, 3303
161. Dehlinger, P. J. and Schimke, R. T. (1971). *J. Biol. Chem.*, **246**, 2574
162. Bock, K. W., Siekevitz, P. and Palade, G. E. (1971). *J. Biol. Chem.*, **246**, 188
163. Gurd, J. W. and Evans, W. H. (1973). *Eur. J. Biochem.*, **36**, 273
164. Korn, E. D. (1969). *Ann. Rev. Biochem.*, **38**, 263
165. Singer, S. J. (1971). *Structure and Function of Biological Membranes*, 145 (L. I. Rothfield, editor) (New York: Academic Press)
166. Singer, S. J. and Nicolson, G. L. (1972). *Science*, **175**, 720
167. Frye, L. D. and Edidin, M. (1970). *J. Cell Sci.*, **7**, 319
168. Taylor, R. B., Duffus, W. P. H., Raff, M. C. and De Petris, S. (1971). *Nature (London), New Biol.*, **233**, 225
169. Edelman, G. M., Yahara, I. and Wang, J. L. (1973). *Proc. Nat. Acad. Sci. USA*, **70**, 1442
170. Nicolson, G. L. (1973). *Nature (London), New Biol.*, **243**, 218
171. Moscona, A. A. (1957). *Proc. Nat. Acad. Sci. USA*, **43**, 184
172. Steinberg, M. S. (1970). *J. Exp. Zool.*, **173**, 395
173. Moscona, A. A. (1968). *Develop. Biol.*, **18**, 250
174. Lüderitz, O., Staub, A. M. and Westphal, O. (1966). *Bacteriol. Rev.*, **30**, 192
175. Bray, D. and Robbins, P. W. (1967). *J. Mol. Biol.*, **30**, 457
176. Losick, R. and Robbins, P. W. (1967). *J. Mol. Biol.*, **30**, 445
177. Robbins, P. W. and Uchida, T. (1965). *J. Biol. Chem.*, **240**, 375
178. Choppin, P. W., Compans, R. W., Scheid, A., McSharry, J. J. and Lazarowitz, S. G. (1972). *Membrane Research*, 163 (C. F. Fox editor) (New York: Academic Press)
179. Kates, M., Allison, A. C., Tyrell, D. A. J. and James, A. T. (1961). *Biochim. Biophys. Acta*, **52**, 455
180. Laver, W. G. and Valentine, R. C. (1969). *Virology*, **38**, 105
181. Compans, R. W., Klenk, H. D., Caliguiri, L. A. and Choppin, P. W. (1970). *Virology*, **42**, 880
182. Compans, R. W. (1971). *Nature (London) New Biol.*, **229**, 114
183. Chen, C., Compans, R. W. and Choppin, P. W. (1971). *J. Gen. Virol.*, **11**, 53
184. McSharry, J. J., Compans, R. W. and Choppin, P. W. (1971). *Bacteriol. Proc.*, 215
185. Rifkin, D. B. and Compans, R. W. (1971). *Virology*, **46**, 485
186. Gahmberg, C. G., Utermann, G. and Simons, K. (1972). *FEBS Lett.*, **28**, 179
187. Noll, H., Aoyagi, T. and Orlando, J. (1962). *Virology*, **18**, 154
188. Laver, W. G. and Webster, R. G. (1966). *Virology*, **30**, 104
189. Strandli, O. K., Mortensson-Egnund, K. and Harboe, A. (1964). *Acta Path. Microbiol. Scand.*, **60**, 265
190. Wiese, L. (1965). *J. Phycol.*, **1**, 46
191. Wiese, L. and Shoemaker, D. W. (1970). *Biol. Bull.*, **138**, 88
192. Crandall, M. A. and Brock, T. D. (1968). *Science*, **161**, 473
193. Gerisch, G. (1968). *Current Topics in Developmental Biology*, Vol. 3, 157 (A. Monroy and A. A. Moscona, editors) (New York: Academic Press)
194. Raper, K. B. and Thom, C. (1941). *Amer. J. Bot.*, **28**, 69
195. Beug, H., Gerisch, G., Kempff, S., Riedel, V. and Cremer, G. (1970). *Exp. Cell Res.*, **63**, 147
196. Rosen, S. D., Kafka, J. A., Simpson, D. L. and Barondes, S. H. (1973). *Proc. Nat. Acad. Sci. USA*, **70**, 2554
197. Weeks, G. (1973). *Exp. Cell Res.*, **76**, 467
198. Humphreys, T. (1967). *The Specificity of Cell Surfaces*, 195 (B. D. Davis and L. Warren, editors) (Englewood Cliffs, New Jersey: Prentice-Hall
199. Turner, R. S. and Burger, M. M. (1973). *Rev. Physiol.*, **68**, 121
200. Wilson, H. V. (1907). *J. Exp. Zool.*, **5**, 245
201. Galtsoff, P. S. (1925). *J. Exp. Zool.*, **42**, 223
202. Humphreys, T. (1963). *Develop. Biol.*, **8**, 27
203. Humphreys, T. (1965). *Exp. Cell Res.*, **40**, 539
204. Humphreys, T. (1970). *Nature (London)*, **228**, 685
205. Turner, R. S. and Burger, M. M. (1973). *Nature (London)*, **244**, 509
206. Henkart, P., Humphreys, S. and Humphreys, T. (1973). *Biochemistry*, **12**, 3045

207. Cauldwell, C. B., Henkart, P. and Humphreys, T. (1973). *Biochemistry*, **12**, 3051
208. Humphreys, T. (1972). *The Comparative Molecular Biology of Extracellular Matrices*, 108 (H. C. Slavkin, editor) (New York: Academic Press)
209. Gasic, G. J. and Galanti, N. L. (1966). *Science*, **151**, 203
210. Müller, W. E. G. and Zahn, R. K. (1973). *Exp. Cell Res.*, **80**, 95
211. Margoliash, E., Schenck, J. R., Hargie, M. P., Burokas, S., Richter, W. R., Barlow, G. H. and Moscona, A. A. (1965). *Biochem. Biophys. Res. Commun.*, **20**, 383
212. MacLennon, A. P. (1970). *Symp. Zool. Soc.*, **25**, 299
213. Weinbaum, G. and Burger, M. M. (1973). *Nature (London)*, **244**, 510
214. Leith, A. G. and Steinberg, M. S. (1972). *Biol. Bull.*, **143**, 468
215. John, H. A., Campo, M. S., Mackenzie, A. M. and Kemp, R. B. (1971). *Nature (London), New Biol.*, **230**, 126
216. Lilien, J. E. (1969). *Current Topics in Developmental Biology*, Vol. 4, 169 (A. Monroy and A. A. Moscona, editors) (New York: Academic Press)
217. Kleinschuster, S. J. and Moscona, A. A. (1972). *Exp. Cell Res.*, **70**, 397
218. Moscona, A. A. (1971). *Science*, **171**, 905
219. Weiser, M. M. (1972). *Science*, **177**, 525
220. Lallier, R. (1972). *Exp. Cell Res.*, **72**, 157
221. Steinberg, M. S. and Gepner, I. A. (1973). *Nature (London) New Biol.*, **241**, 249
222. Abelev, G. I. (1971). *Advances in Cancer Research*, Vol. 14, 295 (G. Klein and S. Weinhouse, editors) (New York: Academic Press)
223. Alexander, P. (1972). *Nature (London)*, **235**, 137
224. Ting, C.-C., Lavrin, D. H., Shiu, G. and Herberman, R. B. (1972). *Proc. Nat. Acad. Sci.* USA, **69**, 1664
225. Baldwin, R. W., Glaves, D. and Vose, B. M. (1972). *Int. J. Cancer*, **10**, 233
226. Iype, P. T., Baldwin, R. W. and Glaves, D. (1973). *Brit. J. Cancer*, **27**, 128
227. Stonehill, E. H. and Bendich, A. (1970). *Nature (London)*, **228**, 370
228. Artzt, K., Dubois, P., Bennet, D., Condamine, H., Babinet, C. and Jacob, F. (1973). *Proc. Nat. Acad. Sci. USA*, **70**, 2988
229. Toole, B. P., Jackson, G. and Gross, J. (1972). *Proc. Nat. Acad. Sci. USA*, **69**, 1384
230. Toole, B. P. (1973). *Science*, **180**, 302
231. Toole, B. P. and Gross, J. (1971). *Develop. Biol.*, **25**, 57
232. Bernfield, M. R. and Wessells, N. K. (1970). *Changing Synthesis in Development, Develop. Biol. Suppl.*, Vol. 4, 195 (M. N. Runner, editor) (New York: Academic Press)
233. Bernfield, M. R. and Banerjee, S. D. (1972). *J. Cell Biol.*, **52**, 664
234. Bernfield, M. R., Banerjee, S. D. and Cohn, R. H. (1972). *J. Cell Biol.*, **52**, 674
235. Curtis, A. S. G. (1962). *Biol. Rev.*, **37**, 82
236. Oppenheimer, S. B., Edidin, M., Orr, C. W. and Roseman, S. (1969). *Proc. Nat. Acad. Sci. USA*, **63**, 1395
237. Oppenheimer, S. B. (1973). *Exp. Cell Res.*, **77**, 175
238. Roth, S., McGuire, E. J. and Roseman, S. (1971). *J. Cell Biol.*, **51**, 525
239. Vicker, M. G. and Edwards, J. G. (1972). *J. Cell Sci.*, **10**, 759
240. Kemp, R. B. (1968). *Nature (London)*, **218**, 1255
241. Matsuzawa, T. (1973). *Exp. Cell Res.*, **80**, 377
242. Cook, G. M. W., Heard, D. H. and Seaman, G. V. F. (1962). *Exp. Cell Res.*, **28**, 27
243. Curtis, A. S. G. (1967). *The Cell Surface: Its Molecular Role in Morphogenesis* (New York: Academic Press)
244. Gottschalk, A. (1960). *Nature (London)*, **186**, 949
245. Weiss, L. (1965). *J. Cell Biol.*, **26**, 735
246. Chipowsky, S., Lee, Y. C. and Roseman, S. (1973). *Proc. Nat. Acad. Sci. USA*, **70**, 2309
247. Roth, S., McGuire, E. J. and Roseman, S. (1971). *J. Cell Biol.*, **51**, 536
248. Bosmann, H. B. (1971). *Biochem. Biophys. Res. Commun.*, **43**, 1118
249. Barber, A. J. and Jamieson, G. A. (1971). *Biochim. Biophys. Acta*, **252**, 533
250. Weiser, M. M. (1973). *J. Biol. Chem.*, **248**, 2542
251. Roth, S. and White, D. (1972). *Proc. Nat. Acad. Sci. USA*, **69**, 485
252. Weiser, M. M. (1973). *J. Biol. Chem.*, **248**, 2536
253. Podolsky, D. K. and Weiser, M. M. (1973). *J. Cell Biol.*, **58**, 497
254. Spaet, T. H. and Zucker, M. B. (1964). *Amer. J. Physiol.*, **206**, 1267
255. Butler, W. T. and Cunningham, L. W. (1966). *J. Biol. Chem.*, **241**, 3882

256. Spiro, R. (1969). *J. Biol. Chem.*, **244,** 602
257. Steiner, M. and Baldini, M. (1969). *Blood*, **33,** 628
258. Booyse, F. M. and Rafelson, M. E. (1968). *Biochem. Biophys. Acta*, **166,** 689
259. Jamieson, G. A., Urban, C. L. and Barber, A. J. (1971). *Nature (London) New Biol.*, **234,** 5
260. Katzman, R. L., Kang, A. H. and Beachy, E. H. (1973). *Science*, **181,** 670
261. Merrell, R. and Glaser, L. (1973). *Proc. Nat. Acad. Sci. USA*, **70,** 2794
262. Ashwell, G. and Morell, A. G. (1971). *Glycoproteins of Blood Cells and Plasma*, 173 (G. A. Jamieson and T. J. Greenwalt, editors) (Philadelphia: J. B. Lippincott)
263. Morell, A. G., Gregoriadis, G., Scheinberg, I. H., Hickman, J. and Ashwell, G. (1971). *J. Biol. Chem.*, **246,** 1461
264. Pricer, W. E. and Ashwell, G. (1971). *J. Biol. Chem.*, **246,** 4825
265. Van Lenten, L. and Ashwell, G. (1972). *J. Biol. Chem.*, **247,** 4633
266. Morell, A. G. and Scheinberg, I. H. (1972). *Biochem. Biophys. Res. Commun.*, **48,** 808
267. Aronson, N. N., Tan, L. Y. and Peters, B. P. (1973). *Biochem. Biophys. Res. Commun.*, **53,** 112
268. Gesner, B. M. and Ginsburg, V. (1964). *Proc. Nat. Acad. Sci. USA*, **52,** 750
269. Woodruff, J. J. and Gesner, B. M. (1969). *J. Exp. Med.*, **129,** 551
270. Fox, T. O., Sheppard, J. R. and Burger, M. M. (1971). *Proc. Nat. Acad. Sci. USA*, **68,** 244
271. Shoham, J. and Sachs, L. (1972). *Proc. Nat. Acad. Sci. USA*, **69,** 2479
272. Burger, M. M. and Goldberg, A. R. (1967). *Proc. Nat. Acad. Sci. USA*, **57,** 359
273. Sela, B.-A., Lis, H., Sharon, N. and Sachs, L. (1970). *J. Mem. Biol.*, **3,** 267
274. Pollack, R. E. and Burger, M. M. (1969). *Proc. Nat. Acad. Sci. USA*, **62,** 1074
275. Inbar, M., Rabinowitz, Z. and Sachs, L. (1969). *Int. J. Cancer*, **4,** 690
276. Ozanne, B. and Sambrook, J. F. (1971). *The Biology of Oncogenic Viruses*, 248, *2nd Lepetit Colloquium on Biology and Medicine* (E. Verwey, editor) (Amsterdam: North Holland)
277. Culp, L. A. and Black, P. H. (1972). *J. Virol.*, **9,** 611
278. Benjamin, T. L. and Burger, M. M. (1970). *Proc. Nat. Acad. Sci. USA*, **67,** 929
279. Biguard, J.-M. and Vigier, P. (1972). *Virology*, **47,** 444
280. Gantt, R. R., Martin, J. R. and Evans, V. J. (1969). *J. Nat. Cancer Inst.*, **42,** 369
281. Eckhart, W., Dulbecco, R. and Burger, M. M. (1971). *Proc. Nat. Acad. Sci. USA*, **68,** 283
282. Renger, H. C. and Basilico, C. (1972). *Proc. Nat. Acad. Sci. USA*, **69,** 109
283. Noonan, K. D., Renger, H. C., Basilico, C. and Burger, M. M. (1973). *Proc. Nat. Acad. Sci. USA*, **70,** 347
284. Albrecht-Bühler, G., Noonan, K. D. and Burger, M. M. (1975). *J. Cell Biol.*, in press
285. Burger, M. M. and Noonan, K. D. (1970). *Nature (London)*, **228,** 512
286. Cline, M. J. and Livingston, D. C. (1971). *Nature (London), New Biol.*, **232,** 155
287. Arndt-Jovin, D. J. and Berg, P. (1971). *J. Virol.*, **8,** 716
288. Ben-Bassat, H., Inbar, M. and Sachs, L. (1971). *J. Mem. Biol.*, **6,** 183
289. Ozanne, B. and Sambrook, J. (1971). *Nature (London) New Biol.*, **232,** 156
290. Noonan, K. D. and Burger, M. M. (1973). *J. Biol. Chem.*, **248,** 4286
291. Rosenblith, J. Z., Ukena, T. E., Yin, H. H., Berlin, R. D. and Karnovsky, M. J. (1973). *Proc. Nat. Acad. Sci. USA*, **70,** 1625
292. Nicolson, G. L. (1973). *Nature (London) New Biol.*, **243,** 218
293. Noonan, K. D. and Burger, M. M. (1975). *J. Cell Biol.*, in press
294. Inbar, M., Shinitzky, M. and Sachs, L. (1973). *J. Mol. Biol.*, **81,** 245
295. Inbar, M., Ben-Bassat, H. and Sachs, L. (1973). *Exp. Cell Res.*, **76,** 143
296. Inbar, M. and Sachs, L. (1973). *FEBS Lett.*, **32,** 124
297. Porter, K., Prescott, D. and Frye, J. (1973). *J. Cell Biol.*, **57,** 815
298. Buck, C. A., Glick, M. C. and Warren, L. (1973). *Biochemistry*, **12,** 85
299. Buck, C. A., Glick, M. C. and Warren, L. (1971). *Biochemistry*, **10,** 2176
300. Muramatsu, T., Atkinson, P. H., Nathenson, S. G. and Ceccarini, C. (1973). *J. Mol. Biol.*, **80,** 781
301. Patt, L. M. and Grimes, W. J. (1974). *J. Biol. Chem.*, **249,** 4157
302. (1975). *MTP International Review of Science, Biochemistry Series One*, Vol. 2, *Biochemistry of Cell Walls and Membranes* (C. F. Fox, editor) (London: Butterworths)

# 3
# Biosynthesis of Connective Tissue Proteoglycans

**L. RODÉN**
University of Alabama in Birmingham
and
**N. B. SCHWARTZ**
University of Chicago

---

---

## 3.1 INTRODUCTION

The connective tissue proteoglycans possess certain characteristic structural features which justify their classification as a separate category of glycoconjugates, and a comprehensive review of the current status of knowledge regarding their structures and properties is presented elsewhere in this volume by Muir and Hardingham (see Chapter 4). A major difference between proteoglycans and glycoproteins is the presence, in the former, of polysaccharide chains composed of repeating disaccharide units, which have a high charge density due to their uronic acid carboxyl groups and/or sulphate groups. Despite these unique characteristics, the differences between the two classes of compounds appear no greater than those which exist between individual members of the glycoprotein group, e.g. collagen and the blood group substances. Indeed, the only distinctive structural feature shared by all glycoproteins is the very presence of one or more carbohydrate prosthetic groups in a covalent linkage to the protein moiety, and a wide range of variation in the structures of the protein as well as the carbohydrate components makes this a rather superficial basis for classification.

Studies of the biosynthesis of proteoglycans have benefited greatly from concurrent developments in the area of glycoprotein biosynthesis and, as might have been expected, it is now evident that the carbohydrate moieties of both classes of compounds are formed by essentially the same routes. In its simplest form, the biosynthesis of a carbohydrate–protein conjugate occurs as follows: the protein core is first synthesised on the ribosomes in the same manner as for simple proteins, and addition of the carbohydrate groups then takes place in a stepwise fashion, monosaccharide by monosaccharide, by transfer of the appropriate glycosyl group from a nucleotide sugar; alternatively, some oligosaccharides are first assembled in the same stepwise manner on polyprenol lipids, and the entire oligosaccharide is then transferred to the protein core. The latter mechanism is presently under active investigation, and it appears that many or all glycoproteins containing an *N*-acetylglucosamine–asparagine linkage may be synthesised by this route (see Section 3.2.3). Since corneal keratan sulphate (keratan sulphate I) is linked to protein by an *N*-acetylglucosamine–asparagine linkage (Chapter 4), formation of this polysaccharide may well be initiated by a lipid–oligosaccharide intermediate, but experimental evidence to this effect has not yet been obtained. The majority of the connective tissue proteoglycans, however, are probably synthesised by the first, more direct route, but it should be emphasised that the available information is not sufficient to preclude the participation of lipid intermediates in this process at some stage or other.

Despite the close analogy with the patterns of glycoprotein biosynthesis, there are naturally some areas in which problems peculiar to the proteoglycans are encountered. An example in this category is the formation of the iduronic acid residues of heparin, which has been shown to occur at the polymer level by epimerisation of glucuronic acid (see Sections 3.2.1 and 3.2.3). Also, the recent finding that the proteoglycan aggregates of cartilage consist of several subunits held together by a specific interaction with hyaluronic acid and one or two 'link proteins' (Chapter 4) has at the same time opened up new vistas in the area of biosynthesis. It will obviously be of great interest to investigate the biosynthetic aspects of this physiologically important phenomenon with a view to increasing our understanding of the factors which influence the formation and organisation of the extracellular matrix *in vivo*.

Interest in various aspects of connective tissue polysaccharide biosynthesis has grown considerably in the past few years, and it is not possible to cover the entire field in depth within the scope of this presentation. The authors have therefore chosen to place the main emphasis of this review on a description of the various glycosyl transfer steps involved in chondroitin sulphate biosynthesis, with a discussion of the general characteristics of mammalian glycosyl transfer reactions as a background. A promising beginning has recently been made in the purification of the individual chondroitin sulphate glycosyltransferases, and the properties of these enzymes will also be described in some detail. Finally, the biosynthesis of chondroitin sulphate in the course of cartilage cell development, and some recent studies of factors which influence the biosynthetic process in cells in tissue culture, will be discussed.

It should be mentioned that several areas of connective tissue polysaccharide biosynthesis, which are closely related to the present subject, have been covered in other recent reviews: the biosynthesis of heparin, heparan sulphate and dermatan sulphate has been reviewed by Lindahl elsewhere in this series[1], and more detailed information has been presented earlier[2,3] concerning the biosynthesis of nucleotide sugars, the process of sulphation, the subcellular localisation of glycoprotein and proteoglycan formation, and certain aspects of the regulation of biosynthesis.

## 3.2 GENERAL MECHANISMS OF COMPLEX CARBOHYDRATE FORMATION

### 3.2.1 Introduction

The mechanisms of complex carbohydrate formation are presently under active investigation[1-12], and seemingly well established principles are currently being challenged and re-evaluated. In the past, a clear distinction was made between two major groups of carbohydrate-containing macromolecules, i.e. those composed exclusively of carbohydrate components, such as glycogen and other homo- or hetero-polysaccharides, and those in which an oligo- or poly-saccharide is linked to a protein or lipid moiety (glycoproteins, proteoglycans, glycolipids, lipopolysaccharides). This distinction has now become somewhat hazy, and an increasing number of substances which have been regarded as purely carbohydrate in nature should more properly be classified among the conjugated macromolecules. A pertinent illustration of this trend is provided by recent studies of the structure and synthesis of heparin[1]. This substance was for many years considered as a sulphated heteropolysaccharide without other major constituents; however, presently available evidence favours the notion that heparin is synthesised as a multichain proteoglycan, and that subsequent modifications occur which result in the formation of single polysaccharide chains, some of which contain covalently bound peptide whereas others have a free reducing terminus. This occurs through (i) proteolytic degradation of the protein core together with (ii) cleavage of certain glucuronosidic linkages by an endo-β-glucuronidase. It is thus apparent that greater caution should be exercised in making assumptions regarding biosynthetic pathways, and the possibilities that 'pure' polysaccharides may contain a small but significant amount of covalently bound peptide, and that they may actually be synthesised as proteoglycans, must be kept in mind. Even glycogen may belong in this category, as some tentative evidence suggests that glycogen chains are initiated by glucosyl transfer to a protein acceptor[13, 14].

In mammalian tissues all complex carbohydrate groups are formed via nucleotide sugars which, in turn, originate from glucose. (Occasionally, sugars other than glucose may serve as alternative precursors for more direct routes to the respective nucleotide derivatives[2].) As has been pointed out by Leloir[15], even polysaccharides which are not formed directly from nucleotide sugars are in all known instances derived from these compounds in an indirect manner, e.g. the levansucrase of *Leuconostoc mesenteroides* requires sucrose as a substrate[16], but this disaccharide is itself produced by a nucleotide sugar-dependent reaction in other organisms.

For each monosaccharide component of the complex carbohydrates of mammalian tissues there is a corresponding nucleotide sugar containing this monosaccharide, and the general pattern of synthesis involves the transfer of the glycosyl group to a suitable acceptor in a reaction catalysed by a specific glycosyltransferase. One exception to this basic pattern has recently been discovered by Lindahl and collaborators[1, 17, 18]. They showed that the iduronic acid component of heparin is formed by epimerisation of glucuronic acid after incorporation into the polymer. Although this reaction appears to be unique in the biosynthesis of mammalian carbohydrates, it has a precedent in the epimerisation of mannuronic acid to guluronic acid, which occurs in the course of alginic acid formation[19, 20]. These findings leave unexplained the occurrence of nucleotide derivatives of iduronic acid and guluronic acid (UDP-L-iduronic acid[21] and GDP-L-guluronic acid[22, 23], respectively) which have previously been regarded as the obvious precursors of the polymer-bound forms of these two uronic acids. The problem of iduronic acid formation falls largely outside the scope of the present review, and the reader is referred to a comprehensive discussion of this subject by Lindahl[1].

Following the discovery of lipid-bound intermediates in the biosynthesis of bacterial complex carbohydrates[4, 12], it became necessary to consider the possibility that similar mechanisms might be operative in mammalian tissues as well. The dolichols, discovered by Hemming in 1963[24], were recognised by Leloir and his collaborators[10, 12, 25] as possible mammalian counterparts to the shorter-chain isoprenoid lipids which participate in microbial polysaccharide biosynthesis. In the past couple of years the assumption that dolichol-bound sugars may be involved in mammalian glycoprotein biosynthesis has received substantial experimental support and, in particular, it appears that some or all glycoproteins containing a core region composed of *N*-acetylglucosamine and mannose are formed via such intermediates. This problem will be discussed in somewhat more detail below in relation to the process of carbohydrate chain initiation.

### 3.2.2 Interconversion of monosaccharides

A total of 10 different monosaccharide components are found in mammalian complex carbohydrates: D-galactose, D-mannose, *N*-acetyl-D-glucosamine, *N*-acetyl-D-galactosamine, D-glucose, *N*-acetyl-D-neuraminic acid (and several other sialic acids), L-fucose, D-xylose, D-glucuronic acid and L-iduronic acid. In addition, the presence of L-arabinose in hyaluronic acid has been reported[26, 27], but this finding remains to be confirmed[28]. Nucleotide derivatives of all 10 sugars have been isolated from a variety of mammalian tissues and, with the exception of UDP-L-iduronic acid, these compounds have been shown to serve as the direct precursors of their respective glycosyl residues in complex carbohydrates[2, 3, 29, 30]. It is obvious that a number of modifications in structure are required in the course of biosynthesis of the nucleotide sugars from glucose, and most of these occur at the nucleotide sugar level. Prior to the nucleotide sugar stage, the mannose configuration is formed by isomerisation of fructose 6-phosphate to mannose 6-phosphate[31], and fructose 6-phosphate is also a key intermediate in glucosamine formation[30], serving as the substrate for transfer of the amide group of glutamine to form glucosamine 6-phosphate, which is subsequently acetylated to *N*-acetylglucosamine 6-phosphate. Interconversions at the nucleotide level include 4′-epimerisation of UDP-glucose and UDP-*N*-acetylglucosamine to yield, respectively, UDP-galactose[32-34] and UDP-*N*-acetylgalactosamine[35-37], 5′-epimerisation of UDP-glucuronic acid[21], oxidation of UDP-glucose to UDP-glucuronic acid[38-41] and decarboxylation of the latter to UDP-xylose[42, 43].

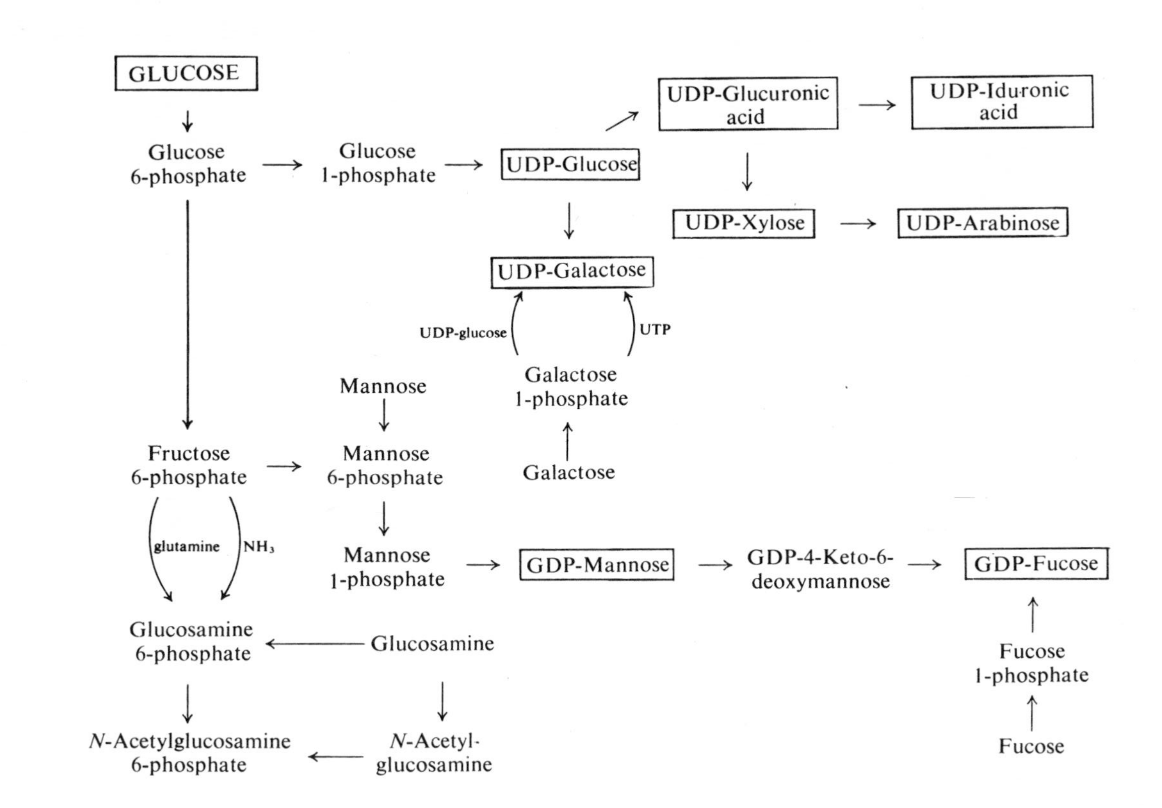

GLUCOSE
Glucose 6-phosphate
Glucose 1-phosphate
UDP-Glucose
UDP-Glucuronic acid
UDP-Iduronic acid
UDP-Xylose
UDP-Arabinose
UDP-Galactose
UDP-glucose
UTP
Mannose
Galactose 1-phosphate
Fructose 6-phosphate
Mannose 6-phosphate
Galactose
glutamine
$NH_3$
Mannose 1-phosphate
GDP-Mannose
GDP-4-Keto-6-deoxymannose
GDP-Fucose
Glucosamine 6-phosphate
Glucosamine
Fucose 1-phosphate
*N*-Acetylglucosamine 6-phosphate
*N*-Acetyl-glucosamine
Fucose

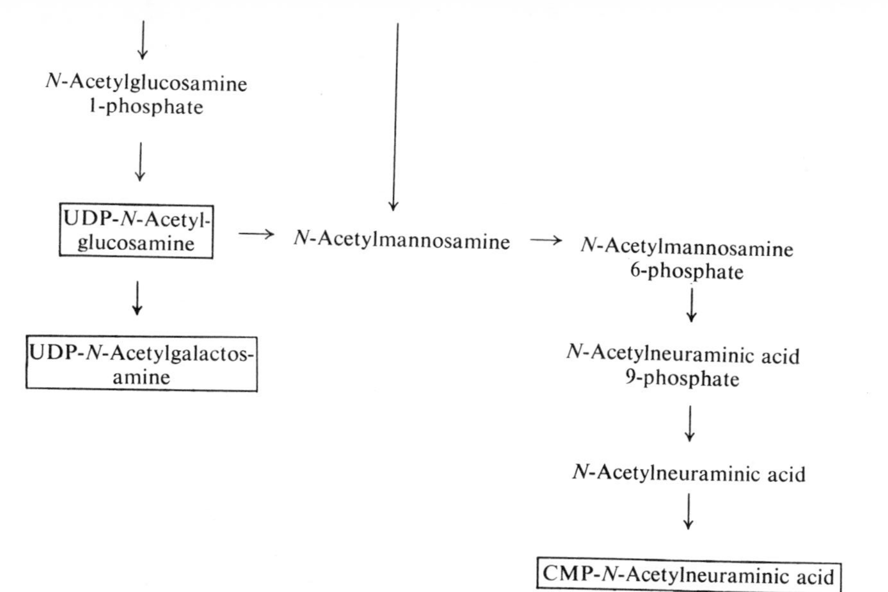


**Figure 3.1** Pathways of nuclestide sugar formation from glucose. (From Schachter and Rodén[2], by courtesy of Academic Press.)

Formation of GDP-fucose from GDP-mannose requires two reactions, 4′-epimerisation and formation of the 6-deoxy group[44-47]. Sialic acid is synthesised by a more complex route[2,5,30], passing through *N*-acetylmannosamine, which may be formed either by epimerisation of *N*-acetylglucosamine or by a reaction in which UDP-*N*-acetylglucosamine is converted into *N*-acetylmannosamine and UDP. This is followed by introduction of a three-carbon fragment from phosphoenolpyruvate which is transferred to *N*-acetylmannosamine 6-phosphate, and subsequent reactions yield CMP-*N*-acetylneuraminic acid. The reactions leading to the nucleotide sugars have all been reviewed extensively elsewhere[2,3,5,7,8,29,30,43] and are therefore only summarised briefly in Figure 3.1. It may be added that part of the regulation of proteoglycan biosynthesis undoubtedly occurs at the level of nucleotide sugar biosynthesis, and this aspect of nucleotide sugar metabolism has been discussed previously[2,3,30].

### 3.2.3 Chain initiation

In protein biosynthesis, chain initiation is a complicated and specific event which requires the participation of a number of initiation factors[48]. In contrast, initiation of a carbohydrate chain may occur in its simplest form by transfer of the first sugar of the chain from the corresponding nucleotide sugar to a suitable acceptor, e.g. the protein moiety of a glycoprotein or a proteoglycan. Alternatively, the acceptor may be a monosaccharide or oligosaccharide, e.g. D-xylose or simple xylosides may serve as the initial acceptor for the formation of chondroitin sulphate chains[49-56] (this is not, however, a physiologically occurring process), and maltose may be utilised as a primer in glycogen synthesis, albeit much less effectively than high molecular weight glycogen[57,58]. As has already been indicated, yet another mode of chain initiation is now known which involves the participation of lipid intermediates[12] and operates on a fundamentally different principle, inasmuch as an entire oligosaccharide is transferred from the lipid intermediate to the acceptor. At our present stage of knowledge it seems that the lipid intermediate route is characteristic of the glycoproteins in which the carbohydrate prosthetic group is linked via an *N*-acetylglucosamine–asparagine linkage and that compounds with other carbohydrate–protein linkages are formed without the participation of lipid intermediates.

From a biosynthetic as well as a structural point of view, it is reasonable to classify the mammalian proteoglycans and glycoproteins according to their carbohydrate–protein linkage types. The following linkage types are presently known to occur in mammalian protein-bound complex carbohydrates: (i) an *O*-glycosidic linkage between xylose and serine hydroxy groups; (ii) an *O*-glycosidic linkage between *N*-acetylgalactosamine and the hydroxy groups of serine or threonine; (iii) an *N*-glycosylamine linkage between *N*-acetylglucosamine and the amide groups of asparagine residues; and (iv) an *O*-glycosidic linkage between galactose and the hydroxy group of hydroxylysine.

Some of these linkages which are present in connective tissue proteoglycans have been discussed in detail by Muir and Hardingham (Chapter 4). The first linkage type is specific for the connective tissue proteoglycans and has so far not been detected in any other type of carbohydrate–protein compounds. Specifically, it is found in the chondroitin sulphates, dermatan sulphate, heparin and heparan sulphate[60]. The second type is involved in the linkage of skeletal keratan sulphate (keratan sulphate II) to protein, and the third is present in corneal keratan sulphate (keratan sulphate I). The *N*-acetylglucosamine–asparagine linkage is in all likelihood the most prevalent carbohydrate–protein linkage in mammalian glycoproteins[59], but many substances containing the second linkage type have also been described. The galactose–hydroxylysine linkage is typical of the collagens of connective tissues and will not be discussed here in any detail.

Whatever the mode of chain initiation, the addition of the carbohydrate prosthetic groups is largely a post-ribosomal event, although it is possible that glycosyl transfer in some cases commences already before the release of the completed peptide from the ribosomes[2].

The three linkage types involving xylose, *N*-acetylgalactosamine and galactose may all be formed by transfer of the individual monosaccharides to appropriate acceptors. The protein core from cartilage proteoglycan serves as an excellent acceptor for xylose transfer from UDP-xylose[61]; collagen in unmodified form, or after removal of galactose units already present, is a substrate for galactosyl transfer[62], and the core protein of submaxillary mucin is an acceptor for transfer of *N*-acetylgalactosaminyl residues[63-67]. In none of these cases is there any evidence for the participation of lipid intermediates, but it is not yet possible to exclude such a mechanism with absolute certainty. For instance, although xylosyltransferase has been purified to apparent homogeneity[68], the enzyme is associated with lipids at an earlier stage of purification[69], and the possibility that some lipid may be retained after extensive purification has not been ruled out. The primary structure of the protein acceptors is obviously an important determinant for the glycosyl transfer reactions, and some generalisations can be made in regard to the structural features required for the various reactions, as will be discussed in somewhat more detail in Section 3.4.1.1.

In view of the relative ease with which transfer of the three sugars xylose, *N*-acetylgalactosamine and galactose to protein acceptors can be demonstrated, it has been somewhat of an enigma why, in analogous experiments, transfer of *N*-acetylglucosamine to asparagine residues cannot be detected. The participation of lipid-bound oligosaccharide intermediates in the formation of the glycoproteins containing an asparagine–*N*-acetylglucosamine linkage now provides a logical explanation for the failure to observe direct transfer of single *N*-acetylglucosamine residues. Through the work of Behrens *et al.*[70] and Hsu *et al.*[71] the following sequence of reactions has been suggested:

$$\text{UDP-Glc}N\text{Ac} + \text{DMP} \rightarrow \text{DDP-Glc}N\text{Ac} + \text{UMP}$$
$$\text{GDP-Man} + \text{DMP} \rightarrow \text{DMP-Man} + \text{GDP}$$
$$\text{UDP-Glc}N\text{Ac} + \text{DDP-Glc}N\text{Ac} \rightarrow \text{DDP-Glc}N\text{Ac}_2 + \text{UDP}$$
$$\text{DDP-Glc}N\text{Ac}_2 + \left\{\begin{array}{l}\text{DMP-Man}_x\\ \text{and/or}\\ \text{GDP-Man}_x\end{array}\right. \rightarrow \text{DDP-Glc}N\text{Ac}_2\text{Man}_x + \left\{\begin{array}{l}\text{DMP}\\ \text{and/or}\\ \text{GDP}\end{array}\right.$$
$$\text{DDP-Glc}N\text{Ac}_2\text{Man}_x + \text{protein} \rightarrow \text{protein-Glc}N\text{Ac}_2\text{Man}_x + \text{DDP}$$

(DMP, dolichol monophosphate; DDP, dolichol diphosphate)

According to this scheme, the *N*-acetylglucosamine group destined to be linked to an asparagine residue of the protein core is initially transferred to dolichol monophosphate and, after addition of a second *N*-acetylglucosaminyl group and several mannose residues, the entire oligosaccharide is transferred to the protein acceptor.

Further details concerning the chain initiation in proteoglycan synthesis will be discussed in Section 3.4.1.

### 3.2.4 Specificity of glycosyltransferases

In the formation of biological macromolecules, the synthetic apparatus must generally be capable of assembling the component parts in a specific order; however, the degree of accuracy which is required in this process may vary from one group of compounds to another. In protein synthesis, a high fidelity of reproduction is necessary, as illustrated by the fact that a mutation resulting in a change of only a single amino acid may drastically alter the properties of the molecule, possibly to the point of complete loss of biological activity. On the other hand, the structural requirements for proper functioning of a complex carbohydrate molecule are often less stringent, and, e.g. the oligosaccharide chains of a particular glycoprotein may show considerable heterogeneity without any obvious adverse effects on the functions of the molecule. This situation is exemplified by ovalbumin which displays a special type of heterogeneity termed 'paucidispersity'. Five different glycopeptides have been isolated from this glycoprotein, which are composed of *N*-acetylglucosamine and mannose[72, 73], varying in ratio from 5:6 to 2:5. The oligosaccharide moieties are all structurally similar, and the available evidence indicates that the smaller ones are the products of incomplete biosynthesis rather than *in vivo* degradation. It has also long been recognised that the connective tissue polysaccharides exhibit a certain degree of molecular weight polydispersity, and an exact mechanism for chain termination at a precise molecular weight apparently does not exist[3]. Even if the function of a carbohydrate prosthetic group is not always absolutely dependent upon the integrity of a single well-defined structure, it is equally clear that a difference of a single monosaccharide unit may be crucial to the behaviour and function of the glycoprotein. We need only be reminded of the elegant work of Ashwell and collaborators[74-78] which has spotlighted the effect of a single sialic acid residue on the physiological behaviour of serum glycoproteins. Removal of the terminal sialic acid uncovers galactose residues which serve as a recognition signal for galactose receptors in the liver cells, and the asialoglycoproteins are therefore, with few exceptions, removed from the circulation at a much higher rate than the sialylated parent substances. The relationship between structure and blood group specificity is yet another classical example of the importance of a single terminal monosaccharide for the function of a glycoprotein, as shown by the conversion of H substance into A or B substance upon addition of *N*-acetylgalactosamine or galactose, respectively, to the non-reducing terminus of the oligosaccharide chain[79].

Whereas the structure of a protein is encoded in the nucleotide sequence of the corresponding messenger RNA, from which the information is directly translated to a specific amino acid sequence, the synthesis of complex carbohydrates is not under direct template control by the genes. Nevertheless, the reproduction of a particular carbohydrate structure occurs by a rigorously controlled mechanism, and whether a certain glycoprotein has a precisely defined carbohydrate structure or contains a paucidisperse oligosaccharide moiety, the monosaccharide components are reliably assembled in a specific order and with proper configurations and positions of the glycosidic linkages. The regulatory basis for this process is not yet fully understood, but at least two important factors are involved, i.e. the specificity of the glycosyltransferases catalysing the individual transfer steps, and their subcellular organisation. A considerable body of information is also available regarding some of the properties of the glycosyltransferases. In most cases, the glycosyltransferases behave in accordance with the 'one enzyme–one linkage' concept, formulated by Hagopian and Eylar[64], which postulates that an enzyme is specific with regard to (i) the sugar transferred, (ii) the acceptor and (iii) the position and anomeric configuration of the linkage formed. These factors will be discussed in some detail below.

### *3.2.4.1 Donor specificity*

On theoretical grounds alone, an extremely high specificity for the sugar transferred may be considered a virtual necessity for proper assembly of the complex carbohydrates and, e.g., a galactosyltransferase may be expected to transfer no sugar other than galactose. A high specificity for the donor sugar has been verified in several instances[80-90], but this aspect of specificity has not received much attention and should be more properly re-assessed with pure enzymes, as these become available. Occasional errors in synthesis do occur, and it has been shown, e.g., that *in vivo* administration of radioactive galactosamine results in a low level of incorporation into glycogen[91-94]. In all likelihood this is due to a lack of absolute donor specificity on the part of glycogen synthase. A low level of mistakes in synthesis is apparently of little practical significance, and accumulative effects are not likely to be observed, since any erroneously inserted glycosyl groups can presumably be removed by the appropriate glycosidases in the course of normal catabolism.

In contrast to the nearly absolute specificity with regard to the glycosyl group of the donor, the requirements for a specific nucleotide component are not as precise. UDP-Glucose is the physiological donor in glycogen synthesis by mammalian liver, yeasts and fungi, but may be substituted by ADP-glucose, which is approximately 50% as effective as the uridine derivative[95,96]. Conversely, ADP-glucose is the natural donor for starch biosynthesis in plants, and glycogen formation in bacteria, but UDP-glucose may also serve as substrate in this reaction[97,98]. Since the normal donors are the best substrates and are also the only ones present in significant concentrations *in vivo*, the lack of absolute specificity for the nucleotide portion of the molecule is of little consequence from a physiological point of view.

### *3.2.4.2 Acceptor specificity*

The second important aspect of specificity in glycosyl transfer reactions concerns the acceptors in these reactions. According to the 'one enzyme–one linkage' concept, a glycosyl transferase is specific for a particular acceptor. This assumption has proven generally valid but for a few exceptions, notably the ability of lactose synthase to utilise either *N*-acetylglucosamine or glucose as acceptors.[99-109]. In discussing acceptor specificity we must consider the following points: (a) nature of the acceptor (it is not always a sugar); (b) size of the acceptor (more than a single monosaccharide may be needed for recognition by the enzyme); (c) the influence of linkage position on acceptor activity in the cases where an acceptor larger than a monosaccharide is required; (d) location of the acceptor unit within the chain (transfer to the non-reducing terminus yields a linear molecule, whereas transfer to an internal glycosyl group yields a branched structure); and (e) the use of endogenous acceptors for assay of glycosyltransferase activities.

It is quite clear from the above that three broad categories of acceptors exist: (i) the core proteins of glycoproteins and proteoglycans; (ii) the growing carbohydrate prosthetic groups of glycoproteins, proteoglycans and glycolipids; and (iii) lipids, including ceramide, which constitutes the lipid portion of the many glycosphingolipids[110], and dolichol phosphate, which serves as an intermediate in glycoprotein biosynthesis. Glycosyl transfer to protein acceptors has been dealt with briefly in the preceding Section (3.2.3), and will be discussed further in Section 3.4.1. The biosynthesis of glycolipids[111-115] has many features in common with the formation of protein-bound carbohydrates but falls outside the scope of the present review, and in the following we shall be concerned primarily with the formation of the carbohydrate groups of glycoproteins and proteoglycans.

After chain initiation, the continued growth of the carbohydrate groups occurs by stepwise addition of glycosyl units from the appropriate nucleotide sugars and, in the majority of cases, the non-reducing terminus of the growing chain serves as the acceptor; however, transfer may also occur to an internal residue (most often, and possibly always, the penultimate group) with the formation of branch points. Although the natural acceptors are invariably glycosylated proteins, a smaller acceptor structure may sometimes be utilised, and in a few cases transfer to free monosaccharides has been observed *in vitro*. The following examples of such reactions may be cited: (i) galactosyl transfer to glucose, *N*-acetylglucosamine or xylose[116], catalysed by lactose synthase[99-109]; (ii) galactosyl transfer to xylose, catalysed by a galactosyltransferase involved in formation of chondroitin sulphate and other glycosaminoglycans[49]; (iii) galactosyl transfer to *N*-acetylgalactosamine, catalysed by an enzyme preparation from submaxillary gland[117], and representing one of the reactions in submaxillary mucin formation; (iv) mannosyl transfer to mannose by one or more enzymes from liver[118]; (v) glucuronosyl transfer to galactose, one of the six glycosyl transfer steps in chondroitin sulphate synthesis[119]; and (vi) glucuronosyl transfer to *N*-acetylgalactosamine[119]. Transfer to monosaccharide acceptors may be used for convenient assay of some of the glycosyltransferases, especially when the physiological acceptors are difficult to prepare. Comparisons of the acceptor activities of monosaccharides and the glycosylated protein acceptors which represent the physiological substrates have only been possible in a few instances, and the available evidence suggests that the 'natural' acceptors are in general far better than the monosaccharides.

Whereas the acceptor monosaccharide itself is sometimes the smallest acceptor structure necessary for transfer, larger structures are usually required. The next higher order of specificity in regard to acceptor size is represented by a simple monosaccharide glycoside in which the glycosidic linkage has a specific anomeric configuration. This situation is exemplified by the sialyltransferase from colostrum, which can utilise aryl β-galactosides but not free galactose or α-galactosides as acceptors[120]. In the majority of cases, however, a disaccharide unit (free or protein-bound) is the smallest group that can be recognised as a substrate by a particular glycosyltransferase. Some enzymes have a nearly absolute requirement for recognition of the correct penultimate sugar and the position of its linkage to the non-reducing terminal acceptor group (see also Section 3.4.2), while others may require a disaccharide structure but are not as specific as to the nature of the penultimate sugar or its linkage position (see Section 3.4.3). In the latter instance, however, it is often observed that the acceptor activity of the 'unnatural' substrate is much lower than that of the proper disaccharide and may be insignificant from a physiological point of view.

It may well be that a terminal disaccharide structure is sufficient to provide most of the glycosyltransferases with a proper recognition signal which is specific enough to permit accurate and reproducible assembly of a particular oligosaccharide prosthetic group. There is as yet no single example of a mammalian glycosyltransferase which has an absolute requirement for a trisaccharide unit. However, the difficulties in preparing naturally occurring trisaccharides and synthetic model compounds for comparative purposes have precluded in-depth investigations of the potential role of the antepenultimate glycosyl residue in the glycosyl transfer reactions. At this point it may also be emphasised that the qualitative information regarding acceptor specificity should be complemented with a quantitative evaluation, especially since it is commonly observed that larger substrates are preferred, even though the minimum acceptor size for a particular glycosyltransferase may be that of a monosaccharide. The effect of acceptor size is strikingly illustrated by the relative activities of oligosaccharides serving as primers in glycogen synthesis: over the range from maltose to a high molecular weight glycogen, the acceptor activity increases[57,58] as much as 14 000-fold. Similarly, the acceptor activity for transfer of glucuronic acid and *N*-acetylgalactosamine, respectively, to two homologous series of chondroitin sulphate oligosaccharides increases with increasing molecular weight[119,121-123]. Whereas a monosaccharide and a disaccharide are sufficiently different in structure to be regarded as qualitatively different substrates (e.g. galactose has only a vestige of the activity exhibited by the disaccharide substrate, 3-*O*-β-D-galactosyl-D-galactose, in the glucuronosyl transfer reaction mentioned above[119]), a ready explanation for the variation in acceptor activity within a homologous oligosaccharide series is not immediately obvious. In addition to a better understanding of the effect of 'bulk' on acceptor activity, we also need further information regarding the influence of the protein moieties of glycoproteins and proteoglycans on glycosyl transfer steps other than the chain-initiating transfers to the protein itself. As rapid progress is now being made in the isolation of various glycosyltransferases, we shall soon be in a position to approach these and other problems relating to their mechanism of action in a more meaningful fashion by the use of purified enzyme preparations.

A specific position of the linkage between the non-reducing terminus and the penultimate residue sometimes appears to be absolutely essential for acceptor activity. The sialyltransferases in pork liver[124] and goat colostrum[5, 20], which catalyse transfer to galactose (to form a 2 → 6 linkage), have a marked preference for β-galactosyl-(1→4)-*N*-acetylglucosamine over the other position isomers (and lactose), and catalyse transfer to both high and low molecular weight acceptors. In contrast, a different sialyltransferase present in mammary gland (catalysing transfer to galactose to form a 2→3 linkage), shows identical activity towards lactose and all three position isomers of galactosyl-*N*-acetylglucosamine; this enzyme utilises low molecular weight acceptors but not glycoprotein acceptors[5, 120]. A substrate specificity identical to that of the liver sialyltransferase is shown by a fucosyltransferase from pork liver, i.e. transfer occurs to sialidase-treated $\alpha_1$-acid glycoprotein (orosomucoid) and *N*-acetyl-lactosamine, but little activity is observed toward lactose or the 1→3 and 1→6 isomers of *N*-acetyl-lactosamine[125]. (Since the liver fucosyl- and sialyl-transferases have identical specificities, i.e. both enzymes require a terminal β-galactosyl residue bound in a 1→4 linkage to a penultimate *N*-acetylglucosamine residue, the ratio of sialic acid to fucose in plasma glycoproteins is due, in part at least, to competition between the two transferases for available galactosyl sites. There is no evidence that the structures internal to this disaccharide play a major role in either fucose or sialic acid transfer). Examples of variability in the specificity with regard to the linkage position of the acceptor are also found among the glycosyltransferases involved in chondroitin sulphate synthesis. UDP-D-Galactose: 4-*O*-β-D-galactosyl-D-xylose galactosyltransferase is highly specific for the 4-linked galactosylxylose isomer, yielding 3-*O*-β-D-galactosyl-4-*O*-β-D-galactosyl-D-xylose[49], but, on the other hand, glucuronosyl transfer to this product or 3-*O*-β-D-galactosyl-D-galactose is catalysed by a less specific enzyme which may also utilise 4-*O*-β-D-galactosyl-D-galactose, 6-*O*-β-D-galactosyl-D-galactose and lactose, although transfer to the latter two acceptors is considerably less than that observed with the 'natural' 3-linked substrates (see further, Sections 3.4.2 and 3.4.3.)[119].

Although the majority of the glycosyl transfer reactions are additions to the non-reducing terminus of a growing chain, a considerable number of branched structures are found among the mammalian glycoproteins, e.g. in blood group specific substances, fetuin, orosomucoid, keratan sulphate and heparin. The formation of a branch point must involve glycosyl transfer to the penultimate sugar or a residue located in an even more internal position. The blood group A and B specific substances are formed in this manner by transfer of either *N*-acetylgalactosamine or galactose to a terminal 2′-fucosyl-lactose group in which the penultimate galactosyl moiety is the acceptor for the addition of the group specific sugar[79, 126-133]. (Beside the growing macromolecule, the trisaccharide 2′-fucosyl-lactose can also serve as acceptor.) A second fucose residue may be present, resulting in Le$^a$ rather than H specificity, and, if this is the case, the addition of *N*-acetylgalactosamine or galactose is blocked, possibly because the molecule then becomes too bulky to be accommodated on the active site of the enzyme. In many glycoproteins, mannose residues participate in the formation of branch points, but comparatively little is known concerning the details of mannosyl transfer reactions in mammalian tissues. Another example of glycosyl transfer to a penultimate residue has been observed in the course of studies of submaxillary gland mucin formation. Following transfer of *N*-acetylgalactosamine and galactose, sialic acid may be added to the *N*-acetylgalactosamine residue, and the product may be further converted into the pentasaccharide characteristic of ovine mucin by addition of fucose and *N*-acetylgalactosamine to the galactose residue[134, 135]. However, if sialic acid is transferred prior to galactose, no further chain growth can occur, and the product remains a disaccharide, i.e. a mucin of the bovine type. Although the sweeping generalisation is often made that chain elongation in the course of glycoprotein synthesis occurs by glycosyl transfer to the non-reducing end of the growing chain, it is thus apparent that the penultimate residue is often the target for glycosyl transfer reactions and that branch formation is by no means an uncommon event in the formation of the complex carbohydrates.

#### *3.2.4.3 Specificity of linkage formation*

We have seen from the preceding discussion that neither the acceptor nor the donor specificity of glycosyltransferases is absolute. In contrast, these enzymes are apparently extremely specific for the anomeric configuration and position of the linkages formed. There is as yet no example of a single mammalian glycosyltransferase which can synthesise more than one type of linkage. An absolute specificity in this regard is not surprising, since an exact reproduction of a complex carbohydrate structure in the course of biosynthesis would otherwise not be possible. Furthermore, it will be realised that whenever a certain lack of specificity as to the nature and linkage of the penultimate group is observed with artificial substrates, this has little bearing on the physiological mode of synthesis, since *in vivo* a glycosyltransferase is always likely to be presented with a substrate of the 'correct' structure resulting from the preceding glycosyl transfer reaction.

### 3.2.5 Assay of glycosyltransferases

The many glycosyl transfer reactions that have now been demonstrated in mammalian tissues have almost always been detected initially by experiments of the same type, i.e. incubation of tissue homogenates or subcellular fractions with radioactive nucleotide sugars and measurement of incorporation into endogenous acceptors present in the crude enzyme preparations[2]. Ideally, these physiological substrates should be used to assay the various glycosyltransferases and to study their properties, particularly since there is reason to believe that they are better acceptors than many exogenous substrates presently employed for these purposes. However, in practice this meets with formidable problems, and there are also certain disadvantages to the use of endogenous acceptors which may be summarised as follows. First, the endogenous acceptor is usually not physically separate from the enzyme preparation, thereby making it impossible to study the effects of independent variation of enzyme concentration and substrate concentration on enzyme activity. In practical terms the lack of proper kinetic data makes accurate measurement of enzyme concentration difficult, since it may be impossible to determine whether the enzyme or substrate concentration is limiting in the reaction. Second, the endogenous acceptor is usually present in rather low concentration. Thus, it is often necessary to work at acceptor concentrations well below saturation. The low levels of acceptor require the use of radioactive sugar nucleotides of high specific activity to detect the enzymic reaction. This in turn means that the nucleotide sugar is present in low concentration. Both substrates of the reaction may therefore be present at concentrations well below saturation, resulting in a non-linear enzyme assay. Third, neither the endogenous acceptor nor the resulting product are usually characterised chemically. There may, for example, be more than one acceptor for a particular sugar donor and consequently more than one enzyme might contribute to the assay. Alternatively, owing to similarities in structure between the carbohydrate prosthetic groups of various glycoproteins, one and the same glycosyltransferase may participate in the synthesis of a host of different products, and the specific acceptor activities and relative quantitative contributions of the various substrate species to the total endogenous acceptor activity cannot be readily assessed. Furthermore, such factors as the multichain structure of the connective tissue proteoglycans, and the presence of several identical oligosaccharides in a single glycoprotein, introduces additional heterogeneity among the endogenous acceptors, since molecules in intermediary stages of synthesis will be present in the tissues at all times. The separation and characterisation of such a multitude of closely related acceptor molecules is largely beyond the limits of present methodology, and even in the cases where this could be accomplished, the preparation of sufficient quantities of the various acceptor species is generally not a practically feasible proposition. Finally, it should also be pointed out that the substrate specificity of the glycosyltransferases cannot easily be studied with the use of endogenous acceptors. This is an especially important shortcoming, since substrate specificity is the major factor controlling the sequence of sugars in polysaccharide and glycoprotein prosthetic groups. In short, if one relies exclusively on the use of endogenous acceptors, the scope of the information that can be obtained concerning the reactions under study is by necessity severely restricted in several important respects.

It is not surprising then that one of the first steps in the study of a particular glycosyltransferase has often been a search for a suitable exogenous substrate which could help overcome the problems associated with an assay using endogenous acceptors. In many cases, oligosaccharides from naturally occurring glycoproteins or polysaccharides have been used successfully for this purpose, and another useful group of substrates comprises several modified glycoproteins from which one or more sugar residues have been removed by digestion with an appropriate glycosidase (e.g. orosomucoid or fetuin treated with neuraminidase, β-galactosidase and *N*-acetylglucosaminidase). Substrates of the latter type may actually bear a great resemblance to the natural endogenous acceptors and are sometimes distinctly superior to small oligosaccharides in their acceptor activity. In a more extensive review of glycoprotein biosynthesis, a number of exogenous substrates for various glycosyl transfer reactions have been discussed in some detail[2].

In concluding our discussion of the problems encountered in the assay of glycosyltransferases it should be emphasised that although the use of exogenous substrates offers many advantages, one of the important goals in the study of these enzymes is to determine the mode of glycoprotein synthesis *in vivo*. This can only be accomplished by studying systems which are likely to reflect the physiological situation with a fair degree of accuracy, and many a time this is certainly not true for the artificial systems consisting of purified enzymes and exogenous substrates. Beautiful examples of the importance of employing physiologically relevant systems are found in the work of Lindahl and collaborators[1] on the biosynthesis of heparin, and of Silbert and co-workers[136, 138] on the mechanism of polymerisation and sulphation of chondroitin sulphate (Section 3.4.5).

### 3.2.6 Survey of known glycosyltransferases

From the 'one enzyme–one linkage' concept it may be inferred that each disaccharide unit which can be visualised as a component of a complex carbohydrate molecule corresponds to a specific glycosyltransferase capable of transferring the non-reducing terminal unit of this particular disaccharide. Provided that this basic premise is valid, it is possible to make predictions as to the types and numbers of the glycosyltransferases present in mammalian tissues by surveying known carbohydrate structures, and the following considerations then emerge.

As has already been indicated, a total of ten different monosaccharides have been distinguished with certainty as components of mammalian glycoproteins and polysaccharides. From these monosaccharides can be constructed exactly 700 different disaccharides, which vary in regard to the position and anomeric configuration of the glycosidic linkages (the sugars are all assumed to be in pyranose form, and only one sialic acid structure has been included, disregarding the variation in the number of acetyl groups and the possibility of substitution of an *N*-glycolyl group for the *N*-acetyl group[139]). The potential for variability in the structure of oligosaccharides larger than a disaccharide is therefore almost unlimited. In actuality, however, the number of oligosaccharide structures which have been encountered so far is reasonably small, and it is often observed that different glycoproteins have identical or closely related carbohydrate prosthetic groups. Furthermore, certain limitations seem to be imposed on some of the monosaccharides in regard to their position within an oligosaccharide group (e.g. terminal *vs.* internal); in addition, some theoretically possible combinations of monosaccharides have never been encountered in naturally occurring mammalian substances and may, at least for the time being, be regarded as 'forbidden' (e.g. disaccharide structures composed of mannose and galactose have never been found). We can therefore make the following generalisations with respect to the structural behaviour of the individual sugars.

*Sialic acid* is located almost exclusively at a non-reducing terminal and has been found in an internal position only in disialyl-lactosylceramide[140], which contains a disaccharide structure composed of two sialic acid residues (see Table 3.2, p. 114). In contrast, a sialic acid polymer, colominic acid, has been described in micro-organisms[141-143]. Sialic acid is not necessarily positioned at the terminus of a linear chain, but may be part of a branch, e.g. as in porcine mucin[144] or skeletal keratan sulphate (keratan sulphate II)[145, 146]. In the latter polysaccharide, one of the sialic acid residues is presumably part of a sialyl-galactose disaccharide which is linked to position-6 of the *N*-acetyl-galactosamine residue involved in the carbohydrate–protein linkage, while the main polysaccharide chain is linked to position-3 of the same sugar.

*Fucose* is similar to sialic acid in regard to its position within oligosaccharide prosthetic groups, despite the great structural differences between these two sugars. It has been found in mammalian tissues exclusively in non-reducing terminal location and mostly as a branch rather than as a terminal residue of a linear chain. Like sialic acid, fucose may also participate in homopolysaccharide formation in organisms other than vertebrates, and is the sugar component of the fucoidan of the brown alga, *Fucus vesiculosus*[147, 148].

*Xylose* occupies a unique position as the monosaccharide mediating the linkage of most of the connective tissue polysaccharides to the protein moieties of the respective proteoglycans (see Chapter 4). Several reports have described the presence of xylose in other complex carbohydrate molecules, such as ribonuclease[149, 150] and a glycoprotein from placenta[151]; also, xylose has been found in an oligosaccharide–lipid from hen's oviduct[152]. In no case has the exact position and linkage type of the xylose residues in these latter compounds been unambiguously established. The xylose content of the mammalian glycoproteins is usually less than one mole per mole of compound (a finding which is entirely compatible with the known microheterogeneity of glycoproteins in general), and it would therefore appear that xylose is not an essential part of these substances, as it is in the proteoglycans. The significance of the xylose-containing oligosaccharide–lipid is not yet clear. The position of xylose at the non-reducing terminus of this substance would seem to preclude a role as an intermediate in the synthesis of connective tissue polysaccharides, and it has been suggested that the oligosaccharide–lipid may rather be a precursor of some hitherto unknown xylose-containing glycoprotein in the oviduct.

*Galactose* and *mannose* have not as yet been found linked to each other in any mammalian glycoproteins. Whereas galactose may be bound to several other monosaccharides, mannose is restricted to combinations with *N*-acetyl-glucosamine or other mannose residues.

N-*Acetylgalactosamine* is rarely found as an internal sugar in glycoproteins, and its most prominent positions are (i) at the non-reducing terminus of blood group substance A and similar glycoproteins, and (ii) as the terminal residue at the proximal end of a chain, linked to hydroxy groups of serine or theronine residues in the protein core (e.g. in salivary mucins and in keratan sulphate II). In addition, *N*-acetylgalactosamine is a prominent constituent of the connective tissue polysaccharides and is then present in internal positions as part of the repeating disaccharide units of the chondroitin sulphates and dermatan sulphate (Chapter 4).

D-*Glucuronic acid* and L-*iduronic acid* are limited to the rather small group of connective tissue polysaccharides which comprises no more than seven or eight different compounds. In these polysaccharides the two uronic acids are linked either to *N*-acetylglucosamine or to *N*-acetylgalactosamine and, in addition, glucuronic acid is part of the specific carbohydrate–protein linkage region, where it is bound to galactose (Chapter 4).

D-*Glucose* is not a common constituent of mammalian complex carbohydrates and is found essentially only in three types of compounds: (i) glycogen, (ii) the collagens, particularly those of basement membranes (in the collagens, glucose is part of the characteristic disaccharide 2-*O*-α-D-glucosyl-D-galactose, which is bound to hydroxylysine residues of the protein), and (iii) lactose and other milk oligosaccharides which are not bound to protein but are structurally related to certain glycoproteins in which *N*-acetylglucosamine replaces the glucose moieties of the oligosaccharides. Many reports of the presence of glucose in various glycoproteins have appeared in the literature, but such findings must be evaluated with great caution in view of the ease with which contamination with glucose can occur through the use of cellulose ion exchangers, Sephadex and other such materials.

**Table 3.1 Possible combinations of naturally occurring monosaccharides into disaccharide units†**

* Sialic acid-Sialic acid
* Sialic acid-Galactose
Sialic acid-Mannose
Sialic acid-*N*-Acetylglucosamine
* Sialic acid-*N*-Acetylgalactosamine
Sialic acid-Glucose

* Fucose-Galactose
* Fucose-Mannose
* Fucose-*N*-Acetylglucosamine
Fucose-*N*-Acetylgalactosamine
* Fucose-Glucose

* Glucose-Glucose
* Glucose-Galactose
Glucose-Mannose
Glucose-*N*-Acetylglucosamine
Glucose-*N*-Acetylgalactosamine

* Galactose-Galactose
Galactose-Mannose
* Galactose-*N*-Acetylglucosamine
* Galactose-*N*-Acetylgalactosamine
* Galactose-Glucose

* Mannose-Mannose
Mannose-Galactose
* Mannose-*N*-Acetylglucosamine
Mannose-*N*-Acetylgalactosamine
Mannose-Glucose

* *N*-Acetylglucosamine-*N*-Acetylglucosamine
* *N*-Acetylglucosamine-Galactose
* *N*-Acetylglucosamine-Mannose
*N*-Acetylglucosamine-*N*-Acetylgalactosamine
* *N*-Acetylglucosamine-Glucose

* *N*-Acetylgalactosamine-*N*-Acetylgalactosamine
* *N*-Acetylgalactosamine-Galactose
*N*-Acetylgalactosamine-Mannose
*N*-Acetylgalactosamine-*N*-Acetylglucosamine
*N*-Acetylgalactosamine-Glucose

* Naturally occurring combination

† Sialic acid and fucose are linked in non-reducing terminal positions only, and disaccharides with reducing terminal fucose and sialic acid have therefore been omitted with exception of the disaccharide sialic acid-sialic acid, which is found in disialyl-lactosylceramide ($G_{D3}$)[140]

Leaving aside, for the moment, the three monosaccharides which are specific for connective tissue polysaccharides (D-glucuronic acid, L-iduronic acid and D-xylose) we can formulate a list of disaccharides—not specified as to linkage configuration and position—which could all theoretically be components of mammalian complex carbohydrate structures. Omitting structures containing fucose and sialic acid at the reducing terminus (with the one exception of sialylsialic acid), we arrive at the 36 disaccharides which are shown in Table 3.1. Of these disaccharide types, 21 are presently known to occur (marked by asterisks) and, as a result of differences in linkage position and anomeric configuration of the glycosidic linkages, they actually represent 44 individual disaccharides, which are listed in Table 3.2 together with their parent compounds. The assumption that each of these disaccharides corresponds to a specific glycosyltransferase is amply verified by a survey of the known enzymes in this class (see Ref. 2 for a detailed review).

**Table 3.2 Naturally occurring disaccharide sequences in glycoproteins*†**

| | *Disaccharide* | *Source* |
|---|---|---|
| 1. | α-Fuc-(1→4)-Glc*N*Ac | Le$^a$, Le$^b$ Type 1 chains[79, 153], $\gamma_G$-myeloma globulin[154] |
| 2. | α-Fuc-(1→3)-Glc*N*Ac | Le$^a$, Le$^b$ Type 2 chains[79, 153], $\gamma_A$-myeloma globulin[154, 155], ribonuclease[156], ceramide-lacto-*N*-fucopentaose III[157] |
| 3. | β-Fuc-(1→4)-Glc*N*Ac | $\gamma_G$-myeloma globulin[154] |
| 4. | α-Fuc-(1→2)-Gal | A, B, H, Le$^b$ Type 1 and 2 chains[79, 153] |
| 5. | α-Fuc-(1→3)-Glc | Lacto-difucotetraose and lacto-*N*-difucohexaose II[79, 153] |
| 6. | α-Fuc-(1→4)-Man | $\gamma_G$-myeloma globulin[154] |
| 7. | β-Gal-(1→3/4)-Glc*N*Ac | Orosomucoid[158], $\gamma_A$-myeloma globulin[155], transferrin[159], blood group substances[79, 154], keratan sulphate[170] |
| 8. | β-Gal-(1→6)-Glc*N*Ac | $\gamma_G$-myeloma globulin[154], ribonuclease[156] |
| 9. | β-Gal-(1→3)-Gal*N*Ac | Pig submaxillary gland mucin[144] |
| 10. | β-Gal-(1→4)-Gal*N*Ac | Gangliosides ($G_{MI}$, $G_{DIa}$, etc.)[140] |
| 11. | β-Gal-(1→6)-Gal*N*Ac | $\gamma_A$-myeloma globulin[155] |
| 12. | α-Gal-(1→4)-Gal | Human digalactosylceramide[160], trihexosylceramide[161], globoside (cytolipin K)[161], digalactosyldiglycerides[162] |
| 13. | α-Gal-(1→3)-Gal | Rat trihexosylceramide[163], cytolipin R[164], rabbit erythrocyte pentaglycosylceramide[165], blood group B substance |
| 14. | β-Gal-(1→4)-Glc | Many glycosphingolipids[140], lactose |
| 15. | α-Gal*N*Ac-(1→3)-Gal*N*Ac | Forssman hapten[166] |
| 16. | α-Gal*N*Ac-(1→3)-Gal | Blood group A substance[79, 153], rabbit pentaglycosylceramide[165], pig submaxillery mucin Type A[144] |
| 17. | β-Gal*N*Ac-(1→3)Gal | Cytolipin K[161], R[164] |
| 18. | β-Gal*N*Ac-(1→4)-Gal | Gangliosides[140], $\gamma_A$-myeloma globulin[155] |
| 19. | β-Gal*N*Ac-(1→4)-Gal*N*Ac | $\gamma_A$-myeloma globulin[155] |
| 20. | α-Glc(1→4)-Glc | Glycogen[167, 168] |

**Table 3.2** *continued*

| | *Disaccharide* | *Source* |
|---|---|---|
| 21. | α-Glc-(1→6)-Glc | Glycogen[167, 168] |
| 22. | α-Glc-(1→2)-Gal | Collagen (glomerular basement membrane)[169] |
| 23. | β-Glc*N*Ac-(1→3)-Gal | Blood group substances[154, 157, 165], keratan sulphate[170] |
| 24. | β-Glc*N*Ac-(1→4)-Gal | Some bovine spleen gangliosides[171] |
| 25. | β-Glc*N*Ac-(1→2)-Man | $\gamma_G$-myeloma globulin[154] |
| 26. | β-Glc*N*Ac-(1→3)-Man | Transferrin[159], orosomucoid[158], ribonuclease[156] |
| 27. | β-Glc*N*Ac-(1→4)-Man | Ovalbumin[172, 173], orosomucoid[158], ribonuclease[156], $\gamma_G$-myeloma globulin[155] |
| 28. | β-Glc*N*Ac-(1→4)-Glc*N*Ac | $\gamma_G$-myeloma globulin[155], chitin[174] |
| 29. | β-Glc*N*Ac-(1→6)-Gal | Group specific H glycoprotein from ovarian cyst[79, 175] |
| 30. | β-Glc*N*Ac-1→6)-Glc | H glycoprotein[79, 175] |
| 31. | α-Man-(1→6)-Man | Ribonuclease[156], $\gamma_G$-myeloma globulin[154] |
| 32. | α-Man-(1→4)-Man | Ovalbumin[172, 173] |
| 33. | α-Man-(1→3)-Man | $\gamma_G$-myeloma globulin[154], ovalbumin[172, 173] |
| 34. | α-Man-(1→2)-Man | $\gamma_A$-myeloma globulin[155] |
| 35. | α-Man-(1→4)-Glc*N*Ac | Ribonuclease[156], orosomucoid[158], $\gamma_A$-myeloma globulin[155] |
| 36. | α-Man-1→3)-Glc*N*Ac | Ovalbumin[172, 173], orosomucoid[158], $\gamma_A$-[155], $\gamma_G$-myeloma globulins[156] |
| 37. | β-Man-(1→3)-Glc*N*Ac | Ovalbumin[172, 173], ribonuclease[156] |
| 38. | α-*N*ANA-(2→2/4)-Gal | $\gamma_G$-heavy chain glycoprotein[176] |
| 39. | α-*N*ANA-(2→3)-Gal | Gangliosides ($G_{M3}$, $G_{M2}$, $G_{M1}$, etc.)[140], sialyl-lactose, $\gamma_A$-myeloma globulin[155] |
| 40. | α-*N*GNA-(2→3)-Gal | Gangliosides ($G_{M3}$, $G_{M2}$, $G_{M1}$, etc.)[140], sialyl-lactose, $\gamma_A$-myeloma globulin[155] |
| 41. | α-*N*ANA-(2→6)-Gal | Fetuin[177], orosomucoid[158] |
| 42. | α-*N*ANA-(2→6)-Gal*N*Ac | Submaxillary gland glycoproteins[111] |
| 43. | α-*N*ANA-(2→8)-*N*ANA | Gangliosides[140] |
| 44. | β-*N*GNA-(2→8)-*N*ANA | Gangliosides[140] |

* Modified from Dawson[315]

† Abbreviations: Fuc, L-fucose (6-deoxy-L-galactose); Gal, D-galactose; Glc, D-glucose; Man, D-mannose; Glc*N*Ac, *N*-acetyl-D-glucosamine; Gal*N*Ac, *N*-acetyl-D-galactosamine; *N*ANA, *N*-acetylneuraminic acid; *N*GNA, *N*-glycolylneuraminic acid *N*ANA and Fuc invariably occupy non-reducing terminal positions; Gal is often the next sugar and Gal*N*Ac, Glc*N*Ac, Man and Glc are usually non-terminal, although many exceptions do exist

In a few cases, the postulated glycosyltransferases have not yet been demonstrated experimentally, largely because of lack of suitable substrates, but it seems safe to predict that the continued search for such enzymes will eventually be successful.

In a complete overview of all mammalian glycosyltransferases, we should also include those participating in connective tissue polysaccharide biosynthesis and in carbohydrate–protein linkage formation. An accurate estimate of the number of glycosyltransferases required for these processes cannot be given at this time*, but it seems likely that approximately 15 enzymes are necessary, and the total number of mammalian glycosyltransferases would then be close to 60. Whereas a great many of these have been partially purified and characterised as to substrate specificity and certain other properties only four have as yet been purified to homogeneity or near homogeneity, i.e glycogen synthase[178-180], lactose synthase[103, 104, 106], UDP-D-xylose:core protein xylosyltransferase[68] and UDP-*N*-acetyl-D-galactosamine: H substance *N*-acetylgalactosaminyltransferase[126].

## 3.3 GLYCOSYLTRANSFERASES OF CONNECTIVE TISSUE POLYSACCHARIDE BIOSYNTHESIS

From the known structures of the connective tissue polysaccharides (Chapter 4) we can derive 15 different disaccharide components (Table 3.3). This list should be regarded as only partial, since knowledge concerning the structure of some of the polysaccharides is still incomplete. For example, heparin may have a branched structure and would then contain at least one additional disaccharide component[1]. Similarly, the detailed structure of keratan sulphate has not yet been determined, and this polysaccharide may conceivably contain disaccharides other than those listed in Tables 3.2 and 3.3. In any event we could postulate, in the vein of the 'one enzyme–one linkage' hypothesis that 15 or more glycosyltransferases participate in formation of the various disaccharide components of the connective tissue polysaccharides, and that an additional three or four enzymes would be required for the chain-initiating glycosyl transfer reactions. However, the validity of the postulate that each disaccharide structure reflects the existence of a specific glycosyltransferase rests on the additional premise that the sugars do not undergo any interconversions following their incorporation into an oligo- or poly-saccharide chain. This is indeed generally true, but in connective tissue polysaccharide biosynthesis we encounter the only known exception to this rule, inasmuch as the iduronic acid residues of heparin and dermatan sulphate are formed by epimerisation of glucuronic acid at the polymer level[1]. Another deviation from the anticipated pattern concerns the *N*-acetylgalactosaminyl transfer in the course of dermatan sulphate biosynthesis. The 'one enzyme–one linkage' concept suggests the existence of two separate enzymes catalysing transfer to iduronic acid and glucuronic acid residues, respectively; however both of these reactions are catalysed by a single enzyme with broad specificity[196]. Taking into account these exceptions from the basic rule, we can now

* The uncertainties in making such an estimate result from lack of knowledge in several important areas: the structure of keratan sulphate is not yet fully known; the detailed mechanism of hyaluronic acid biosynthesis has not been established and might involve the participation of lipid intermediates; the anomalous route of iduronic acid formation raises several problems, e.g. in regard to the acceptor specificity in the *N*-acetylhexosaminyl transfer reactions participating in heparin and dermatan sulphate biosynthesis; formation of *N*-acetylgalactosaminyl–serine and *N*-acetylgalactosaminyl–threonine linkages would be expected to require two enzymes, were the acceptor specificities absolute, but it is possible that only one enzyme is needed; finally, the details of the formation of *N*-acetylglucosaminyl–asparagine linkages have not yet been elucidated.

**Table 3.3 Disaccharide components of connective tissue polysaccharides***

| | *Disaccharide component* | *Parent polysaccharide* |
|---|---|---|
| 1. | β-GlcUA-(1→3)-Glc*N*Ac | Hyaluronic acid[170, 181, 182] |
| 2. | β-Glc*N*Ac-(1→4)-GlcUA | |
| 3. | β-GlcUA-(1→3)-Gal*N*Ac | Chondroitin 4- and 6-sulphate[170, 183], dermatan sulphate[170, 184-186] |
| 4. | β-Gal*N*Ac-(1→4)-GlcUA | |
| 5. | α-IdUA-(1→3)-Gal*N*Ac | Dermatan sulphate[186-190] |
| 6. | β-Gal*N*Ac-(1→4)-IdUA | |
| 7A. | α-IdUA-(1→4)-Glc*N*$SO_4$ | Heparin[170] |
| 7B. | α-IdUA-(1→4)-Glc*N*Ac | Heparan sulphate[170] |
| 8A. | β-GlcUA-(1→4)-Glc*N*$SO_4$ | |
| 8B. | β-GlcUA-(1→4)-Glc*N*Ac | |
| 9A. | α-Glc*N*$SO_4$-(1→4)-IdUA | |
| 9B. | α-Glc*N*Ac-(1→4)-IdUA | |
| 10A. | α-Glc*N*$SO_4$-(1→4)-GlcUA | |
| 10B. | α-Glc*N*Ac-(1→4)-GlcUA | |
| 11. | β-Gal-(1→4)-Glc*N*Ac | Keratan sulphate[170] |
| 12. | β-Glc*N*Ac-(1→3)-Gal | |
| 13. | β-Gal-(1→4)-Xyl | Carbohydrate–protein linkage regions of chondroitin 4- and 6-sulphate, dermatan sulphate, heparin and heparan sulphate[170, 191-195] |
| 14. | β-Gal-(1→3)-Gal | |
| 15. | β-GlcUA-(1→3)Gal | |

* Sulphate groups have been omitted with the exception of *N*-sulphate groups which replace most of the *N*-acetyl groups in the course of biosynthesis of heparin and are also present to a varying extent in heparan sulphate

formulate the following list of enzymes which participate in the formation of the various connective tissue polysaccharides:

*Hyaluronic acid*: one *N*-acetylglucosaminyltransferase and one glucuronosyltransferase;

*Chondroitin sulphate:* (i) a chain-initiating xylosyltransferase catalysing transfer of xylose to the core protein of the proteoglycan; (ii and iii) two galactosyltransferases which transfer, consecutively, the two galactosyl residues of the linkage region; (iv) a glucuronosyltransferase which completes the formation of the specific carbohydrate–protein linkage region; (v) an *N*-acetylgalactosaminyltransferase and (vi) a second glucuronosyltransferase, which jointly synthesise the bulk of the polysaccharide chain, composed of 30–50 repeating disaccharide units;

*Dermatan sulphate:* four linkage-region enzymes; one glucuronosyltransferase catalysing transfer to *N*-acetylgalactosamine units; one epimerase; and one *N*-acetylgalactosaminyltransferase catalysing transfer to iduronic acid and glucuronic acid residues. With the exception of the unique epimerase, these enzymes may all be identical or similar to those involved in chondroitin sulphate synthesis;

*Heparin and heparan sulphate:* four linkage-region enzymes; one glucuronosyltransferase catalysing transfer to *N*-acetylglucosamine residues; one epimerase; and one *N*-acetylglucosaminyltransferase catalysing transfer to glucuronic acid residues (*N*-acetylglucosaminyl transfer is assumed to occur before epimerisation, since oligosaccharides with non-reducing terminal iduronic acid do not serve as acceptors[1]);

*Keratan sulphate:* one galactosyltransferase (possibly identical with lactose synthase), and one *N*-acetylglucosaminyltransferase; since keratan sulphate contains several sugars other than galactose and *N*-acetylglucosamine, i.e. sialic acid, fucose, mannose and galactosamine (keratan sulphate II), as well as galactose in excess over the amount present in the repeating disaccharide units, glycosyltransferases catalysing the transfer of these sugars must also be considered.

The existence of many of the enzymes mentioned above has already been well documented. However, the mechanism of formation of hyaluronic acid remains incompletely understood and, although cell-free synthesis of this polysaccharide from UDP-glucuronic acid and UDP-*N*-acetylglucosamine can readily be obtained in both bacterial and mammalian systems[3], the two separate transfer reactions have not yet been demonstrated. Likewise, the mode of chain initiation is not known, and there is a strong possibility that lipid intermediates are involved in the biosynthesis of this polysaccharide.

Little progress has also been made in the study of keratan sulphate biosynthesis and, since the iduronic acid-containing polysaccharides (dermatan sulphate, heparin and heparan sulphate) are covered elsewhere in this series, the remainder of this presentation will be devoted largely to a discussion of the enzymatic aspects of chondroitin sulphate biosynthesis.

## 3.4 BIOSYNTHESIS OF CHONDROITIN SULPHATE

It is seen from Table 3.4 that six different glycosyltransferases are required for the biosynthesis of chondroitin sulphate. The properties of these enzymes

**Table 3.4 Chondroitin sulphate glycosyltransferases and some of their exogenous substrates**

| *Enzyme* | *Exogenous acceptors* |
|---|---|
| Xylosyltransferase | Smith-degraded cartilage proteoglycan, L-serylglycylglycine |
| Galactosyl-transferase I | Xylose<br>*O*-β-D-Xylosyl-L-serine |
| Galactosyl-transferase II | 4-*O*-β-D-Galactosyl-D-xylose<br>4-*O*-β-D-Galactosyl-*O*-β-D-xylosyl-L-serine |
| Glucuronosyl-transferase I | 3-*O*-β-D-Galactosyl-D-galactose<br>*O*-β-D-Galactosyl-(1→3)-*O*-β-D-galactosyl-(1→4)-D-xylose |
| *N*-Acetylgalactos-aminyltransferase | GlcUA-Gal*N*Ac-GlcUA-Gal*N*Ac-GlcUA-Gal*N*Ac (chondroitin hexasaccharide)<br>GlcUA-(Gal*N*Ac-4S)-GlcUA-(Gal*N*Ac-4S)-GlcUA-(Gal*N*Ac-4S) (chondroitin 4-sulphate hexasaccharide) |
| Glucuronosyl-transferase II | Gal*N*Ac-GlcUA-Gal*N*Ac-GlcUA-Gal*N*Ac (chondroitin pentasaccharide)<br>(Gal*N*Ac-6S)-GlcUA-(Gal*N*Ac-6S)-GlcUA-(Gal*N*Ac-6S) (chondroitin 6-sulphate pentasaccharide) |

and the characteristics of the reactions which they catalyse will be reviewed in some detail below. In addition, we shall consider certain aspects of the process of sulphation, which is a unique and important part of the biosynthetic process and distinguishes proteoglycan formation from glycoprotein synthesis in general.

### 3.4.1 Chain initiation by xylosyltransferase

The xylosyl transfer reaction, which initiates the formation of most of the connective tissue polysaccharides, was first demonstrated by Grebner *et al.*[197], who showed that hen's oviduct contains an enzyme which transfers xylose from UDP-xylose to endogenous acceptors present in the enzyme preparation. This enzyme has subsequently also been found in a mouse mastocytoma[198], embryonic chick cartilage[199, 200], brain[69], and, in lower concentration, in some other tissues.

Like many other endogenous substrates, the xylosyl acceptor has not been properly characterised. While admittedly an extremely difficult task, a thorough analysis of its relevant structural features would help answer a number of simple but important questions concerning the process of chain initiation and chain growth *in vivo*, and among problems which call for a solution we may mention the following. Does xylosyl transfer begin while the protein core is still being synthesised on the polyribosomes, or does it take place only after the polypeptide chain has been completed? Are there any intermediates in this reaction? Are all the polysaccharide chains initiated simultaneously, or do they grow in a random fashion, or is there a strictly controlled progression from either the *N*-terminal or *C*-terminal end of the protein core? In the latter event, is one polysaccharide chain completed before the growth of the next one starts, or is there a whole spectrum of polysaccharide molecules in varying stages of chain elongation? By and large, information concerning these aspects of chondroitin sulphate biosynthesis is still sketchy and lacking in precise detail, but some answers are beginning to emerge from several physiologically oriented studies. The work of Silbert and collaborators[136-138] has shed considerable light on the nature of the endogenous acceptors for repeating disaccharide formation and sulphation, and is also highly relevant to the problems of chain initiation. In an interesting recent preliminary communication, Kimura and Caplan[201] reported that xylose transfer to polysomes from cultured embryonic chick chondrocytes occurs in the presence of added xylosyltransferase. The sizes of the polysomes which served as substrates indicated a molecular weight range of 10 000–20 000 dalton for the acceptor polypeptides, a figure which is considerably lower than that determined for the protein moieties of most proteoglycans which have been studied so far. It might be suggested on the basis of these findings that the larger size of the finished core proteins can be attributed to postribosomal events resulting in covalent association of a number of smaller units.

### *3.4.1.1 Exogenous xylose acceptors*

In view of the need for an exogenous substrate to be used in purification of xylosyltransferase, a search for suitable xylose acceptors was undertaken by Baker *et al.*[61], and a number of potential acceptors were tested, ranging from serine and simple serine derivatives to the entire core protein of the chondroitin sulphate proteoglycan. Since the primary structure of the core protein has not been determined it was not possible to select small peptides of known structure with sequences identical to those present in the native acceptor. However, since a glycine residue always appears to be present on the *C*-terminal side of the polysaccharide-substituted serine residues, it was of particular interest to test peptides containing a Ser-Gly sequence. Whereas the dipeptide serylglycine was not itself an acceptor, the tripeptide serylglycylgylglycine had some acceptor activity but a relatively high $K_m$ value of 20.0 mmol $l^{-1}$. Larger peptides from the native carbohydrate–protein linkage region had higher acceptor activity, and isolation of the entire core protein by Smith degradation (periodate oxidation, followed by reduction with borohydride and mild acid cleavage) yielded the best acceptor obtained so far. The $K_m$ value of this material was 0.064 mmol $l^{-1}$, expressed in terms of the concentration of total serine residues. That transfer did indeed occur to serine residues of the core protein was indicated by isolation of [$^{14}$C]xylosylserine in 45% yield from a proteolytic digest of the reaction product.

The finding that the macromolecular substrate had the highest acceptor activity is consistent with observations on other transferase reactions in which a polypeptide is glycosylated, such as the *N*-acetylgalactosaminyl transfer to serine and threonine residues in the formation of salivary gland mucins, and the galactosyl transfer to hydroxylysine residues of collagen. The only substrate known for the former reaction is the core protein obtained after removal of the sialic acid–*N*-acetylgalactosamine disaccharide, and peptides from a Pronase digest were completely inactive[63]. Similarly, galactosyl transfer to hydroxylysine proceeds most readily with a macromolecular substrate—intact collagen or collagen from which the glucosylgalactose disaccharide has been removed by Smith degradation—but tryptic peptides could also serve as acceptors, although their activity was only 23% of that observed for the macromolecular substrate[62].

In the native proteoglycan, approximately half of the serine residues are bound to polysaccharide chains, and half are not substituted. Nevertheless, the intact proteoglycan is not a xylose acceptor. The failure to obtain xylose transfer to the free serine residues could conceivably be due to the presence of the completed chondroitin sulphate chains, but the products of hyaluronidase[61] or chondroitinase[202] digestion, in which the unsubstituted serines are probably more readily accessible, are also inactive. Although the possibility still exists that the oligosaccharide chains which remain attached to the protein prevent the glycosylation of the unsubstituted residues, it is more reasonable to assume that only specific serine residues can serve as acceptors. As indicated above, a Ser-Gly sequence is always needed for recognition by the xylosyltransferase, but the exact structural requirements for the acceptor are not yet known.

In this context, we may also consider briefly the presence of keratan sulphate in the cartilage proteoglycans. Some of the keratan sulphate chains are bound to serine, whereas others are linked to threonine, in both instances by glycosidic linkages to *N*-acetylgalactosamine[203-205]. Consequently, although initiation of keratan sulphate chains has not yet been investigated experimentally, we may postulate the existence of one or two *N*-acetylgalactosaminyltransferases which catalyse transfer to these hydroxyamino acids. Furthermore, it may be suggested that by analogy with the situation observed for the initiation of chondroitin sulphate chains, only specific serine residues are recognised by the enzyme and that the primary structure in the vicinity of these residues is quite different from that in the chondroitin sulphate regions. Despite the lack of direct experimental evidence, this notion receives support from studies of the structure of the cartilage proteoglycans which have shown that a particular region of the molecule is rich in keratan sulphate and hydrophobic amino acids (see Chapter 4), and it will be of interest to correlate further structural information with investigations of the substrate specificity of the chain-initiating enzyme.

A description of the properties of the Smith-degraded proteoglycan would be incomplete without emphasis of the fact that approximately one-third of this material consists of keratan sulphate. Since this polysaccharide is resistant to oxidation with periodate (with the exception of its terminal residues), most or all of the keratan sulphate in the original proteoglycan still remains after the Smith degradation. It is possible that the keratan sulphate chains influence considerably the acceptor activity of the Smith-degraded proteoglycan, but we have no means of determining their effect at the present time, since methods for the complete removal of keratan sulphate from the protein without extensive degradation of the latter do not currently exist.

#### *3.4.1.2 Purification and properties of xylosyltransferase*

Xylosyltransferase is unique among chondroitin sulphate glycosyltransferases in being more loosely bound to the membranes of the endoplasmic reticulum than the other enzymes of this group and, on thorough homogenisation, 80–90% of the total activity of a homogenate of embryonic chick cartilage is found in the supernatant fraction after high-speed centrifugation[69]. More than the apparent proportion of the activity is left in the particulate fraction, however, and repeated digestion of the pellet with ribonuclease[206, 207] (see Section 3.4.1.3) or treatment with detergent at relatively high ionic strength[208] (e.g. 0.5% Nonidet P-40 in 0.5 M KCl) yields a considerable amount of soluble enzyme approaching that present in the initial high speed supernatant fraction.

Once an exogenous substrate for xylosyltransferase was available, the enzyme could be reliably quantitated and was purified 40- to 50-fold by a procedure involving ammonium sulphate fractionation and gel chromatography on Sephadex G-200, or centrifugation in a sucrose density gradient[69]. Further purification has been obtained by chromatography on an affinity matrix consisting of Sepharose-bound core protein (Smith-degraded proteoglycan)[68]. After removal of endogenous acceptors by the purification steps used by Stoolmiller *et al.*[69], the partially purified enzyme was adsorbed to the affinity matrix and was subsequently eluted by a solution of core protein, yielding quantitative recovery of activity. However, since the degree of purification could not readily be estimated in the presence of the protein substrate, alternative methods of elution were sought. It was found that the enzyme was also eluted, in lower yield (30–40%), by buffer containing 0.25 M KCl. Interestingly, it was also observed that once a column had been saturated with enzyme and eluted with salt, further batches could be adsorbed and quantitatively eluted with salt alone. The binding capacity of such a preconditioned column was correspondingly lower—approximately one-third of the initial capacity—indicating that two-thirds of the binding sites were still occupied by enzyme which could only be desorbed by the substrate itself. The purification obtained in the affinity step was in the order of 160-fold, resulting in an overall purification over the crude homogenate of approximately 4000-fold. Rechromatography of some preparations yielded enzyme of even higher specific activity, representing close to 7000-fold purification.

The affinity-purified enzyme was homogenous on analysis by polyacrylamide gel electrophoresis, immunodiffusion and immunoelectrophoresis. The molecular weight was 95 000–100 000 dalton, as determined by gel chromatography, and the enzyme consisted of two pairs of non-identical subunits, as indicated by the finding that gel electrophoresis after pretreatment with sodium dodecyl sulphate and mercaptoethanol yielded two bands with migration rates corresponding to molecular weights of 23 000 and 27 000 dalton, respectively.

The behaviour of xylosyltransferase on affinity chromatography suggested that two types of interaction are involved in its binding to the matrix: (i) the enzyme–substrate interaction which can be dissociated only by the substrate itself, and (ii) an interaction between individual enzyme molecules which can be dissociated by salt. The latter type of interaction is in keeping with the observation that the enzyme readily aggregates at low ionic strength even at relatively low protein concentration to form cloudy suspensions which may again be clarified by an increase of the ionic strength. It is of interest to note that the xylosyltransferase may also interact with UDP-galactose:D-xylose galactosyltransferase which catalyses the second step in chondroitin sulphate synthesis[208-210]. This interaction will be discussed in more detail below.

### 3.4.1.3 *Effect of ribonuclease and lysozyme on xylosyltransferase*

In a study of xylosyltransferase from mouse mastocytoma, Grebner and Neufeld[206] observed that lysozyme and ribonuclease stimulated the activity of aged preparations of the enzyme. The effect was not due to glycosylation of the added proteins, and the reasons for the activating effect are not yet entirely clear. In the chick cartilage system, ribonuclease caused a three-fold increase in xylose incorporation into the endogenous acceptors of a 100 000 × *g* pellet fraction, but had only slight effect on the soluble transferase in the supernatant fluid[207]. The ribonuclease effect is due, in part at least, to solubilisation of the particle-bound xylosyltransferase, which occurs concomitantly with the digestion of the ribonucleic acid present in the particulate enzyme preparation. After repeated treatments with ribonuclease, the total solubilised activity is close to that initially found in the supernatant fraction of a homogenate. As a corollary to these observations, it was also found that ribonucleic acid inhibits xylosyltransferase activity when added to a solution of the purified enzyme[211]. Although we cannot with any certainty ascribe a physiological significance to these findings, it is tempting to speculate that xylosyltransferase is in some way closely interacting with the ribosomes, since xylosyl transfer probably occurs soon after the synthesis of the protein is completed or possibly even while the peptide is attached to the ribosomes. In any event, it appears likely that xylosyltransferase is in part associated with ribonucleic acid in the particulate enzyme preparation and may be released by digestion with ribonuclease.

### 3.4.2 Galactosyl transfer reactions I and II

The second reaction in chondroitin sulphate synthesis is the transfer of galactose from UDP-galactose to the xylosyl-protein formed by xylosyltransferase (Figure 3.2). This reaction was first demonstrated by Robinson *et al.*[199, 200], who showed that a particulate enzyme preparation from embryonic chick cartilage catalyses incorporation of galactose into endogenous acceptors. The product was characterised by isolation of radioactively labelled galactosylxylitol after cleavage of the xylosyl–serine linkage with alkaline borohydride. Similar evidence for the existence of a second galactosyltransferase, catalysing transfer to the galactosylxylosyl-protein formed by the previous reaction, was obtained by Helting and Rodén[49], who isolated 3-*O*-β-D-galactosyl-D-galactose from a partial acid hydrolysate of the same type of reaction mixture.

In view of the difficulties in studying glycosyl transfer reactions with endogenous acceptors, which have been previously discussed, suitable exogenous substrates for these reactions were sought. Several substrates are now available for the assay of galactosyltransferase I, including D-xylose, *O*-β-D-xylosyl-L-serine, methyl β-D-xylopyranoside and *p*-nitrophenyl β-D-xylopyranoside[49-56]. The finding that D-xylose is an acceptor was somewhat surprising, and it may be noted that galactosyltransferase I is one of the few glycosyltransferases which do not require an acceptor larger than a monosaccharide. If xylose were a common constituent of glycoproteins and could be bound to sugars other than galactose, such a situation could conceivably result in frequent mistakes during the assembly of carbohydrate prosthetic groups. However, this is not the case, and any xylosyl residues that the galactosyltransferase encounters *in vivo* are almost certainly part of growing connective tissue proteoglycan molecules. The ability of D-xylose and *p*-nitrophenyl β-D-xylopyranoside to serve as chain initiators has been used to advantage in studies of cartilage differentiation in cell culture, as will be discussed in Section 3.7.2.

4-*O*-β-D-Galactosyl-D-xylose and 4-*O*-β-D-galactosyl-*O*-β-D-xylosyl-L-serine have been used as exogenous substrates for the assay of galactosyltransferase II and are presently the only known small substrates of well-defined structure[49]. In view of the multitude of glycoprotein structures that contain

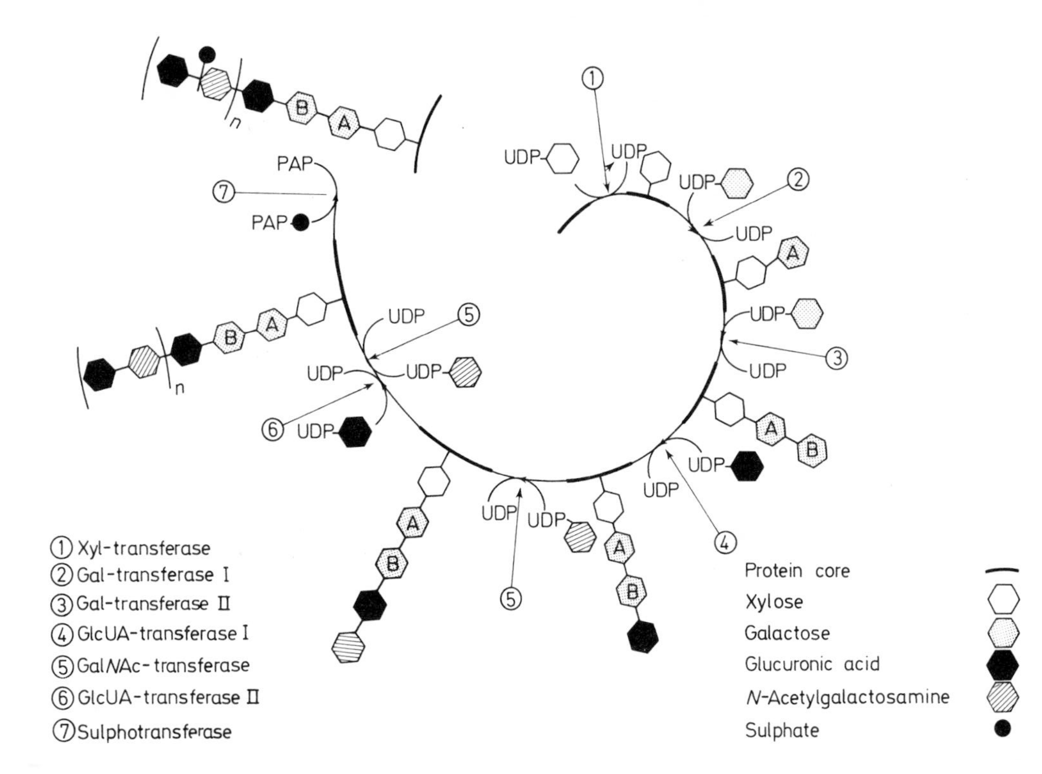


**Figure 3.2** Biosynthesis of chondroitin 4-sulphate chain. (From Rodén[3], by courtesy of Academic Press.)

galactose, it is not surprising that galactosyltransferase II has a higher degree of specificity than does galactosyltransferase I. Apparently, the enzyme needs to recognise both the terminal monosaccharide acceptor and also the penultimate sugar, i.e. D-xylose. Lactose exhibits slight acceptor activity, but it should be remembered that this disaccharide differs from the more natural substrate, 4-*O*-β-D-galactosyl-D-xylose, only by the presence of a primary alcohol group instead of a hydrogen on C-5 of the reducing terminal monosaccharide. Transfer does not occur to free D-galactose, nor do oligosaccharides with non-reducing terminal galactose other than lactose serve as acceptors. The reaction catalysed by galactosyltransferase II is somewhat unusual inasmuch as it involves the addition of a glycosyl group to an acceptor of the same kind. An enzyme of low specificity, recognising only the non-reducing terminus, would obviously be totally unsatisfactory for this particular reaction, as its action would not be limited to the addition of a single galactose residue but would rather result in the formation of a galactose homopolymer.

The substrate specificities of the two galactosyltransferases are summarised in Table 3.5; it should be pointed out that the data shown in this table have been obtained with the use of a crude, particulate enzyme preparation from embryonic chick cartilage and that similar studies have not yet been carried out with purified enzymes.

**Table 3.5 Substrate specificity of galactosyltransferases***

| *Acceptor* | *Activity (relative to* D-*xylose)* |
|---|---|
| D-Xylose | 100 |
| L-Xylose | 4 |
| Methyl β-D-xylopyranoside | 99 |
| *O*-β-D-Xylosyl-L-serine | 206 |
| 4-*O*-β-D-Galactosyl-D-xylose | 50 |
| 3-*O*-β-D-Galactosyl-D-xylose | 70† |
| 4-*O*-β-D-Galactosyl-*O*-β-D-L-serine | 40 |
| Lactose | 9 |
| Raffinose | 5 |
| D-Arabinose | 15 |
| D-Glucose | 9 |

* Inactive substrates: galactose, 3-*O*-β-D-galactosyl-D-galactose, 2-*O*-β-D-galactosyl-D-lyxose, methyl β-D-xylofuranoside, D-lyxose, D-ribose, D-mannose, maltose, sucrose, L-fucose

† Characterisation of the reaction product with this substrate showed that galactosyl transfer had occurred to C-4 of the xylose residue and not to the non-reducing terminal galactose

### 3.4.3 Glucuronosyl transfer reaction I

The specific carbohydrate–protein linkage region is completed by the addition of a glucuronic acid residue[119, 212]. The enzyme catalysing this reaction requires at least a disaccharide for maximal acceptor activity, although detectable transfer also occurs to free D-galactose (at a level of approximately

2% of that observed for the optimal substrate). In addition to the 'natural' disaccharide, 3-*O*-β-D-galactosyl-D-galactose, several other galactose-containing disaccharides may serve as substrates for this reaction, including 4- and 6-*O*-β-D-galactosyl-D-galactose, lactose and 4-*O*-β-D-galactosyl-D-xylose (Table 3.6).

**Table 3.6 Substrate specificity of glucuronosyltransferase I***

| *Acceptor* | *Activity (relative to 3-O-β-D-galactosyl-D-galactose)* |
|---|---|
| 3-*O*-β-D-Galactosyl-D-galactose | 100 |
| 4-*O*-β-D-Galactosyl-D-galactose | 126 |
| 6-*O*-β-D-Galactosyl-D-galactose | 27 |
| 3-*O*-β-D-Galactosyl-4-*O*-β-D-galactosyl-D-xylose | 97 |
| 3-*O*-β-D-Galactosyl-4-*O*-β-D-galactosyl-*O*-β-D-xylosyl-L-serine | 167 |
| Lactose | 27 |
| 4-*O*-β-D-Galactosyl-D-xylose | 9 |
| *N*-Acetyl-D-galactosamine | 7 |
| D-Galactose | 2 |

* Inactive substrates: raffinose, D-glucose, *N*-acetyl-D-glucosamine, D-xylose, L-fucose

The nature of the penultimate group of the acceptor is apparently of less importance for the glucuronosyl transfer reaction than for the preceding galactosyl transfer step. As far as the three isomeric galactosylgalactoses are concerned, this lack of specificity is of little consequence, since the 4- and 6-linked isomers are not known to occur in mammalian tissues, and the terminal galactose unit of the natural acceptor will always be linked to the adjacent galactose residue by a 1→3 β-linkage, owing to the generally absolute specificity of glycosyltransferases with regard to the linkage formed. In contrast, the finding that 4-*O*-β-D-galactosyl-D-xylose is a substrate, albeit a poor one, is somewhat disturbing, since glucuronosyl transfer to this disaccharide *in vivo* could lead to irregularities in the structure of the carbohydrate–protein linkage region. However, glycopeptides with only one galactose residue were never observed in the course of structural studies of the linkage region, and it therefore appears unlikely that mistakes in synthesis occur under physiological conditions. Factors other than the specificity of the glycosyltransferases probably contribute to the orderly growth of the nascent polysaccharide chains, and it has been suggested by Horwitz and Dorfman[69, 122, 123] that the glycosyltransferases are arranged on the membranes of the endoplasmic reticulum in a sequential fashion so that, in this particular instance, the growing chain will not encounter the glucuronosyltransferase until both galactose residues have already been added. Such an arrangement would add another dimension of specificity to the assembly of the polysaccharide chains, and would prevent any irregularity in structure that might potentially arise from lack of specificity on the part of the individual glycosyltransferases.

It should again be emphasised that substrate specificities which have been determined with crude enzyme preparations (as in Table 3.6) do not necessarily reflect the behaviour of the purified enzymes. Glucuronosyltransferase I is a case in point, insofar as the relative acceptor activities of 3-*O*-β-D-galactosyl-D-galactose and lactose depend on the nature of the enzyme preparations used. With a 10 000 × *g* supernatant fraction from a homogenate of embryonic chick cartilage as an enzyme source, both 3-*O*-β-D-galactosyl-D-galactose and lactose were utilised as acceptors for glucuronosyl transfer, although the latter was only about 25% as active as the optimal substrate[69, 119]. After centrifugation at 100 000 × *g*, virtually all of the enzyme was found in the pellet, as determined with both substrates, and little or no activity was present in the supernatant liquid. Interestingly, the total activity in the pellet, with 3-*O*-β-D-galactosyl-D-galactose as substrate, was 130–180% of the initial activity, whereas the activity with lactose as acceptor had decreased drastically, to 10–20% of the original value. A similar observation has been made by Helting[213], who reported that a solubilised and partially purified preparation of glucuronosyltransferase I from a mouse mastocytoma did not catalyse transfer to lactose, although a crude enzyme fraction could utilise this sugar as an acceptor. Since available evidence indicates that 3-*O*-β-D-galactosyl-D-galactose and lactose are both substrates for the same enzyme, these results are mutually exclusive, unless we postulate that a 'modifier' is present in the soluble fraction of the homogenate. Such a modifier would promote transfer to lactose and, when removed, transfer to 3-*O*-β-D-galactosyl-D-galactose would increase and the utilisation of lactose would decrease. While clearly reminiscent of the well-known effect of lactalbumin on the substrate specificity of lactose synthase, these findings cannot as yet be given a physiologically meaningful interpretation and have not been explored further.

### 3.4.4 Repeating disaccharide formation

In the first investigations of the cell-free biosynthesis of chondroitin sulphate it was shown that a low-sulphated polysaccharide was formed on incubation of a particulate enzyme preparation from embryonic chick cartilage with UDP-glucuronic acid and UDP-*N*-acetylgalactosamine[36, 214]. These initial studies provided no information concerning the detailed mechanism of repeating disaccharide formation, but through a series of elegant studies by Dorfman[11] and his collaborators, many fundamental characteristics of the polymerisation process have now been elucidated, and the design of the experimental approach has served as a model for the more recent work on the biosynthesis of the linkage region which has been described above. Thus, it is now well documented that the repeating disaccharides are formed by alternating additions of *N*-acetylgalactosamine and glucuronic acid to the non-reducing terminus of the growing chain and that this process is catalysed by two distinct enzymes, an *N*-acetylgalactosaminyltransferase and a glucuronosyltransferase (for simplicity, this enzyme is called glucuronosyltransferase II to distinguish it from the enzyme involved in linkage region formation). The substrate specificities of the polymerising enzymes have been explored by the use of a great many oligosaccharides, prepared from chondroitin, chondroitin 4- and 6-sulphates, hyaluronic acid and dermatan sulphate, and the major conclusions emerging from these studies are summarised in Table 3.7.

*N*-Acetylgalactosaminyl transfer occurs to non-reducing terminal glucuronic acid residues of sulphated as well as non-sulphated oligosaccharides derived from the polysaccharides mentioned above. The finding that hyaluronic acid hexasaccharide is an acceptor for *N*-acetylgalactosaminyl transfer is particularly interesting, since this compound has an *N*-acetylglucosamine residue in the penultimate position. Although the presence of an *N*-acetylgalactosamine unit in this position is apparently not obligatory, it may well be that an oligosaccharide containing a sugar other than an *N*-acetylhexosamine would be inactive, but a suitable selection of potential acceptors has not been available for examination of this possibility. The constancy in structure of the products of proteoglycan biosynthesis has previously been commented upon, and in the polymerisation process we have another example of the fact that a certain lack of specificity on the part of an individual glycosyltransferase does not result in irregularities in the physiological assembly of the polysaccharide. Thus, despite the theoretical possibility that a hybrid structure containing both glucosamine and galactosamine could be formed, this does not occur, and we must assume that a strict compartmentalisation of the formation of the various polysaccharides exists within the cell, which may have its basis in the organisation of multienzyme complexes specific for each polysaccharide. (We must, of course, recognise the existence of glucuronic acid–iduronic acid hybrid molecules, i.e. dermatan sulphate and heparin, but these are in a category by themselves in view of the exceptional mode of synthesis of iduronic acid.)

As expected, glucuronosyltransferase II catalyses transfer to terminal *N*-acetylgalactosamine residues of oligosaccharides from chondroitin and chondroitin 6-sulphate, but not to the *N*-acetylglucosamine unit of a hyaluronic acid pentasaccharide. Surprisingly, however, a pentasaccharide from chondroitin 4-sulphate does not serve as an acceptor, and it may therefore be concluded that sulphation *in vivo* must occur at least one step behind the polymerisation, although the exact relationship between these two processes is not yet entirely clear.

The size of the acceptor is of great importance for the acceptor activity, and although a comprehensive and systematic study of this aspect of the specificity of the two polymerising enzymes has not yet been undertaken, it is evident from a few isolated observations that the larger oligosaccharides are vastly better acceptors than the smaller members of the respective homologous series[119, 121-123]. This is in keeping with the behaviour of other glycosyltransferases, such as glycogen synthase, which exhibits a distinct preference for larger acceptor molecules[57, 58, 167, 168].

With increasing awareness of the participation of polyprenol intermediates in the biosynthesis of some complex carbohydrates, we must consider the possibility that the connective tissue polysaccharides may be formed in a similar manner. So far, no evidence has been obtained to this effect, and presently available information is entirely consistent with the notion that the growth of the chondroitin sulphate chains occurs by stepwise addition

**Table 3.7 Acceptor and donor specificities of chondroitin sulphate 'polymerase'**

| *Donor UDP-nucleotide sugar* | *Oligosaccharide acceptors with non-reducing terminal glucuronic acid** | | | | | | |
|---|---|---|---|---|---|---|---|
| | Chondroitin Hexa-saccharide | Chondroitin Tetra-saccharide | Chondroitin 4-sulphate Hexa-saccharide | Chondroitin 4-sulphate Tetra-saccharide | Chondroitin 6-sulphate Hexasaccharide | Hyaluronic acid Hexasaccharide | Desulphated dermatan sulphate Tetrasaccharide |
| UDP-*N*-Acetylgalactosamine | + | + | + | + | + | + | — |
| UDP-Glucuronic acid | — | — | — | — | — | — | |
| UDP-*N*-Acetylglucosamine | | | — | | — | — | |

| | *Oligosaccharide acceptors with non-reducing terminal* N-*acetylhexosamine** | | | |
|---|---|---|---|---|
| | Chondroitin Pentasaccharide | Chondroitin 4-sulphate Pentasaccharide | Chondroitin 6-sulphate Pentasaccharide | Hyaluronic acid Pentasaccharide |
| UDP-*N*-Acetylgalactosamine | — | — | — | — |
| UDP-Glucuronic acid | + | — | + | — |

* Even-numbered oligosaccharides from chondroitin, the chondroitin sulphates and hyaluronic acid were prepared by digestion with testicular hyaluronidase, which yields homologous oligosaccharides with glucuronic acid at the non-reducing end. Odd-numbered oligosaccharides with *N*-acetylhexosamine at the non-reducing end were obtained from the even-numbered compounds by digestion with β-glucuronidase. The desulphated dermatan sulphate tetrasaccharide was isolated from an acid hydrolysate of the polysaccharide

of single monosaccharide units directly from the nucleotide sugars. Never theless, since much of the work described above has been carried out with crude enzyme preparations, the participation of intermediates could have gone undetected, and it will be of importance to confirm some of the previou studies with the use of purified enzymes. Furthermore, it is not yet known how the formation of the repeating disaccharide units is initiated, although it has been assumed that this would occur by transfer of *N*-acetylgalacto samine to the galactose-bound uronic acid residue of the linkage region, in a reaction catalysed by the same *N*-acetylgalactosaminyltransferase that i involved in the formation of the remainder of the repeating units. It is a matter of some concern that, in their studies of heparin biosynthesis, Helting and Lindahl[215] were unable to detect the analogous *N*-acetylglucosaminy transfer reaction, using GlcUA-Gal-Gal-Xyl-Ser as a potential acceptor This negative result might have a trivial explanation, but it is equally possible that repeating disaccharide formation is initiated by a specific reaction which has so far escaped detection, and it must still be considered possible that lipid intermediates are involved in this phase of chondroitin sulphate biosynthesis

### 3.4.5 Endogenous acceptors, polymerisation and sulphation

In a series of ingenious studies, Silbert and his co-workers[136-138] have investi gated the mechanism of chondroitin sulphate biosynthesis with a view to determining the mode of synthesis *in vivo*. Using the microsomal fraction of an embryonic chick cartilage homogenate as the best available approximation to an *in vivo* system, they have addressed themselves to questions such as the following. (i) What are the characteristics of the polysaccharide molecules present in the microsome fraction? (ii) What are the primers for polymerisa tion? (iii) What are the endogenous acceptors for sulphation? (iv) What is the time course of polymerisation and sulphation?

Initially it may be noted that most of these studies have been carried out at pH 6.5 and that chondroitin 6-sulphate exclusively is formed at this pH whereas, at pH 7.8, 30–40% of the newly synthesised polysaccharide is 4 sulphated. Characterisation of the radioactive products was carried out largely by two procedures, i.e. chromatography on DEAE-cellulose to achieve separation according to charge and charge density, and gel chroma tography on Sepharose 4B for purposes of determining approximate mole cular weights.

The nature of the endogenous acceptors was studied by radioactive labelling of the microsomal fraction with either sulphate, UDP-glucuronic acid, or UDP-*N*-acetylgalactosamine. Since the endogenous levels of the nucleotide sugars in the microsomal fraction are too low to sustain polymerisation, incubation with either of the two nucleotide sugars will usually result in addition of a single glycosyl group only to the non-reducing terminus of the chain. Characterisation of the sulphate-labelled material showed that the molecular weights of the acceptors ranged from 2000 to 40 000 dalton, with 90% of the total polysaccharide having a molecular weight of 16 000 dalton or larger. Incorporation of sulphate appeared to occur only into occasional non-sulphated galactosamine residues in predominantly sulphated molecules, as indicated by a similarity in chromatographic patterns between the sulphate-labelled and the sugar-labelled products. No over-sulphation—yielding 4,6-disulphated hexosamine residues—was observed, nor was any endogenous polysaccharide of low sulphate content found. It should be emphasised that the characterisation of the acceptors on the basis of sulphate incorporation alone does not necessarily reflect the actual proportion of molecules in a particular size class, since completely sulphated molecules would not be detected. Therefore, through the additional labelling of the sugar components, a more reliable overall assessment of the nature of the endogenous substrates is obtained.

Extending this work to a study of the polymerisation process, Richmond *et al.*[137] showed the presence of two types of primer, one of molecular weight less than 8000 dalton, and a second class with a molecular weight of 25 000 dalton or more. The growth of the chains apparently occurred rapidly, and the chain length after incubation for 2 minutes was about half that observed after 4 hours. Since the amount of polysaccharide synthesised at the latter time was 10 times higher, it was concluded that polymerisation did not occur by simultaneous growth of all existing primer molecules, but rather that a portion of the chains was rapidly completed and that, subsequently, a new set of primer molecules was utilised for repetition of the same process. It is important to note that no primers of intermediate size were observed in these experiments, nor were any endogenous non-sulphated chains present. It was furthermore suggested that the utilisation of the low-molecular-weight primer represented the physiological mode of synthesis and that the addition to the larger primers might be an artifact of the *in vitro* system which merely resulted from continued polymerisation of chains which would not have grown further under *in vivo* conditions.

In the presence of 3′-phosphoadenylylsulphate under properly chosen incubation conditions, Richmond *et al.*[138] obtained sulphation of as much as 85% of the newly formed chondroitin sulphate. Sulphation occurred in an 'all or nothing' fashion and, maximally, 75% of the polysaccharide chains were sulphated to the extent of 96%, while approximately 5% of the chains were essentially free of sulphate, and only 20% of the newly formed chondroitin sulphate (representing less than 11% of the total sulphate incorporated) might have contained intermediate amounts of sulphate. Time studies confirmed the 'all or nothing' pattern and suggested that the sulphotransferase is located close to the polymerising enzymes so that sulphation of the chain proceeds during polymerisation or immediately following this process.

## 3.5 GENERAL PROPERTIES OF CHONDROITIN SULPHATE GLYCOSYLTRANSFERASES

It has been indicated previously that mammalian tissues contain approximately 60 different glycosyltransferases which are involved in the formation of a great many complex carbohydrates. Although a fair proportion of the reactions catalysed by these enzymes have been studied in considerable detail, relatively little information is available concerning the individual enzymes. Similarly, knowledge regarding the glycosyltransferases involved in the biosynthesis of the connective tissue polysaccharides is mostly limited to information relating to their substrate specificity and kinetic parameters, and these properties have often been determined with crude enzyme preparations. This situation is due, in large part, to certain qualities which are characteristic of many glycosyltransferases, such as an often firm association with cellular membranes (solubilisation is therefore an obligatory first step in purification and may be difficult to achieve), and a tendency of solubilised and partially purified preparations to aggregate, sometimes irreversibly and with loss of activity, if the ionic strength is lowered beyond a certain critical level or if the enzyme concentration is raised above 0.5–1.0 mg $ml^{-1}$. In addition, some glycosyltransferases are dependent on phospholipids or detergents for their enzymatic activity, and this will be discussed separately in Section 3.6.

Thus, it is not surprising that only a few glycosyltransferases have been purified to homogeneity so far, i.e. glycogen synthase, lactose synthase, xylosyltransferase and an *N*-acetylgalactosaminyltransferase which catalyses transfer of the *N*-acetylgalactosamine unit responsible for blood group A specificity. It should be noted that these enzymes are all among the more readily solubilised members of the group or occur in a naturally soluble form in body fluids (serum, milk) in sufficient quantities to permit convenient preparation from these sources.

### 3.5.1 Solubilisation and purification of chondroitin sulphate glycosyltransferases

In attempting to solubilise and purify the chondroitin sulphate glycosyltransferases, we encounter much the same problems which are common to investigations of many other membrane-bound proteins, and glycosyltransferases in particular. Rapid advances have been made in the general area of membrane biochemistry in the recent past, and a variety of procedures are now available which, alone or in combination, can be successfully applied to the solubilisation and subsequent purification of many membrane components (for review, see Ref. 216). Briefly, a number of proteins have been extracted from membranes by procedures based on changes of a single factor in the milieu of the native membrane; such methods include extraction with hypotonic or hypertonic salt solutions, treatment with buffers of alkaline pH, solubilisation with detergents, or extraction with organic solvents. However, as it has become apparent that several types of forces are instrumental in maintaining the stability of membrane architecture, including ionic and hydrophobic interactions, van der Waals forces and hydrogen bonding, it is to be expected that certain membrane components are held firmly *in situ* by the combined actions of several of these forces. It may not then be possible to solubilise such compounds by any one method which is based on perturbation of only a single milieu factor, and it becomes necessary to design a procedure based on the concerted action of several agents (or a single multifunctional agent) which are capable of overcoming all existing interactions. A particular requirement for enzyme purification is of course that the integrity of the molecule should be maintained to the extent that it is still fully enzymatically active and, e.g., treatment with sodium dodecyl sulphate, albeit an excellent solubilisation procedure, most often results in complete loss of enzymatic activity owing to dissociation into subunits.

With the exception of xylosyltransferase, the chondroitin sulphate glycosyltransferases appear to be unusually firmly bound to the membranes of the endoplasmic reticulum, and previous attempts to solubilise these enzymes by 'monofunctional' procedures have met with only limited success. However, it was recently shown by Helting[217] that a combination of detergent (Tween 20) and alkali solubilised a large proportion of the two galactosyltransferases from a heparin-producing mouse mastocytoma, which are identical or analogous to the enzymes involved in chondroitin sulphate biosynthesis. This method is similar to procedures which have previously been used for the solubilisation of mitochondrial enzymes[218].

Helting's method has also been applied successfully to the solubilisation of the chondroitin sulphate glycosyltransferases of embryonic chick cartilage, and, furthermore, it was found that substitution of Nonidet P-40 for Tween 20 considerably increased the efficiency of the procedure[208]. Since the detergent–alkali method involves a substantial increase in ionic strength, it seemed possible that this could be the determining factor for solubilisation rather than the brief exposure to an alkaline pH. Indeed, when the alkali treatment was replaced by addition of salt, the glycosyltransferases were solubilised to the extent of 70–90%, with some variation between the individual enzymes. At optimal ionic strength, the efficiency of the detergent–salt procedure was equal to or somewhat greater than that of the detergent–alkali method. Freezing and thawing in the presence of detergent and salt increased the yield of soluble enzymes and this step might be particularly useful in dealing with enzymes which are even more resistant to solubilisation than the chondroitin sulphate glycosyltransferases. It is suggested that the Nonidet–salt procedure can serve as a general method for the solubilisation of many or all of the particulate glycosyltransferases found in mammalian tissues.

The term 'solubilisation' is often used only in an operational sense to indicate that a particular substance is no longer sedimentable under certain conditions, e.g. 100 000 × *g* for 1 h. However, this does not by any means guarantee that complete dissociation into a monomolecular species has occurred, and there are many examples to illustrate that 'solubilisation' may not have gone as far as is necessary for a successful pursuit of continued purification. A particular solubilisation method would not be useful as an initial step in purification unless it resulted in a substantial degree of dissociation of the individual molecules from each other. In the studies of the cartilage glycosyltransferases, it was therefore important to determine whether this had indeed been achieved by the solubilisation procedure discussed above. Some evidence that dissociation into single molecules had occurred to a significant extent was obtained by gel chromatography on Sephadex G-200. A bimodal pattern was observed for three of the chondroitin sulphate glycosyltransferases (galactosyltransferase I and II and *N*-acetylgalactosaminyltransferase), with the bulk of the activity emerging in well retarded positions and minor portions appearing in the void volume; in contrast, xylosyltransferase appeared as a single peak, overlapping with the included part of galactosyltransferase II, at an effluent volume corresponding to a molecular weight of approximately 100 000 dalton. This finding suggested that the 'solubilised' preparation consisted largely of monomolecular enzyme species, but that a small proportion of multimolecular aggregates was also present[208]. It should be pointed out, however, that a portion of the aggregates could have arisen in the course of gel chromatography, since it was not established that the conditions of elution would have prevented reaggregation.

Despite the lack of precise and complete information, some general conclusions can now be drawn concerning the properties of the chondroitin sulphate glycosyltransferases and the forces that control their behaviour in the intact membranes. Thus, it seems evident that both hydrophobic and ionic forces contribute to the stabilisation of the glycosyltransferases within the membranes and that interactions of this type occur not only between the glycosyltransferases and other membrane constituents but also between individual glycosyltransferases and between molecules of the same enzyme species (Section 3.5.2). Furthermore, the relative strengths of the ionic and hydrophobic interactions apparently differ significantly from one enzyme to another, e.g. xylosyltransferase and galactosyltransferase I represent opposite poles in this regard. A number of observations indicate that the ionic milieu is of prime importance for the behaviour of xylosyltransferase. The enzyme is readily extracted from the tissues by buffers without detergent; a high salt concentration yields virtually complete dissociation of the enzyme into single molecules, and a decrease in ionic strength results in a readily reversible aggregation. In contrast, galactosyltransferase I is distinctly hydrophobic and is presumably dependent on lipids for its enzymatic activity. It is not extracted by salt and, after solubilisation by detergent and salt, it remains in large part as a multimolecular aggregate. Similarly, previous studies of the naturally soluble fraction of chondroitin sulphate glycosyltransferases showed that, even in the presence of 1 M KCl, the bulk of the galactosyltransferase activity emerged with the void volume from Sephadex G-200[69, 123]. In addition, the activity of the membrane-bound enzyme is drastically reduced by treatment with phospholipase C and can be restored by phospholipids (see Section 3.6).

Chondroitin sulphate glycosyltransferases other than xylosyltransferase and galactosyltransferase I exhibit properties intermediate between these two; thus *N*-acetylgalactosaminyltransferase could be partially solubilised by extraction of the tissues by buffer alone (to the extent of 30% of the total activity) and, whereas gel chromatography of this material at low ionic strength yielded a single peak in the void volume, half of the total activity appeared[69, 123] in a retarded position on elution with 1 M KCl.

### 3.5.2 Interactions between glycosyltransferases

It has previously been suggested that the chondroitin sulphate glycosyltransferases may be arranged on the membranes of the endoplasmic reticulum in such a fashion that enzymes catalysing consecutive transfer steps are located in adjacent positions[69, 122, 123]. An arrangement of this type could conceivably result from specific interactions between the various glycosyltransferases, and evidence in support of this idea has recently been obtained[208-210]. It was observed that an affinity matrix designed to adsorb xylosyltransferase was also capable of binding galactosyltransferase I, after having been equilibrated with xylosyltransferase, and the galactosyltransferase could be eluted by Nonidet P-40 in the presence of salt but not by salt alone, suggesting that hydrophobic binding forces are involved. A high degree of specificity in the binding between the two enzymes is indicated by a considerable increase in the specific activity of the galactosyltransferase, and the use of a similar matrix with xylosyltransferase covalently bound to the resin has resulted in approximately 1100-fold purification of galactosyltransferase I. It may be envisioned that these or analogous affinity systems will continue to be of great value not only as tools for the purification of glycosyltransferases but, perhaps more importantly, as simplified model systems for membrane assembly. As highly purified preparations of the various chondroitin sulphate glycosyltransferases become available, the use of such systems will make it possible to establish the existence and nature of specific interactions more conclusively, and the simplicity afforded by model systems with a limited number of variables will obviate many ambiguities and difficulties in the interpretation of the behaviour of the more complex native membranes.

## 3.6 EFFECT OF LIPIDS AND DETERGENTS ON GLYCOSYLTRANSFERASES

As discussed in Section 3.2.3, lipids may participate as intermediate carriers in the biosynthesis of complex carbohydrates in bacterial and mammalian systems. In addition, lipids may function in the regulation of polysaccharide formation indirectly by affecting the glycosyltransferases. This has been most elegantly demonstrated in bacterial systems by Rothfield *et al.*[219]. Comparable studies in mammalian systems have not been carried out because of the difficulties encountered in the purification of membrane-bound glycosyltransferses. However, a lipid- or detergent-dependence for a number of glycosyltransferases has been demonstrated. Some of these findings are summarised in the following discussion.

Many glycosyltransferases which are localised in microsomal, Golgi or plasma membrane fractions have been shown to have a lipid- or detergent-dependence. For instance, a mannosyltransferase[220] and a galactosyltransferase[221] involved in glycoprotein synthesis in developing rat liver both have an absolute dependence on the detergent Triton X-100. Likewise, a distinguishing characteristic of the particulate sialyltransferase, galactosyltransferase and *N*-acetylglucosaminyltransferase participating in glycoprotein synthesis in embryonic chicken brain is that they have an absolute requirement for detergent (Cutscum)[222].

An analogous situation exists for the synthesis of complex glycolipids; Cutscum was absolutely required for UDP-galactose:glucosylceramide galactosyltransferase activity in embryonic chick brain[110, 233] and adult rat spleen[224]. Similarly, an *N*-acetylgalactosaminyltransferase involved in the synthesis of gangliosides in rat spleen required a detergent mixture composed of Triton CF-54 and Tween 80 for activity[225]. Many additional examples of the detergent dependence of glycosyltransferase activity can be cited; however, in none of these examples has the actual mechanism by which detergents activate the glycosyltransferases been demonstrated. It is usually suggested that detergent releases the glycosyltransferase from an inaccessible membrane compartment or activates a cryptic form of the enzyme. Alternatively, especially in the case of glycolipid biosynthesis, it has been postulated that detergents may help to solubilise the hydrophobic substrates and thus facilitate enzyme–substrate interactions.

A few studies on perturbation of membrane lipids by agents like phospholipase C and phospholipase A, whose site of action on membrane phospholipids are fairly well defined, have been conducted in relation to glycosyltransferase activity. Treatment of a microsomal fraction from rat brain with phospholipase C, phospholipase A or detergents (Triton X-100 or deoxycholate) resulted in complete loss of UDP-glucose:ceramide glucosyltransferase activity[226]. No attempt was made in this instance to reverse the inactivation by addition of lipid.

Recently, the first evidence for activation of a membrane-bound enzyme required for serum glycoprotein synthesis by added phospholipid was presented in a study on the effect of phospholipids and other lipid fractions on UDP-galactose:glycoprotein galactosyltransferase from rat liver[227]. The enzyme was markedly stimulated by incubation with lysolecithin, phospholipase A or Triton X-100. There was no change in the $K_m$ value for UDP-galactose in the presence or absence of lysolecithin; however, the $V_{max}$ of the enzyme was increased six-fold in the presence of the phospholipid. Thus it was suggested that this membrane-bound glycosyltransferase may be amenable to physiological regulation by membrane lipid components.

### 3.6.1 Phospholipid dependence of UDP-glucuronosyltransferase

Most of the information on the effects of detergents and phospholipids on glycosyltransferases has come from studies on a complex enzyme system involved in detoxification via glucuronidation of carboxylic, phenolic, alcoholic, amino and possibly SH compounds[228].

Early work showed that detergents (Triton X-100 or digitonin) activated UDP-glucuronosyltransferase 5–10 fold[229-232]. The activation was irreversible and not accompanied by solubilisation[229]. Attempts to explain the activation by kinetic analysis were unsuccessful, and the results were interpreted as an exposition of active sites on the latent UDP-glucuronosyltransferase in the inactive tissue homogenate[229-231].

As indicated previously, specific degradation by phospholipases is important in the study of membrane-bound enzymes; however, the results with UDP-glucuronosyltransferase were ambiguous. Graham and Wood[233] and Atwood *et al.*[234] showed that specific degradation of the microsomal phospholipid membrane by phospholipase A or phospholipase C produced a concomitant inactivation of UDP-glucuronosyltransferase. The inactivation was reversed by phosphatidylcholine and mixed microsomal phospholipid micelles at concentrations similar to those present in the intact microsome preparations[234]. It was concluded, therefore, that the activity of the enzyme was dependent on phospholipids and hence, probably, on the structural integrity of the microsomal membrane.

On the other hand, Hänninen and Puukka[235] and Vessey and Zakim[236] found that phospholipase digestion activated this enzyme. Treatment of liver microsomal preparations from normal rats with lysophosphatidylcholine or with phospholipase A, which generates lysophosphatides, greatly stimulated UDP-glucuronosyltransferase[235]. The stimulation of activity by phospholipase A was due to a six-fold increase of $V_{max}$, with a progressive decrease in binding affinity for both UDP-glucuronic acid and *p*-nitrophenol as substrates[236]. Similar effects, but to a lesser extent, were observed with phospholipase C, Triton X-100, sonication and exposure to pH 9.8. Activation in these experiments was attributed to phospholipid-induced alterations of enzyme conformation, rather than compartmentation of the enzyme[236].

The apparent discrepancy in the observed effects of phospholipase digestions on UDP-glucuronosyltransferase activity has in part been clarified by the work of Winsnes[237], who showed that both activation and inhibition can occur, depending on substrate concentration in the enzyme assay. This is because the apparent $K_m$ values towards both acceptor and donor substrates are strikingly increased by digestion with phospholipase C. Lecithin micelles largely reversed the change in $K_m$ for UDP-glucuronate, the donor substrate, but not for *p*-nitrophenol, the acceptor substrate[237].

In all of these examples the enzyme requirements for a given level of activity appeared to be satisfied by a bulk property of the lipid moiety rather than an absolute requirement for an individual phospholipid molecule. In an attempt to delineate which properties of the lipid moieties are of particular importance for enzyme activity, Eletr *et al.*[238] examined whether the fluidity of lipids in hepatic microsomal membranes correlated in any way with the activity levels of two tightly bound membrane enzymes: UDP-glucuronosyltransferase and glucose 6-phosphatase. The activities of the enzymes were determined at assay temperatures in the range between 5 and 40 °C. Arrhenius plots of the activity of UDP-glucuronosyltransferase displayed abrupt changes at *ca.* 19 and 32 °C. Lipophilic spin probes were introduced into samples of the same microsomal preparations and the corresponding electron spin resonance spectra were recorded over the same temperature range. Temperature dependence of an empirical spectral parameter, related to the fluidity of the matrix solubilising the molecular probes, revealed apparent breaks at 19 and 32 °C in intact microsomes. The authors concluded that the correlation between the enzyme data and the data obtained from the lipophilic spin-probes was indicative of the dependence of the tightly membrane-bound UDP-glucuronosyltransferase on the physical state of membrane lipids[238].

In some of these studies it was speculated that the effect *in vitro* on liver microsomal UDP-glucuronosyltransferase by membrane-perturbing agents may reflect a regulatory mechanism of importance in the intact cell[237], although no direct evidence for such mechanism was presented. Recently, Graham *et al.*[239, 240] have obtained evidence which establishes a physiological regulatory mechanism based on membrane phospholipid perturbation which might exist in the cell. Previously it was shown that protein deficiency markedly increased the activity of rat liver UDP-glucuronosyltransferase[241, 242]. Likewise, protein deficiency profoundly altered the phospholipid composition of rat liver microsomal membranes[240]. The most striking difference in phospholipid composition between control and protein-deficient rats was their content of lysophosphatides. Whereas microsomal membranes from protein-deficient rats contained significant amounts of lysophophatidylcholine and lysophosphatidylethanolamine, very little or no lysophosphatides were detected in control preparations. Pretreatment of microsomal fractions from normal rats with phospholipase A markedly increased their UDP-glucuronosyltransferase activity, as did pretreatment with lysophosphatidylcholine. It was concluded that the quantities of lysophosphatide present in microsomal membranes from protein-deficient rats were sufficient to have caused the increased UDP-glucuronosyltransferase activity of these preparations. It was further suggested that the changes that occur in phospholipid composition reflect changes that occur *in vivo*, caused by protein deficiency, which regulate the activity of UDP-glucuronosyltransferase[239, 240].

### 3.6.2 Role of lipids in activity of UDP-galactose:xylose galactosyltransferase

Since the glycosyltransferases involved in chondroitin sulphate biosynthesis are firmly bound to the membranes of the endoplasmic reticulum, it is of interest to examine the effects of membrane-perturbing agents on the activities of these enzymes. As previously discussed, solubilisation of the chondroitin sulphate glycosyltransferases (with the exception of xylosyltransferase) required a detergent in combination with salt or alkali[208, 243-246]. When the detergent and salt concentrations of the solubilised preparation were rapidly lowered by dialysis, a portion of each glycosyltransferase enzyme was no longer soluble, and in particular the UDP-galactose:xylose galactosyltransferase (galactosyltransferase I) became largely insoluble[208, 246]. If, on the other hand, the glycosyltransferases of the solubilised preparation were separated by gel filtration on Sephadex, three of the enzymes (xylosyl-, galactosyl II- and *N*-acetylgalactosaminyl-transferase), and a portion of galactosyltransferase I were found to be included[208]. In addition, a portion of galactosyltransferase I was found in the void volume fraction, and was most likely associated with large molecular weight membrane components[208].

Although some activation of each of the glycosyltransferases was always observed in the presence of detergent, it would appear that the main function of the detergent was to dissociate the enzyme from the membrane matrix. Additional experiments, using phospholipase digestions, indicated that galactosyltransferase I may be dependent on lipids for activity[245, 246]. A $100\,000 \times g$ pellet fraction from embryonic chick cartilage was treated with phospholipase C and four glycosyltransferases were studied (galactosyl I-, xylosyl-, *N*-acetylgalactosaminyl- and glucuronosyl I-transferase). The latter three were not significantly affected by phospholipase C, whereas a substantial amount of galactosyltransferase I was lost following digestion (Figure 3.3). The phospholipase C inactivation of galactosyltransferase I was accompanied by a substantial loss of phospholipids in the form of ethanol

–ether extractable phosphate. Analysis of lipid extracts by thin-layer chromatography revealed that phosphatidylcholine and phosphatidylethanolamine were the predominant phospholipids in the particulate fraction of embryonic chick cartilage[246]. These phospholipids were barely detectable in phospholipase C-treated samples[246].

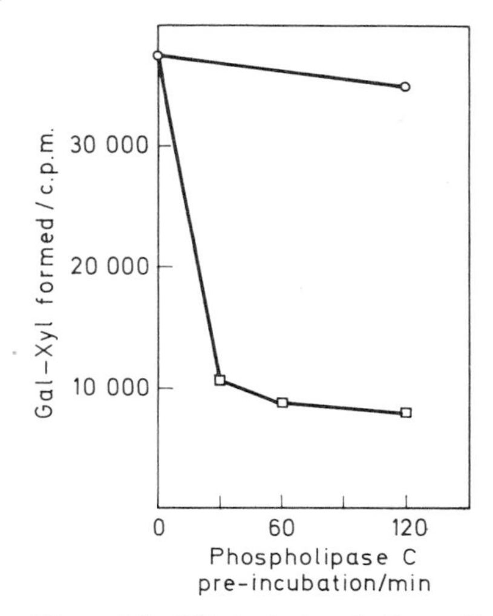


**Figure 3.3** Effect of phospholipase C on galactosyltransferase I. ○, control; □, phospholipase-treated enzyme. (From Schwartz[246], by courtesy of *J. Biol. Chem.*)

The inactivation of galactosyltransferase I by phospholipase C was reversed to varying degrees by the addition of phospholipds and Nonidet P-40[245, 246]. At low concentrations (2–4 μg $ml^{-1}$), lysophosphatidylcholine and lysophosphatidylethanolamine were more effective in restoring activity than phosphatidylcholine and phosphatidylethanolamine. However, the latter phospholipids did restore activity at higher concentrations. There was no significant activation of control particulate enzymes by addition of the phospholipids, but phospholipase A, which generates lysophosphatides, did stimulate galactosyltransferase I activity approximately two-fold[246].

Following partial purification of the solubilised preparation by gel filtration, only that portion of galactosyltransferase I which was excluded on Sephadex was susceptible to the action of phospholipase C[246]. Presumably, the retarded fractions did not contain substantial amounts of phospholipids. However, the enzymes in these included fractions were now probably associated with Nonidet P-40, which may substitute in some way for the lipid.

Further evidence for an obligate detergent substitution for some lipid property was obtained from experiments with more highly purified preparations of galactosyltransferase I. Additional purification of the enzyme was achieved by chromatography on one of a number of affinity matrices[208, 244]. Galactosyltransferase I was eluted from these columns with 1% Nonidet P-40–0.25 M KCl, yielding preparations with about 1100-fold purification over the original homogenate[208]. The galactosyltransferase I in the detergent–salt elute was inactivated by dialysis against detergent-free buffer (Table 3.8); activity was restored upon addition of detergent. These results indicate that galactosyltransferase I is distinctly hydrophobic and is presumably dependent on lipids or detergent for its enzyme activity. As in the case of UDP-glucuronosyltransferase, the activity requirements of the chondroitin sulphate galactosyltransferase appear to be satisfied by a general property of lipids or detergents, rather than an absolute requirement for a particular phospholipid molecule. Thus a certain hydrophobic micro-environment may be necessary for maximal galactosyltransferase I activity.

**Table 3.8 Inactivation of galactosyltransferase I by dialysis and reactivation by NP-40 and KCl**

| *Treatment* | *Activity** /c.p.m. | *Recovery of activity*/% |
|---|---|---|
| 1. Control | 1175 | — |
| 2. Dialysis *vs.* 0.05 M KCl, 1 h | 448 | 38 |
| 3. Dialysis *vs.* 0.05 M KCl, 1 h, followed by dialysis *vs.* 0.25 M KCl–1 % NP-40, 18 h | 1281 | 109 |
| 4. Dialysis *vs.* 0.05 M KCl, 18 h | 185 | 16 |
| 5. Dialysis *vs.* 0.05 M KCl, 18 h, followed by dialysis *vs.* 0.05 M KCl–1 % NP-40, 24 h | 286 | 24 |
| 6. Same as 5, followed by dialysis *vs.* 0.25 M KCl–1 % NP-40, 24 h | 875 | 74 |
| 7. Same as 5, followed by dialysis *vs.* 0.25 M KCl–1 % NP-40, 48 h | 1014 | 86 |

* Activity is expressed as radioactivity incorporated into product (4-*O*-β-D-galactosyl-D-xylose)

## 3.7 CHONDROITIN SULPHATE SYNTHESIS IN CARTILAGE CELLS

Most of the preceding discussion pertains to the examination of complex carbohydrate formation in the *in vitro* situation. In addition, a number of studies concerning the composition and regulation of heteropolysaccharides in intact cells have been conducted. However, the volume of information in this area is too large to review in its entirety here, and will therefore be limited to a discussion of some experiments dealing with chondroitin sulphate synthesis in cells of connective tissue origin.

Increasing knowledge of the biochemistry of cartilage has indicated that there are two identifiable biochemical markers for this tissue: large quantities of chondroitin 4- and/or 6-sulphate and a specific collagen of the type $[\alpha_1(II)]_3$. Chondrocytes are differentiated cells present in cartilage which synthesise these two specialised macromolecules.

Chondrocytes presumably arise from predifferentiated mesenchymal cells. However, the sequence of events that results in this transition is not understood. For the purpose of this discussion, evidence for the synthesis of chondroitin sulphate at different stages of cartilage cell development and studies on factors which affect this biosynthetic process will be presented.

### 3.7.1 Somites

Embryonic notochord and spinal cord have been shown by many workers to promote or 'induce' chondrogenesis of somitic mesoderm[247]. However, Franco-Browder *et al.*[248] have shown that pre-induced somites, spinal cord and notochord all contain chondroitin sulphate, indicating that the enzymatic machinery for synthesis of chondroitin sulphate exists before cartilage induction. Furthermore, under suitable nutritive conditions, embryonic chick somites can synthesise chondroitin sulphate and undergo chondrogenesis *in vitro* in the absence of spinal cord or notochord[249]. It is established, therefore, that in cell culture, chondrogenesis is dependent on cultural conditions as well as specific inducers. The latter greatly increase the rate and amount of matrix accumulation[250-252] and may have a role in enhancing or stabilising a pre-existing chondrogenic bias of somites.

The mechanism by which these inducer tissues may enhance chondrogenesis remains unexplained; however, a number of studies have suggested that certain matrix components may be required for differentiation of somite mesenchyme to cartilage. Collagen is synthesised during the interaction of spinal cord and somitic mesenchyme, and the importance of this finding has been stressed[253, 254]. Minor[255] has demonstrated that, in addition to collagen, proteoglycan granules are synthesised and secreted by the inducer tissue. Similar findings have been observed *in vivo*[256]. Most recently, Kosher *et al.*[257] reported that exogenous embryonic chick chondromucoprotein consisting mainly of chondroitin 4- and 6-sulphate greatly stimulated somite chondrogenesis *in vitro*. Evidence is thus accumulating that extracellular materials influence differentiation; however, the mechanisms involved are still in the realm of speculation.

### 3.7.2 Limb chondrogenesis

Zwilling[258] has described the events occurring in cartilage formation in the limb which serve as a basis for separating limb development into morphogenetic and cyto-differential phases. In this scheme, cells are not differentiated as myoblasts or chondroblasts through stage 25.

Chondroitin sulphate is synthesised by stage 19 bud cells and is probably even produced by the prelimb region of stage 15 chick embryos[259]. Later, during stage 22–23, the rate of $^{35}SO_4$ incorporation into chondroitin 4- and 6-sulphate increases in the central, prechondrogenic area, whereas the peripheral, myogenic region decreases its incorporation of $^{35}SO_4$[248, 260, 261]. Therefore, prior to the appearance of metachromatic extracellular material (cartilage) the limb tissue does possess the machinery for synthesis of chondroitin sulphate.

A process which has been shown to effect limb chondrogenesis *in vivo* is the synthesis of hyaluronate and its subsequent removal by endogenous hyaluronidase. The latter phase of this process is accompanied by differentiation of the cells and formation of cartilage[262, 263]. Subsequently, it was shown that addition of small amounts of hyaluronate to the media of stage 26 chondrocytes inhibited cartilage nodule formation[264]. Conditioned media from limb bud cultures at different stages and from muscle and neural retina cultures significantly suppressed chondrogenic expression of limb bud cultures[265, 266]. However, specific components of the media, such as glycosaminoglycans, were not tested.

The effects of the thymidine analogue 5-bromo-2′-deoxyuridine (BUdR) on limb bud differentiation *in vitro* have recently been summarised by Levitt and Dorfman[267, 268], and the reader is referred to these excellent reviews for a further discussion of this topic.

Briefly, when limb bud cells are grown at high density on plastic dishes or cultured over agar for a brief period and subsequently subcultured at low density on plastic dishes, chondrogenesis occurs as monitored by morphology, metachromasia and incorporation of $^{35}SO_4$ into glycosaminoglycans[269].

If the limb bud cells are exposed to BUdR during the brief period of culture over agar, inhibition of cartilage formation is observed when the cells are subsequently cultured on plastic dishes at low density[270]. The nature of this inhibition has been examined in detail, since it appears to be a specific effect only on the synthesis of chondroitin sulphate proteoglycan while other parameters of cell metabolism remain unaffected.

Of the number of possible ways that chondroitin sulphate proteoglycan synthesis may be diminished, two were shown to be untenable. In order to determine the effect of BUdR treatment on nucleotide precursor pools, the activities of two critical enzymes involved in the synthesis of nucleotide sugar precursors were measured and found to be depressed only slightly[268]. Furthermore, pool sizes of UDP-glucuronic acid, UDP-*N*-acetylgalactosamine and UDP-*N*-acetylglucosamine in control cells and progeny of BUdR-treated cells did not differ significantly.

A decreased level of activity of the enzymes involved in polysaccharide chain formation was also considered a possible explanation for BUdR-inhibition of chondroitin sulphate proteoglycan synthesis. However, the levels of xylosyltransferase, the chain-initiating glycosyltransferase, and *N*-acetylgalactosaminyltransferase, one of the polymerising enzymes, were found to be reduced only 35–40%, whereas overall glycosaminoglycan synthesis was reduced 85–90% after BUdR treatment. In addition, an indirect procedure was used to measure polysaccharide chain synthesis while obviating the need for core protein. Xylose, which appears to act as a chondroitin sulphate chain initiator, reversed the BUdR inhibition by stimulating chondroitin sulphate chain synthesis almost to control levels[270]. These results are consistent with the previous findings of Brett and Robinson[51], who showed that high concentrations of xylose partially overcame the inhibition of chondroitin sulphate synthesis by puromycin in minced cartilage.

Thus, the results indicate that the potential exists in BUdR-treated mesenchymal cells for the formation of chondroitin sulphate chains, and this leads to the postulation that BUdR treatment affects the synthesis of core protein.

In this respect, the nature of the small amount of proteoglycan produced by BUdR-treated cells was compared with that produced in control cells by chromatography on Bio-gel A 50. The elution patterns indicated the presence of at least two kinds of material: a large molecular weight fraction which appeared with the void volume and was the predominant component in the control cells, and a second smaller peak of included material which was present in approximately equal amounts in control and BUdR-treated cells. The larger molecular weight material, which was nearly absent in BUdR-treated cells, was postulated to be cartilage-specific[268].

Similar results have been obtained for cultured chondrocytes treated with BUdR, and sternal cartilage from nanomelic chick embryos[271]. Although not conclusive, there is a suggestion that the inhibition of chondroitin sulphate proteoglycan synthesis by treatment of cartilage cells with BUdR, and in nanomelic cartilage, may be due to a failure to synthesise core protein, or the synthesis of an abnormal core protein which cannot be glycosylated.

### 3.7.3 Chondrocytes

A number of studies have been concerned with the culture of chondrocytes, the postmitotic differentiated cells of cartilage, which are characterised as producing chondroitin sulphate proteoglycan and the cartilage-type collagen. In early studies the cartilage phenotype could be maintained after liberating chondrocytes from their matrix, and growing them in pellets or as monodisperse cells plated on plasma clots[272]. Coon[273] then demonstrated that with appropriate media, in low density cultures, it was possible to obtain stable clones of polygonal cells which formed metachromatic matrix. Eventual loss of the cartilage phenotype with prolonged culture was suggested by Bryan[274, 275] as due to partial overgrowth by non-cartilage cells present in the original population and/or suppression of the cartilage phenotype by non-cartilage cells. However, true cartilage clones still undergo an alteration of morphology after long periods in culture. Methods have also been developed for the growth of cartilage cells in soft agar and in liquid suspension[276]; however, these procedures still allow only short-term culture. Since a permanent line of chondrocytes is difficult to establish, most studies have been performed on primary or secondary cell cultures, which, if grown appropriately, express and maintain the cartilage phenotype.

The synthesis of a glycosaminoglycan identified as chondroitin sulphate was demonstrated in cultured chondrocytes by Glick and Stockdale[277] and confirmed by Nameroff and Holtzer[278]. The matrix produced in chondrocyte cell cultures was shown to contain a protein–polysaccharide with alkali-labile linkages of chondroitin 4-sulphate to the protein core[279]. An additional fraction was isolated with many properties of keratan sulphate. The amino acid composition of chondroitin sulphate proteoglycan isolated from suspension cultures of epiphyseal cartilage was similar to that from chick embryo epiphyseal cartilage and bovine nasal septum[280]. Thus the proteoglycan material produced in culture by chondrocytes appears identical to that found in homologous cartilage tissue.

A number of studies have been directed toward the examination of factors which influence chondrocyte growth and metabolism in culture. The importance of medium composition and density of initial inoculum in formation of cartilage nodules by dissociated chick chondrocyte cultures has been investigated[281]. Ascorbic acid, which is required for extensive collagen synthesis[282], enhances both matrix accumulation and the organisation of matrix around cultured cartilage cells[281, 283]. Thyroxine and low oxygen tension have been shown to promote chondrocyte-like morphology and high levels of $^{35}SO_4$ uptake in embryonic chick cartilage cultures[284].

The effect of hormones on cartilage growth and function has been the object of a number of reports. A potent circulating serum growth factor called sulphation factor, or more recently, somatomedin[285], was shown to mediate growth hormone-enhanced sulphate incorporation by cartilage *in vitro*[286]. The possible modulation of somatomedin action on embryonic chick cartilage by cyclic AMP[287, 288] and fatty acids and ketones[289, 290] was investigated. Direct treatment of cultures of embryonic chick chondrocytes with ovine and some other mammalian growth hormones resulted primarily in increased acid mucopolysaccharide synthesis[291, 292]. This first demonstration of a mammalian growth hormone effect on cartilage expression was explained on the basis of the hormone's ability to alter the availability of the essential amino acid valine, which may play a regulatory role in the synthesis of sulphated mucopolysaccharides[293].

Factors released into the media of growing cells have been postulated to play important roles in regulating cellular activities such as contact inhibition[294], cell aggregation[295], growth[296] and differentiation. Studies on the effects of chondrocyte conditioned media (i.e. media that have been conditioned by exposure to cells) on cartilage differentiation has indicated that there may be specialised factors present that promote differentiation of this cell type[297, 298]. In other studies by these authors, chick chondromucoprotein, chondroitin sulphate preparations from whale and shark cartilage, and soluble calf skin collagen did not affect sulphate incorporation, as did conditioned media[298]. To an extent this finding is in contrast to the observations of Nevo and Dorfman[299] that the synthesis of chondroitin sulphate by suspension-cultured chondrocytes was stimulated by chondroitin sulphate proteoglycan and other polyanions. The enhancement of chondroitin sulphate production by exogenous chondroitin sulphate has also been demonstrated in monolayer cultures of chick sternal chondrocytes[300, 301]. In one of these studies[300], the stimulatory effect on sulphated mucopolysaccharide in chondrocyte monolayers was found to be time-specific with regard to the growth cycle, and time-dependent in that a latent period was required for maximal effect.

Another glycosaminoglycan, hyaluronic acid, has been reported to cause the opposite effect on chondrocyte expression. Wiebkin and Muir[302] observed an inhibition of sulphate incorporation in isolated pig laryngeal chondrocytes by hyaluronic acid. Similarly, added hyaluronic acid specifically inhibited the synthesis of chondroitin sulphates A and C as well as the synthesis of collagen in cultured chick embryo chondrocytes[303]. Under the conditions of culture employed, a significant inhibition of sulphated mucopolysaccharide synthesis by exogenous hyaluronic acid in suspension cultures[299] and monolayer cultures of chondrocytes[304] has not been observed.

Manipulation of the environment around cells by the action of enzymes has also been suggested to influence the synthetic functions of chondrocytes. There was a marked increase in proteoglycan synthesis after the removal of chondroitin sulphate by the action of hyaluronidase on the cartilaginous portions of chick embryo tibiae grown in organ culture[305]. This same response probably occurs *in vivo* in cartilage of young animals, since intravenous injection of papain caused a loss of chondroitin sulphate from cartilage, which was rapidly replaced[306]. In culture, phospholipase C suppressed cartilage matrix synthesis by chondrocytes, as measured by failure to exhibit metachromasia and a marked inhibition of $^{35}SO_4$ incorporation into chondroitin sulphate[307].

All of these studies indicate that the matrix or microenvironment around cells exerts some influence over the synthetic and secretory processes of the chondrocyte. The mechanisms by which the processes are regulated remain to be elucidated.

As previously mentioned, BUdR selectively interferes with either differentiation or the expression of the differentiated state in several systems. In chondrocytes, differentiated characteristics are reversibly suppressed by treatment with BUdR. In addition to changes in cell morphology[308, 309], the production of cartilage matrix by cultured chondrocytes is reduced after exposure to BUdR for varying lengths of time[310, 311].

As previously discussed in the case of the predifferentiated limb bud cultures, the mechanism of BUdR inhibition of glycosaminoglycan synthesis is unknown. It has been postulated that the presence of BUdR alters the cell membrane, by direct or indirect modification of the glycoproteins and glycolipids of the cell surface. In this regard, Chi *et al.*[312] have shown that BUdR-suppressed chondrocytes accumulate glycoprotein(s) not found in normal chondrocytes, which may be similar to the glycoproteins detected on the surfaces of viral transformed cells. The significance of this finding in relation to the inhibition of differentiated functions by BUdR in chondrocytes (other than an additional consequence of the BUdR-treatment), remains to be established.

The activity levels of enzymes involved in the production of nucleotide sugars and sulphate activation are reduced in chondrocytes exposed to BUdR[313, 314]. A decrease in the activity of some of the glycosyltransferases that catalyse chondroitin sulphate chain synthesis has been observed. However, the degree of inhibition and the length of time required to achieve maximal inhibition indicate that reduced glycosyltransferase activity is not the initial event responsible for inhibition of overall glycosaminoglycan synthesis by BUdR[268, 304]. In addition, with the use of xylose or β-xylosides as chain-initiators in BUdR-treated chondrocytes, it was found that the cells still retain a high capacity for the synthesis of chondroitin sulphate chains[55, 268]. These results tend to indicate that the primary site of BUdR inhibition of glycosaminoglycan production in chondrocytes may also involve the synthesis of core protein.

## References

1. Lindahl, U. (1976). In *International Review of Science, Organic Chemistry Series Two*, Vol. 7, *Carbohydrates* (G. O. Aspinall, editor) (London: Butterworths)

2. Schachter, H. and Rodén, L. (1972). In *Metabolic Conjugation and Metabolic Hydrolysis*, Vol. 3, 1 (W. H. Fishman, editor) (New York and London: Academic Press)
3. Rodén, L. (1970). In *Metabolic Conjugation and Metabolic Hydrolysis*, Vol. 2, 34 (W. H. Fishman, editor) (New York and London: Academic Press)
4. Robbins, P. W., Bray, D., Dankert, M. and Wright, A. (1967). *Science*, **158,** 1536
5. Roseman, S. (1968). In *Biochemistry of Glycoproteins and Related Substances—Cystic Fibrosis, Part II*, 244 (E. Rossi and E. Stoll, editors) (Basel: Karger)
6. Stoolmiller, A. C. and Dorfman, A. (1969). In *Comprehensive Biochemistry*, Vol. 17 241 (M. Florkin and E. H. Stotz, editors) (Amsterdam, London, New York: Elsevier
7. Hassid, W. Z. (1970). In *The Carbohydrates*, Vol. IIA, 301 (W. Pigman and D. Horton editors) (New York and London: Academic Press)
8. Nikaido, H. and Hassid, W. Z. (1971). *Adv. Carbohyd. Chem. Biochem.*, **26,** 351
9. Heath, E. C. (1971). *Ann. Rev. Biochem.*, **40,** 29
10. Behrens, N. H., Parodi, A. J. and Leloir, L. F. (1972). *In Biochemistry of the Glycosidic Linkage*, 195 (R. Piras, and H. G. Pontis, editors) (New York and London: Academic Press)
11. Dorfman, A. (1974). *Mol. Cell. Biochem.*, **4,** 45
12. Lennarz, W. J. (1975). *Science*, **188,** 986
13. Krisman, C. R. (1973). *Ann. N.Y. Acad. Sci.*, **210,** 81
14. Krisman, C. R. (1972). In *Biochemistry of the Glycosidic Linkage*, 629 (R. Piras and H. G. Pontis, editors) (New York and London: Academic Press)
15. Leloir, L. F. (1972). In *Biochemistry of the Glycosidic Linkage*, 1 (R. Piras and H. G Pontis, editors) (New York and London: Academic Press)
16. Hestrin, S., Shapiro, A. and Aschner, M. (1953). *Biochem. J.*, **37,** 450
17. Lindahl, U., Bäckström, G., Malmström, A. and Fransson, L.-Å. (1972). *Biochem. Biophys. Res. Commun.*, **46,** 985
18. Höök, M., Lindahl, U., Bäckström, G., Malmström, A. and Fransson, L.-Å. (1974) *J. Biol. Chem.*, **249,** 3908
19. Larsen, B. and Haug, A. (1971). *Carbohyd. Res.*, **17,** 287
20. Haug, A. and Larsen, B. (1971). *Carbohyd. Res.*, **17,** 297
21. Jacobson, B., and Davidson, E. A. (1972). *J. Biol. Chem.*, **237,** 638
22. Lin, T. C. and Hassid, W. Z. (1966). *J. Biol. Chem.*, **241,** 5284
23. Su, J. C. and Hassid, W. Z. (1962). *Biochemistry*, **1,** 468
24. Burgos, J. Hemming, F. W., Purnock, J. F. and Morton, R. A. (1963). *Biochem. J.* **88,** 470
25. Behrens, N. H. and Leloir, L. F. (1970). *Proc. Nat. Acad. Sci. USA*, **66,** 153
26. Wardi, A. H., Allen, W. S., Turner, D. L. and Stary, Z. (1966). *Arch. Biochem. Biophys.*, **177,** 44
27. Wardi, A. H., Allen, W. S., Turner, D. L. and Stary, Z. (1969). *Biochim. Biophys. Acta* **192,** 151
28. Katzman, R. L. (1971). *J. Neurochem.*, **18,** 1187
29. Hassid, W. Z. (1972). In *Biochemistry of the Glycosidic Linkage*, 315 (R. Piras and H. G. Pontis, editors) (New York and London: Academic Press)
30. Warren, L. (1972). In *Glycoproteins*, 1097 (A. Gottschalk, editor) (Amsterdam London, New York: Elsevier)
31. Bruns, F. H., Noltman, E. and Willemsen, A. (1958). *Biochem. J.*, **330,** 411
32. Maitra, U. S. and Ankel, H. (1971). *Proc. Nat. Acad. Sci. USA*, **68,** 2660
33. Nelsestuen, G. L. and Kirkwood, S. (1971). *J. Biol. Chem.*, **246,** 7533
34. Maitra, U. S. and Ankel, H. (1973). *J. Biol. Chem.*, **248,** 1477
35. Maley, F. and Maley, G. F. (1959). *Biochim. Biophys. Acta*, **31,** 577
36. Perlman, R. L., Telser, A. and Dorfman, A. (1964). *J. Biol. Chem.*, **239,** 3623
37. Jacobson, B. and Davidson, E. A. (1963). *J. Biol. Chem.*, **237,** 638
38. Strominger, J. L., Kalckar, H. M., Axelrod, J. and Maxwell, E. S. (1954). *J. Amer. Chem. Soc.*, **76,** 79
39. Strominger, J. L., Maxwell, E. S., Axelrod, J. and Kalckar, H. M. (1957). *J. Biol. Chem.*, **224,** 79
40. Zalitis, J. and Feingold, D. S. (1969). *Arch. Biochem. Biophys.*, **132,** 457
41. Nelsesteun, G. L. and Kirkwood, S. (1971). *J. Biol. Chem.*, **246,** 3828
42. Schutzbach, J. S. and Feingold, D. S. (1970). *J. Biol. Chem.*, **245,** 2476

43. Feingold, D. S. (1972). In *Biochemistry of the Glycosidic Linkage*, 79 (R. Piras and H. G. Pontis, editors) (New York and London: Academic Press)
44. Ginsburg, V. (1960). *J. Biol. Chem.*, **235,** 2196
45. Ginsburg, V. (1961). *J. Biol. Chem.*, **236,** 2389
46. Ginsburg, V. (1964). *Advan. Enzymol.*, **26,** 35
47. Foster, E. W. and Ginsburg, V. (1961). *Biochim. Biophys. Acta*, **54,** 376
48. Haselkorn, R. and Rothman-Denes, L. B. (1973). *Ann. Rev. Biochem.*, **42,** 397
49. Helting, T. and Rodén, L. (1969). *J. Biol. Chem.*, **244,** 2790
50. Okayama, M., Kimata, K. and Suzuki, S. (1971). *Seikagaku*, **43,** 454 Abstr.
51. Brett, M. J. and Robinson, H. C. (1971). *Proc. Aust. Biochem. Soc.*, **4,** 92 Abstr.
52. Okayama, M. and Lowther, D. A. (1972). *Proc. Aust. Biochem. Soc.*, **6,** 75 Abstr.
53. Okayama, M., Kimata, K. and Suzuki, S. (1973). *J. Biochem. (Tokyo)*, **74,** 1069
54. Levitt, D. and Dorfman, A. (1973). *Proc. Nat. Acad. Sci. USA*, **70,** 2201
55. Schwartz, N. B., Ho, Pei-Lee, Levitt, D. and Dorfman, A. (1974). *Fed. Proc.*, **33,** 1553 Abstr.
56. Schwartz, N. B., Galligani, L., Ho, Pei-Lee, and Dorfman, A. (1974). *Proc. Nat. Acad. Sci. USA*, **71,** 4047
57. Goldemberg, S. H. (1962). *Biochim. Biophys. Acta*, **56,** 357
58. Leloir, L. F. (1964). *Plenary Lecture, 6th International Congress of Biochemistry, New York, I.U.B.*, 15
59. Neuberger, A., Gottschalk, A., Marshall, R. D. and Spiro, R. G. (1972). In *Glycoproteins*, 450 (A. Gottschalk, editor) (Amsterdam, London, New York: Elsevier)
60. Lindahl, U. and Rodén, L. (1972). In *Glycoproteins*, 491 (A. Gottschalk, editor) (Amsterdam, London, New York: Elsevier)
61. Baker, J. R., Rodén, L. and Stoolmiller, A. C. (1972). *J. Biol. Chem.*, **247,** 3838
62. Spiro, R. G. (1972). In *Glycoproteins*, 964 (A. Gottschalk, editor) (Amsterdam, London, New York: Elsevier)
63. McGuire, E. J. and Roseman, S. (1967). *J. Biol. Chem.*, **242,** 3745
64. Hagopian, A. and Eylar, E. H. (1968). *Arch. Biochem. Biophys.*, **126,** 785
65. Hagopian, A. and Eylar, E. H. (1968). *Arch. Biochem. Biophys.*, **128,** 422
66. Hagopian, A. and Eylar, E. H. (1969). *Arch. Biochem. Biophys.*, **129,** 447
67. Hagopian, A. and Eylar, E. H. (1969). *Arch. Biochem. Biophys.*, **129,** 515
68. Schwartz, N. B. and Rodén, L. (1974). *Carbohyd. Res.*, **37,** 167
69. Stoolmiller, A. C., Horwitz, A. L. and Dorfman, A. (1972). *J. Biol. Chem.*, **247,** 3525
70. Behrens, N. H., Carminatti, H., Staneloni, R. J., Leloir, L. F. and Cantarella, A. I. (1973). *Proc. Nat. Acad. Sci. USA*, **70,** 3390
71. Hsu, A. F., Baynes, J. W. and Heath, E. C. (1974). *Proc. Nat. Acad. Sci. USA*, **71,** 2391
72. Montgomery, R. (1970). In *The Carbohydrates*, Vol. IIB, 627 (W. Pigman and D. Horton, editors) (New York and London: Academic Press)
73. Montgomery, R. (1972). In *Glycoproteins*, 518 (A. Gottschalk, editor) (Amsterdam, London, New York: Elsevier)
74. Van den Hamer, C. J. A., Morell, A. G., Scheinberg, I. H., Hickman, J. and Ashwell, G. (1970). *J. Biol. Chem.*, **245,** 4397
75. Morell, A. G., Gregoriadis, G., Scheinberg, I. H., Hickman, J. and Ashwell, G. (1971). *J. Biol. Chem.*, **246,** 1461
76. Pricer, W. E., Jr. and Ashwell, G. (1971). *J. Biol. Chem.*, **246,** 4825
77. Morell, A. G. and Scheinberg, I. H. (1972). *Biochem. Biophys. Res. Commun.*, **48,** 808
78. Hudgin, R. L., Pricer, W. E., Jr., Ashwell, G., Stockert, R. J. and Morell, A. G. (1974). *J. Biol. Chem.*, **249,** 5536
79. Watkins, W. M. (1972). In *Glycoproteins*, 830 (A. Gottschalk, editor) (Amsterdam, London, New York: Elsevier)
80. Bosmann, H. B. (1972). *J. Neurochemistry*, **19,** 763
81. Bosmann, H. B. (1970). *Eur. J. Biochem.*, **14,** 33
82. Bosmann, H. B. and Eylar, E. H. (1968). *Biochem. Biophys. Res. Commun.*, **30,** 89
83. Bosmann, H. B. and Eylar, E. H. (1968). *Biochem. Biophys. Res. Commun.*, **33,** 340
84. Bosmann, H. B., Hagopian, A. and Eylar, E. H. (1968). *Arch. Biochem. Biophys.*, **128,** 470
85. Bosmann, H. B. and Martin, S. S. (1969). *Science*, **164,** 190
86. Hagopian, A. and Eylar, E. H., (1969). *Arch. Biochem. Biophys.*, **129,** 515

87. Johnston, I. R., McGuire, E. J., Jourdian, G. W. and Roseman S., (1966). *J. Biol. Chem.*, **241,** 5735
88. McGuire, E. J., Jourdian, G. W., Carlson, D. M. and Roseman, S., (1965). *J. Biol. Chem.*, **240,** PC4112
89. McGuire, E. J. and Roseman, S., (1967). *J. Biol. Chem.*, **242,** 3745
90. Spiro, M. J. and Spiro, R. G., (1968). *J. Biol. Chem.*, **243,** 6529
91. Maley, G., and Maley, G. F., (1959). *Biochim. Biophys. Acta*, **31,** 577
92. Maley, F., McGarrahan, J. F. and DelGiacco, R., (1966). *Biochem. Biophys. Res Commun.*, **23,** 85
93. McGarrahan, J. F. and Maley, F., (1962). *J. Biol. Chem.*, **237,** 2458
94. Maley, F., Tarentino, A. L., McGarrahan, J. F. and DelGiacco, (1968). *J. Biol. Chem.* **107,** 637
95. Kornfeld, R. and Brown, D. H. (1963). *Biochem. J.*, **238,** 1604
96. Rabinowitz, M. and Goldberg, I. H., (1963). *J. Biol. Chem.*, **238,** 1801
97. Preiss, J. (1972). In *Biochemistry of the Glycosidic Linkage*, 517 (R. Piras and H. G Pontis, editors) (New York and London: Academic Press)
98. Hawker, J. S., Ozbun, J. L. and Preiss, J., (1972). In *Biochemistry of the Glycosidic Linkage*, 529 (R. Piras and H. G. and Pontis, editors) (New York and London Academic Press)
99. Brodbeck, U., Denton, W. L., Tanahashi, N. and Ebner, K. E. (1967). *J. Biol. Chem* **242,** 1391
100. Brew, K., Vanaman, T. C. and Hill, R. L. (1968). *Proc. Nat. Acad. Sci. USA*, **59,** 491
101. Hudgin, R. L. and Schachter, H. (1971). *Can. J. Biochem.*, **49,** 838
102. Schanbacher, F. L. and Ebner, K. E. (1970). *J. Biol. Chem.*, **245,** 5057
103. Fitzgerald, D. K., Brodbeck, U., Kiyosawa, I., Mawal, R., Colvin, B. and Ebner K. E. (1970). *J. Biol. Chem.*, **245,** 2103
104. Barker, R., Olsen, K. W., Shaper, J. H. and Hill, R. L. (1972). *J. Biol. Chem.*, **247,** 713
105. Klee, W. A. and Klee, C. B. (1972). *J. Biol. Chem.*, **247,** 2336
106. Magee, S. C., Mawal, R. and Ebner, K. E. (1974). *Biochemistry*, **13,** 99
107. Magee, S. C., Mawal, R. and Ebner, K. E. (1973). *J. Biol. Chem.*, **248,** 7565
108. Magee, S. C. and Ebner, K. E. (1973). *Fed. Proc.* **32,** 541
109. Magee, S. C. and Ebner, K. E. (1974). *J. Biol. Chem.*, **249,** 6992
110. Basu, S., Kaufman, B. and Roseman, S. (1968). *J. Biol. Chem.*, **243,** 5802
111. Roseman, S. (1970). *Chem. Phys. Lipids*, **5,** 270
112. Cumar, F. A., Fishman, P. H. and Brady, R. O. (1971). *J. Biol. Chem.*, **246,** 5075
113. Cumar, F. A., Tallman, J. F. and Brady, R. O. (1972). *J. Biol. Chem.*, **247,** 2322
114. Caputto, R., Maccioni, H. J. and Arce, A. (1974). *Mol. Cell. Biochem.*, **4,** 97
115. Brady, R. O. and Fishman, P. H. (1974). *Biochim. Biophys. Acta*, **355,** 121
116. Hill, R. L. (1974). Unpublished results
117. Schachter, H., McGuire, E. J. and Roseman, S. (1971). *J. Biol. Chem.*, **246,** 5321
118. Raizada, M. K. and Schutzbach, J. S. (1973). *Abstr. 166th Amer. Chem. Soc. Meeting Biol.*, 214
119. Helting, T. and Rodén, L. (1969). *J. Biol. Chem.*, **244,** 2799
120. Roseman, S., Carlson, D. M., Jourdian, G. W., McGuire, E. J., Kaufman, B., Basu S. and Bartholomew, B. (1966). *Methods Enzymol.*, **8,** 354
121. Telser, A., Robinson, H. C. and Dorfman, A. (1966). *Arch. Biochem. Biophys.*, **116,** 45
122. Horwitz, A. L. and Dorfman, A. (1968). *J. Cell. Biol.*, **38,** 258
123. Horwitz, A. L. (1972). *Ph.D. thesis*, University of Chicago
124. Hudgin, R. L. and Schachter, H. (1972). *Can. J. Biochem.*, **50,** 1024
125. Jabbal, I. and Schachter, H. (1971). *J. Biol. Chem.*, **246,** 5154
126. Whitehead, J. S., Bella, A., Jr. and Kim, Y. S. (1974). *J. Biol. Chem.*, **249,** 3442
127. Whitehead, J. S., Bella, A., Jr. and Kim, Y. S. (1974). *J. Biol. Chem.*, **249,** 3448
128. Ziderman, D., Gompertz, S., Smith, Z. G. and Watkins, W. M. (1967). *Biochem Biophys. Res. Commun.*, **29,** 56
129. Race, C., Ziderman, D. and Watkins, W. M. (1968). *Biochem. J.*, **107,** 733
130. Race, C. and Watkins, W. M. (1969). *Biochem. J.*, **114,** 86P
131. Race, C. and Watkins, W. M. (1970). *FEBS Lett.*, **10,** 279
132. Kobata, A., Grollman, E. F. and Ginsburg, V. (1968). *Biochem. Biophys. Res. Commun.* **32,** 272
133. Schenkel-Brunner, H. and Tuppy, H. (1970). *Eur. J. Biochem.*, **17,** 218

134. Carlson, D. M., Iyer, R. N. and Mayo, J. (1970). In *Blood and Tissue Antigens*, 229 (D. Aminoff, editor) (New York: Academic Press)
135. McGuire, E. J. (1970). In *Blood and Tissue Antigens*, 461 (D. Aminoff, editor) (New York: Academic Press)
136. Richmond, M. E., DeLuca, S. and Silbert, J. E. (1973). *Biochemistry*, **12**, 3898
137. Richmond, M. E., DeLuca, S. and Silbert, J. E. (1973). *Biochemistry*, **12**, 3904
138. DeLuca, S., Richmond, M. E. and Silbert, J. E. (1973). *Biochemistry*, **12**, 3911
139. Tuppy, H. and Gottschalk, A. (1972). In *Glycoproteins*, 403 (A. Gottschalk, editor) (Amsterdam, London, New York: Academic Press)
140. Svennerholm, L. (1964). *J. Lipid Res.*, **5**, 145
141. Barry, G. T. and Goebel, W. F. (1957). *Nature*, **179**, 206
142. Barry, G. T. (1958). *J. Exp. Med.*, **107**, 507
143. Barry, G. T., Abbot, V. and Tsai, T. (1962). *J. Gen. Microbiol.*, **29**, 335
144. Carlson, D. M. (1968). *J. Biol. Chem.*, **243**, 616
145. Hopwood, J. J. (1972). *Ph.D. thesis*, Monash University
146. Hopwood, J. J. and Robinson, H. C. (1975). *Biochem. J.*, **141**, 517
147. Percival, E. and McDowell, R. H. (1967). In *Chemistry and Enzymology of Marine Algal Polysaccharides* (London and New York: Academic Press)
148. Percival, E. (1970). In *The Carbohydrates*, Vol. IIB, 537 (W. Pigman and D. Horton, editors) (New York: and London Academic Press)
149. Weicker, H. and Jeanloz, R. W. (1967). *Fed. Proc.*, **26**, 607
150. Weicker, H. and Grässlin, D. (1966). *Nature*, **212**, 715
151. Tomoda, M. and Murayama, K. (1965). *Jap. J. Exp. Med.*, **35**, 159
152. Waechter, C. J., Lucas, J. J. and Lennarz, W. J. (1974). *Biochem. Biophys. Res. Commun.*, **56**, 343
153. Marcus, D. M. (1969). *New Engl. J. Med.*, **280**, 944
154. Kornfeld, R., Keller, J., Baenziger, J. and Kornfeld, S. (1971). *J. Biol. Chem.*, **246**, 3259
155. Dawson, G. and Clamp, J. R. (1968). *Biochem. J.*, **107**, 341
156. Kabasawa, R. and Hirs, C. (1972). *J. Biol. Chem.*, **247**, 1610
157. Yang, H.-J. and Hakomori, S.-I. (1971). *J. Biol. Chem.*, **246**, 1192
158. Wagh, P. V., Bornstein, I. and Winzler, R. J. (1969). *J. Biol. Chem.*, **244**, 658
159. Jamieson, G. A., Jett, M. and DeBernardo, S. L. (1971). *J. Biol. Chem.*, **246**, 3686
160. Li, Y.-T., Li, S.-C. and Dawson, G. (1967). *Biochim. Biophys. Acta*, **260**, 1163
161. Hakomori, S.-I., Siddiqui, B., Li, Y.-T., Li, S.-C. and Hellerquist, C. G. (1971). *J. Biol. Chem.*, **246**, 2271
162. Rouser, G. and Yamamoto, A. (1969). In *Handbook of Neurochemistry*, Vol. 1, 121 (A. Lajtha, editor) (New York: Plenum)
163. Stoffyn, P., Stoffyn, A. and Hauser, G. (1972). *Trans. Amer. Soc. Neurochem.*, **3**, 125
164. Laine, R. A., Sweeley, C. C., Li, Y.-T., Kisic, A. and Rapport, M. M. (1972). *Fed. Proc.*, **31**, 437
165. Eto, T., Ichikawa, Y., Nishimura, K., Ando, S. and Yamakawa, T. (1968). *J. Biochem (Tokyo)*, **64**, 205
166. Siddiqui, B. and Hakomori, S.-I. (1971). *J. Biol. Chem.*, **246**, 5766
167. Smith, E. E., Taylor, P. M. and Whelan, W. J. (1968). In *Carbohydrate Metabolism and its Disorders*, Vol. 1, 89 (F. Dickens, P. J. Randle and W. J. Whelan, editors) (London and New York: Academic Press)
168. Ryman, B. E. and Whelan, W. J. (1971). *Adv. Enzymol.*, **34**, 285
169. Spiro, R. G. (1967). *J. Biol. Chem.*, **242**, 4813
170. Jeanloz, R. W. (1970). In *The Carbohydrates*, Vol. IIB, 589 (W. Pigman and D. Horton, editors) (New York and London: Academic Press)
171. Wiegandt, H. (1970). *Chem. Phys. Lipids*, **5**, 192
172. Sukeno, T., Tarentino, A. L., Plummer, T. H., Jr. and Maley, F. (1971). *Biochem. Biophys. Res. Commun.*, **45**, 219
173. Makino, M. and Yamashina, I. (1966). *J. Biochem. (Tokyo)*, **60**, 262
174. Ward, K., Jr., and Seib, P. A. (1970). In *The Carbohydrates*, Vol. IIA, 413 (W. Pigman and D. Horton, editors) (New York and London: Academic Press)
175. Aston, W. P., Donald, A. S. R. and Morgan, W. T. J. (1968). *Biochem. Biophys. Res. Commun.*, **30**, 1

176. Clamp, J. R., Dawson, G. and Franklin, E. C. (1968). *Biochem. J.*, **110**, 385
177. Spiro, R. G. (1964). *J. Biol. Chem.*, **239**, 567
178. Soderling, T. R., Hickenbottom, J. P., Reimann, E. M., Hunkeler, F. L., Walsh, D. A and Krebs, E. G. (1970). *J. Biol. Chem.*, **245**, 6317
179. Sanada, Y. and Segal, H. L. (1971). *Biochem. Biophys. Res. Commun.*, **45**, 1159
180. Issa, H. A. and Mendicino, J. (1973). *J. Biol. Chem.*, **248**, 685
181. Rapport, M. M., Weissmann, B., Linker, A. and Meyer, K. (1951). *Nature*, **168**, 99
182. Weissmann, B. and Meyer, K. (1954). *J. Amer. Chem. Soc.*, **76**, 1753
183. Hebting, J. (1914). *Biochem. Z.*, **63**, 353
184. Fransson, L.-Å. and Rodén, L. (1967). *J. Biol. Chem.*, **242**, 4161
185. Fransson, L.-Å. and Rodén, L. (1967). *J. Biol. Chem.*, **242**, 4170
186. Hoffman, P., Linker, A. and Meyer, K. (1956). *Science*, **124**, 1252
187. Cifonelli, J. A., Ludowieg, J. and Dorfman, A. (1958). *J. Biol. Chem.*, **233**, 541
188. Satake, M. and Masamune, H. (1958). *Tohoku J. Exp. Med.*, **68**, 114
189. Satake, M. (1958). *Tohoku J. Exp. Med.*, **68**, 385
190. Stoffyn, P. J. and Jeanloz, R. W. (1960). *J. Biol. Chem.*, **235**, 2507
191. Lindahl, U. and Rodén, L. (1966). *J. Biol. Chem.*, **241**, 2113
192. Rodén, L. and Smith, R. (1966). *J. Biol. Chem.*, **241**, 5949
193. Rodén, L. and Armand, G. (1966). *J. Biol. Chem.*, **241**, 65
194. Lindahl, U. (1966). *Biochim. Biophys. Acta*, **130**, 361
195. Lindahl, U. (1966). *Arkiv. Kemi*, **26**, 101
196. Malmström, A. and Fransson, L.-Å. (1971). *FEBS Lett.*, **16**, 105
197. Grebner, E. E., Hall, C. W. and Neufeld, E. F. (1966). *Biochem. Biophys. Res. Commun.*, **22**, 672
198. Grebner, E. E., Hall, C. W. and Neufeld, E. F. (1966). *Arch. Biochem. Biophys.*, **116**, 391
199. Robinson, H. C., Telser, A. and Dorfman, A. (1966). *Proc. Nat. Acad. Sci. USA*, **56**, 1859
200. Telser, A. (1968). *Ph.D. thesis*, University of Chicago
201. Kimura, J. H. and Caplan, A. I. (1974). *Cell Biol. Abstr.*, **63**, 167A
202. Rodén, L. (1972). Unpublished results
203. Seno, N., Meyer, K., Anderson, B. and Hoffmann, P. (1965). *J. Biol. Chem.*, **240**, 1005
204. Mathews, M. B. and Cifonelli, J. A. (1965). *J. Biol. Chem.*, **240**, 4140
205. Anderson Bray, B., Lieberman, R. and Meyer. K. (1967). *J. Biol. Chem.*, **242**, 3373
206. Grebner, E. E. and Neufeld, E. F. (1969). *Biochim. Biophys. Acta*, **192**, 347
207. Baker, J. R., Rodén, L. and Yamagata, S. (1971). *Biochem. J.*, **125**, 93P
208. Schwartz, N. B. and Rodén, L. (1975). *J. Biol. Chem.*, **250**, 5200
209. Schwartz, N. B., Rodén, L. and Dorfman, A. (1974). *Biochem. Biophys. Res. Commun.*, **56**, 717
210. Schwartz, N. B. (1975). *FEBS Lett.*, **49**, 342
211. Yamagata, S. and Rodén, L. (1974). Unpublished results
212. Brandt, E. A., Distler, J. and Jourdian, G. W. (1969). *Proc. Nat. Acad. Sci. USA*, **64**, 374
213. Helting, T. (1972). *J. Biol. Chem.*, **247**, 4327
214. Silbert, J. E. (1964). *J. Biol. Chem.*, **239**, 1310
215. Helting, T. and Lindahl, U. (1972). *Acta Chem. Scand.*, **26**, 3515
216. Steck, T. L. (1972). In *Membrane Molecular Biology*, 27 (C. F. Fox and A. Keith, editors) (Stamford, Conn.: Sinauer Associates)
217. Helting, T. (1971). *J. Biol. Chem.*, **246**, 815
218. Yamazaki, M. and Hayaishi, O. (1968). *J. Biol. Chem.*, **243**, 2934
219. Rothfield, L. and Romeo, D. (1971). In *Structure and Function of Biological Membranes*, 251 (L. I. Rothfield, editor) (New York and London: Academic Press)
220. Pradal, M. B., Louisot, P. and Got, R. (1971). *Z. Naturforsch.* **26b**, 625
221. Jato-Rodriguez, J. J. and Mookerjea, S. (1974). *Arch. Biochem. Biophys.*, **162**, 281
222. Den, H., Kaufman, B. and Roseman, S. (1970). *J. Biol. Chem.*, **245**, 6607
223. Basu, S., Kaufman, B. and Roseman, S. (1973). *J. Biol. Chem.*, **248**, 1388
224. Hauser, G. (1967). *Biochem. Biophys. Res. Commun.*, **28**, 502
225. Yip, M. C. M. (1972). *Biochim. Biophys. Acta*, **273**, 374
226. Shah, S. N. (1973). *Arch. Biochem. Biophys.*, **159**, 143
227. Mookerjea, S. and Yung, J. W. M. (1974). *Biochem. Biophys. Res. Commun.*, **57**, 815

228. Gregory, D. H., II, and Strickland, R. D. (1973). *Biochim. Biophys. Acta*, **327**, 36
229. Leuders, K. K. and Kuff, E. L. (1967). *Arch. Biochem. Biophys.*, **120**, 198
230. Winsnes, A. (1969). *Biochim. Biophys. Acta*, **191**, 279
231. Mulder, G. J. (1970). *Biochem. J.*, **117**, 319
232. Graham, A. B. and Wood, G. C. (1972). *Biochim,. Biophys. Acta*, **276**, 392
233. Graham, A. B. and Wood, G. C. (1969). *Biochem. Biophys. Res. Commun.*, **37**, 567
234. Attwood, D., Graham, A. B. and Wood, G. C. (1971). *Biochem. J.*, **123**, 875
235. Hänninen, O. and Puukka, R. (1971). *Chem.–Biol. Interact.*, **3**, 282
236. Vessey, D. A. and Zakim, D. (1971). *J. Biol. Chem.*, **246**, 4649
237. Winsnes, A. (1972). *Biochim. Biophys. Acta*, **284**, 394
238. Eletr, S., Zakim, D. and Vessey, D. A. (1973). *J. Mol. Biol.*, **78**, 351
239. Graham, A. B., Wood, G. C. and Woodcock, B. G. (1972). *Biochem. J.*, **129**, 22P
240. Graham, A. B., Woodcock, B. G. and Wood, G. C. (1974). *Biochem. J.*, **137**, 567
241. Wood, G. C. and Woodcock, B. G. (1970). *J. Pharm. Pharmacol.*, **22**, 605
242. Woodcock, B. G. and Wood, G. C. (1971). *Biochem. Pharmacol.*, **20**, 2703
243. Rodén, L., Baker, J. R., Helting, T., Schwartz, N. B., Stoolmiller, A. C., Yamagata, S. and Yamagata, T. (1972). *Methods Enzymol.*, **28**, 638
244. Schwartz, N. B., Rodén, L. and Dorfman, A. (1973). *Abstr. Ninth Int. Congr. Biochem.*, 427 Abst.
245. Schwartz, N. B. and Rodén, L. (1972). *Fed. Proc.*, **31**, 434Abst.
246. Schwartz, N. B. (1976). *J. Biol. Chem.*, in press
247. Lash, J. W. (1968). *J. Cell. Physiol.*, **72**, 35
248. Franco-Browder, S., DeRydt, J. and Dorfman, A. (1963). *Proc. Nat. Acad. Sci. USA*, **49**, 643
249. Lash, J. W., Glick, M. C. and Madden, J. W. (1964). *Nat. Cancer Inst. Monogr.*, **13**, 39
250. Lash, J. W. (1967). *J. Exp. Zool.*, **165**, 47
251. Ellison, M. L. and Lash, J. W. (1971). *Develop. Biol.*, **26**, 486
252. Gordon, J. S. and Lash, J. W. (1974). *Develop. Biol.*, **36**, 88
253. Cohen, A. M. and Hay, E. D. (1971). *Develop. Biol.*, **26**, 578
254. Trelstad, R. L., Kang, A. H., Cohen, A. M. and Hay, E. D. (1973). *Science*, **173**, 295
255. Minor, R. R. (1973). *J. Cell Biol.*, **56**, 27
256. Ruggeri, A. (1972). *Z. Anat. Entwicklungsgesch.*, **138**, 20
257. Kosher, R. A., Lash, J. W. and Minor, R. R. (1973). *Develop. Biol.*, **35**, 210
258. Zwilling. E. (1968). *Develop. Biol. Suppl.*, **2**, 184
259. Medoff, J. (1967). *Develop. Biol.*, **16**, 118
260. Searls, R. L. (1965). *Develop. Biol.*, **11**, 155
261. Searls, R. L. (1965). *Proc. Soc. Biol. Exp. Med.*, **118**, 1172
262. Toole, B. P. and Gross, J. (1971). *Develop. Biol.*, **25**, 57
263. Toole, B. P. (1972). *Develop. Biol.*, **29**, 321
264. Toole, B. P., Jackson, G. and Gross, J. (1972). *Proc. Nat. Acad. Sci. USA*, **69**, 1384
265. Schacter, L. P. (1970). *Exp. Cell Res.*, **63**, 19
266. Schacter, L. P. (1970). *Exp. Cell Res.*, **63**, 33
267. Levitt, D. and Dorfman, A. (1973). In *Biology of Fibroblasts*, 79 (E. Kulonen and J. Pikkarainen, editors) (New York and London: Academic Press)
268. Levitt, D. and Dorfman, A. (1974). In *Current Topics in Developmental Biology*, Vol. 8, 103 (A. Moscona, editor) (New York and London: Academic Press)
269. Levitt, D. and Dorfman, A. (1972). *Proc. Nat. Acad. Sci. USA*, **69**, 1253
270. Levitt, D. and Dorfman, A. (1973). *Proc. Nat. Acad. Sci. USA*, **70**, 2201
271. Palmoski, M. J. and Goetinck, P. F. (1972). *Proc. Nat. Acad. Sci. USA*, **69**, 3385
272. Abbot, J. and Holtzer, H. (1966). *J. Cell Biol.*, **28**, 473
273. Coon, H. G. (1966). *Proc. Nat. Acad. Sci., USA*, **55**, 66
274. Bryan, J. (1968). *Exp. Cell Res.*, **52**, 319
275. Bryan, J. (1968). *Exp. Cell Res.*, **52**, 327
276. Horwitz, A. L. and Dorfman, A. (1970). *J. Cell Biol.*, **45**, 434
277. Glick, M. and Stockdale, F. E. (1964). *Develop. Biol.*, **83**, 61
278. Nameroff, M. and Holtzer, H. (1967). *Develop. Biol.*, **16**, 250
279. Shulman, H. S. and Meyer, K. (1970). *Biochem. J.*, **120**, 689
280. Nevo, Z., Horwitz, A. L. and Dorfman, A. (1972). *Develop. Biol.*, **28**, 219
281. Levietes, B. (1971). *Exp. Cell Res.*, **68**, 43
282. Robertson, W. B. (1961). *Ann. N.. Acad. Sci.*, **92**, 159

283. Levenson, G. E. (1970). *Exp. Cell Res.*, **62,** 271
284. Pawelek, J. (1969). *Develop. Biol.*, **19,** 52
285. Daughaday, W. H., Hall, K., Raben, M. S., Salmon, W. D., Jr. Van den Brande, J. L. and Van Wyk, J. J. (1972). *Nature (London)*, **235,** 107
286. Salmon, W. D., Jr. and Daughaday, W. H. (1957). *J. Lab. Clin. Med.*, **49,** 825
287. Rendall, J. L., Delcher, K. H. and Lebovitz, H. E. (1972). *Biochem. Biophys. Res. Commun.*, **46,** 1425
288. Birch, B. M., Delcher, H. K., Rendall, J. L., Eisenbarth, G. S. and Lebovitz, H. E. (1973). *Biochem. Biophys. Res. Commun.*, **52,** 1184
289. Eisenbarth, G. S., Beuttel, S. C. and Lebovitz, H. E. (1973). *Biochim. Biophys. Acta*, **331,** 397
290. Delcher, H. K., Eisenbarth, G. S. and Lebovitz, H. E. (1973). *J. Biol. Chem.*, **248,** 1901
291. Meier, S. and Solursh, M. (1972). *Gen. Comp. Endocrinol.*, **18,** 89
292. Meier, S. and Solursh, M. (1972). *Edocrinology*, **90,** 1447
293. Meier, S. and Solursh, M. (1973). *Develop. Biol.*, **30,** 290
294. Yeh, J. and Fisher, H. W. (1969). *J. Cell. Biol.*, **40,** 382
295. Roth, S. (1968). *Develop. Biol.*, **18,** 602
296. Rein, A. and Rubin, H. (1968). *Exp. Cel Res.*, **49,** 666
297. Solursh, M., Meier, S. and Vaerwyck, S. (1973). *Amer. Zool.*, **13,** 1051
298. Solursh, N. and Meier, S. (1973). *Develop. Biol.*, **30,** 279
299. Nevo, Z. and Dorfman, A. (1972). *Proc. Nat. Acad. Sci. USA*, **69,** 2069
300. Schwartz, N. B. and Dorfman, A. (1975). *Conn. Tiss. Res.*, **3,** 115
301. Huang, D. J. (1974). *J. Cell Biol.*, **62,** 881
302. Wiebkin, O. W. and Muir, H. (1973). *FEBS Lett.*, **37,** 42
303. Vaerewyck, S. A. and Solursh, M. (1973). *J. Cell Biol.*, **59,** Abst. 351a
304. Schwartz, N. B. and Dorfman, A. (1974). Unpublished results
305. Hardingham, T. E., Fitton-Jackson, S. and Muir, H. (1972). *Biochem. J.*, **129,** 101
306. Bryant, J. H., Leder, I. G. and Stetten, D. W. (1958). *Arch. Biochem. Biophys.*, **76,** 122
307. Nameroff, M., Trotter, J. A., Keller, J. M. and Munar, E. (1973). *J. Cell Biol.*, **58,** 107
308. Abbott, J. and Holtzer, H. (1968). *Proc. Nat. Acad. Sci. USA*, **59,** 1144
309. Holtzer, H. and Abbott, J. (1968). In *The Stability of the Differentiated State*, 1 (H. Ursprung, editor) (Berlin and New York: Springer-Verlag)
310. Lasher, R. and Cahn, R. D. (1969). *Develop. Biol.*, **19,** 415
311. Coleman, A. W., Coleman, J. R., Kankel, D. and Werner, I. (1970). *Exp. Cell Res.*, **59,** 319
312. Chi, J. C. H., Holtzer, S. and Holtzer, H. (1973). *J. Cell Biol.*, **59,** Abst. 52a
313. Schulte-Holthausen, H. S., Chacko, S., Davidson, E. A. and Holtzer, H. (1969). *Proc. Nat. Acad. Sci. USA*, **63,** 864
314. Marzullo, G. (1972). *Develop. Biol.*, **27,** 20
315. Dawson, G. (1973). Table in paper by Touster, O. In *Birth Defects—Enzyme Therapy in Genetic Diseases*, Vol. IX, 9 (National Foundation)

# 4
# Structure of Proteoglycans

**H. MUIR and T. E. HARDINGHAM**
Mathilda and Terence Kennedy Institute of Rheumatology, London

---

---

## 4.1 INTRODUCTION

The glycosaminoglycans are characteristic components of vertebrate connective tissues. They are long, straight-chain carbohydrates which contain many acidic (carboxyl or sulphate) groups (see Section 4.2). They do not normally occur as free polysaccharide chains *in vivo*, however, but as proteoglycans in which many chains are linked at the terminal reducing sugar residue to a protein molecule. The initial stages in the biosynthesis of proteoglycans (see Chapter 3) have been elucidated and show that they are formed in an analogous way to glycoproteins; the polysaccharide chains are synthesised by stepwise addition of monosaccharide units only after the initiation of chain synthesis by the addition of specific sugar residues to the protein core. The name, proteoglycan, is of recent origin (see Balazs[1]). It was devised to replace several terms such as protein–polysaccharide (complex) and chondromucoprotein that had been applied to various preparations containing glycosaminoglycans and protein in ill-defined forms of covalent and non-covalent association (Table 4.1). The conclusive evidence that glycosaminoglycans, with the possible exception of hyaluronate, exist covalently linked to protein *in vivo* led to the introduction of the term proteoglycan as the name for this family of molecules in which protein and glycosaminoglycan were attached by covalent bonds.

**Table 4.1 Nomenclature**

| *Current name* | *Previous name* |
|---|---|
| Proteoglycan | Protein–polysaccharide (complex)<br>Chondromucoprotein |
| Glycosaminoglycan | Mucopolysaccharide |
| Hyaluronic acid | No change |
| Chondroitin 4-sulphate | Chondroitin sulphate A |
| Chondroitin 6-sulphate | Chondroitin sulphate C |
| Dermatan sulphate | Chondroitin sulphate B |
| Heparan sulphate | Heparitin sulphate<br>Heparin monosulphate |
| Heparin | No change |
| Keratan sulphate | Keratosulphate |

Before the nature of their association with protein was determined, the glycosaminoglycans were investigated as structural carbohydrates. The direction taken by research was largely a result of the difficulty experienced in preparing purified proteoglycans both in high yield and in native form; this contrasted with the isolation of glycosaminoglycan chains after proteolytic digestion, which was both simple and quantitative. An extensive technology was thus developed, associated with the isolation and identification of glycosaminoglycans[2], and their detailed distribution and biosynthesis in many tissues has been documented. It is not yet possible to do this for the intact proteoglycans, although it is becoming increasingly apparent that it is the physical properties and interactions entered into by the proteoglycans that are of physiological importance and these cannot be predicted from the glycosaminoglycan analysis alone. Much has thus to be done before the precise role of proteoglycans in tissues can be determined.

Proteoglycans are primarily molecules of the extracellular space and they thus occur in greatest abundance in those tissues in which the extracellular space is large. These are often tissues with a structural function such as cartilage, nucleus pulposus, cornea, skin and blood vessel wall. In other 'soft' tissues such as liver, kidney and brain, proteoglycans are also present, but in smaller amounts. Many mammalian cell types examined in culture have been shown to synthesise glycosaminoglycans[3] and this may be a property common to most eukaryotic cells[4].

The proteoglycans are synthesised intracellularly and secreted into the extracellular space, as is the fibrous protein collagen, which also has a major structural role in connective tissue. Collagen is laid down as fibres which form a network and provides the tissue with tensile strength. The function of proteoglycans is complementary to that of collagen and can be related to their physicochemical properties in solution. The intact proteoglycan such as that from cartilage is an extremely large molecule. Not only has it a high molecular weight but it has a highly expanded structure in solution. The relative sizes of a proteoglycan molecule, collagen and several globular proteins are shown in Figure 4.1. Because proteoglycans are very large

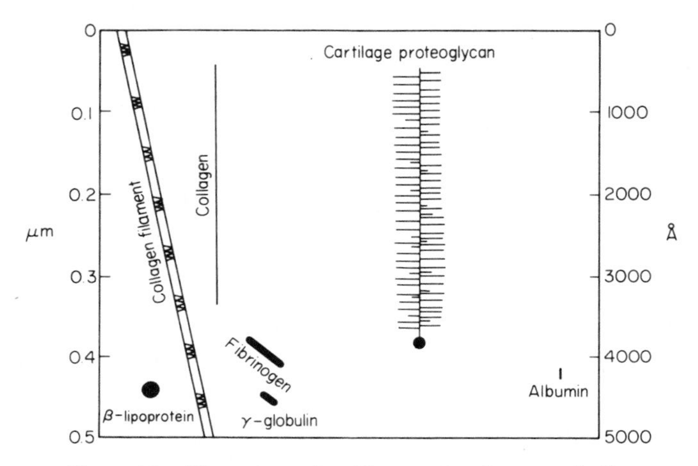


**Figure 4.1** Dimensions of cartilage proteoglycans and other common fibrous and globular proteins. (With the exception of proteoglycan, the figure was supplied by Dr. E. A. Balazs)

expanded molecules they become entangled and immobilised in the collagen network as their concentration increases. They impede the flow of interstitial water when an external force is applied, thus giving the tissue elasticity and resilience to compressive forces[5]. The hydrophilic nature of proteoglycans and collagen induces a swelling pressure in the tissue to which osmotic pressure also contributes. Thus proteoglycans are immobilised polyanions and mobile counter ions therefore become more concentrated in their vicinity, creating a Donnan equilibrium and a large positive osmotic pressure within the tissue. The swelling pressure of connective tissue thus increases with the proteoglycan content.

Proteoglycans also act as molecular sieves as a consequence of their steric exclusion of other solutes. This is particularly evident in cartilage where they are most concentrated and where the permeability of solutes decreases sharply with molecular size[6] such that it is virtually impenetrable to molecules the size of immunoglobulins[7]. The exclusion of other large solutes such as proteins from the molecular domain of proteoglycans increases the effective concentration and hence the activity of solutes in interstitial fluids. This may be of particular importance during tissue growth and development when collagen fibres are being laid down, as proteoglycans affect fibril formation (Section 4.4.1), which may be related to both exclusion effects and to specific interactions between the proteoglycans and collagen.

In mature tissues the turnover of proteoglycans is generally much faster than that of collagen. The physical properties of tissues thus depend upon the constant renewal of the proteoglycan matrix. There is evidence that in cartilage the extracellular content of proteoglycans is subject to 'feedback' control, such that if the matrix is depleted this is immediately balanced by increased synthesis[8,9]. The relative rates of synthesis of proteoglycans and collagen must also in some way be coordinated, although it has been shown that they are not interdependent[10]. The mechanisms by which the synthetic activities of the cell are controlled and modified by its environment have yet to be deduced.

These properties relate to the function of connective tissue proteoglycans in the extracellular space, but there are several examples of proteoglycans that are located within cells in organelles or membrane fractions (Section 4.7.4). These proteoglycans are generally of much lower molecular weight than those of the extracellular space and as they are strong polyanions they often function in binding and acting as a carrier of cationic material in cells.

Relatively little is known of the detailed structure of the protein cores of the proteoglycans. They are synthesised by the normal mechanism of protein synthesis and are presumably of precisely determined amino acid sequence. However, the properties of the proteoglycan are largely determined by the subsequent attachment within the cell of the large number of glycosaminoglycan chains. Although the enzymes involved in this process are specific (see Chapter 3), analysis of proteoglycans suggests that there can be large variation in the number and type of glycosaminoglycan chains attached and variation also in their length and degree of sulphation, such that the observed heterogeneity and polydispersity of proteoglycan preparations can largely be related to the variations amongst the attached glycosaminoglycan chains. It has not yet been possible to identify and characterise proteoglycans according to the structure of their protein cores. Only when this is achieved will it be possible to assess the significance of the morphological distribution of different proteoglycans in terms of the synthesis of genetically distinct protein cores.

## 4.2 GLYCOSAMINOGLYCANS

### 4.2.1 General features

There are seven different types of glycosaminoglycan that are commonly found in vertebrate tissues (Tables 4.1 and 4.2). They are all long unbranched polysaccharides in which the chain is made up of disaccharide repeating units consisting of a hexosamine and a hexuronic acid (except in keratan sulphate where galactose replaces hexuronic acid). The amino groups of the hexosamine residues are generally substituted with *N*-acetyl groups, but can also contain *N*-sulphate groups in heparin and heparan sulphate. The presence of sulphate and/or carboxyl groups on each disaccharide unit makes the chains strong polyanions and this dominates their physical properties and to a large extent their interaction with other molecules. The charge density varies from one per disaccharide in hyaluronic acid,

**Table 4.2 Composition of glycosaminoglycans**

| | *Disaccharide repeating unit* | | *Sulphate* | *Other sugar residues, including those in the linkage region* |
|---|---|---|---|---|
| | *Hexuronic acid* | *Hexosamine* | | |
| Hyaluronic acid | D-glucuronic acid | D-glucosamine | — | ? |
| Chondroitin 4-sulphate | D-glucuronic acid | D-galactosamine | *O*-sulphate | D-xylose, D-galactose |
| Chondroitin 6-sulphate | D-glucuronic acid | D-galactosamine | *O*-sulphate | D-xylose, D-galactose |
| Dermatan sulphate | L-iduronic acid or D-glucuronic acid | D-galactosamine | *O*-sulphate | D-xylose, D-galactose |
| Keratan sulphate | D-galactose | D-glucosamine | *O*-sulphate | D-mannose, D-fucose, sialic acid, D-galactosamine |
| Heparan sulphate | D-glucuronic acid or L-iduronic acid | D-glucosamine | *O*-sulphate and *N*-sulphate | D-xylose, D-galactose |
| Heparin | D-glucuronic acid or L-iduronic acid | D-glucosamine | *O*-sulphate and *N*-sulphate | D-xylose, D-galactose |

which contains one carboxyl group on each disaccharide, to a maximum of four in heparin which can contain up to three sulphate groups and one carboxyl group on each disaccharide. The original evidence for the assignment of the structures shown in Figures 4.2–4.5 and 4.7 can be found in several reviews[11-13] and will not be discussed here. In most cases the structures have been confirmed by subsequent investigation, but they should not be regarded as being exclusive since there is evidence that even with a single glycosaminoglycan preparation there is considerable variation in structural detail. It is also apparent that the distinctions between the different types of glycosaminoglycan are not in every case as clear-cut as is implied by these idealised structures.

The general features of all the glycosaminoglycans are described in the following sections. More detailed information of those glycosaminoglycans containing iduronic acid are presented in Table 4.2, in Ref. 13a and the biosynthesis of all glycosaminoglycans is described in Chapter 3.

### 4.2.2 Hyaluronic acid

Hyaluronic acid is the only non-sulphated glycosaminoglycan and has a wide distribution in both vertebrate and invertebrate connective tissues. It is also present in the capsule of some bacteria such as Staphylococcus and Group A and C Streptococcus. In mammals it is abundant in many embryonic tissues and it occurs in large amounts in synovial fluid, the vitreous humour of the eye, umbilical cord and in cock's comb[14], and recently it has also been shown to be present in considerable amounts in cartilage[15,16].

The repeating disaccharide unit is *O*-β-D-glucopyranosyluronic (1→3)-*O*-β-D-2-acetamido-2-deoxy-D-glucopyranose which is 1→4 β-linked (see Figure 4.2). Hyaluronic acid has a molecular weight in the range from $10^5$–$10^7$ dalton and is thus 10 times larger than other glycosaminoglycans. The

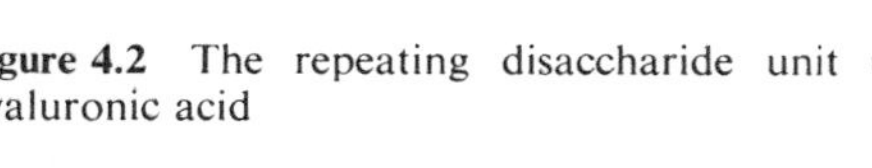

**Figure 4.2** The repeating disaccharide unit of hyaluronic acid

molecular weight depends on the source from which it is prepared and can also be limited as a result of degradation during preparation[14] as it is sensitive to 'oxido-reductive depolymerisation' catalysed by various agents such as ascorbate and cysteine[17-21]. Thus methods used in the routine preparation of glycosaminoglycans, such as papain digestion in the presence of cysteine and EDTA, can produce a degraded preparation of hyaluronic acid[17]. The extent of depolymerisation, however, is limited because after extended treatment with ascorbate the average size of the molecules was still 65 000 dalton[21]. This led to the suggestion that the hyaluronate chain might be made up of subunits of 65 000 molecular weight linked by ascorbate-sensitive bonds. As the reaction mechanism involved in oxido-reductive depolymerisation remains unknown, but probably involves the generation of free radicals, there is as yet no conclusive evidence for this hypothesis. The other glycosaminoglycans are very much less sensitive to this type of depolymerisation.

Hyaluronic acid does not occur as a multi-chain proteoglycan and this is shown by its viscosity in solution being unaffected by proteolytic digestion[17] and electron micrographs of undegraded molecules from human synovial fluid show separate single chains[22]. The question whether single chains are linked to peptide still remains in doubt. If hyaluronic acid is prepared from synovial fluid by ultrafiltration through sintered glass filters, the product still contains up to 20% protein[23]. However, using more effective methods of purification, such as equilibrium density gradient centrifugation, the protein content can be reduced[14,24] to less than 2%. Scher and Hammerman[25] purified hyaluronic acid from synovial fluid by ion-exchange chromatography on ECTEOLA-cellulose and it contained only 0.35% protein. By labelling the tyrosine in the protein with $^{125}I$ it was shown that the protein remained bound to the hyaluronic acid when subjected to caesium chloride density gradient fractionation in the presence of 4 M guanidinium chloride. This treatment should have removed any protein that was not bound by covalent bonds. The label could only be released by digestion with bacterial hyaluronidase or proteolytic enzymes. However, proof of the covalent nature of the association of protein with hyaluronic acid awaits the isolation of a carbohydrate–protein linkage fragment. There has been no conclusive evidence for the suggested presence of arabinose in the linkage of hyaluronic acid to protein[26,27].

### 4.2.3 Chondroitin sulphates

Chondroitin sulphate occurs in very large amounts in all types of cartilage, where it can account for up to 40% of the dry weight[28], but it also has a broad distribution in other connective tissue such as skin, cornea, blood vessel wall and nucleus pulposus. The disaccharide repeating unit has a structure similar to that of hyaluronic acid (Figure 4.3), but contains galactosamine instead of glucosamine and the galactosamine has an ester sulphate group attached at the 4 or 6 position. The position of the two ester sulphate groups give them distinct infrared spectra. Chondroitin 4-sulphate has strong absorption bands at 928, 852 and 725 $cm^{-1}$, whereas chondroitin 6-sulphate has bands at 1000, 820 and 775 $cm^{-1}$ [29]. The molecular weight of chondroitin sulphate is in general between 10 000 and 60 000 dalton, and most preparations are polydisperse in size. The degree of sulphation can vary within a single preparation and from one tissue site to another. Some sources, such as shark cartilage, characteristically contain oversulphated chondroitin 6-sulphate[30] with the extra sulphate on the uronic acid residue, whereas cornea contains an undersulphated form referred to as chondroitin which has a molar ratio of sulphate to hexosamine as low as 0.12[31]. Preparations from mammalian cartilage are often 'undersulphated' by as much as 10%, and in bovine nasal septal cartilage fragments isolated from the region close to the linkage to protein had less sulphate than the remainder of the chain[32]. Non-sulphated and disulphated disaccharide units can be identified after digestion with chondroitinase[33] (see Section 4.2.9.2), the former decreasing and the latter increasing with age in human knee cartilage[34]. Some of the extra sulphate on the chondroitin sulphate from squid cartilage is attributable to 2,6-disulphated galactosamine[33], which has also been found in chondroitinase digests of preparations from human rib cartilage[35].

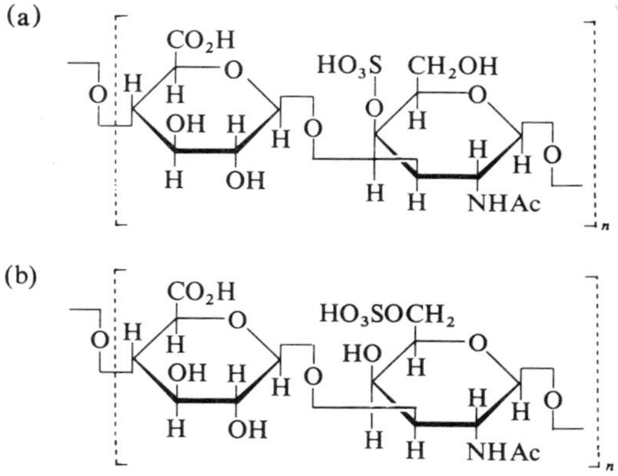

**Figure 4.3** The repeating disaccharide units of (a) chondroitin 4-sulphate; and (b) chondroitin 6-sulphate

The isomeric chondroitin 4- and 6-sulphates occur in tissues both independently of each other or as a mixture. Thus pig laryngeal cartilage contains chondroitin 4-sulphate, umbilical cord and bovine aorta contain chondroitin 6-sulphate, and rat skin, human costal cartilage and rabbit ear cartilage all contain a mixture of the two isomers. Where they occur as a mixture they are not separated by the more simple methods of glycosaminoglycan fractionation such as CPC precipitation or ion-exchange chromatography[2]. The relative amounts of both isomers in a mixture can be assessed using testicular hyaluronidase digestion[36] or by a rather more effective method using the bacterial chondroitinase enzymes[37] (see Section 4.2.9.2). A method was also developed for their isolation on a preparative scale using a CPC-cellulose column[38], by exploiting the preferential solubility of the chondroitin 4-sulphate–cetylpyridinium chloride complex in an organic solvent mixture. The technique also separated dermatan sulphate and chondroitin 6-sulphate complexes as a result of the higher solubility of the latter at low pH. However, the separation achieved with standards is often difficult to reproduce with natural mixtures, and this probably results from a larger variation in the degree of sulphation and greater polydispersity of crude mixtures compared with highly purified material. There is also, in some instances, the possibility of copolymeric structure as demonstrated with dermatan sulphate (see Section 4.2.4) in which both condroitin 4-sulphate and chondroitin 6-sulphate disaccharide units may be present within the same chain[39]. The distribution and frequency of such copolymeric structures has not yet been assessed.

### 4.2.4 Dermatan sulphate

Dermatan sulphate is an isomer of the chondroitin sulphates in which a large proportion of the D-glucuronate residues is replaced by L-iduronate residues (Figure 4.4). It occurs in large amounts in skin and in other connective tissues, such as blood vessel walls, heart valves, umbilical cord and scar tissue, and it accumulates in tissues and is excreted in large amounts by patients with Hurler's or Hunter's syndrome (Chapter 8). All preparations are digested by testicular hyaluronidase to a limited extent, the chains being cleaved where glucuronate residues occur[40,41]. In a preparation from pig skin[42] these were mainly near the linkage region, but were more randomly distributed as clusters along the chains in material from umbilical cord[43].

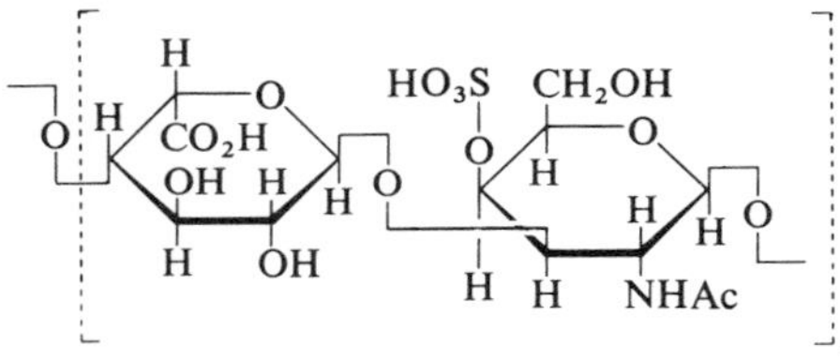

H O H $CO_2H$ OH H H O $HO_3S$ $CH_2OH$ O O H H H O H OH H H NHAc n

**Figure 4.4** The repeating disaccharide unit of dermatan sulphate. All preparations contain some D-glucuronate residues

Dermatan sulphate from pig skin is thus a copolymer having disaccharide units typical of both dermatan sulphate and chondroitin 4-sulphate. The relative proportion of iduronic acid varies but is generally in the range 10–20% of the glucuronic acid[40]. Dermatan sulphate from umbilical cord was shown to be a copolymer containing some chondroitin 6-sulphate disaccharides[43], and that of a preparation from horse aorta was even more complex, containing chondroitin 4-sulphate, chondroitin 6-sulphate and dermatan sulphate units[44]. This type of structure would obviously not be resolved into different isomeric glycosaminoglycans by such procedures as described in Section 4.2.3. In addition to the sulphate groups, characteristic of the different disaccharide units, dermatan sulphate also has extra sulphate groups on the uronate residues[45] which appear to be primarily on the iduronate residues[46].

### 4.2.5 Keratan sulphate

Keratan sulphate has many characteristics that distinguish it from other glycosaminoglycans. It has a repeating disaccharide (Figure 4.5) containing *N*-acetylglucosamine and galactose, but no uronic acid[47]. It is also often of relatively low molecular weight even in undegraded preparations and is of limited distribution, as it has been characterised only in cornea, cartilage and nucleus pulposus.

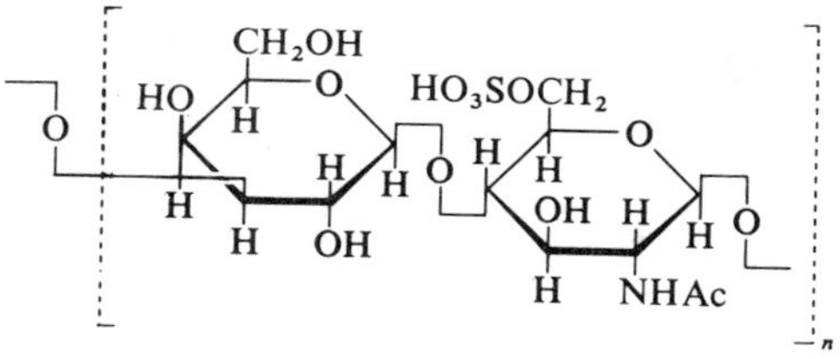

**Figure 4.5** The repeating disaccharide unit of keratan sulphate

On the basis of its linkage to protein it has been classified into two types. KS-I occurs in cornea and contains an alkali-stable linkage to protein. From the analysis of preparations that had been digested with proteolytic enzymes or degraded by alkali it was suggested to be either an amide linkage from glucosamine to aspartic or glutamic acid or a glycosylamine linkage from glucosamine to asparagine or glutamine[48]. The isolation of the linkage fragment 2-acetamido-1-(β-L-aspartamido)-1,2-dideoxy-β-D-glucose from mild acid hydrolysates of keratan sulphate[49] confirmed the linkage (Figure 4.6a) as that found in several types of glycoprotein[50]. KS-II is found in cartilage and nucleus pulposus. It is invariably linked to the same protein as some chondroitin sulphate[51,52] and although the linkage is more resistant to alkaline cleavage than that of chondroitin sulphate, a loss of serine and threonine residues[48] is accompanied by the appearance of a fragment that reacted with Erlich's reagent when KS–protein was treated with 0.5 M alkali at room temperature[53,54]. It was suggested that the linkage was between

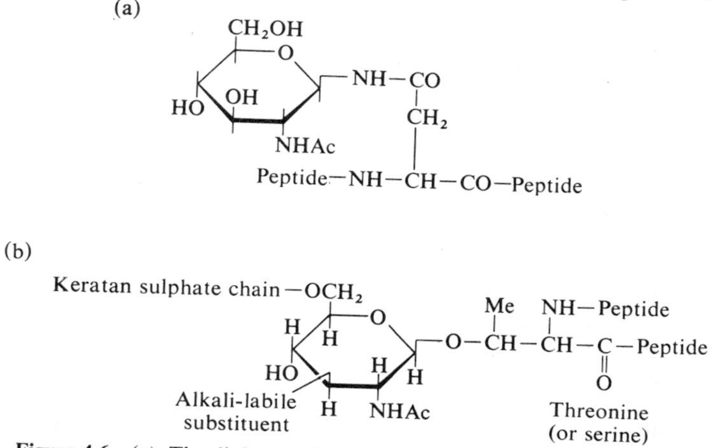

**Figure 4.6** (a) The linkage of corneal keratan sulphate (KS-I) to protein, from *N*-acetylgalactosamine to asparagine. (b) The possible linkage of skeletal keratan sulphate (KS-II) to protein from *N*-acetylgalactosamine to threonine or serine. Mannose is also present in the linkage region

galactosamine, which decreased in content during the treatment with alkali, and serine and threonine, which were converted by the β-elimination reaction into dehydroalanine and α-aminocrotonic acid, respectively. It was concluded that the galactosamine contained both a 3-substituent which facilitated its conversion to a Kuhn's chromogen[55], and an alkali-stable linkage at C-6 by which it was attached to a higher molecular weight fragment[53], which it was suggested was the remaining part of the keratan sulphate chain (Figure 4.6b)[54]. Preliminary evidence has also been presented that some KS-II may be linked to glutamate (or glutamine) residues in an alkali-stable linkage[56]. In addition to galactosamine in KS-II, preparations of the two types of keratan sulphate have been shown to contain traces of other sugars, mannose, fucose and sialic acid, that are not characteristic of the repeating disaccharide unit[48,53,57-62] and also a molar excess of galactose over glucosamine[57,59]. A study of the acid hydrolysis products of permethylated KS-I[59] suggested that a small amount of galactose and fucose occupied terminal positions, with galactose terminal on the main chain and fucose attached to branch points off the main chain. KS-II gave similar results[60], but the degree of branching appeared to be considerably greater. The results also suggested that about half the galactose residues and rather more of the glucosamine residues in KS-I and KS-II contained 6-*O*-sulphate groups, although the degree of sulphation varied greatly from one preparation to another[58]. The sialic acid in a preparation of KS-II from whale cartilage[62], and in KS-II attached to proteoglycan from bovine nasal cartilage[63], was shown to occupy terminal positions as it was removed by the action of neuraminidase. It thus appears that both fucose and sialic acid may occupy terminal positions in a branched keratan sulphate structure in much the same way as they occur at the non-reducing terminals of the oligosaccharides attached to many glycoproteins[64].

The results of enzymic digestion of preparation of KS-I and KS-II showed that the majority of the mannose residues in both types remained associated with an oligosaccharide–peptide fragment that was resistant to digestion[65,66]. Most of the mannose would thus seem to occur close to the linkage to protein. The structure of keratan sulphate determined so far has several characteristics that liken it to a glycoprotein oligosaccharide. KS-I contains a glycosylamine linkage to asparagine that is common to many glycoproteins and both types of KS contain a repeating disaccharide that contains no uronic acid, while the evidence for a branched structure and the presence of mannose, fucose and sialic acid all suggest further strong similarities with glycoproteins.

### 4.2.6 Heparin and heparan sulphate

Heparin and heparan sulphate are glycosaminoglycans of closely related structure but of different morphological distribution and function. Although heparin is found in many connective tissues, it is not present as a structural component but as an intracellular component in mast cells. It occurs in skin, lung and umbilical cord and in particularly large amounts in bovine liver capsule and pig gastric mucosa which are both used as commercial sources of heparin.

Heparin contains glucosamine and uronic acid in its disaccharide repeating unit (Figure 4.7) and is highly sulphated. A large proportion of the glucosamine residues contain *N*-sulphate groups instead of *N*-acetyl groups. Both glucuronic acid and iduronic acid are present in the molecule[67,68] and the latter has been estimated to account for at least half of the total uronate[69-71]. It was thought that both linkages in the disaccharide unit had the 1→4 α-configuration[72], but this has been recently challenged as some glucuronic acid residues in nitrous acid-degraded fragments of heparin were shown to be substrates for β-glucuronidase[73] and x-ray fibre diffraction patterns suggested that the structure was not an entirely 1→4 α-linked polymer[74].

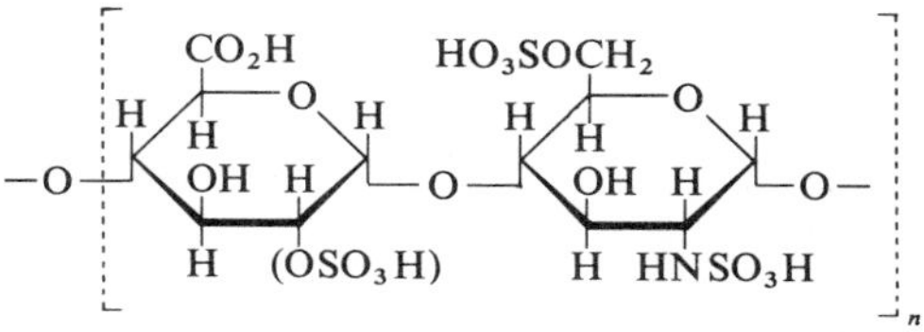

$CO_2H$ $HO_3SOCH_2$
H H O H H H O H
—O— OH H —O— OH H —O—
H $(OSO_3H)$ H $HNSO_3H$
n

**Figure 4.7** The repeating disaccharide unit of heparin. Heparan sulphate contains a similar disaccharide unit, but with more *N*-acetyl and less *N*-sulphate groups, and a lower degree of *O*-sulphation. Both heparin and heparan sulphate contain some L-iduronate residues

The chain may thus contain some regular 1→4 α- and β-linkages. Most of the glucosamine residues contain *N*-sulphate groups and a small proportion contain *N*-acetyl groups[75], which occur particularly in the region of the linkage to protein[76]. About 30% of the glucosamine of heparin has been isolated after enzymic digestion as 2,6-disulphate[77] and about half of the uronic acid residues were shown to be sulphated[78]. It was suggested that this sulphate group was primarily on the iduronate residue[69], in the 2-position[79]. The heparin chains thus contain up to three sulphate groups per disaccharide, but a considerable variation in the degree of sulphation is evident within preparations from different sources[75].

The structure of heparan sulphate is based upon the same oligosaccharide repeating unit as that of heparin, but it differs markedly in its sulphate content. It contains as much *N*-acetyl as *N*-sulphate substituents, and a lower degree of *O*-sulphation than heparin[79], such that there is only, on average, one sulphate group per disaccharide. However, the sulphate is not evenly distributed, as both non-sulphated disaccharides and disulphated glucosamine residues have been isolated from enzymic digests of heparan sulphate[80]. The chain appears to contain a block structure in which some regions have a structure comparable with that of heparin and others contain no *N*-sulphate and very little *O*-sulphate[81,82]. The relative amounts of the two types of structure vary with the source of the heparan sulphate and also within each sample. There has thus been much speculation that heparin and heparan sulphate may be related forms of the same biosynthetic unit, which has been synthesised and exported in the case of heparan sulphate, or retained within the cell and modified in the case of heparin[81,83]. However, any evidence on this point awaits the isolation of the intact proteoglycans for comparison of their protein components.

### 4.2.7 Linkage of glycosaminoglycans to protein

The linkage between chondroitin 4-sulphate and protein was shown to involve an alkali-labile linkage to serine[84]. It was subsequently established that the linkage region consisted of a characteristic trisaccharide unit (Gal-Gal-Xyl) in which the xylose residue was glycosidically linked to serine (Figure 4.8)[85,86], a sequence also found in chondroitin 6-sulphate[87], heparin[76], heparan sulphate[88] and dermatan sulphate[42,89]. The biosynthesis of the linkage region is discussed in Chapter 3. In 0.5 N NaOH at room temperature the xylosyl–serine bond is cleaved by a β-elimination reaction in which the serine is converted into dehydroalanine (Figure 4.8)[90]. This reaction has been used for the isolation of glycosaminoglycans free of protein[2]. The reaction is most quantitative with the intact proteoglycans because low yields are obtained when the carboxyl group or amino group of serine are free[91].

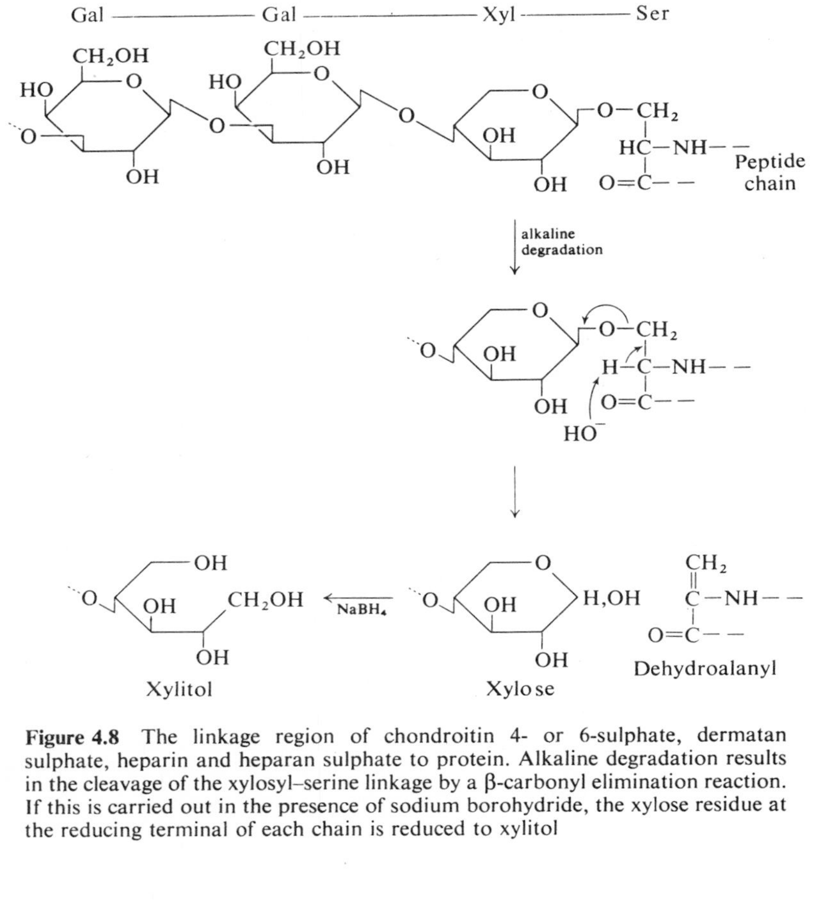

**Figure 4.8** The linkage region of chondroitin 4- or 6-sulphate, dermatan sulphate, heparin and heparan sulphate to protein. Alkaline degradation results in the cleavage of the xylosyl–serine linkage by a β-carbonyl elimination reaction. If this is carried out in the presence of sodium borohydride, the xylose residue at the reducing terminal of each chain is reduced to xylitol

KS-I and KS-II do not contain this sequence of linkage region. KS-I is linked by a glycosylamine linkage to asparagine, which is stable to alkali, whereas the linkage of KS-II is mainly labile to alkali. It has not been fully characterised although there is evidence that it is a glycosidic link between galactosamine and serine or threonine (see Section 4.2.5).

## 4.2.8 Molecular weight determinations

### *4.2.8.1 Physical methods*

Most physical methods developed for the determination of the molecular weight of proteins have also been used successfully to determine the molecular weight of the glycosaminoglycans. Sedimentation velocity and sedimentation equilibrium methods[92-94] can be used to provide accurate measurements of the weight average ($M_w$) and number average ($M_n$) molecular weights[95-97]. Light scattering[94,98] has also been used to determine $M_n$ but is effective mainly with high molecular weight material such as hyaluronic acid[99,100], whereas osmometry[94] can be used to determine $M_n$ for material of lower molecular weight such as chondroitin 4-sulphate[101,102]. The swelling of a bead of cross-linked dextran (Sephadex), suitably calibrated, may be used to record the osmotic pressure of non-penetrating macromolecules in the solution surrounding it[103]. The sensitivity of this micro-method has been greatly increased, making it comparable with conventional osmometry, by measuring the separation between the ends of a loop of polyacrylamide gel[104].

The viscosity of a linear polymer in solution is related to its molecular weight by an empirical relationship, the Mark–Houwink equation:

$$[\eta] = KM^{\alpha}$$

where $[\eta]$ is the limiting viscosity number and $K$ and $a$ are constant for each polymer under fixed conditions of solvent, temperature and pH. Values for these constants have been determined for all the glycosaminoglycans except KS-II and heparan sulphate[2] and they thus provide a useful method of molecular weight determination that does not require expensive equipment or standards of known molecular weight. The average molecular weight determined by viscosity[94] has a value close to that of $M_w$.

Gel chromatography and polyacrylamide gel electrophoresis have been developed to provide secondary methods of molecular weight determination. A relationship was shown to exist between the position of elution from the column and the Stokes radii of the solute[105-107]. For a homologous series of polymers such as chondroitin 4-sulphate, the position of elution from the column was related to the logarithm of the molecular weight[95,108]. The distribution of the sample in the eluate enabled both $M_n$ and $M_w$ to be calculated and the degree of polydispersity to be assessed[95]. Although chondroitin 4-sulphate standards of known molecular weight were required to calibrate the column, samples of heparan sulphate, dermatan sulphate and chondroitin 6-sulphate were found to be on the same calibration line[108].

However, a thorough comparison of the behaviour of the different glycosaminoglycans on gel chromatography has not yet been reported.

Polyacrylamide gel electrophoresis provides a micro-method that can be used to determine the molecular weights of glycosaminoglycans. The mobility was proportional to the logarithm of the molecular weight[109,110]. In low porosity gels (6% cross-linked), the mobility appeared to be largely dependent upon size, probably reflecting filtration through the gel matrix, as chondroitin 4- and 6-sulphates, dermatan sulphate and heparin of different molecular weight all conformed to the same straight-line relationship[109]. In higher porosity gels (3–5% cross-linked) the effect of charge density was more pronounced, but could be compensated for by comparison of the electrophoretic mobility in gels of different porosity. Hyaluronic acid was too large to penetrate normal polyacrylamide gels[109].

#### *4.2.8.2 Chemical methods*

The identification of the atypical trisaccharides that link several types of glycosaminoglycan to protein (Figure 4.8) has lead to the development of several chemical methods of molecular weight determination. The preparation of glycosaminoglycans by proteolytic digestion leaves the chains attached to a short peptide with the linkage region intact. As there is only one linkage region on each polysaccharide chain, the xylose content of a glycosaminoglycan preparation is a measure of the number of chains and the molar ratio of hexosamine or hexuronic acid to xylose is therefore a measure of the average number of disaccharide units in the chain and hence its molecular weight. Xylose may be assayed by a sensitive colorimetric method that does not require prior hydrolysis of the sample[111]. The assay is specific for pentose or methylpentose and the small amount of interference from other sugars (less than 5%) may be allowed for by including appropriate blanks[112]. Xylose can also be measured quantitatively by gas–liquid chromatography after hydrolysis by a procedure such as that described by Clamp *et al.*[113]. Both methods have been used to determine molecular weights of chondroitin sulphate from various sources and the values were found to be in agreement with those determined by other methods[112,114-116]. Galactose has often been assayed at the same time as xylose as a further measure of chain length, but its measurement alone might give unreliable results as it also occurs in many glycopeptides and as part of the disaccharide unit of keratan sulphate.

The alkaline β-elimination of chondroitin sulphate chains from protein has also been used as the basis of a method to determine the molecular weight of the chains (Figure 4.8)[117]. Proteoglycans from cartilage were treated with 0.5 M KOH at 4°C for 4 days in the presence of tritiated sodium borohydride of known specific radioactivity. The xylose at the reducing terminal of the released chains was reduced to xylitol, which was labelled with tritium. As there was no evidence of further degradation of the linkage region by peeling reactions[118] or of random chain cleavage, the radioactivity incorporated into the chains was a measure of chain length. The molecular weight ($M_n$) thus calculated was comparable with that determined by equilibrium sedimentation methods. The procedure was also effective when carried out on whole cartilage, when it was estimated that at least 95% of the total chondroitin sulphate was accounted for. Keratan sulphate (KS-II) was also released and radioactively labelled by this method. The labelling appeared to be specific, and enabled the molecular weight to be calculated, although the actual residue labelled (presumably galactosamine, see Section 4.2.5) was not stated. The method has not yet been applied to other glycosaminoglycans, but borotritiide labelling of terminal reducing sugars has been shown to be an accurate method of estimating molecular weights of neutral mono- and oligo-saccharides[119], and of maltose oligosaccharides and partially hydrolysed amylose chains[120].

The determination of molecular weight based upon the assay of galactose and xylose in the linkage region does not apply to keratan sulphate or hyaluronic acid which do not have this linkage. It would also give a misleading estimate where the glycosaminoglycan chains are partially degraded. This has been reported with heparan sulphate from the liver and urine of patients with Hurler's syndrome[88] and a recent study shows that some linkage region sugars were partially lacking from the degraded heparan sulphate in the liver and spleen of similar patients[121].

### 4.2.9 X-Ray fibre diffraction

Over the past three years considerable evidence has been obtained from x-ray diffraction of fibrous crystalline structures in several forms of the principal glycosaminoglycans. It has been established that in spite of their undoubted irregularities in structure already discussed, they showed a high degree of order when dried films of the glycosaminoglycans were aligned by stretching[122]. The degree of stretching (about $1\frac{1}{2}$ times the length) was insufficient to align completely all the chains and the type of orientation achieved is shown in Figure 4.9. The quality of the x-ray fibre patterns produced by irradiating perpendicular to the stretch axis suggested a high degree of order and therefore that the 'loop' regions (Figure 4.9) were not greatly

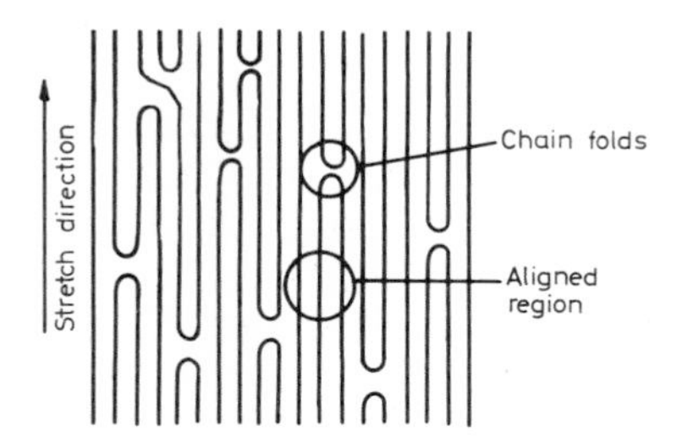


**Figure 4.9** The orientation of single polysaccharide chains achieved by stretching dried films of glycosaminoglycans. The degree of stretching (10–300%) was insufficient to align all the chains. The folds are in general sharp, as the diffraction patterns obtained suggested the degree of crystallinity to be quite high. (Diagram provided by Dr. E. D. T. Atkins)

disturbing the packing of the 'aligned' portions of the chain. It was found that the degree of crystallinity in the sample could be increased, or the form altered, by stretching under conditions of high humidity (70–95%) and by annealing at high temperature (60–70°C). The effect of humidity suggested the importance of water in the structures obtained, and calculations revealed that considerable spaces were present in those structures when examined in detail that may well be occupied by water[123].

Hyaluronic acid was the first glycosaminoglycan to be studied in this way. Initial results[124,125] showed that sodium hyaluronate readily forms ordered structures that can be interpreted as single stranded threefold helices, i.e. containing three disaccharide units per twist of the chain, and models suggested that these were stabilised by hydrogen bonding between the acetamido groups of neighbouring chains. In the free acid state the chains formed two-fold helices with 180° rotation between each disaccharide unit. Subsequently a further form which contained double-stranded four-fold helices was obtained by drying samples of a viscoelastic 'putty' formed by high molecular weight hyaluronic acid at pH 2.5 under conditions of low ionic strength[126]. The two strands of the double helix were antiparallel and it was suggested that the existence of double helices in solution may contribute to the formation of the viscoelastic 'putty'[127]. The double helix was also shown to be formed from potassium hyaluronate[122] which on annealing was converted into the single stranded three-fold helix. It was suggested that the different forms of hyaluronate represented states of rather similar free energy and interchange between forms was possible by varying the environment of the fibres[122]. The results obtained with the other glycosaminoglycans have been in many ways comparable with those obtained with hyaluronic acid with the notable exception that no other double stranded helices have been observed, possibly because of the bulky sulphate side groups in other glycosaminoglycans. Chondroitin 6-sulphate[123] formed single-stranded three-fold helices as the sodium salt and two-fold helices as the free acid. Similar structures were also present in stretched films of chondroitin 4-sulphate. Evidence was also obtained for an eight-fold helix in chondroitin 6-sulphate in addition to the three-fold form[129], and both types were also identified in a preparation of dermatan sulphate[130] in addition to a two-fold helix formed on annealing the three-fold helix. Although the types of helices formed by chondroitin 4-sulphate and chondroitin 6-sulphate were similar, axial projections showed that sulphate groups in the 6-position projected much further from the helix than in the 4-position[131]. In the L-iduronate residue in dermatan sulphate the 1C conformation is not as energetically unfavourable as it would be in D-glucuronate residues, but calculations of the disaccharide repeat distances in all three types of helix observed showed that they were too long for the iduronate to be in the 1C chair form with both glycosidic bonds axial. The iduronate residues thus appeared to be in the same C1 conformation as the glucuronate and hexosamine residues.

The data for heparan sulphate[128] and heparin[74] showed some characteristics that were at variance with the accepted structures of these glycosaminoglycans. The results were inconsistent with an entirely 1→4 α-linked polymer and were more in keeping with a tetrasaccharide repeat sequence containing alternating glucosamine-glucuronate and glucosamine-iduronate disaccharides. It was suggested that the glucuronate residue in this sequence might be 1→4 β-linked, which was also indicated by the susceptibility of nitrous acid degraded fragments of heparin to β-glucuronidase[73]. In this instance the iduronate was possibly in the 1C chair conformation. However, none of the model structures suggested satisfy all the observations and more detailed information is required before the problem can be resolved.

### 4.2.10 Enzymic digestion of glycosaminoglycans

#### *4.2.10.1 Testicular hyaluronidase*

Testicular hyaluronidase (EC 3.2.1.35) is the most widely available enzyme which degrades glycosaminoglycans. It functions as an endohexosaminidase and acts on hyaluronate, chondroitin and chondroitin 4- and 6-sulphates (Table 4.3). It releases even-numbered oligosaccharides with *N*-acetylhexosamine at the reducing terminal end. The enzyme has some specificity for the hexuronate residues involved, as hexosaminyl linkages to iduronate in dermatan sulphate are not cleaved[43]. Both hydrolysis and transglycosylation are catalysed, the former predominating in the early stages of digestion of high molecular weight substrate, while transglycosylation increases as the digestion proceeds[132]. The major end products of digestion are tetrasaccharides, which cannot be degraded further, together with smaller amounts of larger oligosaccharides resulting from transglycosylation[132].

Tetrasaccharides containing a reducing terminal hexosamine group may be assayed by the Morgan–Elson reaction, except for those having substituents on C-4, such as 4-*O*-sulphate. The estimation of Morgan–Elson-positive material in the digest of a mixture of chondroitin sulphates thus enables the relative amounts of 4- and 6-sulphate to be determined[36]. At the same time, gel chromatography of the digest products may be used to detect dermatan sulphate in the same mixture[133]. This method tends to overestimate the proportion of chondroitin 4-sulphate in the mixture as it is usual to assume that all the digest products are reduced to tetrasaccharides, whereas the average size will be slightly larger than this (see above), and, moreover, whenever dermatan sulphate is present a longer digestion time is needed because dermatan sulphate inhibits the enzyme.

#### *4.2.10.2 Chondroitinases ABC and AC*

Chondroitinase ABC (EC 4.2.2.4) has been purified from *Flavobacterium heparinum* and from *Proteus vulgaris* grown in the presence of chondroitin 6-sulphate[33]. The enzyme digests chondroitin 6-sulphate, chondroitin 4-sulphate, dermatan sulphate, hyaluronic acid and desulphated chondroitin sulphate, but not heparin, heparan sulphate or keratan sulphate. The enzyme acts as an eliminase, producing unsaturated disaccharides (Figure 4.10) with hexosamine at the reducing end, and the absence of digest products of intermediate size[33] suggests that it is an exo-enzyme, but this has yet to be confirmed. Chondroitin 4-sulphate and dermatan sulphate give rise to the same unsaturated disaccharide because the formation of a double bond between C-4 and C-5 of the uronate residue removes the asymmetry at C-5 and hence the distinction between D-glucuronate and L-iduronate (Figure 4.10).

**Table 4.3 Enzymes degrading glycosaminoglycans**

| | *Substrates* | *Type of action* |
|---|---|---|
| 1. Testicular hyaluronidase | Hyaluronate<br>Chondroitin 4-sulphate<br>Chondroitin 6-sulphate | *endo*-β-hexosaminidase |
| 2. Chondroitinase ABC | Hyaluronate<br>Chondroitin 4-sulphate<br>Chondroitin 6-sulphate<br>Dermatan sulphate | (*exo*?)-β-hexosaminyl eliminase |
| 3. Chondroitinase AC | Hyaluronate<br>Chondroitin 4-sulphate<br>Chondroitin 6-sulphate | (*endo*?)-β-hexosaminyl eliminase |
| 4. Bacterial hyaluronidase | Hyaluronate | β-glucosaminyl eliminase |
| 5. Leech hyaluronidase | Hyaluronate | *endo*-β-glucuronidase |
| 6. Heparinase (crude preparation) | Heparin<br>Heparan sulphate | hexosaminyl eliminase and α-glycuronidase |
| 7. Keratanase (crude preparation) | Keratan sulphate I<br>Keratan sulphate II | *endo*-β-galactosidase and others |

Chondroitinase AC (EC 4.2.2.5) has been purified from *Proteus vulgaris* and is very similar in action to chondroitinase ABC except that dermatan sulphate is not degraded, but is a strong competitive inhibitor of the enzyme[33] as are, to a lesser extent, heparan sulphate and keratan sulphate. It was not shown whether the enzyme functioned as an exo- or endo-eliminase. The unsaturated disaccharides produced by the action of chondroitinase ABC or AC absorb strongly at 232 nm ($E_{max}$ 5.1 for ΔDi-4S and 5.5 for ΔDi-6S) and thus the progress of digestion can be monitored with a spectrophotometer[37] and a

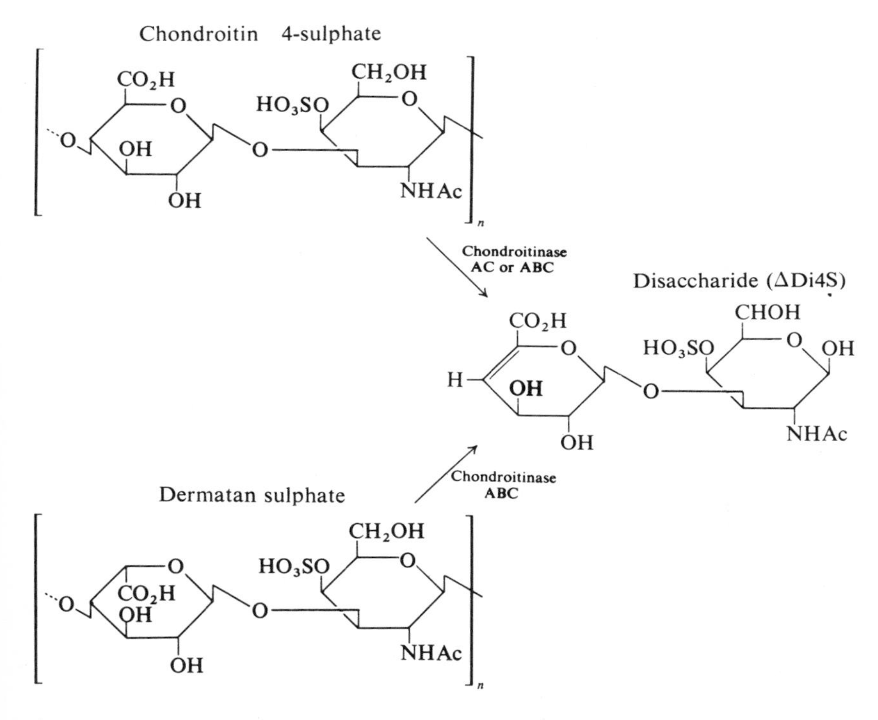


**Figure 4.10** The action of chondroitinases ABC and AC on chondroitin 4-sulphate and dermatan sulphate. Chondroitinase ABC digests both glycosaminoglycans, producing the same reduced disaccharide product (ΔDi4S). Chondroitinase AC is unable to degrade dermatan sulphate

colorimetric assay was developed for use when the presence of protein makes monitoring at 232 nm no longer possible[134]. The products of chondroitin sulphate digestion contain non-sulphated (ΔDi-0S), monosulphated (ΔDi-4S or ΔDi-6S) and occasionally disulphated (ΔDi-diS) disaccharides. These can be separated by paper chromatography or electrophoresis[33]. The respective disaccharides from each chondroitin sulphate isomer can be desulphated by specific chondro 4-sulphatase (EC 3.1.6.9) or chondro 6-sulphatase (EC 3.1.6.10) which are also obtained from *Proteus vulgaris*[33]. These sulphatases act only upon the corresponding unsaturated disaccharides or their saturated analogues. The release of inorganic sulphate can be used as a measure of the enzyme activity.

Chondroitinase ABC, AC and chondro 4- and 6-sulphatase are all available commercially and several procedures have been developed, using them in various combinations to estimate chondroitin 4-sulphate, 6-sulphate and dermatan sulphate in unknown samples. These methods offer the most accurate way of carrying out these determinations and they can be applied to extremely small samples[37].

#### *4.2.10.3 Other enzymes degrading hyaluronic acid*

Hyaluronic acid is digested by both chondroitinase ABC and AC and it is also degraded by a hyaluronate lyase (EC 4.2.2.1) that has been isolated from cultures of Group A and C Streptococcus, *Staphylococcus aureus* and *Clostridium welchii*. This enzyme has a similar eliminase action to the chondroitinases but has no activity against chondroitin 4- or 6-sulphate[135], and chondroitin sulphate and heparin were both found to be competitive inhibitors of hyaluronate digestion[136]. The leech (*Hirudino medicinalis*) also contains a hyaluronidase that has a unique action as it functions as an endoglucuronidase (Table 4.3). The major product of digestion of hyaluronic acid was a tetrasaccharide containing glucuronate at the reducing terminal[137]. The enzyme appeared highly specific and did not attack other glycosaminoglycans.

#### *4.2.10.4 Enzymes degrading heparin*

An enzyme system degrading heparin can also be induced in *Flavobacterium heparinum*[138]. The crude extract was shown to degrade heparin mainly to monosaccharides, amongst which glucosamine 2,6-disulphate was a major fraction[80,139,140]. Although it was suggested that the principal degrading enzyme was a hydrolase[140,141], there was evidence that most of the activity was due to an eliminase[80]; three major enzyme fractions were subsequently resolved[77], of which two were eliminases with maximum activity against heparin and heparan sulphate, respectively, and the third was an α-glycuronidase which cleaved the unsaturated uronate residue from the digest products formed by the eliminases. Specific sulphatases for the removal of *N*-sulphate residues and *O*-sulphate residues have also been demonstrated[142]. The action of the glycuronidase was accompanied by a decrease in ultraviolet absorbance which resulted from the isomerisation of the $\Delta^{4,5}$-unsaturated uronate to the corresponding α-keto acid[80]. The rapid conversion of unsaturated uronate into α-keto acid in the presence of high glycuronidase activity would thus explain why no increase in $\Delta^{4,5}$-unsaturated uronate or u.v. absorbance accompanied the action of the crude enzyme mixture[140,141]. In the absence of glycuronidase the eliminase degraded heparin to a series of even-numbered oligosaccharides. The major product was disaccharide, together with smaller amounts of larger fragments which were resistant to further digestion, probably because of the presence of 2-*O*-sulphated uronate residues that may block further degradation[77]. The crude enzyme mixture has been used in conjunction with cellulose acetate electrophoresis to identify heparin and heparan sulphate in unknown mixtures of glycosaminoglycans[143].

#### *4.2.10.5 Keratan sulphate degrading enzymes*

A bacterial enzyme system from Coccabacillus which degrades keratan sulphate was first reported in 1960[144], but attempts to purify the enzymes involved were not reported until more recently[65,66]. The major part of the activity was due to an endo-β-galactosidase that reduced keratan sulphate to a series of oligosaccharides. KS-I was hydrolysed at about twice the rate of KS-II, and desulphated KS-I was hydrolysed even faster. The preparation also showed activity against a sulphated glycoprotein from chick allantoic fluid and a neutral glycopeptide from colostrum, but other glycosaminoglycans were not attacked. Both KS-I and KS-II contained oligosaccharide peptide fragments that were resistant to digestion and contained up to 80% of the mannose of each preparation. That from KS-II was also enriched in galactosamine, which is probably the linkage sugar (see Section 5.2.5) and also contained extra sulphate. The yield of other oligosaccharides was 50–55% and a tetrasaccharide containing galactose at the reducing terminal was a major fraction, but at least 15 other oligosaccharide fractions were produced. The complexity of the mixture may have reflected the branched structure of keratan sulphate, and also the presence of trace amounts of other glycosidases in the enzyme preparation[66].

## 4.3 CARTILAGE PROTEOGLYCANS

### 4.3.1 General introduction

Most studies of the structure of proteoglycan have been restricted to those from cartilage, and the relatively slow progress made over the past 20 years is related to various problems associated with the composite structure of proteoglycans, that arise only to a limited extent with other proteins:
(a) Preparation. They are difficult to extract from many tissues unless either extensive mechanical disruption or strong denaturing solvents (e.g. 4 M guanidinium chloride and 8 M urea) are used in their preparation.
(b) Physical properties. Their very high molecular weight and non-ideal solute behaviour has hindered progress in the determination of their size and shape by classical methods of biophysics.
(c) Polydispersity. Most preparations are found to be very polydisperse in size, often with a 10-fold variation in molecular weight.
(d) Heterogeneity. Within each preparation the molecules contain variations in the number and type of glycosaminoglycan attached to the protein core and there may be different core proteins.

These problems have led to a large variety of techniques being used both to extract and isolate proteoglycans and subfractionate them in order to determine the extent of their heterogeneity and polydispersity. However, when methods of extraction are not quantitative, they may select different proportions of the various proteoglycans present in the same tissue and so give conflicting evidence of molecular weight polydispersity and composition[145]. In assessing results it is thus necessary to consider carefully the methods used in obtaining them. The problem is also made more difficult, as there is no simple physiological function of proteoglycans that can be measured and used as a criterion of native structure in the way that an enzyme's activity can be determined and used as a measure of its structural integrity. Partial degradation as a result of autolysis or chemical cleavage during extraction and isolation is thus not easily detected, especially where purified proteoglycans appear to be both polydisperse in size and heterogeneous in composition.

The proteoglycans in cartilage are not a single molecular species, but a population of closely related molecules. They contain chondroitin 4-sulphate or chondroitin 6-sulphate chains attached to the protein core which usually also contains some keratan sulphate. The general structure of cartilage proteoglycans was proposed by Mathews and Lozaityte[146] and Partridge, Davis and Adair[147] (Figure 4.11). In this structure the protein core forms a central

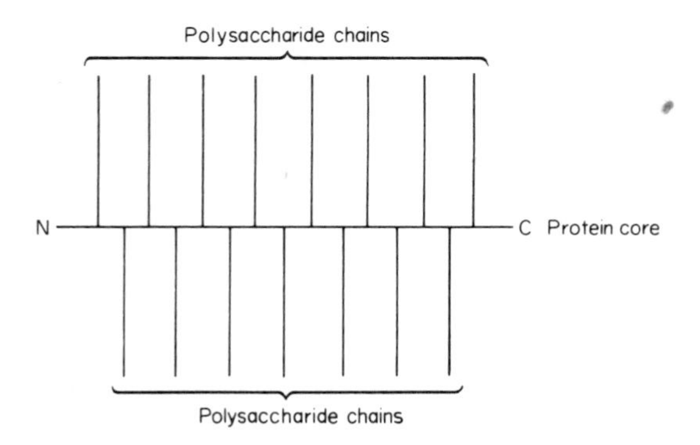


**Figure 4.11** Diagrammatic representation of the 'bottle-brush' model of proteoglycan structure proposed by Mathews and Lozaityte (1958)[146] and Partridge *et al.* (1961)[147]

backbone to which the glycosaminoglycan chains are attached by their reducing terminal ends and has often been referred to as a 'bottle brush' model. Subsequent analysis of the rates of degradation of the molecule by proteolytic attack and by digestion with testicular hyaluronidase gave support to this model[148] and electron micrographs of proteoglycan molecules are also in keeping with this structure[149,150]. The molecule typically contains 90–95% chondroitin sulphate and keratan sulphate and only 5–10% protein.

In the following sections recent developments in the understanding of proteoglycan structure are described. Improved techniques have enabled proteoglycans to be extracted and purified by methods avoiding high shear forces or precipitation (Section 4.3.2.2). It is thus more likely that the proteoglycans prepared in this way are in the native state and they have been shown to aggregate specifically into very high molecular weight complexes, suggesting that a much greater degree of order exists within the cartilage matrix than had previously been supposed (Section 4.3.5). It has also been possible to distinguish proteoglycans that aggregate from those that do not, making it necessary to reassess the heterogeneity of proteoglycan preparations (Section 4.3.7.1), and reconsider the evidence for the existence of more than one variety of core protein. Many questions remain, but these developments are a major advance in the understanding of proteoglycan structure.

## 4.3.2 Extraction

### *4.3.2.1 Disruptive extraction*

Although proteoglycans can account for up to 40% of the dry weight of cartilage, they do not readily diffuse out of the sliced tissue in solutions of low ionic strength[151] unless the collagen network is subjected to mechanical disruption by high-speed homogenisation[152]. Up to 50% of the proteoglycans may then be extracted, the final yield depending upon the type of cartilage and the extent of disruption achieved.

An extensive procedure for the fractionation of proteoglycans extracted from cartilage by high-speed homogenisation in water was developed by Schubert and co-workers[153-156]. The extracted proteoglycans were subjected to high-speed centrifugation in 0.15 M KCl, which sedimented a collagen-rich proteoglycan fraction (PPH) leaving in the supernatant a large proportion of proteoglycan (PPL) relatively free of collagen[153]. Centrifugation of the PPL in 2 M potassium acetate and subsequently in 2 M $CaCl_2$ produced further sedimentable and non-sedimentable subfractions, while treatment of PPH and the cartilage residue with hydroxylamine yielded more soluble fractions. Finally, collagenase digestion of the residue brought the remaining proteoglycan into solution. In this way all the proteoglycan of cartilage was extracted in a soluble form. The technique revealed large differences in the composition and extractability of cartilage proteoglycans from different sources[28] and of different age[154,156] (see Section 4.3.7.2). However, it was known that the treatment with hydroxylamine partially degraded the proteoglycan[157] so that it was not released in its native form. The polypeptide chain of collagen has been shown to be cleaved by hydroxylamine at asparagine–glycine linkages[158] and its action on proteoglycans may be similar. The digestion with collagenase may also produce further degradation as it is difficult to obtain purified bacterial collagenase that is free of non-specific proteolytic activity[159]. The task of solubilising all the proteoglycans was thus achieved at the expense of some limited degradation. Although the various fractions produced by differential centrifugation showed consistent differences in composition, Campo *et al.*[160] have questioned whether the fractions were discreet. They showed that when the proteoglycans from cartilage were homogenised in the presence of a radioactive preparation of PPL and subsequently fractionated according to the Schubert procedure, a significant proportion of the radioactivity was distributed in other fractions besides the main PPL fraction. It was concluded that the fractions separated were not of physiological significance. However, some of the fractionation achieved has been more recently attributed to the presence of aggregated and disaggregated proteoglycans[161] (see Section 4.3.5) and therefore some of the randomisation of labelled PPL added to a preparation may have resulted from its equilibrium with aggregated and disaggregated species.

The use of procedures involving high shear has been shown to degrade high molecular weight polymers[162] and during the preparation of proteoglycans it disrupts naturally occurring aggregates[163,164] (see Section 4.3.5) and may thus involve an alteration in their native structure. The use of high-speed homogenisation and differential centrifugation in the preparation of proteoglycans has thus largely been superseded by dissociative extraction methods and equilibrium density gradient centrifugation (see Sections 4.3.2.2 and 4.3.3.2).

### *4.3.2.2 Dissociative extraction*

The extraction of proteoglycans from sliced bovine nasal cartilage by salt solutions of different ionic strength was examined systematically by Sajdera and Hascall[165]. It was found that the solutions of several metal chlorides were effective at optimum concentrations in extracting a large proportion of the proteoglycans without homogenisation (Figure 4.12). Thus 3 M $MgCl_2$, 2 M $CaCl_2$ and 3–4 M guanidinium chloride extracted 80–85% of the total uronic acid in the cartilage within 24 hours at room temperature. *S*-Methylisothiouronium chloride, LiCl, $SrCl_2$[165], $LaCl_3$ and $CeCl_3$[166] also showed

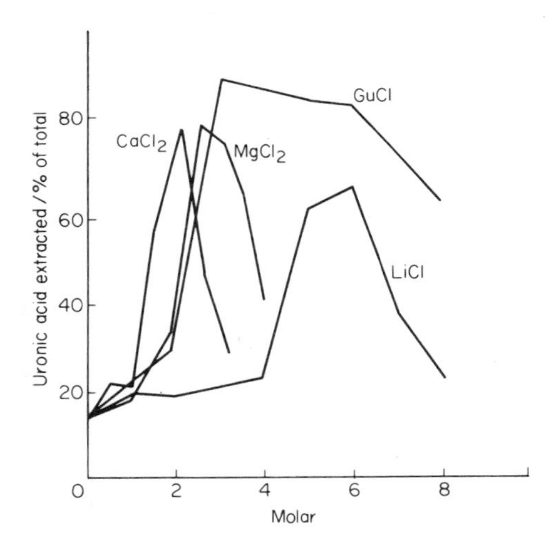


**Figure 4.12** Results of the extraction of sliced bovine nasal cartilage with solutions of various electrolytes for 20 h at 25 °C. Guanidinium chloride (GuCl), LiCl and $CaCl_2$ were buffered with 0.05 M Tris-HCl at pH 7.5. $MgCl_2$ solutions were unbuffered (pH 5–6). (From Sajdera and Hascall[165], by courtesy of the American Society of Biological Chemists.)

xtraction optima, but KCl[165] NaCl and CsCl[166] did not. After extraction he cartilage lost its rigidity but remained otherwise intact, showing that the emoval of most of the proteoglycan was not critical to its overall shape[165]. At salt concentrations significantly above the optimum, such as 4 M $CaCl_2$, ery little proteoglycan was extracted; subsequently when optimum concentrations of $CaCl_2$ or guanidinium chloride were used, the yields were greatly educed. The effect of high salt concentration thus appeared irreversible, ossibly because of denaturation of the collagen fibres[167]. Electron microscopy upported this contention as the normal banding pattern of collagen fibres n cartilage was lost after treatment[168] with 4 M $CaCl_2$. The importance of effects upon the collagen network is emphasised by the results of Herberge *t al.*[169], which showed that the optimum salt concentration for extraction of proteoglycans decreased with temperature and could be related to the lenaturation of collagen fibres revealed by x-ray diffraction. At temperatures rom 4 to 50°C the optimum concentration of $CaCl_2$ for the extraction of proteoglycan was just below that required to denature the collagen fibres. This observation implies that progressive changes in the collagen meshwork which precede denaturation are important for the extraction of proteoglycans.

Mason and Mayes[166] noted that a logarithmic relationship exists between he enthalpy of hydration of a metal ion and its optimum concentration for he extraction of proteoglycan (Figure 4.13). As the enthalpy of hydration

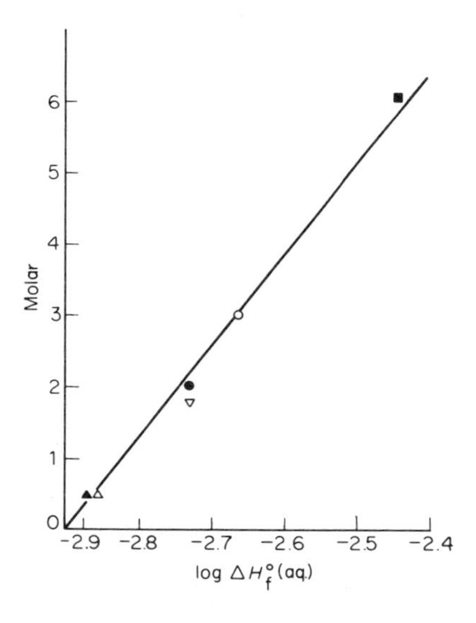


**Figure 4.13** Plot of the log values of the enthalpy of hydration [$\Delta H_f^\circ$(aq)] for various cations against the molar concentration of their chloride salts maximally effective in extracting proteoglycans from bovine nasal cartilage ■, LiCl; ○, $MgCl_2$; ●, $CaCl_2$; ▽, $BaCl_2$; ▲, $LaCl_3$; △, $CeCl_3$. (From Mason and Mayes[166], by courtesy of the Biochemical Society.)

of a cation is a measure of its affinity for water, it was proposed that, at the ery high concentration of ions necessary for extraction, there was competiion between charged groups for the limited amount of water available for ydration, which altered the interaction and conformation of matrix compoents and facilitated the diffusion of proteoglycans out of the tissue. This ypothesis also explained why $Na^+$, $K^+$ and $Cs^+$, which all have a low nthalpy of hydration, did not show extraction optima within their range of olubility. The efficient extraction achieved with guandinium chloride and

*S*-methylisothiouronium chloride is difficult to explain, however, solely on the basis of a selective decrease in solvation of charged groups[166]. These extractants have other well known properties in destroying tertiary and quaternary protein structure that may induce changes in the collagen network and other components, even more effective than those produced by high concentrations of metal ions. Thus guanidinium chloride has been shown to extract a larger proportion of proteoglycans from bovine articular cartilage than $MgCl_2$ or $CaCl_2$[169,170]. Furthermore, 6 M urea in the presence of other metal chlorides lowered their optimum concentration of extraction, although it is a very poor extractant on its own[165]. It thus had a synergistic affect, perhaps owing to its ability to disrupt protein structure.

Disaggregation of proteoglycans may be necessary for efficient extraction since conditions of extraction, which give good yields, dissociate aggregates[165]. On the other hand, aggregates are dissociated in 2 M guanidinium chloride or after reduction with dithiothreitol, although neither is effective in extracting proteoglycans from cartilage[167], so that disaggregation is not the only factor governing the extraction of proteoglycans.

#### 4.3.2.3 *Sequential extraction*

Prior to the introduction of dissociative extraction procedures, it was realised that when using mild homogenisation which did not disrupt the cartilage extensively, yields of proteoglycan were increased by raising the salt concentration of the extractant[171]. When the same cartilage was extracted sequentially with solutions of increasing salt concentration, the proteoglycans extracted by each solution differed in molecular size and composition[112]. This showed that the proteoglycans were not all bound in the same way in the tissue and provided a method of fractionating proteoglycans during their extraction. The recognition of optimum salt concentrations for extraction of proteoglycans has extended this method of fractionation to include a large proportion of the total proteoglycan in cartilage which has helped to show the extent of their heterogeneity (see Section 4.3.7)[115,172-175].

#### 4.3.2.4 *Inextractable proteoglycans*

The extraction of proteoglycans from hyaline cartilage such as bovine nasal cartilage or pig laryngeal cartilage is extremely efficient. Only about 15% of the total uronic acid remains associated with the residue after extracting for 24 hours, but prolonged extraction does not bring this fraction into solution. It can only be solubilised by treatment with hydroxylamine[157] or collagenase[156]. Although the product from either of these treatments is partially degraded, it was comparable in general composition with the salt-extractable proteoglycans[28] and hence may not be of unusual structure. Whether it remains bound to the residue because of covalent linkages to collagen or to other matrix components, or because of physical entrapment within the network of collagen fibres, remains to be established. A much larger proportion of the total proteoglycan remains in the residue of articular cartilage after dissociative extraction and this may be related to the higher collagen content of this type of cartilage. The proportion bound to the residue also increases with age (Section 4.3.7.2), which may be related to the increase in collagen content that occurs with age or the increased intermolecular and intramolecular cross-links in collagen. A proteoglycan fraction containing some hydroxyproline has been detected in a residue fraction of human costal cartilage[176]. The proteoglycan was extracted by collagenase digestion, isolated by equilibrium density gradient centrifugation and fractionated with cetyltrimethylammonium bromide on a cellulose column. The hydroxyproline remained associated with the proteoglycan throughout this procedure and with the protein core after digestion with hyaluronidase. A possible covalent linkage between collagen and the residual proteoglycan which is suggested by this evidence requires confirmation by rigorous methods.

### 4.3.3 Purifications of proteoglycans of cartilage

#### *4.3.3.1 Precipitation and ion-exchange procedures*

Cartilage contains collagen, proteoglycans and considerable amounts of other proteins including lysozyme[177], most of which have not been characterised[5]. Proteoglycans in aqueous extracts may be partially purified by precipitating collagen and other proteins by the addition of two volumes of ethanol. In the absence of added salt, the proteoglycans remain in solution[152,178]; however, this method is not sufficiently selective. Several procedures originally developed for the isolation and purification of free glycosaminoglycan chains[2] have also been applied to proteoglycans. Thus proteoglycans have been precipitated with cetylpyridinium chloride (CPC)[165], cetyltrimethylammonium bromide (CTAB)[179], and with 9-aminoacridine[171]. The risk of coprecipitation of other material can be reduced by repeated precipitation and this has been carried out effectively at different pH with 9-aminoacridine[171]. Column procedures using CPC and CTAB have also been described[116,176,180,181]. The proteoglycan–CPC or –CTAB complexes behave as the corresponding chondroitin sulphate complexes, but have higher critical electrolyte concentrations at which they become soluble. Other procedures of ion-exchange chromatography on columns of DEAE-cellulose[182] and DEAE-Sephadex[183] have also been described. Chromatography on DEAE-cellulose has also been performed in the presence of 8 M urea to reduce interaction between proteoglycans and other components. It can also be used with crude extracts which contain material insoluble at low ionic strength in the absence of urea[184].

#### *4.3.3.2 Equilibrium density gradient centrifugation*

Equilibrium density gradient centrifugation in CsCl solutions was developed for the isolation of nucleic acids and it was first applied to the purification of proteoglycans by Franek and Dunstone[185]. The technique fractionates molecules according to their buoyant density in solution. The buoyant density of a molecule is inversely related to its partial specific volume ($\bar{V}$), which for polysaccharide is *ca.* 0.60–0.65 ml $g^{-1}$ and the presence of sulphate residues give glycosaminoglycans even lower values (e.g. 0.53 ml $g^{-1}$ for chondroitin 4-sulphate)[186]. On the other hand, the partial specific volume of protein is *ca.* 0.70–0.75 ml $g^{-1}$. Thus cartilage proteoglycans, which contain about 90% glycosaminoglycans, have a buoyant density of *ca.* 1.80 g $ml^{-1}$, whereas proteins, such as collagen, which contains less than 5% of carbohydrate, have a buoyant density of *ca.* 1.35 g $ml^{-1}$ and they can thus be easily separated from each other in a density gradient.

This method was developed by Hascall and Sajdera[163] in conjunction with dissociative extraction methods (Section 4.3.2.2) to purify aggregated and disaggregated proteoglycans. It was established that proteoglycan aggregates were dissociated in the presence of the optimum concentrations of salt solutions used for extraction, such as 4 M guanidinium chloride[165], and were reformed when the concentration of salt was lowered.

Two stages of purification were devised whereby proteoglycan aggregates were separated from collagen and other protein contaminants under 'associative' conditions and then disaggregated and separated from non-covalently bound hyaluronate[16] and protein-link fractions[163] (Section 4.3.5) in a second density gradient under 'dissociative' conditions (Figure 4.14).

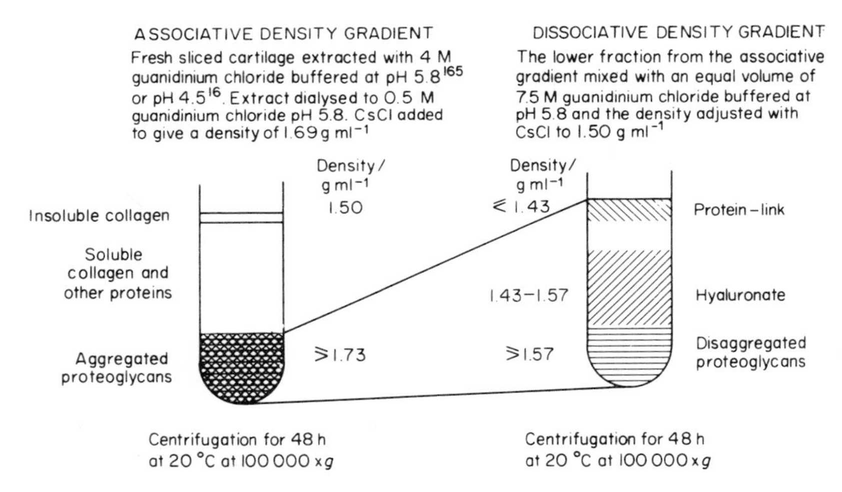
ASSOCIATIVE DENSITY GRADIENT
Fresh sliced cartilage extracted with 4 M guanidinium chloride buffered at pH 5.8[165] or pH 4.5[16]. Extract dialysed to 0.5 M guanidinium chloride pH 5.8. CsCl added to give a density of 1.69 g ml⁻¹
DISSOCIATIVE DENSITY GRADIENT
The lower fraction from the associative gradient mixed with an equal volume of 7.5 M guanidinium chloride buffered at pH 5.8 and the density adjusted with CsCl to 1.50 g ml⁻¹
Density/ g ml⁻¹
1.50
Density/ g ml⁻¹
≤ 1.43
Insoluble collagen
Soluble collagen and other proteins
Aggregated proteoglycans
≥1.73
Protein – link
1.43–1.57
Hyaluronate
≥1.57
Disaggregated proteoglycans
Centrifugation for 48 h at 20 °C at 100 000 x g
Centrifugation for 48 h at 20 °C at 100 000 x g

**Figure 4.14** Preparation and purification of cartilage proteoglycans by equilibrium density gradient centrifugation as developed by Hascall and Sajdera[163]

(a) The 'associative' density gradient. Proteoglycan extracts in 4 M guanidinium chloride are dialysed to 0.5 M guanidinium chloride to allow reaggregation to occur. Solid CsCl is then added to the dialysed extract to a density of 1.69 g $ml^{-1}$ and the solution is centrifuged at 100 000 × *g* at 20 °C for 48 h. A density gradient forms during the first 24 h of centrifugation and after 48 h essentially all the proteoglycan separates at the bottom of the gradient at densities above 1.73 g $ml^{-1}$, while the collagen and other proteins float at the top of the gradient[165] at a density below 1.55 g ml$^{-1}$. Increasing the time of centrifugation to 72 or 96 h did not improve the separation[187]. When the proteoglycans are of higher protein content than in hyaline cartilage, and as a consequence will have lower buoyant densities, the starting density of the gradient may need to be decreased to *ca.* 1.50 g ml$^{-1}$ in order to achieve comparable separations. This has been found necessary with proteoglycans from canine articular cartilage[188].

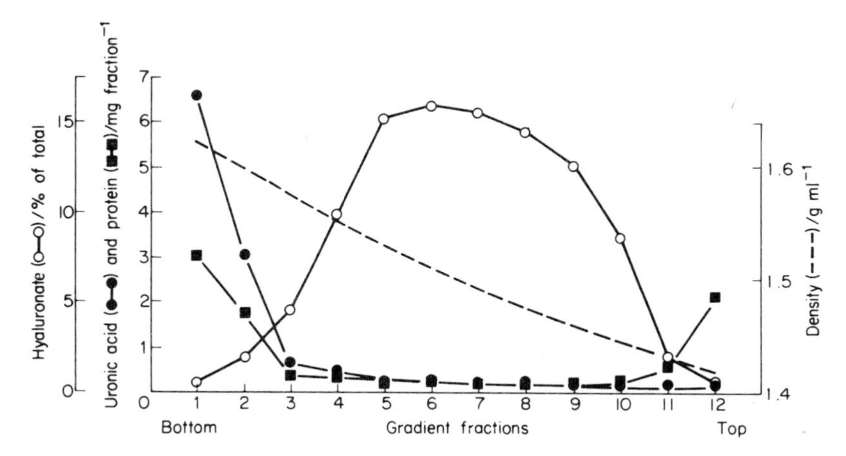


**Figure 4.15** The detailed distribution of disaggregated proteoglycan, hyaluronate and protein-link components when purified proteoglycans are fractionated by equilibrium density gradient centrifugation under dissociative conditions. (From Hardingham and Muir[16], by courtesy of the Biochemical Society.)

(b) The 'dissociative' density gradient. The purified aggregated proteoglycan fractions from the first density gradient are mixed with an equal volume of 7.5 M guanidinium chloride and the density adjusted with CsCl to 1.50 g ml$^{-1}$. The solution is then centrifugated[163] at 100 000 $\times$ *g* at 20 °C for 48 h. When the components of dissociated aggregate separate in the gradient, the proteoglycans are at the bottom, the hyaluronic acid which is of lower of buoyant density in the middle and the protein-link fraction of lowest buoyant density near the top (see Section 5.3.5) (Figure 4.14). The distribution of these components is shown in detail in Figure 4.15.

### 4.3.4 General structure of proteoglycans

Proteolytic digestion of proteoglycans cleaves the protein core and liberates free glycosaminoglycan chains attached to small peptides. The action of papain has been shown to release single chondroitin sulphate chains, but trypsin and chymotrypsin cannot degrade the proteoglycan as far. Their action produces a large proportion of doublets in which two chondroitin sulphate chains remain linked by a peptide[189]. Papain is thus able to cleave the peptide linking the chains. From the analysis of the 'doublets' produced by digestion with trypsin and chymotrypsin it appeared that the linking peptide contained less than 10 amino acids, whereas the distance between adjacent doublets was calculated to be much larger, about 35 amino acid residues[190]. The composition of the linking peptide was found to be similar in preparations of proteoglycan from cartilage from a wide variety of animals of different species and from different phyla. This suggested that there was some homology of protein structure between the proteoglycan protein cores in these different preparations[190]. Comparable evidence on the distribution of keratan sulphate chains along the protein core is not available. Keratan sulphate is of much lower molecular weight than chondroitin sulphate; thus even in bovine nasal cartilage proteoglycan, where it accounts for less than 10% of the glycosaminoglycan, it has been calculated that there are 60 keratan sulphate chains per molecule compared with about 100 chondroitin sulphate chains[63]. These are bound to a protein core of 200 000 molecular weight, which would thus contain one substituent glycosaminoglycan chain per 12 amino acid residues.

The molecular weight of proteoglycans has been estimated to be in the range of $1 \times 10^5$ to $4 \times 10^6$, varying with the methods of preparation and the source used. All preparations are polydisperse; thus the proteoglycan from bovine nasal cartilage varies in molecular weight from $1 \times 10^6$ to $4 \times 10^6$ with a weight average[163] of $2.5 \times 10^6$. Small amounts of proteoglycans of much lower molecular weight ($2 \times 10^5$ to $3 \times 10^5$) have been shown to be present in pig laryngeal cartilage[112]. These determinations of molecular weight have been by sedimentation velocity and sedimentation equilibrium methods in the ultracentrifuge. It is evident from their physical properties in solution, which include a high viscosity and excluded volume and entanglement effects, that proteoglycans occupy an extremely large domain in solution[28]. This results from the charge repulsion between the glycosaminoglycan chains favouring an expanded structure with the chains extending away from the protein core. The effect is reduced in the presence of increasing concentrations of counter-ions and it has been shown[191] that the radius of gyration of a proteoglycan measured by light scattering decreased from 1590 to 570 Å in guanidinium chloride as the concentration increased from 0.05 to 0.50 M, but did not decrease further in higher concentrations of guanidinium chloride. Because of the expanded state of proteoglycans in solution there is no rigorous mathematical treatment for the analysis of their sedimentation behaviour in a centrifugal field and the results have to be analysed in terms of equivalent rod, sphere or ellipsoid models.

The general concensus is that the proteoglycans behave in solution as a hydrated sphere or prolate ellipsoid with a low axial ratio[186,192]. However, using sedimentation equilibrium techniques and a different form of analysis developed for associating protein systems it was concluded that the proteoglycans were of basically low molecular weight (some fractions as low as 67 000 molecular weight), but formed a self-associating system producing larger species in solution[193]. However, this very low estimate of molecular weight is not easily reconciled with other evidence of proteoglycan size from light scattering[191], from the electron microscopy of the proteoglycans[149,150], from the analysis of the protein core[63] and from the stoichiometry of the interaction with hyaluronate (see Section 4.3.6)[16,194].

### 4.3.5 Proteoglycan aggregation

In 1969 Hascall and Sajdera[163] produced detailed evidence of aggregation of highly purified proteoglycans. The possibility of proteoglycan aggregation had been raised by several authors over the past 15–20 years, but it was only with the development of efficient and reproducible methods of extraction and purification that unequivocal results were obtained that formed a basis for future work.

The proteoglycans of bovine nasal cartilage, which had a molecular weight of *ca.* $1 \times 10^6$ to $4 \times 10^6$, were shown to form very high molecular weight aggregates (molecular weight $30 \times 10^6$). These aggregates were not formed by the self-association of the proteoglycans but by their binding with other specific non-proteoglycan components. About 25% of the protein in the aggregate was shown to be present in a 'protein-link' fraction (initially called 'glycoprotein-link') that was separated from the proteoglycan by equilibrium density gradient centrifugation in 4 M guanidinium chloride, under conditions where the aggregates were dissociated[163]. The role of linking proteoglycans together to form aggregates was initially attributed to this component[161,163,170]. However, it was shown subsequently that hyaluronic acid accounted for *ca.* 1% of the uronic acid in the aggregate and was also separated from the proteoglycans by the dissociative density gradient procedure (Figures 4.14, 4.15)[15,16], and the proteoglycans were shown to bind directly to hyaluronate molecules even in the absence of the 'protein-link' component[16,194]. However, the formation of aggregates that were stable in the ultracentrifuge required both hyaluronate and the 'protein-link' component to be present[195]. The basis of aggregation was thus concluded to be the binding of many proteoglycan molecules along a single hyaluronate chain with the protein-link components stabilising this interaction[16]. The role of hyaluronate in aggregation was confirmed in pig laryngeal cartilage, articular cartilage, nasal cartilage and nucleus pulposus[196] and in bovine nasal and tracheal cartilage[197]. The protein-link fraction contained two proteins that were separated by disc electrophoresis[197] and which had a molecular weight of 40 000–65 000. They have also been identified immunologically as they appear to contain separate species-specific antigens[198]. There are thus four separate molecular species in proteoglycan aggregates: proteoglycans, hyaluronate and two proteins.

With the identification of the components that are necessary for the formation of proteoglycan aggregates, it is possible to summarise the events that take place in the dissociative extraction and density gradient purification procedures (Figure 4.14). The proteoglycan aggregates present in the cartilage are dissociated in the optimum concentrations of the most effective salts. The crude extract thus contains disaggregated proteoglycans which are reaggregated when the solution is dialysed to low ionic strength. The reformed aggregates then remain stable in spite of the high CsCl concentration (4.6 M) during the first density gradient centrifugation step under 'associative' conditions, and the hyaluronate and protein-link fractions remain bound to the proteoglycans at high buoyant density. It is only in the second density gradient in the presence of a 'dissociative' solvent, 4 M guanidinium chloride,

that these components are separated from the disaggregated proteoglyca
and migrate to their own characteristic density in the gradient. The behaviou
of the purified proteoglycan fractions in the ultracentrifuge and on ge
chromatography is shown in Figure 4.16.

Proteoglycan aggregates were shown to be reversibly dissociated as th
pH was decreased from pH 7 and dissociation was complete[163] at pH 3.
The proportion of proteoglycans that were aggregated in an extract was als
found to depend upon the pH of extraction[163]. An optimum of about 80%

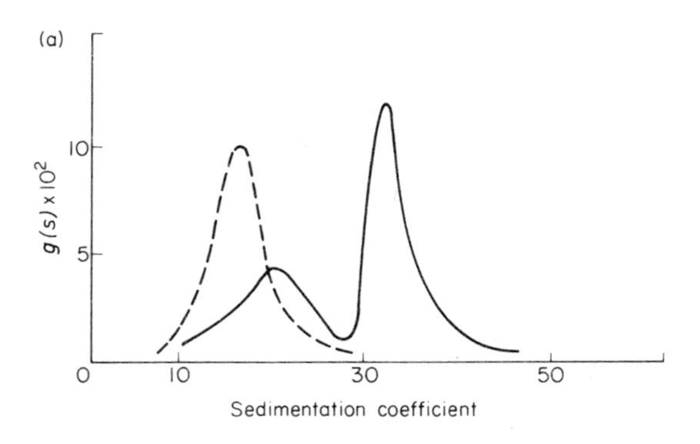


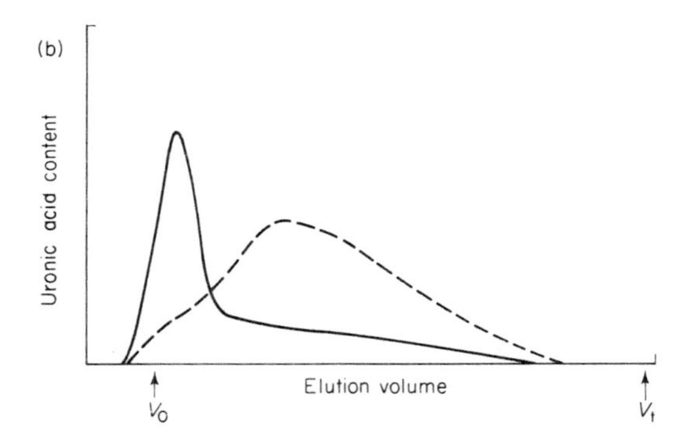


**Figure 4.16** Behaviour of purified proteoglycan fractions in the ultracentrifuge and on gel chromatography. (a) Apparent sedimentation coefficient distributions of aggregated (———) [0.15% (w/v) solution] and disaggregated proteoglycan (----) [0.12% (w/v) solution] in 0.5 M guanidinium chloride buffered with 0.05 M sodium acetate at pH 5.8 determined after centrifugation at 35 500 r.p.m. in a Spinco model E ultracentrifuge at 20°C. (Redrawn from Hascall and Sajdera[163], by courtesy of the American Society of Biological Chemists). (b) Gel chromatography of purified aggregated (———) and disaggregated (----) proteoglycan on a column (165 cm × 1.1 cm) of Sepharose 2B in 0.5 M sodium acetate, pH 6.8 at 4°C. (Redrawn from Hardingham and Muir[16], by courtesy of the Biochemical Society.)

of the proteoglycans from bovine nasal cartilage reformed aggregates when extracted in 4 M guanidinium chloride at pH 5.8, but, in a separate study using different techniques to assess size, an optimum of *ca.* 75% at pH 4.5 was found for pig laryngeal cartilage[16]. The amount of aggregate isolated was not related to the amount of hyaluronate extracted, which remained constant between pH 3.0 and pH 8.5, but it was found that after extraction either above or below pH 4.5 there was an irreversible loss in the ability of some of the proteoglycans to interact with hyaluronate[16]. The choice of pH for extraction thus appeared to be important in maintaining the integrity of the interaction that contributed to the formation of stable aggregates.

The reduction and alkylation of cysteine residues in the proteoglycan and to a lesser extent in the protein-link components prevented the formation of aggregates[163]. It was subsequently shown that the binding of proteoglycans to hyaluronate was abolished by this treatment[16]. The evidence suggested that it was a structure of the protein 'core' of the proteoglycans maintained by disulphide bridges that was essential for the binding of proteoglycans to hyaluronate. The adverse effect of 4 M guanidinium chloride at high pH during extraction may thus reflect denaturation of this binding site which would limit the amount of aggregate isolated. There may also be similar effects upon the protein-link components and perhaps it is more surprising that such a large proportion of the binding sites, both on the proteoglycans and on the protein-link components, appear to survive the extraction procedure which exposes them to such effective denaturing solvents. Some caution is thus needed in the assessment of the yield of aggregates obtained from tissues where the critical effects of the type of extractant, the pH of extraction and the purification procedure used have not been fully evaluated.

Further insight into the structure of the aggregate was provided by the observation that treatment of the aggregate with chondroitinase ABC removed a large proportion of the chondroitin sulphate and thus reduced the size of the complex, but it remained much larger than similarly digested disaggregated proteoglycan cores[134] and could still be reversibly dissociated into smaller components. This suggested that the proteoglycan cores and protein-link components were still able to bind to the hyaluronate[199]. Further experiments confirmed that after chondroitinase ABC digestion proteoglycans were still able to bind reversibly to hyaluronate and also showed that the presence of the chondroitin sulphate was not necessary for binding to take place[197]. Hyaluronate is normally a substrate for chondroitinase ABC, but the reason for its failure to digest hyaluronate in the aggregate may be because the binding of proteoglycans and the 'protein-link' components along the entire length of the chain may make it sterically inaccessible to the digesting enzyme[197]. Partial digestion of proteoglycan aggregates under mild conditions with papain (EC 3.4.22.2) has been shown to remove progressively the region of the proteoglycan containing the chondroitin sulphate, leaving a large protein fraction still bound to the hyaluronate, which may contain the protein-link components and probably the binding region of the proteoglycan[197]. Similar treatment of intact cartilage led to the isolation of hyaluronate with comparable protein fractions bound to it, which suggested that aggregates similar to those isolated after dissociative extraction and purification were indeed present within the cartilage *in vivo*[197].

### 4.3.6 The binding of proteoglycans to hyaluronate

The characteristics of the binding of proteoglycans to hyaluronate have bee studied in detail in our laboratory[16, 196, 200]. The binding produced a comple that could be detected by its large size, measured by gel chromatograph or by its high viscosity (Figure 4.17)[194]. The interaction was detectable at weight ratio of proteoglycan:hyaluronate of 10 000:1 and increased wit hyaluronate concentration until it was optimal at about 150:1. With highe proportions of hyaluronate the effect diminished, which suggested that th proteoglycans did not cross-link hyaluronate chains, and hence containe only a single binding site.

Hyaluronate from a variety of sources such as umbilical cord, cock comb and cartilage all interacted with proteoglycans, as also did hyaluronat that was partially degraded by treatment with ascorbate and $Cu^+$ ion (Section 4.2.2). The increase in viscosity of a proteoglycan solution wa related to the amount and the molecular weight of the hyaluronate chain added, but gel chromatography showed that the proportion of proteoglycan bound was related to the total amount of hyaluronate added and not to it molecular weight. The interaction was specific to hyaluronate, as comparabl effects were not produced by other macromolecular polyanions such a alginate, dextran sulphate or DNA, and the interaction was also unaffecte by the presence of EDTA[194] and was therefore unlikely to depend on divaler alkali metal ions. The complex formed was reversibly dissociated in solution of guanidinium chloride above 2 M and the interaction progressively de creased and was finally eliminated as the pH was lowered from pH 7 to or the temperature was raised from 20 to 60 °C.

Under conditions which favoured binding (in 0.5 M sodium acetate, pH 6.8 at 4 °C), gel chromatography on Sepharose 2B separated the comple from the free proteoglycan[194] and showed that hyaluronate chains boun *ca.* 250 times their weight of proteoglycan when the latter was present in excess. This was a measure of the maximum packing of proteoglycan along the hyaluronate chains. Assuming[16, 161] a molecular weight of protec glycans[161, 163, 201, 202] of $1 \times 10^6$ to $4 \times 10^6$, it was calculated that each protec glycan was bound to a region of the hyaluronate chain, *ca.* 20 nm long of $1 \times 10^4$ equivalent weight. A single hyaluronate chain of $1 \times 10^6$ mole cular weight could thus bind as many as 100 proteoglycans. If a simila calculation is made for the natural aggregate from pig laryngeal cartilage assuming hyaluronate to account for 0.7% of the total uronic acid in th cartilage, and 80% of the extracted proteoglycans to be aggregated[16], th number of proteoglycans bound is considerably less than this, suggestin that each proteoglycan molecule occupies a region of the hyaluronat chain of *ca.* 25 000 equivalent weight. A similar result was suggested fo proteoglycan aggregates from bovine nasal cartilage[197]. The number of protec glycans bound to hyaluronate in the cartilage aggregate is thus less than th maximum number suggested by a study of proteoglycan–hyaluronat interaction. The apparent increased spacing between the proteoglyca molecules on the hyaluronate chain may merely indicate that there is mor hyaluronate in the cartilage than is required to bind all the availabl

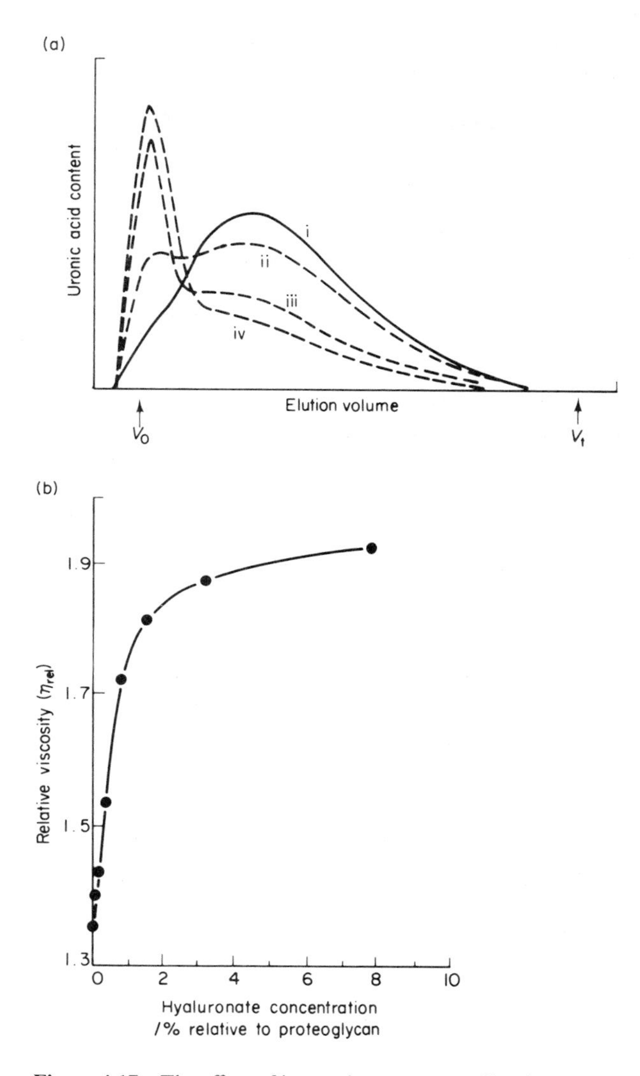


**Figure 4.17** The effect of increasing amounts of hyaluronate on the gel chromatographic profile and viscosity of disaggregated proteoglycans. (a) Disaggregated proteoglycans chromatographed on a column of Sepharose 2B with (i) no hyaluronate; (ii) 0.03% hyaluronate added; (iii) 0.13% hyaluronate added; and (iv) 0.63% hyaluronate added. (b) Viscosity of disaggregated proteoglycan with increasing amounts of hyaluronate added, measured in an Ostwald capillary viscometer at 30°C in 0.5 M guanidinium chloride, 0.05 M sodium acetate, pH 5.8 (Redrawn from Hardingham and Muir[194], by courtesy of Elsevier.)

proteoglycan, or that the protein-link components may also occupy spac
along the chain[197].

Small oligosaccharides derived from hyaluronate by digestion wit
testicular hyaluronidase inhibited the interaction of proteoglycans wit
intact hyaluronate measured in a viscometer[200, 203], and the binding o
chondroitinase ABC digested proteoglycan cores to hyaluronate measure
by gel chromatography[203]. This suggested that the oligosaccharides wer
competing with the hyaluronate for the binding site on the proteoglycan
The inhibition produced by purified oligosaccharide fractions showed tha
the smallest fraction able to bind strongly was a decasaccharide. Its effective
ness in competing with macromolecular hyaluronate showed that it wa
bound almost as strongly. This indicated that although at saturation th
distance between adjacent proteoglycans was at least 20 nm (10 000 equivalen
weight)[200], most of the binding was with a region corresponding to a de
casaccharide which would be only 5 nm long.

From a consideration of the molecular dimensions of proteoglycans, an
assuming a structure similar in outline to the 'bottle brush' model propose
by Mathews and Lozaityte[146] and Partridge, Davis and Adair[147], a model o
the proteoglycan–hyaluronate complex was proposed (Figure 4.18). I

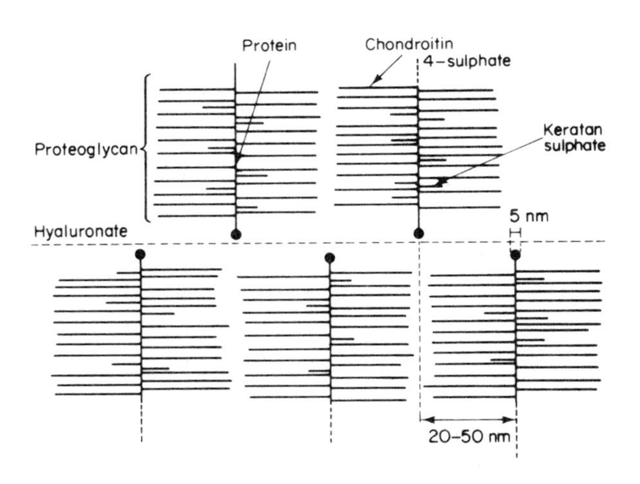


**Figure 4.18** Two-dimensional representation of the proteo-glycan–hyaluronic acid complex. The details and dimensions are taken from the results of Hardingham and Muir[194, 200]. This structure also forms the basis of the proteoglycan aggregate, which contains the protein-link components bound to the hyaluronate–proteoglycan binding region[197]

order for each proteoglycan to occupy only a 20 nm length of chain it wa
necessary that the proteoglycans align themselves with the axis of thei
protein cores perpendicular to the hyaluronate chain, which suggested tha
the binding site on the proteoglycans was located at one end of the protei
core. This is consistent with the appearance of an electron micrograph of

large proteoglycan aggregate which was published before the involvement of hyaluronate in aggregation was known[161]. The overall size of the complex would obviously be directly related to the molecular weight of the hyaluronate chain, and if the latter were polydisperse this would also produce aggregates of different size.

The binding of proteoglycan to hyaluronate appears to be an equilibrium which lies well in favour of complex formation under physiological conditions of ionic strength, pH and temperature. There seems to be no such equilibrium with intact aggregates containing the protein-link components, as there was only a small decrease in viscosity when an excess of oligosaccharides of hyaluronate were added to a solution of aggregates[203]. The proteoglycan–hyaluronate complex differed from the natural aggregate in other ways, as it was unstable both in the ultracentrifuge and on composite agarose–polyacrylamide disc-gel electrophoresis[204].

### 4.3.7 Heterogeneity of cartilage proteoglycans

#### *4.3.7.1 Introduction*

The foregoing discussion of aggregation has been developed without reference to the heterogeneity of the proteoglycans. The ability of equilibrium density gradient centrifugation under 'dissociative' conditions to separate a non-covalently bound protein fraction that is an integral part of the proteoglycan aggregate, but is not separated from it by precipitation with 9-aminoacridine or cetylpyridinium chloride, suggests that it is difficult to assess the chemical heterogeneity amongst the proteoglycan population unless the absence of such protein and hyaluronate components has been demonstrated, or the contribution to heterogeneity made by aggregation is taken into account. Much of the data on heterogeneity reported before the presence of this type of interaction was understood has thus required re-evaluation and in some cases reinvestigation.

Although the concept of reversible aggregation usually implies the binding together of identical subunits[163], there is ample evidence that even after disaggregation the proteoglycans remain heterogeneous in composition and polydisperse in size. However, in assessing heterogeneity it is apparent that the type of preparation studied and the methods used to examine it are most important. Even dissociative extraction and density gradient purification can lead to the preferential selection of proteoglycans of high buoyant density, with proteoglycans of low buoyant density (that are of high protein content) being discarded[205,206]. The effect this has on the representative nature of the purified proteoglycan varies from one tissue to another. Its effect is probably low with proteoglycans from bovine nasal cartilage where the average protein content is low, but could be very significant with proteoglycans from articular cartilage where the average protein content is much higher. A careful choice of the conditions of preparation is therefore required if this sort of selection is to be minimised.

The use of sequential extraction techniques (Section 4.3.2.3) showed that the composition of proteoglycans progressively changed when they were extracted from cartilage by increasing the concentration of the extractin salt or prolonging the time of extraction, and the change in composition wa similar in different types of cartilage. Proteoglycans extracted from pi laryngeal cartilage by mild homogenisation in 0.15 M sodium acetate wa found to be of much lower molecular weight and smaller hydrodynam size than those extracted subsequently with 0.9 M $CaCl_2$ (10% w/v) an analysis showed them to have a lower keratan sulphate and protein content[11] This was also observed with proteoglycans in sequential extracts of pi articular cartilage[172,173]. The changes in composition observed did not onl reflect a higher proportion of large molecules, because in later extracts eve the smallest molecules had a higher keratan sulphate and protein conten Thus even amongst molecules of the same size it appeared that those wit higher keratan sulphate and protein content were bound more firmly withi the cartilage matrix. The later extracts in such studies undoubtedly containe

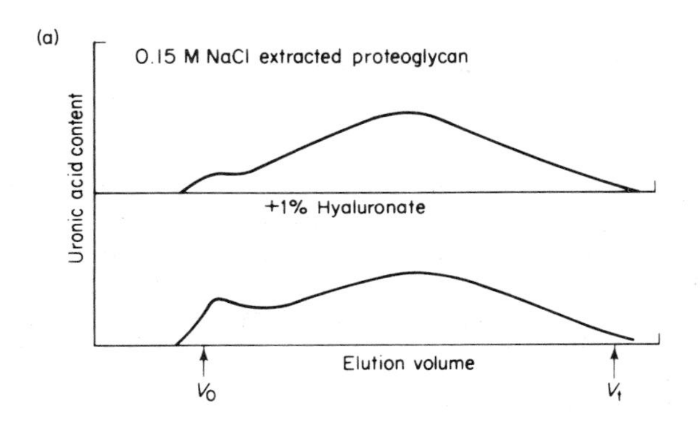


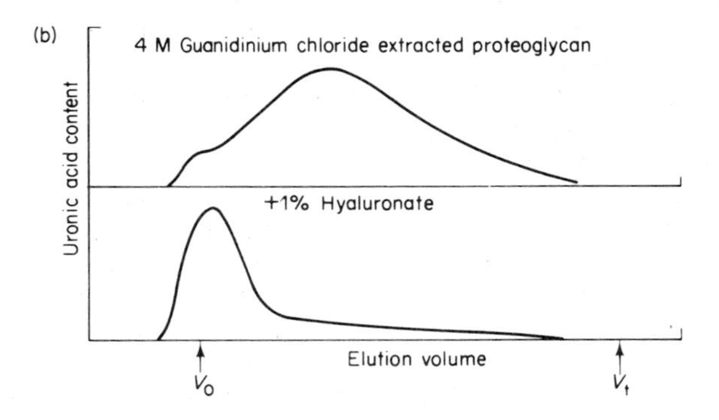


**Figure 4.19** Comparison of the hydrodynamic size and ability to interact with hyaluronic acid of proteoglycans which were disaggregated after sequential extraction from pig laryngeal cartilage with (a) 0.15 M NaCl; and (b) 4 M guanidinium chloride. Gel chromatography on a column (165 cm × 1.1 cm) of Sepharose 2B eluted with 0.5 M sodium acetate, pH 6.8 at 4 °C. (From Hardingham and Muir[16], by courtesy of the Biochemical Society.)

some aggregates, but the differences in size and composition applied to fractions that were too small to be aggregated. This has been exemplified by comparison of the proteoglycans extracted from pig laryngeal cartilage with 0.15 M NaCl and those extracted subsequently with 4 M guanidinium chloride[16]. These two extractants were found to give a significant separation of non-aggregated and aggregated proteoglycans (Figure 4.19). The 0.15 M NaCl extract contained only 12% of the total uronic acid of the cartilage and the proteoglycans were of small size and contained very few aggregates, whereas the guanidinium chloride extract accounted for 71% of the total uronic acid and more than 90% of the proteoglycans in it were aggregated. The ability to extract small non-aggregated proteoglycans from cartilage under mild conditions that would not dissociate aggregates was strong evidence for their co-existence with aggregates in cartilage *in vivo*[16]. After disaggregation, the guanidinium chloride-extracted proteoglycans were still of larger average size and of higher protein and keratan sulphate content than the non-aggregated proteoglycans, which remained unaffected by the disaggregation procedure. From the difference in size and composition of these two proteoglycan fractions it can be implied that they were based upon different protein cores, for if they contained the same protein cores the proteoglycan of smaller size would be expected to have a higher protein content. The presence of dissimilar protein cores was also supported by further evidence. There were significant differences in their amino acid composition as the non-aggregated fraction contained fewer aromatic and basic amino acids and noticeably fewer cysteine residues (Table 4.4). The

**Table 4.4 Amino acid compositions of aggregated and non-aggregated proteoglycans***

| | *Residues per 1000* | |
|---|---|---|
| | *Non-aggregating proteoglycan* | *Aggregating proteoglycan* |
| Asp | 68.3 | 76.3 |
| Thr | 56.2 | 62.3 |
| Ser | 150.0 | 122.7 |
| Glu | 145.8 | 140.0 |
| Pro | 79.8 | 98.5 |
| Gly | 142.9 | 129.5 |
| Ala | 71.9 | 79.4 |
| Cys | Trace | 2.4 |
| Val | 61.8 | 58.9 |
| Met | 3.9 | 3.4 |
| Ileu | 38.8 | 36.5 |
| Leu | 80.7 | 76.7 |
| Tyr | 13.0 | 19.8 |
| Phe | 25.3 | 28.0 |
| Lys | 10.2 | 11.5 |
| His | 8.0 | 9.8 |
| Arg | 29.1 | 44.1 |

* Amino acid compositions of non-aggregated proteoglycan extracted from pig laryngeal cartilage with 0.15 M NaCl, and aggregating proteoglycan from the same cartilage released by further extraction with 4 M guanidinium chloride. Both preparations were purified in a 'dissociative' density gradient[16]

latter may be particularly significant as cysteine disulphide bridges have been identified as being essential for the structure that binds the proteoglycan to hyaluronate (Section 4.3.6). The non-aggregated proteoglycan was found unable to bind to hyaluronate and the presence of cysteine in only trace amounts may thus imply the absence of a hyaluronate binding region in the protein core of these proteoglycans.

Proteoglycan fractions comparable to the non-aggregating fraction have also been observed in pig articular cartilage[115,174] and also in bovine nasal cartilage[187]. About 20% of the proteoglycans from bovine nasal cartilage were extracted in 0.15 M KCl and comparison of the composition of these proteoglycans with those extracted with 4 M guanidinium chloride (representing 85% of the total) showed the former to be of lower protein and keratan sulphate content and they contained significant differences in amino acid composition[187]. This result again suggested more than one type of proteoglycan to be present and contrasts with results from a further detailed study in which it was concluded that only a single polydisperse proteoglycan species was present in this cartilage. Disaggregated proteoglycans from bovine nasal cartilage were fractionated in a dissociative density gradient and produced three major fractions that varied in their relative content of protein, chondroitin sulphate and keratan sulphate, but showed negligible difference in amino acid composition. This, together with their physical properties, suggested that they represented a single polydisperse proteoglycan species[163]. A similar conclusion was drawn after fractionation of proteoglycans following exhaustive digestion with chondroitinase ABC or AC. The majority of the chondroitin sulphate was removed from the proteoglycan, leaving a keratan sulphate-enriched protein core that appeared polydisperse, but not heterogeneous. The variation in composition observed appeared directly related to the molecular weight and was suggested to result from a variable degree of substitution of the protein core with chondroitin sulphate and keratan sulphate chains[63]. These apparently conflicting results may yet be consistent with each other if the aggregating and non-aggregating proteoglycans are both polydisperse species with variations in their carbohydrate/protein ratios that cause them to overlap in composition and fail to be separated by equilibrium density gradient centrifugation.

Further evidence of heterogeneous protein cores has been reported for the proteoglycans from pig larynx[207] and bovine trachea[56]. After digestion with testicular hyaluronidase, fractionation by gel chromatography showed three components of different size and composition. The most abundant 'core' fraction was that of largest size and was also of high keratan sulphate content. Further degradation of this fraction with alkali suggested that some of the keratan sulphate contained an alkali-stable linkage to protein. The high glutamate (or glutamine) content of the fraction, and of the keratan sulphate peptide fraction derived from it, suggested that some keratan sulphate may be linked to this amino acid in the proteoglycan core of largest size, but not in the smaller core fractions.

Electron microscopy, in addition to confirming the general structure of proteoglycans, also suggested that they were based on a protein core of two different sizes[149,150]. A long form of the molecule was almost twice the average length of a short form (3200–3400 Å compared with 1700–2000 Å), and it was noted that only the long forms seemed to be present in electron micrographs of aggregates[149]. On this basis it was proposed that aggregation might proceed by the binding of two short forms to make a long form which could then form multiple aggregates. However, no evidence has been presented that short forms can bind to make long forms, and in the light of more recent information another interpretation is possible. The absence of short forms from aggregates suggests that they may be the non-aggregating proteoglycans that have been shown to be of small size[16]. The fact that they appeared more frequently in the electron micrographs than the long form could result from the use of PPL preparations of proteoglycans which have been subjected to high-speed homogenisation and may be partly degraded (Section 4.3.2.1), and also represent only a minority of the soluble proteoglycans of cartilage. The short forms occurred much less frequently than long forms in a preparation of disaggregated proteoglycan prepared by dissociative extraction techniques (see Section 4.3.2.2)[149], which suggested that their abundance was similar to that of non-aggregating proteoglycans.

The amino acid composition of cartilage proteoglycans from different sources shows a general similarity that has suggested some homology in protein structure[190]. However, this is not inconsistent with the presence of more than one type of proteoglycan core, because the homology could apply equally well to a group of polypeptides as to a single species. The existence of distinct types of proteoglycan suggests that developmental and pathological changes in proteoglycan composition may have at least two origins. One may be the result of preferential synthesis of particular proteoglycan cores, the amino acid sequence of which may determine the average number and type of polysaccharide chains attached[208]. The other may involve some changes in the production and organisation of the enzymes and substrates necessary for polysaccharide chain synthesis (see Chapter 3). Such changes may result in alteration of the physical properties of proteoglycans by limiting the amount or size of the aggregates they form or by affecting their interaction with other components of the extracellular matrix.

#### *4.3.7.2 Changes in composition of proteoglycans with development and ageing*

The changes in proteoglycan composition in cartilage during development and ageing have been shown mainly by analysis of the glycosaminoglycan content, and also by comparison of disruptively and dissociatively extracted (Section 4.3.2) proteoglycans. Although cartilage from different anatomical sites and species contains proteoglycans in which there is a large variation in the relative amounts of chondroitin 4-sulphate, chondroitin 6-sulphate, keratan sulphate and protein, there are some general trends in development that are common to many types of cartilage. The degree of sulphation of chondroitin sulphate increases during development, which is reflected in a decline in non-sulphated disaccharide units that occur in embryonic or young tissue, and an increase in disulphated disaccharide units[34, 209]. There is also frequently an increase in the amount of chondroitin 6-sulphate relative to the amount of chondroitin 4-sulphate[209, 210], but in some cartilages, such as rabbit costal cartilage, the reverse occurs and this may be related to subsequent calcification and loss of proteoglycan[209]. The proportion of keratan sulphate and its degree of sulphation also increases with age[210]. Human costal cartilage is an example of such changes in glycosaminoglycan composition (Figure 4.20)[211], which obviously reflect a profound alteration in the proteoglycans being synthesised. Direct examination of proteoglycans by disruptive[154] or dissociative extraction[174] procedures shows that they become more difficult to extract with age and the proportion associated with the residue increases, which may partly result from progressive changes in the collagen matrix (see Section 4.3.2.4). Thus in foetal pig, articular cartilage

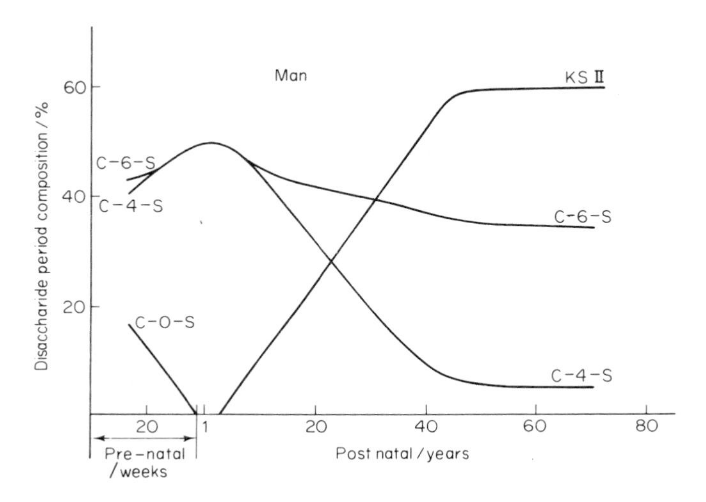


**Figure 4.20** The composition of the glycosaminoglycans of human costal cartilage at various embryonic and postnatal ages shown by the analysis of the disaccharide units: C-4-S, chondroitin 4-sulphate; C-6-S, chondroitin 6-sulphate; C-O-S, non-sulphated chondroitin disaccharides; KS-II, skeletal keratan sulphate. (Smoothed data from Mathews and Glagov[211], by courtesy of the American Society for Clinical Investigation.)

almost no proteoglycan remained in the residue after extraction with 2 M $CaCl_2$, whereas the residue fraction accounted for 25% of the total uronic acid 25 weeks after birth and 30% at 5 years of age[174]. The keratan sulphate content and protein content of the extacted proteoglycan fractions increased with development even amongst fractions of small size that would appear to be non-aggregated[174]. A preliminary study of the proteoglycans from pig laryngeal cartilage suggested that a considerably smaller proportion of the proteoglycans were able to aggregate in cartilage from a 5 year old animal than in that from animals 6–9 months old[196]. However, in pig articular cartilage of comparable ages there was no evidence of a decreased proportion of aggregates[174]. It is not yet possible to relate the changes in proteoglycan composition to the aggregation of the proteoglycans and their interaction with other matrix components. There is some evidence that in human articular cartilage the molecular weight of the chondroitin sulphate chains decreases slightly during development[212], but this type of change could not account for the magnitude of the alterations in composition seen in many tissues, which must reflect a decrease in the number of chondroitin sulphate chains and an increase in the number of keratan sulphate chains attached to the protein core. However, it is only when the detailed structure of the protein cores is determined that the observed alterations in composition can reliably be interpreted on the basis of proteoglycan structure.

## 4.4 INTERACTION OF PROTEOGLYCANS AND GLYCOSAMINOGLYCANS WITH SPECIFIC PROTEINS

### 4.4.1 Interaction with collagen

The interaction of proteoglycans with proteins in connective tissue, particularly collagen, is functionally of great importance. Despite considerable research over more than a decade, the way in which proteoglycans interact with collagen is not yet fully understood, although it depends mainly on electrostatic forces[189, 213-218]. In general, glycosaminoglycans interact more strongly with collagen with increasing charge density and chain length, and heparan sulphate and dermatan sulphate, which contain some iduronic acid, interact more strongly than chondroitin sulphates of similar charge density[218]. This may partly explain why dermatan sulphate proteoglycans are much more difficult to extract from tissues than chondroitin sulphate proteoglycans (see Section 4.7.1), and why they have different effects on the precipitation of soluble collagen.

The ability to form fibrils is an inherent property of collagen, but, at physiological pH and ionic strength, soluble collagen remains in solution at 4 °C; when the solution is warmed to 37 °C, however, it precipitates spontaneously as fibrils. On the other hand, collagen will precipitate at 4 °C in the presence of dermatan sulphate proteoglycan from skin[218] or heart valves, but it will not precipitate in the presence of chondroitin sulphate proteoglycan from heart valves[214]. This suggests that dermatan sulphate may have a special role in the formation of collagen fibres *in vivo*. Meyer[219] has pointed out that dermatan sulphate is found in those tissues where there are coarse or thick collagen fibres and has so far not been identified in tissues where the fibres are much finer such as cartilage or cornea.

What controls how collagen is laid down is not understood. The morphology of collagen fibres, their orientation and distribution, as well as thickness, vary greatly from one tissue to another and even in a comparatively uniform tissue such as articular cartilage there is considerable variation. The fibres towards the surface are mainly much finer than those at greater depths, the fibres becoming thicker with increasing depth from the surface[220, 221].

The formation of fibrils from soluble collagen takes place in two phases, nucleation and growth (Figure 4.21)[222]. The nucleation phase, when molecular aggregates are forming, corresponds to the lag before the turbidity of the solution starts increasing. During the growth phase when fibres are forming

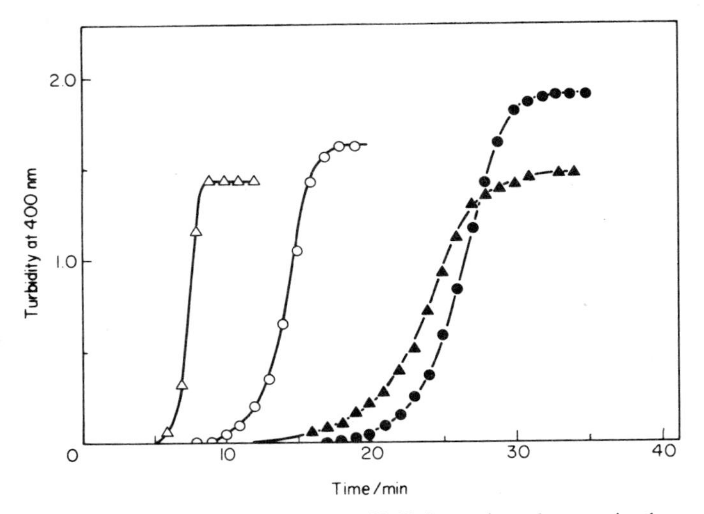


**Figure 4.21** The rate of collagen fibril formation alone or in the presence of glycosaminoglycans or cartilage proteoglycan. Fibril formation was initiated by warming collagen solutions from 4 to 37 °C. Collagen alone (●); collagen + chondroitin sulphate (○); collagen + dermatan sulphate (△); collagen + proteoglycan (▲). (From Obrink[223], by courtesy of Springer-Verlag.)

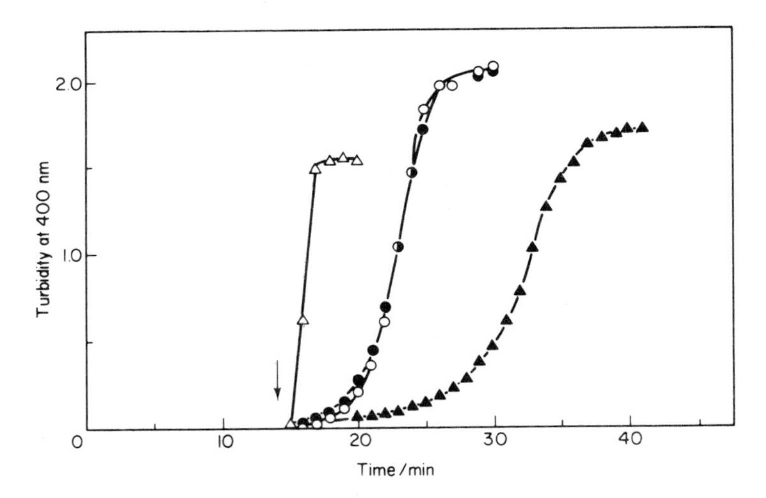


**Figure 4.22** Effects of glycosaminoglycans or cartilage proteoglycan on the rate of collagen fibril formation when added at the end of the lag phase (indicated by arrow). Fibril formation was initiated by warming collagen solutions from 4 to 37 °C. Collagen alone (●); collagen + chondroitin sulphate (○); collagen + dermatan sulphate (△); collagen + proteoglycan (▲). (From Obrink[223], by courtesy of Springer-Verlag.)

the turbidity increases in a sigmoidal manner. The final diameter and heterogeneity of the fibres is determined by the properties of the nuclei, which, as first shown by Wood[213], may be modified by glycosaminoglycans and proteoglycans[214-217, 223]. By adding glycosaminoglycans before or after the nucleation phase, their influence on each phase may be studied[223]. Most glycosaminoglycans and proteoglycans which interact[223] with soluble collagen at 4 °C tend to accelerate fibre formation when present during the nucleation phase (Figure 4.21), whereas when added during the growth phase, chondroitin sulphate, and particularly chondroitin sulphate proteoglycans, delayed fibre formation (Figure 4.22), presumably by inhibiting fibre growth[223]. On the other hand, aggregates of chondroitin sulphate proteoglycans did not delay fibrillogenesis[224]. Keratan sulphate, which does not interact with soluble collagen[217], had no effect. Hyaluronic acid increased the activity of tropocollagen due to mutual steric exclusion and hence accelerated both nucleation and growth[223]. The initial interaction of proteoglycans with collagen and their state of aggregation may thus be critical in determining the morphology of collagen fibrils. Results from different laboratories are not always consistent, however, probably because of differences in the collagen preparations used and whether or not they already contained some aggregated collagen[217].

The interaction with collagen of chondroitin sulphate proteoglycan from cartilage, which is significant under physiological conditions, increases as pH and ionic strength are lowered[218]. This interaction may increase the thermal stability of collagen *in vivo* before it is laid down as fibres, since the thermal stability of soluble collagen was increased by interaction with chondroitin 6-sulphate, as shown by circular dichroism spectroscopy. At an optimum concentration ratio the melting temperature was raised[225] from 38 to 46 °C. Below this ratio a biphasic melting temperature curve showed that some of the collagen was free and some bound, which suggests that there is some form of co-operative binding between collagen and chondroitin 6-sulphate.

In addition to stabilising a soluble protein, interaction with glycosaminoglycans may also influence the conformation of the protein. Thus in the presence of chondroitin 6-sulphate, poly-L-lysine is forced to adopt an α-helix to a much greater extent, as shown by circular dichroism, than in the presence of chondroitin 4-sulphate, whereas in their absence it has a charged coil form[226]. Sulphate groups would appear to be necessary for these effects since no α-helix was detectable in the presence of hyaluronic acid or desulphated chondroitin sulphate. Recent X-ray analyses of oriented films of isomeric chondroitin sulphates show the sulphate group to be further from the polysaccharide chain in the 6-sulphate than in the 4-sulphate (see Section 4.2.9). Each isomer may well have a different biological function because their relative proportions vary in different tissues, as well as during development and ageing (see Section 4.3.7.2). Their different functions might depend on their influence on the conformation of proteins at the time that they are being laid down in the intercellular matrix of connective tissue

It is becoming evident that there are several genetic forms of mammalian collagen, that of cartilage and skin being of different types[227, 228]. It is possible that the interaction of a given proteoglycan with different types of collagen may not be the same, which may be another reason why the morphology of collagen fibres is so varied. Proteoglycans are not visualised in tissue sections by the staining procedures normally used in electron microscopy. Serafini-Fracassini and Smith[229] showed, however, that proteoglycans may be revealed when stained with bismuth nitrate, when they appear as a string of compact dots, each dot being considered to be a coiled glycosaminoglycan side chain. The collagen–proteoglycan complex known as PP-H isolated from cartilage (see Section 4.3.2.1), when stained by this method showed a line of dots at right angles to the fibre axis at regular intervals along the fibre. The proteoglycan appeared to be attached to a certain band of the collagen fibre as revealed by staining first with phosphotungstic acid followed by bismuth nitrate. Since a similar regular arrangement of proteoglycan molecules was seen in cartilage itself when stained with bismuth nitrate, it was concluded that proteoglycans in the tissue were attached tangentially to the collagen fibre axis at certain regions only. Periodic attachment of proteoglycans to collagen fibres has been shown in various types of cartilage[230, 231], nucleus pulposus[232] and also in cornea[233], where the arrangement appears to differ from that in cartilage. Proteoglycans therefore appear to be bound to collagen fibrils at certain definite regions, which implies that the interaction is specific. The exact nature of the interaction remains to be established, however. As yet there is no proof that covalent linkages are involved, although there is evidence that part of the proteoglycans of costal cartilage which cannot be extracted with 4 M guanidinium chloride (see Section 4.3.2.4) may be linked to collagen by covalent bonds[176].

### 4.4.2 Interaction with plasma lipoproteins and platelet factor 4

Proteoglycans may also interact with proteins that are not structural constituents of connective tissue, such as plasma lipoproteins. This interaction is probably an important factor in the development of atherosclerosis. Arterial walls contain several different types of proteoglycan, notably those of dermatan sulphate and heparan sulphate in addition to that of chondroitin sulphate. Using these glycosaminoglycans rather than the corresponding proteoglycans, which are difficult to extract, the affinity of each for plasma lipoproteins has been studied by Iverius[234], using a chromatographic method. Low density (LDL) and very low density (VLDL) but not high density (HDL) lipoproteins interacted to a significant extent with dermatan sulphate at physiological pH and ionic strength. It may be significant that the concentration of dermatan sulphate in fatty streaks of early atherosclerosis of human aorta was much higher than in control tissue[235]. Whole complexes of lipoprotein and glycosaminoglycan have been extracted from fatty streaks[236]. The interaction of plasma lipoproteins with glycosaminoglycans depends mainly on electrostatic forces and increases with charge density of the glycosaminoglycan. Dermatan sulphate and heparan sulphate, which contain some iduronic acid in place of glucuronic acid residues, interacted more strongly than chondroitin sulphate of comparable charge density[234]. From these studies it was calculated that the minimum binding site for VLDL or LDL on glycosaminoglycans appeared to be a trisaccharide having three negative groups. Since iduronic acid and glucuronic acid residues in dermatan sulphate occur in clusters along the chain (see Section 4.2.4), two iduronic acid residues might sometimes occur in the same trisaccharide. Hence, not only would the total iduronic acid content affect binding, but also its distribution, as well as the proportion and distribution of sulphate groups, which vary considerably in heparan sulphate. Since dermatan sulphates from different sources have differing proportions of iduronic acid (see Section 4.2.4), it is possible that the development of atherosclerosis may be accompanied by a change in the iduronic acid content of the dermatan sulphate of affected arteries.

The chondroitin sulphate proteoglycan in the cytoplasmic granules of blood platelets is present as a complex combined with platelet factor 4 (heparin neutralising factor)[237]. The size of the proteoglycan (350 000) and of its constituent chondroitin sulphate chains (12 000) indicate that four chains are attached to the core protein. In the fully saturated complex, four moles of platelet factor 4 appear to be bound to each proteoglycan, the complex itself existing as a dimer[237]. Isolated chondroitin sulphate chains combine with platelet factor 4 at a binding ratio of one mole of platelet factor 4 per carbohydrate chain. The high affinity of the proteoglycan for platelet factor 4 is evident from the fact that it dissociates into its components only at high ionic strength ($I = 0.75$). Heparin, and to a lesser extent heparan sulphate, have higher affinities for platelet factor 4 and will displace chondroitin sulphate from the complex, both chondroitin sulphate isomers and dermatan sulphate having approximately equal affinities[237]. The relative binding affinities of the various glycosaminoglycans for platelet factor 4 are therefore somewhat different from the affinities for β-lipoproteins, when the presence of iduronic acid as well as total charge density increases affinity[234].

## 4.5 TURNOVER OF PROTEOGLYCANS

### 4.5.1 Enzymic attack *in situ*

Isotope experiments *in vivo* show that there is a continuous turnover of proteoglycans, which even in adult tissues may be quite rapid[238-240]. Thus the half-life of $^{35}$S-labelled chondroitin sulphate in the lateral and medial zones of rat costal cartilage was 8.5 and 7 days, respectively[241], which are comparable with those calculated for articular cartilage[239]. The general structure of proteoglycans, where a central protein core holds together a number of carbohydrate chains, makes them particularly vulnerable to attack by proteolytic enzymes, since cleavage of even a few peptide bonds would cause the whole molecule to fall apart. This was shown *in vivo* by the intravenous injection of papain, which resulted in rapid loss of chondroitin sulphate from cartilage[242, 243] (reviewed by Muir[13]). There is considerable evidence that cathepsin D, which can degrade isolated proteoglycans (Figure 4.23)[244-246], is the principal enzyme involved in the catabolism of proteoglycans in cartilage[247]. Cathepsin D may be inhibited *in vitro* by monospecific antiserum[246, 247] or by the peptide pepstatin[248]. During the remodelling that accompanies growth, connective tissue already formed has to be removed. This process is particularly rapid in the developing chick embryo in which cathepsin D has been implicated in the remodelling of cartilage. Using a immunofluorescence technique and monospecific antiserum to chicke cathepsin D, Poole, Hembry and Dingle[249] have demonstrated cathepsin D around the cells of the perichondrium and, to a lesser degree, around chondro cytes of the epiphysial cartilage of normal chick cartilaginous limb-bones i organ culture. In cultured rabbit ear cartilage, cathepsin D was similarl detected around chondroblasts and peripheral chondrocytes. When thes tissues were exposed to doses of retinol (vitamin A) considerably abov

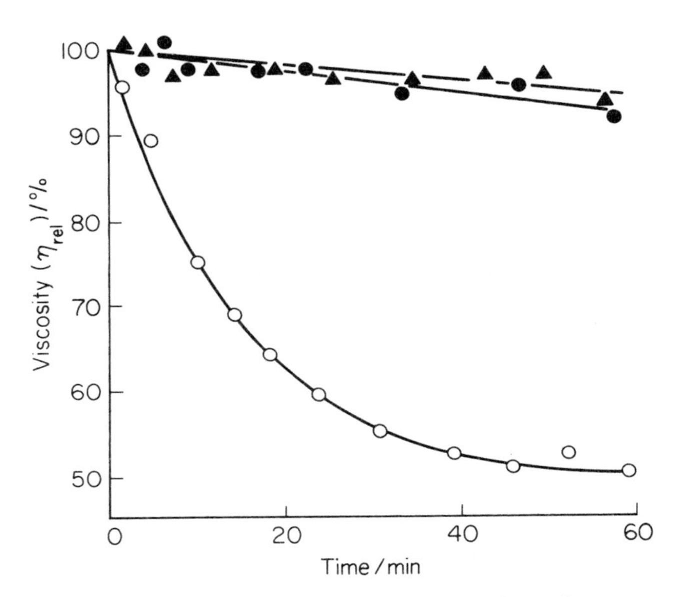


**Figure 4.23** The effect on the viscosity of solutions of proteoglycans of cathepsin D in the presence of antiserum to cathepsin D or in the presence of normal serum. Cathepsin D + normal serum (○); cathepsin D + antiserum to cathepsin D (●); heat inactivated cathepsin D (▲). (From Dingle *et al.*[246], by courtesy of the Biochemical Society.)

physiological levels, a treatment well-known to induce the release of lyso somal enzymes[250], cathepsin D was seen around many of the chondrocytes o rabbit ear cartilage and particularly around chondrocytes of epiphysial an hypertrophic cartilage of chick limb bones. This was accompanied by a los of proteoglycan as shown by histological staining.

Cathepsin D has similarly been demonstrated in ossifying embryoni chick cartilage, where it appears to be derived mainly from osteoblasts[307] It has been suggested that proteoglycans act as inhibitors of calcification[25 and there is now considerable evidence for the degradation and disappearanc of proteoglycans during endochondral ossification[252-254]. It is notable tha a proteoglycan fraction has been isolated from cartilage which inhibit calcification[255].

The product of the action of cathepsin D on proteoglycans is similar to tha produced by trypsin, and consists of a peptide with several chondroitin sulphate chains attached to it[256]. Other proteolytic enzymes present in lysosomes would degrade this peptide further and synergism between degradative proteases seems highly probable (see review by Muir[257]). Cathepsin D is the principal protease in articular cartilage, and in osteoarthrosis the levels per unit of DNA and protein are significantly raised. It is also principally involved in the breakdown of tissues other than cartilage[258], such as the uterus during *post partum* involution[258,259]. Cathepsin D has an acid pH optimum which varies according to the substrate, being about pH 5 for proteoglycans[256]. Acid hydrolase activity is evident in normal, and considerably increased in osteoarthrotic cartilage[260]. A neutral protease is present in polymorphonuclear cells[261-263], which degrades proteoglycans and which may be especially important in the breakdown of tissue during inflammation. The degradation of proteoglycans by extracts of cells from rheumatoid synovial fluids is largely due to this enzyme[264]. Recent work by Fell and her colleagues[7,265] has shown that, if soft connective tissue is included with articular cartilage in organ culture, there is some loss of cartilage matrix in the vicinity of the soft connective tissue. This loss is greatly enhanced by the addition of complement-sufficient antiserum, whereas with cartilage alone, complement-sufficient antiserum had little effect.

Using fluorescein-labelled immunoglobulins it was shown that although immunoglobulin reached the cells of soft connective tissue, it did not penetrate into the matrix of intact cartilage. However, in the presence of complement-sufficient antiserum and soft connective tissue, immunoglobulin was no longer excluded from the matrix. This increase in permeability is presumably due to the local release of lysosomal hydrolases from soft connective tissue in response to complement-sufficient antiserum. These experiments demonstrate the importance of soft connective tissue in the pathology of inflammatory arthritis and the resulting destruction of cartilage. Cartilage is normally almost impermeable to IgG immunoglobulin, but degradation of adult human articular cartilage by cathepsin D increases its permeability to immunoglobulins[266].

### 4.5.2 Clearance and fate of catabolic products

Intact proteoglycans are entrapped in the collagenous framework of connective tissue and cannot be extracted easily (see Section 4.3.2). Once they have been partially degraded, however, they may diffuse out of the tissue into the circulation where small amounts of chondroitin sulphate are present[267], or they may be taken up by cells locally and degraded by lysozomal enzymes. These enzymes are present in embryonic cartilage[268,269], and sulphatase activity was indicated by the release of labelled inorganic sulphate in embryonic[256] and adult costal cartilage[241]. In the cultures of costal cartilage, partial degradation of proteoglycans took place, but the chondroitin sulphate chains themselves remained entirely intact with no desulphation. Hence, the labelled inorganic sulphate[241] must have been released during the complete intracellular degradation of chondroitin sulphate by chondrocytes which are capable of ingesting proteoglycans[270].

The fate of chondroitin sulphate or proteoglycan in the circulation has been followed by injecting $^{35}$S-labelled materials[271]. A proportion of the chondroitin sulphate is rapidly excreted in the urine unchanged with no shortening of chain length. Proteoglycans, however, have to be degraded as far as single polysaccharide chains to pass into the urine. In normal urine the number average molecular weight of chondroitin sulphate and the amount of xylose per chain indicate that single chains are present which are largely intact and have not been attacked by an endoglycosidase[272]. Normal urine, however, contains only a few milligrams of glycosaminoglycan, yet calculations suggest that *ca.* 250 mg of glycosaminoglycan is catabolised each day in an adult human[273]. Most of this must therefore be completely degraded. There is a reciprocal relationship between the disappearance from the plasma of injected $^{35}$S-labelled chondroitin sulphate and the appearance of labelled inorganic sulphate (Figure 4.24)[340]. The organ responsible for

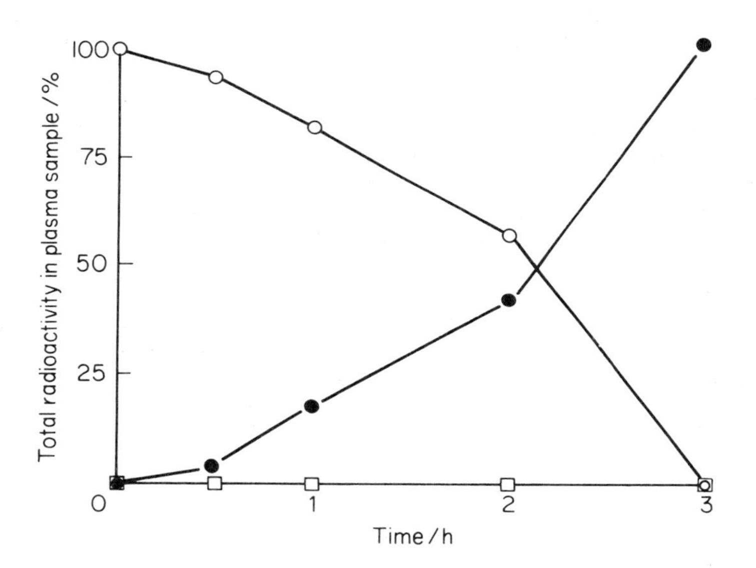


**Figure 4.24** The relationship between chondroitin 4-[$^{35}$S]sulphate injected intravenously and inorganic $^{35}SO_4$ in rat plasma. The injected chondroitin sulphate chains are taken up by the liver before being degraded, and sulphate is then released back into the plasma as inorganic sulphate. Sulphated oligosaccharides were not detected in the plasma. Radioactivity in chondroitin sulphate chains (○), oligosaccharides (□) and inorganic sulphate (●). (From Wood *et al.*[340], by courtesy of the Biochemical Society.)

this breakdown has recently been shown to be the liver[274], as after hepatectomy no detectable degradation of injected chondroitin [$^{35}$S]sulphate occurred[274]. From the calculated half-life of chondroitin sulphate[241] and from the results of liver perfusion experiments, it was found[274] that the liver is capable of entirely degrading all the chondroitin sulphate being turned over in the body.

Liver lysosomes are able to degrade chondroitin sulphate completely[274], presumably by the attack in sequence of hyaluronidase, β-glucuronidase and β-*N*-acetylhexosaminidase. The removal of sulphate is a key step in the breakdown, because the hexosaminidase is inactive until the sulphate has been removed from the hexosamine (reviewed by Muir)[257]. A sulphatase acting on sulphated oligosaccharides has been isolated from liver lysosomes[275]. It is evident that as the glycosaminoglycans in urine normally represent only a very small proportion of the total that is being catabolised, urinary glycosaminoglycans do not provide a sensitive index of glycosaminoglycan catabolism, although age differences and patterns have been noted[276, 277]. A useful method for determining the amounts and proportions of different glycosaminoglycans in small samples of urine has been described recently[278, 279].

On the other hand, in the mucopolysaccharidoses[257, 280], quite large amounts of glycosaminoglycans are present in the urine. These recessively inherited diseases are comparatively rare and result from the lack of factors that are proving to be degradative enzymes specific for particular glycosidic linkages or sulphate groups in different glycosaminoglycans (see review by Neufeld and Cantz[281]). These enzyme deficiencies result in the accumulation of affected glycosaminoglycans in the organs, and excretion in the urine (see review by Muir[257]). The diseases have a progressive downhill course, in which the symptoms probably result from the continuous accumulation of glycosaminoglycans within cells. In a study of an early foetus (20 weeks) with Hurler's syndrome (Type I mucopolysaccharidosis)[280], only the liver was affected at this stage in development[282]. As the liver is principally involved in degrading glycosaminoglycans that are being turned over, it would appear that the catabolic products had been taken up by the liver, but could not be degraded in the normal way owing to the absence of an essential degradative enzyme, which in Hurler's syndrome has been identified as β-iduronidase[283, 284].

### 4.5.3 Maintenance of the intercellular matrix

Since proteoglycans are continuously turned over, to maintain a steady state, proteoglycans must be formed at a rate about equal to their loss. How this is controlled is not known. The effect of the environment on chondrocytes of embryonic chick cartilage in organ culture has been shown by depleting the matrix by the action of hydrolases added to the medium. The cells responded rapidly by increasing the synthesis of matrix constituents[8, 285] and re-establishing their cellular environment within a few days (Figure 4.25). Since the response was dose-dependent[8], there must be a sensitive mechanism for assessing the quality of the matrix. Adult chondrocytes may possess a similar mechanism since chondrocytes in osteoarthrotic cartilage, where there is a loss of chondroitin sulphate[286, 287], apparently incorporate labelled sulphate into chondroitin sulphate at a higher rate than normal[239, 240, 288, 289]. It is possible that the small amount of hyaluronic acid that is present in cartilage may function not only in proteoglycan aggregation (see Section 4.3.5) but also in controlling proteoglycan synthesis. Hyaluronic acid at concentrations as low as $1 \times 10^{-4}$ μg $ml^{-1}$ inhibited the incorporation of $^{35}SO_4$ into polymeric material by isolated adult chondrocytes in suspension cultures[290].

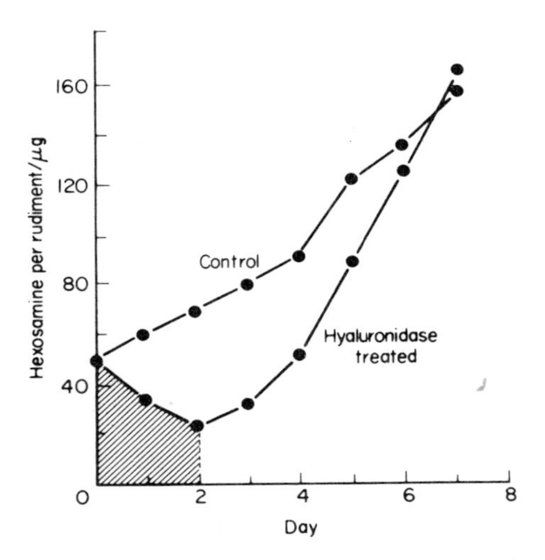


**Figure 4.25** The response of growing chick cartilage in organ culture to the depletion of chondroitin sulphate from the matrix. Hyaluronidase was added to the medium for the first two days (hatched) and the cultures placed in control medium thereafter. Controls were not treated with hyaluronidase. The chondroitin sulphate was determined as total hexosamine per cartilage rudiment. (Redrawn from Fitton-Jackson[8], by courtesy of the Royal Society.)

Proteoglycans formed in response to acute depletion from the matrix of embryonic chick cartilage differed from those in control tissue in chemical composition and in being of smaller hydrodynamic size[9]. Qualitative differences in proteoglycans have been found in natural and experimental osteoarthrosis in dogs[188, 291, 292] and also shown in the knee-joint cartilage of lame and sound pigs being reared by intensive methods[175]. The proteoglycans were more readily extracted and were of smaller hydrodynamic size, even though their total content was unchanged. These results suggest that qualitative changes in proteoglycans precede other pathological changes.

## 4.6 ANTIGENIC PROPERTIES OF PROTEOGLYCANS

Proteoglycans of cartilage are only weakly antigenic but antibodies may be raised against them by repeated immunisation[293-297]. Although cartilage proteoglycans contain only 5–15% of protein, the antigenicity is entirely attributable to the protein moiety since digestion with papain completely destroyed the antigenicity[253, 293-295], whereas digestion with hyaluronidase did not destroy, but rather enhanced, the antigenicity[295, 298]. Moreover, neither chondroitin sulphate nor keratan sulphate are themselves antigenic[299, 300].

Two classes of antigenic determinants have been identified: one is common to proteoglycans from cartilage of a variety of mammals[295-297] and lower vertebrates[190], while the other is specific to each species. In proteoglycan preparations from bovine nasal cartilage which contain aggregates, the species-specific determinants may be separated along with the 'protein-link' fraction (PL) from the rest of the proteoglycan[198, 301, 302] after disaggregation and fractionation in a caesium chloride density gradient according to the procedure of Hascall and Sajdera[163]. Proteoglycans prepared in other ways, such as high-speed homogenisation (Section 3.2.1)[153], retain the determinants due to the protein-link[302]; even when this is followed by an extensive salt fractionation procedure[155] some, but not all, fractions retained these determinants[198]. In the intact native aggregate, the protein-link is 'hidden' because the aggregate does not prevent agglutination of red cells which have been coated with protein-link when these are exposed to antibody to protein-link[302]. That the protein-link is inaccessible in the intact aggregate was confirmed by the finding that on disaggregation it was the only component of the aggregate which precipitated with the lectin concanavalin A, whereas in the intact aggregate it did not react[302]. These results imply that the protein-link is inaccessible and is bound near the centre of the aggregate, which as well as being very large probably has a complex structure. High-speed homogenisation, however, breaks up this structure sufficiently to expose some of the protein-link which is then available to concanavalin A[302]. Whether all species-specific determinants are attributable to the protein-link is not established. Thus in pig laryngeal cartilage there appears to be another species-specific determinant distinct from the protein-link[303].

The determinant common to proteoglycans from different species is attributable to that part of the protein core which resists digestion by trypsin and chymotrypsin[304] and has a very similar amino acid sequence in proteoglycans from different species and even phyla[190]. This region of the core protein has many serine residues carrying chondroitin sulphate chains and it might be expected that the sugars of the linkage region could play some role in the expression of antigenicity. To test this possibility, Baxter and Muir[305] compared the antigenic properties of intact bovine and pig proteoglycans prepared by the procedure of Hascall and Sajdera[163] with the corresponding bovine proteoglycan from which the chondroitin sulphate chains, together with the linkage region, had been removed by the Smith degradation[208], in which the terminal xylose residues are oxidised by periodate. The serine residues involved in the linkage are unaffected by this treatment and the protein core is left with almost no uronic acid attached to it. This product retained its antigenic characteristics and cross-reacted with anti-sera against intact pig proteoglycans. Moreover, antiserum to the Smith-degraded bovine proteoglycan cross-reacted with intact bovine and pig proteoglycans. The sugars of the linkage region are therefore not necessary for the expression of the species-common antigenic determinant, which depends solely on the sequence of amino acids in the region of the protein core which carries the chondroitin sulphate chains and which is resistant to trypsin and chymotrypsin. Indeed, cross reaction was shown more clearly by the fact that this resistant fragment from pig proteoglycan inhibited the reactions of the Smith-degraded bovine proteoglycan with its own antiserum[305].

In studies of their antigenic properties, proteoglycans have usually first been digested with hyaluronidase before they are reacted with antisera. Brandt, Tsiganos and Muir[298], however, compared the behaviour of proteoglycans from articular cartilage before and after digestion with hyaluronidase. Before digestion, a single precipitin line appeared on immunodiffusion with all preparations that contained even a few per cent of proteoglycans small enough to be retarded by gel chromatography on Sepharose 6B (Figure 4.26).

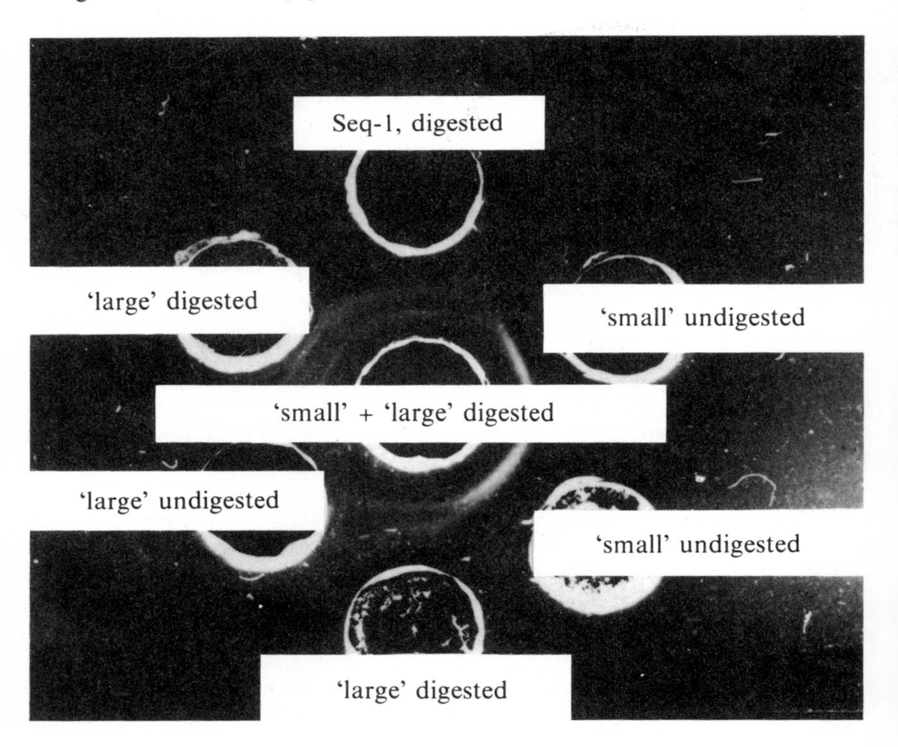


**Figure 4.26** The effect of hyaluronidase treatment on the reactions of proteoglycans of larger or smaller hydrodynamic size (i.e. retarded or excluded on Sepharose 6B) on immunodiffusion against antiserum to the unfractionated mixture of proteoglycans (centre well). Note that the 'small' proteoglycans react without hyaluronidase digestion, whereas the 'large' proteoglycans do not react before digestion. (From Brandt *et al.*[298], by courtesy of Elsevier)

After these were removed, however, the remaining proteoglycans did not react with antiserum, since no precipitin lines developed on immunodiffusion (Figure 4.26) and they were incapable of absorbing specific antibody. These larger proteoglycans could not react with antiserum until after they had been digested with hyaluronidase, when they were effective in absorbing specific antiserum (Figure 4.27) and also gave multiple precipitin lines on immunodiffusion, one of which showed identity with the line formed by the 'small'

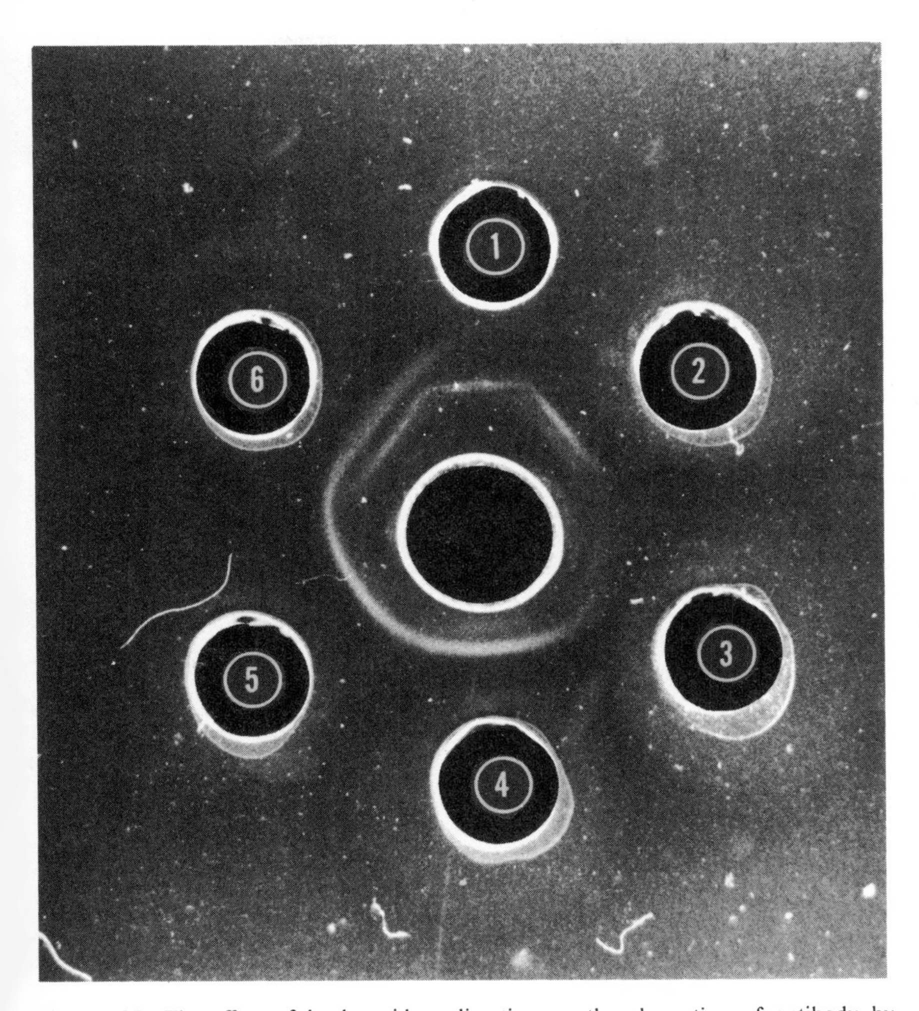


**Figure 4.27** The effect of hyaluronidase digestion on the absorption of antibody by 'large' or 'small' proteoglycans. The antiserum (in the outer wells) was absorbed with various proteoglycan fractions and then diffused against the unfractionated proteoglycans in the centre well. Anti-serum absorbed with: (1) 'small' proteoglycans treated with hyaluronidase, and (2) 'small' proteoglycans undigested with hyaluronidase {Note that only one precipitin line appears in either (1) or (2) and the outer line typical of 'small' proteoglycans does not appear [shown in (4) and (5)]}; (3) 'large' proteoglycans digested with hyaluronidase (note that all reactivity is effectively abolished); (4) and (5) unabsorbed antiserum specific for 'small' proteoglycans only; (6) 'large' proteoglycans undigested with hyaluronidase {Note that all reactivity is retained and the precipitin line typical of 'small' proteoglycans is present [shown in (4) and (5)]}. (From Brandt *et al.*[298], by courtesy of Elsevier.)

proteoglycan (Figure 4.26)[298]. It would appear from these results that there is a common determinant amongst these proteoglycans, but that the larger proteoglycans possess several additional determinants (all determinants being inaccessible to antibody before hyaluronidase digestion). Aggregation of the larger proteoglycans could not explain why they did not react with antibody when they were intact, because even after they were subjected to the disaggregation procedure of Hascall and Sajdera[163], they still did not react with antibody before hyaluronidase digestion[298]. Neither could steric hindrance due to carbohydrate be an explanation, because the 'small' proteoglycans contained more carbohydrate than larger proteoglycans and, since there was no difference between them in the lengths of chondroitin sulphate chains, the number of chains along the protein core would be greater in the 'small' proteoglycans. These contain very little keratan sulphate. Hence the determinant common to different proteoglycans represents the region of the protein core bearing chondroitin sulphate chains.

During the development of articular cartilage the relative proportion of small proteoglycans changes[174]. Since they are not biosynthetic precursors of larger proteoglycans[306], and since they react directly with antibody, unlike larger proteoglycans, they would appear to be distinct from other proteoglycans. This distinction, which illustrates the usefulness of immunological methods in studying the structure of cartilage proteoglycans and their aggregates, has not so far been extended to proteoglycans from other tissues.

## 4.7 PROTEOGLYCANS OF TISSUES OTHER THAN CARTILAGE

### 4.7.1 Introduction

There is much less information about proteoglycans in tissues other than cartilage, partly because these tissues contain less proteoglycan and partly because the proteoglycans are more difficult to extract.

Dermatan sulphate is found in skin, heart valves and aorta, and accounts for 10–15% of the total glycosaminoglycans, of which chondroitin sulphate is the main constituent. These tissues also contain heparan sulphate in about the same amounts as dermatan sulphate. The ease with which the corresponding proteoglycans can be extracted differs considerably. Hyaluronic acid is largely extractable from dried finely milled heart valves by homogenisation in water[308] or 0.2 M NaCl[309]. Chondroitin sulphate proteoglycans can be extracted preferentially from heart valves[310] and aorta[311] with salt solutions. About half the chondroitin sulphate proteoglycan of heart valve was extracted with 0.2 M NaCl and the remainder by M NaCl[310]. In contrast, proteoglycans of dermatan sulphate and heparan sulphate are much more difficult to extract. Thus only a minor proportion of either dermatan sulphate proteoglycan of heart valves was extracted with M NaCl[214, 312, 313], or heparan sulphate proteoglycan of aorta with 2 M NaCl[314]. Much more drastic conditions such as hot concentrated urea were required to extract these proteoglycans to any extent. The physical state of different types of proteoglycan cannot therefore be the same, and they must differ in the way they are bound in the tissues.

Since hyaluronic acid can largely be extracted separately from chondroitin sulphate and dermatan sulphate proteoglycans, it would not appear to be bound to them in a specific way as it is to chondroitin sulphate proteoglycans of cartilage in the formation of aggregates (Section 4.3.6). Moreover, hyaluronic acid which is extracted from heart valves together with some chondroitin sulphate proteoglycan may be isolated by equilibrium density gradient centrifugation[309] under conditions which would not dissociate hyaluronic acid from chondroitin sulphate proteoglycan aggregates of cartilage (Section 4.3.6). Blood vessels therefore probably do not contain such aggregates nor does skin, since the different proteoglycans may be extracted sequentially from it[215, 312]. Furthermore, dermatan sulphate proteoglycans of skin were unaffected by 4 M guanidinium chloride[313] which would dissociate proteoglycan aggregates of cartilage. Since the amino acid compositions of these proteoglycans differ significantly (Table 4.5), differences in their properties arise from differences in their protein cores and hence in their structures.

**Table 4.5 Amino acid compositions of dermatan sulphate proteoglycans***

| | *Residues per 1000* | | |
|---|---|---|---|
| *Amino acid* | *Proteoglycan of pig skin* | *Proteoglycan of bovine heart valves* | *Proteoglycan of pig laryngeal cartilage* |
| Hyp | — | 14 | — |
| Asp | 134 | 122 | 76 |
| Thr | 46 | 56 | 62 |
| Ser | 69 | 67 | 123 |
| Glu | 110 | 139 | 140 |
| Pro | 79 | 75 | 98 |
| Gly | 81 | 90 | 129 |
| Ala | 53 | 77 | 79 |
| Cys | 23 | — | 3 |
| Val | 56 | 50 | 59 |
| Met | 14 | 5 | 3 |
| Ileu | 56 | 37 | 36 |
| Leu | 132 | 96 | 77 |
| Tyr | 24 | 23 | 20 |
| Phe | 34 | 33 | 28 |
| Lys | 29 | 46 | 11 |
| His | 22 | 19 | 10 |
| Arg | 38 | 51 | 44 |

* Amino acid compositions of dermatan sulphate proteoglycans from pig skin[215], bovine heat valves[313], and for comparison, the chondroitin sulphate proteoglycan from pig laryngeal cartilage[16]

### 4.7.2 Dermatan sulphate proteoglycans from skin, heart valve and aorta

Dermatan sulphate proteoglycans which are extracted with hot concentrated urea from heart valves[215] and skin[313] contain as much as 50–60% of protein, whereas cartilage proteoglycans contain only about 5–10%. The molecular weight of dermatan sulphate proteoglycan of skin[313] was $2.9 \times 10^6$, whereas that from heart valves[315] was only $2 \times 10^5$. The amino acid composition was similar in the two preparations[313] but differed from that of cartilage proteoglycans (Table 4.5) and both were polydisperse, although the heterogeneity was not assessed in detail. They accounted for only a fraction of the total dermatan sulphate (30% of that of skin). Differences in the relative proportions of iduronic acid and glucuronic acid in dermatan sulphates from different sources, and differences in the proportion of 4- and 6-sulphate groups, suggest there may be several dermatan sulphate proteoglycans, just as there are several chondroitin sulphate proteoglycans of cartilage. Indeed, biosynthetic studies of dermatan sulphate of aorta would support this suggestion. A proteoglycan from calf aorta consisting of 73% of disaccharides of chondroitin sulphate and 27% of disaccharides of dermatan sulphate differed in the turnover rate of each type of glycosaminoglycan[316], suggesting considerable metabolic heterogeneity amongst the proteoglycans. Moreover, some of the dermatan sulphate of heart valve is particularly resistant to extraction[215] and is presumably in some firm association with collagen. It may function in modifying the tensile properties of collagen which would be particularly important in heart valves and large vessels where pulsatile forces are greatest. Indeed, dermatan sulphate itself interacts with soluble collagen in a different way from chondroitin sulphate (Section 4.4.1).

### 4.7.3 Chondroitin sulphate proteoglycans of aorta and heart valves

About half the glycosaminoglycan of aorta is chondroitin sulphate[317]. The chondroitin sulphate proteoglycans of bovine heart valves[214] and aorta[311] differ from those in cartilage in being of lower molecular weight and in lacking keratan sulphate, but as in cartilage there appears to be more than one type of chondroitin sulphate proteoglycan. When bovine heart valve was extracted sequentially first with 0.2 M NaCl followed by 2 M NaCl, the first extract contained a proteoglycan having a ratio of chondroitin 6-sulphate:chondroitin 4-sulphate of 1.5:1, whereas in the proteoglycan extracted with M NaCl this ratio was 0.6:1 and its molecular weight about twice that of the proteoglycan extracted[310] with 0.2 M NaCl. Nevertheless, the molecular weights of both were lower than molecular weights of chondroitin sulphate proteoglycans of cartilage. There is thus a similar propensity in both tissues for proteoglycans of smaller size to be extracted by less concentrated salt solutions than proteoglycan of larger size (see Section 4.3.2.3). Small amounts of chondroitin sulphate proteoglycans have been obtained from bone after demineralisation and collagenase digestion, which resembled cartilage proteoglycan[318]. When extracted with EDTA, however, the proteoglycan appeared to be bound to a sialo protein that is present in bone[319]. The presence of xylose and the lability to alkali suggest that the xylosyl–serine linkage occurs in this, as in the majority of proteoglycans, even in those from invertebrate tissues, as may be deduced from the properties of a chondroitin sulphate peptide isolated after papain digestion of squid skin. However, unlike that from mammalian sources, the chondroitin sulphate peptide from squid skin appeared also to be bound to short oligosaccharide units by other alkali-stable linkages[320].

### 4.7.4 Proteoglycans of heparan sulphate and heparin

That heparan sulphate exists in the tissues as proteoglycan was presumed from the fact that, when isolated after proteolysis of the tissues, it contained galactose and xylose[88] as well as serine as the main residual amino acid[321]. The proteoglycan has been extracted with 6 M urea from human aorta and partially purified by density gradient centrifugation after digestion of chondroitin sulphate and dermatan sulphate proteoglycans with chondroitinase[314]. There is evidence to suggest that it consists of several polysaccharide chains attached to a polypeptide[322] and is therefore constructed in essentially the same way as cartilage chondroitin sulphate proteoglycans. Mast cells contain heparin, which is presumed to be present as a proteoglycan stored in intracellular granules. Its isolation as a proteoglycan from liver capsule has been reported[323] but not confirmed[13a, 324].

A macromolecular heparin has been prepared from rat skin by proteolytic digestion which could be degraded with ascorbate (Section 4.2.2)[325], but its relationship to a heparin proteoglycan remains to be determined. All the heparin preparations examined contain the same linkage sugars as the chondroitin sulphates. It is reasonable to assume that heparin is synthesised as a multi-chain proteoglycan which is subsequently degraded by a polysaccharidase in conjunction with a protease[81].

### 4.7.5 Intracellular proteoglycans

The majority of proteoglycans occur in the extracellular matrix of connective tissue. Nevertheless, certain specialised cells in the circulation such as granulocytes[326, 327], blood platelets[179, 237] and guinea-pig Kurloff cells[114] contain chondroitin 4-sulphate proteoglycans stored in special organelles. These intracellular proteoglycans resemble those in cartilage[112] and heart valves[310], which are extractable in salt solutions of physiological ionic strength, in being of rather low protein content with very little or no cysteine or methionine and in being of comparatively small size. The molecular weight[237] of the proteoglycan in platelets is only 350 000. The Kurloff cell proteoglycan contains hexose, pentose and serine in molar proportions of 2.2:1.07:1, most of the serine being destroyed on alkaline β-elimination of the carbohydrate chains[114]. Thus the Kurloff cell proteoglycan appears to contain the same linkage region as do cartilage proteoglycans, and, in the proteoglycan from granulocytes, serine is likewise destroyed by alkali[327]. The mature Kurloff cell does not synthesise the proteoglycan[114] which is apparently stored in the inclusion body. In immature myeloid cells, chondroitin sulphate is synthesised in the microsomal fraction and subsequently transferred to cytoplasmic granules for storage[328, 329]. Similarly, the chondroitin 4-sulphate proteoglycan of platelets is located in cytoplasmic granules where it exists in the form of a complex with platelet factor 4 which has the property of neutralising heparin[237].

The apparent function of chondroitin sulphate proteoglycan in platelets is to act as a carrier for heparin neutralising factor[237], since heparin, and to a lesser extent heparan sulphate, displaces chondroitin sulphate from the complex. The complex is released from platelets by a variety of agents, including thrombin[330], and any heparin in the plasma would be neutralised by reacting with platelet factor 4 and displacing the chondroitin 4-sulphate proteoglycan from the complex[237]. Platelets labelled with $^{35}S$ were shown to release labelled chondroitin sulphate when treated with thrombin[331].

The functions of proteoglycans in other types of cell remain obscure, although *in vitro* the proteoglycan of Kurloff cells inhibits the migration of peritoneal macrophages[332]. The proteoglycan appears to be bound in some way to material with characteristic infrared[114] and ultraviolet[334] absorption spectra. Kurloff cells may have some function in pregnancy when they increase in number in the circulation, spleen, bone marrow and vascular channels of the placenta[333]. Although Kurloff cells are unique to the guinea pig, proteoglycans with similar chemical spectral and biological properties have been extracted from the spleen of human and other species during pregnancy[334].

## 4.8 SOME BIOLOGICAL FUNCTIONS OF HYALURONIC ACID

Hyaluronic acid has certain functions that are entirely different from other connective tissue glycosaminoglycans. From observations of hyaluronic acid content, synthesis and degradation during morphogenesis, Toole, Jackson and Gross[335] have suggested that as hyaluronic acid is associated with the morphogenic phase, it controls mesenchymal cell aggregation in embryogenesis and that its removal by endogenous hyaluronidase forms part of the mechanism timing the aggregation and subsequent differentiation of mesenchymal cells. In the regenerating newt limb[336], in the developing chick limb and axial skeleton[337] and during the development of the cornea while transparency increases[338] there is an inverse relationship between the synthesis of hyaluronic acid and the synthesis of chondroitin sulphate (Figure 4.28); maximal synthesis of hyaluronic acid occurs initially, but at the onset of differentiation it is removed by the action of hyaluronidase. At low concentrations, down to 1 ng $ml^{-1}$, hyaluronic acid inhibited chondrogenesis *in vitro* when added to the medium of stationary cultures of chick embryo somite cells[335]. The effect was evident even when the cells were exposed to hyaluronic acid for only a few minutes. It was specific to hyaluronic acid and was not shared by cartilage proteoglycan, the isomers of chondroitin sulphate, heparin, DNA or RNA[339]. The inhibition of chondrogenesis was overcome by thyroxine, growth hormone, calcitonin and adenosine 3′,5′-cyclic monophosphate (cyclic AMP), but not by a number of other hormones[339]. These remarkable effects of hyaluronic acid on embryonic cells are somewhat analogous to its inhibitory effect at comparably low concentrations[290] on the synthesis of chondroitin sulphate proteoglycan by stationary suspension cultures of adult chondrocytes[290], which, like the inhibition of chondrogenesis, was exclusive to hyaluronic acid. Adult chondrocytes were inhibited, but confluent cultures of fibroblasts or synovial cells from adult tissues were not[290]. It is possible that the inhibition may operate through the

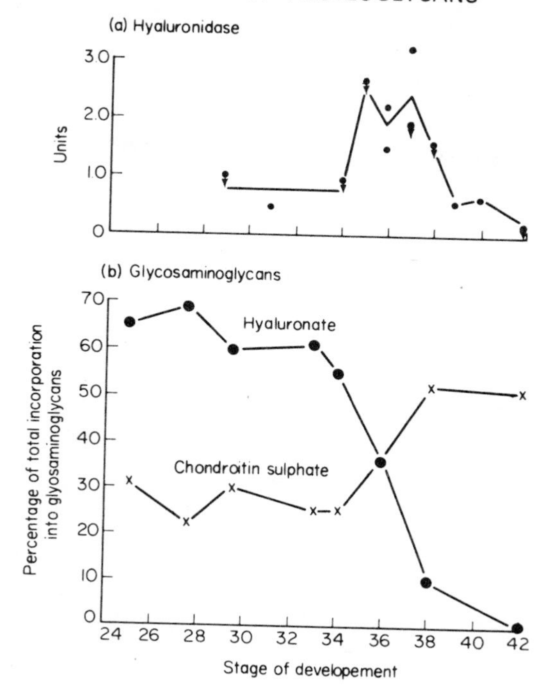


**Figure 4.28** Comparison of (a) hyaluronidase activity with (b) synthesis of glycosaminoglycans (determined by the incorporation of [$^3$H]acetate) in the developing chick cornea. (From Toole and Trelstad[338], by courtesy of Academic Press.)

participation of hyaluronic acid in proteoglycan aggregation[16, 194, 200] (see Section 4.3.6) since proteoglycans that can aggregate are formed by cartilage cells. It is possible that there may well be other specific effects of hyaluronic acid on connective tissue cells.

## References

1. Balazs, E. A. (1970). In *Chemistry and Molecular Biology of the Intercellular Matrix*, Vol. 1, XXX (E. A. Balazs, editor) (London & New York: Academic Press)
2. Rodén, L., Baker, J. R., Cifonelli, J. A. and Mathews, M. B. (1972). In *Methods of Enzymology*, Vol. XXVIII, 73 (V. Ginsburg, editor) (New York & London: Academic Press)
3. Kraemer, P. M. (1971). *Biochemistry*, **10**, 1437
4. Slavkin, H. C. (1972). In *The Comparative Molecular Biology of the Extracellular Matrix*, 35 (H. C. Slavkin, editor) (New York & London: Academic Press)
5. Kempson, G. E., Muir, H., Swanson, S. A. V. and Freeman, M. A. R. (1970). *Biochim. Biophys. Acta*, **215**, 70
6. Maroudas, A. (1970). *Biophys. J.*, **10**, 365
7. Poole, A. R., Barratt, M. E. J. and Fell, H. B. (1973). *Int. Arch. Allergy*, **44**, 469
8. Fitton-Jackson, S. (1970). *Proc. Roy. Soc. (London)*, **B175**, 405
9. Hardingham, T. E., Fitton-Jackson, S. and Muir, H. (1972). *Biochem. J.*, **129**, 101

10. Bhatnager, R. S. and Prockop, D. J. (1966). *Biochim. Biophys. Acta*, **130,** 383
11. Meyer, K. (1957). *Harvey Lect.*, **51,** 88
12. Brimacombe, J. S. and Webber, J. M. (1964). In *Mucopolysaccharides*, BBA Library, Vol. 6 (Amsterdam, London and New York: Elsevier)
13. Muir, H. (1964). *Int. Rev. Connect. Tiss. Res.*, **2,** 101
13a. Lindahl, U. (1976). In *International Review of Science*, *Organic Chemistry Series Two*, Vol. 7, *Carbohydrates* (G. O. Aspinall, editor) (London: Butterworths)
14. Laurent, T. C. (1970). In *Chemistry and Molecular Biology of the Intercellular Matrix*, Vol. 2, 703 (E. A. Balazs, editor) (London & New York: Academic Press)
15. Hardingham, T. E. and Muir, H. (1973). *Biochem. Soc. Trans.*, **1,** 282
16. Hardingham, T. E. and Muir, H. (1974). *Biochem. J.*, **139,** 565
17. Ogston, A. G. and Sherman, T. F. (1959). *Biochem. J.*, **72,** 301
18. Pigman, W. and Rizvi, S. (1959). *Biochem. Biophys. Res. Commun.*, **1,** 39
19. Matsumura, G. and Pigman, W. (1965). *Arch. Biochem. Biophys.*, **110,** 526
20. Niedemeier, W., Laney, R. P. and Dobson, C. (1967). *Biochim. Biophys. Acta*, **148,** 400
21. Swann, D. A. (1969). *Biochem. J.*, **114,** 819
22. Fessler, J. H. and Fessler, L. I. (1966). *Proc. Nat. Acad. Sci. USA*, **56,** 141
23. Ogston, A. G. and Stanier, J. E. (1951). *Biochem. J.*, **52,** 149
24. Silpananta, P., Dunstone, J. R. and Ogston, A. G. (1968). *Biochem. J.*, **109,** 43
25. Scher, I. and Hammerman, D. (1972). *Biochem. J.*, **126,** 1073
26. Wardi, A. H., Allen, W. S., Turner, K. D. and Stary, Z. (1966). *Arch. Biochem. Biophys.*, **117,** 44
27. Wardi, A. H., Allen, W. S., Turner, K. D. and Stary, Z. (1969). *Biochim. Biophys. Acta*, **192,** 151
28. Rosenberg, L. and Schubert, M. (1970). In *The Immunochemistry and Biochemistry of Connective Tissue*, Vol. 3, 1 (J. Rotstein, editor) (Basel, Munchen and New York: Karger)
29. Mathews, M. B. (1958). *Nature*, **181,** 421
30. Suzuki, S. (1960). *J. Biol. Chem.*, **235,** 3580
31. Meyer, K., Linker, A., Davidson, E. A. and Weissman, G. (1953). *J. Biol. Chem.*, **205,** 611
32. Wasteson, Å. and Lindahl, U. (1971). *Biochem. J.*, **125,** 903
33. Yamagata, T., Saito, H., Habuchi, O. and Suzuki, S. (1968). *J. Biol. Chem.*, **243,** 1523
34. Greiling, H. and Baumann, G. (1973). In *Connective Tissue and Ageing*, Vol. 1, 160 (H. G. Vogel, editor) (Amsterdam: Excerpta Medica)
35. Iwata, H. (1969). *J. Jap. Orthop. Ass.*, **43/6,** 455
36. Mathews, M. B. and Inouye, M. (1961). *Biochim. Biophys. Acta*, **53,** 509
37. Saito, H., Yamagata, T. and Suzuki, S. (1968). *J. Biol. Chem.*, **243,** 1536
38. Antonopoulos, C. A. and Gardell, S. (1963). *Acta Chem. Scand.*, **17,** 1474
39. Antonopoulos, C. A., Engfeldt, B., Gardell, S., Hjertquist, S. O. and Solheim, K. (1965). *Biochim. Biophys. Acta*, **101,** 150
40. Fransson, L. Å. and Rodén, L. (1967). *J. Biol. Chem.*, **242,** 4161
41. Fransson, L. Å. and Rodén, L. (1968). *J. Biol. Chem.*, **243,** 1504
42. Fransson, L. Å. (1968). *Biochim. Biophys. Acta*, **156,** 311
43. Fransson, L. Å. (1968). *J. Biol. Chem.*, **243,** 1504
44. Fransson, L. Å. and Havsmark, B. (1970). *J. Biol. Chem.*, **245,** 4770
45. Suzuki, S., Saito, H., Yamagata, T., Anno, K., Seno, N., Kawai, Y. and Furuhashi, T. (1968). *J. Biol. Chem.*, **243,** 1543
46. Fransson, L. Å. (1970). In *Chemistry and Molecular Biology of the Intracellular Matrix*, Vol. 2, 823, (E. A. Balazs, editor) (London and New York: Academic Press)
47. Hirano, S., Hoffman, P. and Meyer, K. (1961). *J. Org. Chem.*, **26,** 5064
48. Seno, N., Meyer, K., Anderson, B. and Hoffman, P. (1965). *J. Biol. Chem.*, **240,** 1005
49. Baker, J. R., Cifonelli, J. A. and Rodén, L. (1969). *Biochem. J.*, **115,** 11P
50. Marshall, R. D. and Neuberger, A. (1970). *Adv. Carbohyd. Chem.*, **25,** 407
51. Hoffman, P., Mashburn, T. A. and Meyer, K. (1967). *J. Biol. Chem.*, **242,** 3805
52. Tsiganos, C. P. and Muir, H. (1967). *Biochem. J.*, **104,** 26c
53. Bray, B. A., Lieberman, R. and Meyer, K. (1967). *J. Biol. Chem.*, **242,** 3373
54. Seno, N. and Toda, N. (1970). *Biochim. Biophys. Acta*, **215,** 544
55. Neuberger, A., Marshall, R. D. and Gottschalk, A. (1966). In *Glycoproteins*, BBA

Library, Vol. 5, 166 (A. Gottschalk, editor) (Amsterdam, London and New York: Elsevier)
56. Heinegård, D. (1972). *Biochim. Biophys. Acta*, **285**, 193
57. Gregory, J. D. and Rodén, L. (1961). *Biochem. Biophys. Res. Commun.*, **5**, 430
58. Mathews, M. B. and Cifonelli, J. A. (1965). *J. Biol. Chem.*, **240**, 4140
59. Bhavanadan, V. P. and Meyer, K. (1967). *J. Biol. Chem.*, **242**, 4352
60. Bhavanadan, V. P. and Meyer, K. (1968). *J. Biol. Chem.*, **243**, 1052
61. Greiling, H. and Stuhlsatz, H. W. (1969). *Z. Physiol. Chem.*, **350**, 449
62. Toda, N. and Seno, N. (1970). *Biochim. Biophys. Acta*, **208**, 227
63. Hascall, V. C. and Riolo, R. L. (1972). *J. Biol. Chem.*, **247**, 4529
64. Neuberger, A. and Marshall, R. D. (1966). In *Glycoproteins*, BBA Library, Vol. 5, 263 (A. Gottschalk, editor) (Amsterdam, London and New York: Elsevier)
65. Hirano, S. and Meyer, K. (1971). *Biochem. Biophys. Res. Commun.*, **44**, 1371
66. Hirano, S. and Meyer, K. (1973). *Conn. Tissue Res.*, **2**, 1
67. Cifonelli, J. A. and Dorfman, A. (1962). *Biochem. Biophys. Res. Commun.*, **7**, 41
68. Radhakrishnamurthy, B. and Berenson, G. S. (1963). *Arch. Biochem. Biophys.*, **101**, 360
69. Perlin, A. S. and Sanderson, G. R. (1970). *Carbohyd. Res.*, **12**, 183
70. Lindahl, U. and Axelsson, O. (1971). *J. Biol. Chem.*, **246**, 74
71. Inouye, S. and Miyawaki, M. (1973). *Biochim. Biophys. Acta*, **320**, 73
72. Wolfrom M. L., Vercellotti, J. R., Tomomatsu, H. and Horton, D. (1964). *J. Org. Chem.*, **29**, 540
73. Helting, T. and Lindahl, U. (1971). *J. Biol. Chem.*, **246**, 5442
74. Nieduszynski, I. A. and Atkins, E. D. T. (1973). *Biochem. J.*, **135**, 729
75. Cifonelli, J. A. and King, J. (1973). *Biochim. Biophys. Acta*, **320**, 331
76. Lindahl, U. (1966). *Biochim. Biophys. Acta*, **130**, 368
77. Linker, A. and Hovingh, P. (1972). *Biochemistry*, **11**, 563
78. Foster, A. B., Harrison, R., Inch, T. D., Stacey, M. and Webber, J. M. (1963). *J. Chem. Soc.*, 2279
79. Linker, A., Hoffman, P., Sampson, P. and Meyer, K. (1958). *Biochim. Biophys. Acta*, **29**, 443
80. Linker, A. and Hovingh, P. (1965). *J. Biol. Chem.*, **240**, 3274
81. Lindahl, U. (1970). In *Chemistry and Molecular Biology of the Intercellular Matrix*, Vol. 2, 943 (E. A. Balazs, editor) (London and New York: Academic Press)
82. Linker, A. and Hovingh, P. (1973). *Carbohyd. Res.*, **29**, 41
83. Cifonelli, J. A. (1970). In *Chemistry and Molecular Biology of the Intercellular Matrix*, Vol. 2, 961 (E. A. Balazs, editor) (London and New York: Academic Press)
84. Muir, H. (1958). *Biochem. J.*, **69**, 195
85. Lindahl, U. and Rodén, L. (1966). *J. Biol. Chem.*, **241**, 2113
86. Rodén, L. and Smith, R. (1966). *J. Biol. Chem.*, **241**, 5949
87. Helting, T. and Rodén, L. (1968). *Biochim. Biophys. Acta*, **170**, 301
88. Knecht, J., Cifonelli, J. A. and Dorfman, A. (1967). *J. Biol. Chem.*, **242**, 4652
89. Stern, E. L., Lindahl, U. and Rodén, L. (1971). *J. Biol. Chem.*, **246**, 5707
90. Anderson, B., Hoffman, P. and Meyer, K. (1965). *J. Biol. Chem.*, **240**, 156
91. Stern, E. L., Cifonelli, J. A., Fransson, L. Å., Lindahl, B., Rodén, L., Schiller, S. and Spach, M. L. (1969). *Arkiv Kemi*, **30**, 583
92. Creeth, J. M. and Pain, R. H. (1967). *Progr. Biophys. Mol. Biol.*, **17**, 217
93. Coates, J. H. (1970). In *Physical Principles and Techniques of Protein Chemistry*, Vol. B, 1 (S. J. Leach, editor) (New York and London: Academic Press)
94. Gibbons, R. A. (1966). In *Glycoproteins*, BBA Library, Vol. 5, 29 (A. Gottschalk, editor) (Amsterdam, London and New York: Elsevier)
95. Wasteson, Å. (1969). *Biochim. Biophys. Acta*, **177**, 152
96. Cleland, R. L. and Wang, J. L. (1970). *Biopolymers*, **9**, 799
97. Lasker, J. E. and Stivala, S. S. (1966). *Arch. Biochem. Biophys.*, **115**, 360
98. Timasheff, S. N. and Townend, R. (1970). In *Physical Principles and Techniques of Protein Chemistry*, Vol. B, 147 (S. J. Leach, editor) (New York and London: Academic Press)
99. Mathews, M. B. (1956). *Arch. Biochem. Biophys.*, **61**, 367
100. Cleland, R. L. (1960). In *Chemistry and Molecular Biology of the Intercellular Matrix*, Vol. 2, 733 (E. A. Balazs, editor) (London and New York: Academic Press)

101. Mathews, M. B. and Dorfman, A. (1953). *Arch. Biochem. Biophys.*, **42,** 41
102. Mason, R. M. and Wusteman, F. S. (1970). *Biochem. J.*, **120,** 777
103. Ogston, A. G. and Wells, J. D. (1970). *Biochem. J.*, **119,** 67
104. Ogston, A. G. and Preston, B. N. (1973). *Biochem. J.*, **131,** 843
105. Laurent, T. C. and Killander, J. (1964). *J. Chromatog.*, **14,** 417
106. Ackers, G. K. (1964). *Biochemistry*, **3,** 723
107. Squires, P. G. (1964). *Arch. Biochem. Biophys.*, **107,** 471
108. Constantopoulos, G., Dekaben, S. and Carrol, W. R. (1969). *Anal. Biochem.*, **31,** 59
109. Hilborn, J. C. and Anastassiadis, P. A. (1971). *Anal. Biochem.*, **39,** 88
110. Mathews, M. B. and Decker, L. (1971). *Biochim. Biophys. Acta*, **244,** 30
111. Tsiganos, C. P. and Muir, H. (1966). *Anal. Biochem.*, **17,** 495
112. Tsiganos, C. P. and Muir, H. (1969). *Biochem. J.*, **113,** 885
113. Clamp, J. R., Dawson, G. and Hough, L. (1964). *Biochim. Biophys. Acta*, **148,** 342
114. Dean, M. F. and Muir, H. (1970). *Biochem. J.*, **118,** 783
115. Brandt, K. D. and Muir, H. (1971). *Biochem. J.*, **121,** 261
116. Kleine, T. O., Kirsig, H. J. and Hilz, H. (1971). *Z. Physiol. Chem.*, **352,** 479
117. Robinson, H. C. and Hopwood, J. J. (1973). *Biochem. J.*, **133,** 457
118. Lloyd, K. O., Kabat, E. A., Layug, E. J. and Gruezo, F. (1966). *Biochemistry*, **5,** 1489
119. McLean, C., Werner, D. A. and Aminoff, D. (1973). *Anal. Biochem.*, **55,** 72
120. Richards, G. N. and Whelan, W. J. (1973). *Carbohyd. Res.*, **27,** 185
121. Dean, M. F., Muir, H. and Ewins, R. J. F. (1971). *Biochem. J.*, **123,** 883
122. Atkins, E. D. T. and Sheehan, J. K. (1973). *Science*, **179,** 562
123. Atkins, E. D. T., Phelps, C. F. and Sheehan, J. K. (1972). *Biochem. J.*, **128,** 1255
124. Atkins, E. D. T. and Sheehan, J. K. (1971). *Biochem. J.*, **125,** 92
125. Atkins, E. D. T. and Sheehan, J. K. (1972). *Nature New Biol.*, **235,** 253
126. Balazs, E. A. (1966). *Fed. Proc.*, **25,** 1817
127. Dea, I. C. M., Moorhouse, R., Rees, D. A., Arnott, S., Guss, J. H. and Balazs, E. A. (1973). *Science*, **179,** 560
128. Atkins, E. D. T. and Laurent, T. C. (1973). *Biochem. J.*, **133,** 605
129. Arnott, S., Guss, M. M., Hukins, D. W. L. and Mathews, M. B. (1973). *Science*, **180,** 743
130. Atkins, E. D. T. and Isaac, D. H. (1973). *J. Mol. Biol.*, **80,** 773
131. Isaac, D. H. and Atkins, E. D. T. (1973). *Nature New Biol.*, **244,** 252
132. Weissman, B., Meyer, K., Sampson, P. and Linker, A. (1954). *J. Biol. Chem.*, **208,** 417
133. Schmidt, M. and Dmowchowski, A. (1964). *Biochim. Biophys. Acta*, **83,** 137
134. Hascall, V. C., Riolo, R. L., Hayward, J. and Reynold, C. C. (1972). *J. Biol. Chem.*, **247,** 4521
135. Linker, A., Meyer, K. and Hoffman, P. (1956). *J. Biol. Chem.*, **219,** 13
136. Greiling, H. and Stuhlsatz, H. W. and Eberhard, T. (1965). *Z. Physiol. Chem.*, **340,** 243
137. Linker, A., Meyer, K. and Hoffman, P. (1960). *J. Biol. Chem.*, **235,** 924
138. Payza, A. N. and Korn, E. D. (1956). *J. Biol. Chem.*, **223,** 853
139. Yosizawa, Z. (1967). *Biochim. Biophys. Acta*, **141,** 600
140. Dietrich, C. P. (1968). *Biochem. J.*, **108,** 647
141. Dietrich, C. P. (1969). *Biochemistry*, **8,** 2089
142. Dietrich, C. P., Silva, M. E. and Michelacci, Y. M. (1973). *J. Biol. Chem.*, **248,** 6408
143. Dietrich, C. P. and Dietrich, S. M. C. (1972). *Anal. Biochem.*, **46,** 209
144. Rosen, O., Hoffman, P. and Meyer, K. (1960). *Fed. Proc.*, **19,** 147
145. Tsiganos, C. P. and Muir, H. (1970). In *Chemistry and Molecular Biology of the Intercellular Matrix*, Vol. 2, 859 (E. A. Balazs, editor) (London and New York: Academic Press)
146. Mathews, M. B. and Lozaityte, I. (1958). *Arch. Biochem. Biophys.*, **74,** 158
147. Partridge, S. M., Davis, H. F. and Adair, G. S. (1961). *Biochem. J.*, **79,** 15
148. Cessi, D. and Bernardi, G. (1965). *Structure and Function of Connective and Skeletal Tissue*, 152 (London: Butterworths)
149. Rosenberg, L., Hellman, W. and Kleinschmidt, A. K. (1970). *J. Biol. Chem.*, **245,** 4123
150. Buddecke, E., Wellauer, P. and Wyler, T. (1973). In *Connective Tissue and Ageing*, Vol. 1, 140 (H. G. Vogel, editor) (Amsterdam: Excerpta Medica)
151. Shatton, J. and Schubert, M. (1954). *J. Biol. Chem.*, **211,** 565
152. Malawista, I. and Schubert, M. (1958). *J. Biol. Chem.*, **230,** 535
153. Gerber, B. R., Franklin, E. C. and Schubert, M. (1960). *J. Biol. Chem.*, **235,** 2870

154. Rosenberg, L., Johnson, B. and Schubert, M. (1965). *J. Clin. Invest.*, **44**, 1647
155. Pal, S., Doganges, P. T. and Schubert, M. (1966). *J. Biol. Chem.*, **241**, 4261
156. Rosenberg, L., Johnson, B. and Schubert, M. (1966). *J. Clin. Invest.*, **48**, 3037
157. Pal, S. and Schubert, M. (1965). *J. Biol. Chem.*, **240**, 3245
158. Bornstein, P. (1970). *Biochemistry*, **9**, 2408
159. Peterkofsky, B. and Diegelmann, R. (1971). *Biochemistry*, **10**, 988
160. Campo, R. D., Bielen, R. J. and Hetherington, J. (1972). *Biochim. Biophys. Acta*, **261**, 136
161. Rosenberg, L. C., Pal, S., Beale, R. and Schubert, M. (1970). *J. Biol. Chem.*, **245**, 4112
162. Harrington, R. E. and Zimm, B. H. (1965). *J. Phys. Chem.*, **69**, 161
163. Hascall, V. C. and Sajdera, S. W. (1969). *J. Biol. Chem.*, **244**, 2384
164. Franek, M. D. and Dunstone, J. R. (1969). *J. Biol. Chem.*, **244**, 3654
165. Sajdera, S. W. and Hascall, V. C. (1969). *J. Biol. Chem.*, **244**, 77
166. Mason, R. M. and Mayes, R. W. (1973). *Biochem. J.*, **131**, 535
167. Gregory, J. D., Sajdera, S. W., Hascall, V. C. and Dziewiatkowski, D. D. (1970). In *Chemistry and Molecular Biology of the Intercellular Matrix*, Vol. 2, 843 (E. A. Balazs, editor) (London and New York: Academic Press)
168. Anderson, C. H. and Sajdera, S. W. (1971). *J. Cell. Biol.*, **49**, 650
169. Herbage, D., Lucas, J. M. and Huc, A. (1973). *Biochim. Biophys. Acta*, **336**, 108
170. Rosenberg, L. C., Pal, S. and Beale, R. J. (1973). *J. Biol. Chem.*, **248**, 3681
171. Tsiganos, C. P. and Muir, H. (1969). *Biochem. J.*, **113**, 879
172. Brandt, K. D. and Muir, H. (1969). *FEBS Lett.*, **4**, 16
173. Brandt, K. D. and Muir, H. (1969). *Biochem. J.*, **114**, 871
174. Simunek, Z. and Muir, H. (1972). *Biochem. J.*, **126**, 515
175. Simunek, Z. and Muir, H. (1972). *Biochem. J.*, **130**, 181
176. Kobayashi, T. K. and Pedrini, V. (1973). *Biochim. Biophys. Acta*, **303**, 148
177. Keuttner, K. E., Guenther, H. L., Ray, R. D. and Schumacher, G. F. B. (1968). *Calc. Tissue Res.*, **1**, 298
178. Jeffrey, P. L. and Reinits, K. G. (1972). *Aust. J. Biol. Sci.*, **25**, 115
179. Olsson, I. and Gardell, S. (1967). *Biochim. Biophys. Acta*, **141**, 348
180. Heinegård, D. and Gardell, S. (1967). *Biochim. Biophys. Acta*, **148**, 164
181. Kleine, T. O. and Hilz, H. (1970). In *Chemistry and Molecular Biology of the Intercellular Matrix*, Vol. 2, 907 (E. A. Balazs, editor) (London & New York: Academic Press)
182. Cöster, L. and Fransson, L.-Å. (1972). *Scand. J. Clin. Lab. Invest.*, **29**, Suppl. 123, 9
183. Berman, E. R. (1970). In *Chemistry and Molecular Biology of the Intercellular Matrix*, Vol. 2, 879 (E. A. Balazs, editor) (London & New York: Academic Press)
184. Antonopoulos, C. A., Axelsson, I., Heinegård, D. and Gardell, S. (1974). *Biochim. Biophys. Acta*, **338**, 108
185. Franek, M. D. and Dunstone, J. R. (1966). *Biochim. Biophys. Acta*, **127**, 213
186. Luscombe, M. and Phelps, C. F. (1967). *Biochem. J.*, **103**, 103
187. Mayes, R. W., Mason, R. M. and Griffin, D. C. (1973). *Biochem. J.*, **131**, 541
188. McDevitt, C. A. and Muir, H. (1974). *Proc. Symp. Bruges, Protides of the Biological Fluids, XXII* (H. Peeters, editor) (London: Pergamon)
189. Mathews, M. B. (1965). *Biochem. J.*, **96**, 710
190. Mathews, M. B. (1971). *Biochem. J.*, **125**, 37
191. Pasternak, S. G. and Veis, A. (1973). *Proc. Ninth Int. Congr. Biochem., Stockholm*, 427
192. Hascall, V. C. and Sajdera, S. W. (1970). *J. Biol. Chem.*, **245**, 4920
193. Woodward, C. B., Hranisaljevic, J. and Davidson, E. A. (1972). *Biochemistry*, **11**, 1168
194. Hardingham, T. E. and Muir, H. (1972). *Biochim. Biophys. Acta*, **279**, 401
195. Gregory, J. D. (1973). *Biochem. J.*, **133**, 383
196. Tsiganos, C. P. and Muir, H. (1973). In *Connective Tissue and Ageing*, Vol. 1, 132 (H. G. Vogel, editor) (Amsterdam: Excerpta Medica)
197. Hascall, V. C. and Heinegård, D. (1974). *J. Biol. Chem.*, **249**, 4232
198. Keiser, H., Shulman, H. J. and Sandson, J. I. (1972). *Biochem. J.*, **126**, 163
199. Gregory, J. D. and Hascall, V. C. (1972). *Fed. Proc.*, **31**, 433
200. Hardingham, T. E. and Muir, H. (1973). *Biochem. J.*, **135**, 905
201. Luscombe, M. and Phelps, C. F. (1967). *Biochem. J.*, **102**, 110
202. Eyring, E. J. and Yang, J. T. (1968). *J. Biol. Chem.*, **243**, 1306
203. Hascall, V. C. and Heinegård, D. (1974). *J. Biol. Chem.*, **249**, 4242

204. Hardingham, T. E. and McDevitt, C. A. (1974). Unpublished results
205. Tsiganos, C. P., Hardingham, T. E. and Muir, H. (1971). *Biochim. Biophys. Acta*, **229**, 529
206. Mashburn, T. A. and Hoffman, P. (1971). *J. Biol. Chem.*, **246**, 6497
207. Baxter, E. (1972). *Proteoglycans of Cartilage: the Isolation and Study of their Protein Moieties, Ph.D. Thesis*, University of London
208. Baker, J. R., Rodén, L. and Stoolmiller, A. C. (1972). *J. Biol. Chem.*, **247**, 3838
209. Mathews, M. B. (1973). In *Connective Tissue and Ageing*, Vol. 1, 151 (H. G. Vogel, editor) (Amsterdam: Excerpta Medica)
210. Hjertquist, S. O. and Lemperg, R. (1972). *Calc. Tissue Res.*, **10**, 223
211. Mathews, M. B. and Glagov, S. (1966). *J. Clin. Invest.*, **45**, 1103
212. Hjertquist, S. O. and Wasteson, Å. (1972). *Calc. Tissue Res.*, **10**, 31
213. Wood, G. C. (1960). *Biochem. J.*, **75**, 605
214. Toole, B. P. and Lowther, D. A. (1968). *Biochem. J.*, **109**, 857
215. Toole, B. P. and Lowther, D. A. (1968). *Arch. Biochem. Biophys.*, **128**, 567
216. Mathews, M. B. and Decker, L. (1968). *Biochem. J.*, **109**, 517
217. Öbrink, B. (1973). *Eur. J. Biochem.*, **33**, 387
218. Öbrink, B. and Wasteson, Å. (1971). *Biochem. J.*, **121**, 227
219. Meyer, K. (1970). In *Molecular Biology*, 69 (D. Nachmansohn, editor) (New York & London: Academic Press)
220. Weiss, C., Rosenberg, L. and Helfet, A. J. (1968). *J. Bone Jt. Surg.*, **50A**, 663
221. Muir, H., Bullough, P. and Maroudas, A. (1970). *J. Bone Jt. Surg.*, **52B**, 554
222. Wood, G. C. and Keech, M. K. (1960). *Biochem. J.*, **75**, 588
223. Öbrink, B. (1973). *Eur. J. Biochem.*, **34**, 129
224. Lowther, D. A. and Natarajan, M. (1972). *Biochem. J.*, **127**, 607
225. Gelman, R. A. and Blackwell, J. (1973). *Connect. Tissue Res.*, **2**, 31
226. Gelman, R. A. and Blackwell, J. (1973). *Biochim. Biophys. Acta*, **297**, 452
227. Miller, E. J. and Matukas, V. J. (1969). *Proc. Nat. Acad. Sci. USA*, **64**, 1264
228. Miller, E. J. (1971). *Biochemistry*, **10**, 3030
229. Serafini-Fracassini, A. and Smith, J. W. (1966). *Proc. Roy. Soc. (London)*, **B165**, 440
230. Smith, J. W., Peters, T. J. and Serafini-Fracassini, A. (1967). *J. Cell Sci.*, **2**, 129
231. Smith, J. W. (1970). *J. Cell Sci.*, **6**, 843
232. Smith, J. W. and Serafini-Fracassini, A. (1968). *J. Cell Sci.*, **3**, 33
233. Smith, J. W. and Frame, J. (1969). *J. Cell Sci.*, **4**, 421
234. Iverius, P. H. (1972). *J. Biol. Chem.*, **247**, 2607
235. Kumar, V., Berenson, G. S., Ruiz, H., Dalferes, E. R. and Strong, J. P. (1967). *J. Atheroscler. Res.*, **7**, 583
236. Srinivasan, S. R., Dolan, P., Radhakrishnamurthy, B. and Berenson, G. S. (1972). *Atherosclerosis*, **16**, 95
237. Barber, A. J., Kaser-Glanzmann, R., Jakabova, M. and Luscher, E. F. (1972). *Biochim. Biophys. Acta*, **286**, 312
238. Schiller, S., Mathews, M. B. and Dorfman, A. (1956). *J. Biol. Chem.*, **227**, 625
239. Mankin, H. J. and Lippiello, L. (1969). *J. Bone Jt. Surg.*, **51A**, 1591
240. Mankin, H. J. and Lippiello, L. (1970). *J. Bone Jt. Surg.*, **52A**, 424
241. Wasteson, Å., Lindahl, U. and Hallén, A. (1972). *Biochem. J.*, **130**, 729
242. Thomas, L. (1956). *J. Exp. Med.*, **104**, 245
243. Spicer, S. S. and Bryant, J. H. (1957). *J. Biol. Chem.*, **227**, 625
244. Woessner, J. F. (1967). In *Cartilage Degradation and Repair*, 99 (C. A. C. Bassett, editor) (Washington, D.C.: National Acad. Sci.)
245. Dziewiatkowski, D. D., Hascall, V. C. and Sajdera, S. W. (1970). *Calc. Tissue Res.*, Suppl. 4, 64
246. Dingle, J. T., Barrett, A. J. and Weston, P. D. (1971). *Biochem. J.*, **123**, 1
247. Weston, P. D., Barratt, A. J. and Dingle, J. T. (1969). *Nature (London)*, **222**, 285
248. Dingle, J. T., Barrett, A. J., Poole, A. R. and Stovin, P. (1972). *Biochem. J.*, **127**, 443
249. Poole, A. R., Hembry, R. M. and Dingle, J. T. (1974). *J. Cell Sci.*, **14**, 139
250. Lucy, J. A., Dingle, J. T. and Fell, H. B. (1961). *Biochem. J.*, **79**, 500
251. Glimcher, M. J. (1959). *Rev, Mod. Phys.*, **31**, 359
252. Campo, R. D. and Dziewiatkowski, D. D. (1963). *J. Cell Biol.*, **18**, 19
253. Hirschman, A. and Dziewiatkowski, D. D. (1966). *Science*, **154**, 393
254. Baylink, D., Wergedal, J. and Thompson, E. (1972). *J. Histochem. Cytochem.*, **20**, 279

255. Pita, J. C., Cuervo, L. A., Madrazo, J. E., Muller, F. J. and Howell, D. S. (1970). *J. Clin. Invest.*, **49,** 2188
256. Morrison, R. I. G. (1970). In *Chemistry and Molecular Biology of the Intercellular Matrix*, Vol. 3, 1683 (E. Balazs, editor) (London & New York: Academic Press)
257. Muir, H. (1973). In *Lysosomes and Storage Diseases*, 79 (H. G. Hers and F. van Hoof, editors) (London & New York: Academic Press)
258. Ali, S. Y. and Evans, L. (1973). *Fed. Proc.*, **32,** 1494
259. Woessner, J. F. (1971). In *Tissue Proteinase*, 291 (A. J. Barrett and J. T. Dingle, editors) (Amsterdam: North Holland)
260. Ehrlich, M. G., Mankin, H. J. and Treadwell, B. V. (1973). *J. Bone Jt. Surg.*, **55A,** 1068
261. Ziff, M., Gribetz, H. J. and Lo Spalluto, L. (1960). *J. Clin. Invest.*, **39,** 405
262. Wiesmann, G. and Spilberg, I. (1968). *Arthritis Rheum.*, **11,** 162
263. Janoff, A. and Blondin, J. (1970). *Proc. Soc. Exp. Biol. Med.*, **135,** 302
264. Wood, G. C., Pryce-Jones, R. J., White, D. D. and Nuki, G. (1971). *Ann. Rheum. Dis.*, **30,** 73
265. Fell, H. B. and Barratt, M. E. J. (1973). *Int. Arch. Allergy*, **44,** 441
266. Lotke, J. C. and Granda, J. C. (1972). *Arth. Rheum.*, **15,** 302
267. Calatroni, A., Donnelly, P. V. and Di Ferrante, N. (1969). *J. Clin. Invest.*, **48,** 332
268. Dingle, J. T. (1961). *Biochem. J.*, **79,** 509
269. Fell, H. B. and Dingle, J. T. (1963). *Biochem. J.*, **87,** 403
270. Saito, H. and Uzman, B. G. (1970). *Exp. Cell. Res.*, **60,** 301
271. Revell, P. A. and Muir, H. (1972). *Biochem. J.*, **130,** 597
272. Wasteson, Å. and Wessler, E. (1971). *Biochim. Biophys. Acta*, **251,** 13
273. Leaback, D. H. (1970). In *Metabolic Conjugation and Metabolic Hydrolysis*, Vol. 2, 443 (W. H. Fishman, editor) (London & New York: Academic Press)
274. Wood, K. M., Wusteman, F. S. and Curtis, C. C. (1973). *Biochem. J.*, **134,** 1009
275. Tudball, N. and Davidson, E. A. (1969). *Biochim. Biophys. Acta*, **171,** 113
276. Taniguchi, N. (1972). *Clin. Chim. Acta*, **37,** 225
277. Allalouf, D. and Ber, A. (1970). *Biochim. Biophys. Acta*, **201,** 61
278. Whiteman, P. (1973). *Biochem. J.*, **131,** 351
279. Whiteman, P. (1973). *Biochem. J.*, **131,** 343
280. McKusick, V. A. (1966). *Heritable Disorders of Connective Tissue* (St. Louis: C. V. Mosby Co.)
281. Neufeld, E. F. and Cantz, M. (1973). In *Lysosomes and Storage Diseases*, 262 (H. G. Hers and F. van Hoof, editors) (London & New York: Academic Press)
282. Crawfurd, M. d'A., Dean, M. F., Hunt, D. M. Johnson, D. R., Macdonald, R. R., Muir, H., Payling-Wright, E. A. and Payling-Wright, C. R. (1973). *J. Med. Genetics*, **10,** 144
283. Matalon, R. and Dorfman, A. (1972). *Biochem. Biophys. Res. Commun.*, **47,** 959
284. Bach, G., Friedman, R., Weissmann, B. and Neufeld, E. F. (1972). *Proc. Nat. Acad. Sci. USA*, **69,** 2048
285. Bosmann, H. B. (1968). *Proc. Roy. Soc.* (*London*), **B169,** 399
286. Bollett, A. J., Handy, J. R. and Sturgill, B. C. (1963). *J. Clin. Invest.*, **42,** 853
287. Bollett, A. J. and Nance, J. L. (1966). *J. Clin. Invest.*, **45,** 1170
288. Collins, D. H. and McElligott, T. F. (1960). *Ann. Rheum. Dis.*, **19,** 318
289. Mankin, H. J., Dorfman, H., Lippiello, L. and Zarins, A. (1971). *J. Bone. Jt. Surg.*, **53A,** 523
290. Wiebkin, O. W. and Muir, H. (1973). *FEBS Lett.*, **37,** 42
291. McDevitt, C. A., Muir, H. and Pond, M. J. (1973). *Biochem. Soc. Trans.*, **1,** 287
292. McDevitt, C. A., Muir, H. and Pond, M. J. (1974). In *Symposium on Normal and Osteoarthritic Cartilage, Stanmore* (S. Y. Ali, M. W. Elves and D. H. Leaback, editors) (London: Institute of Orthopaedics)
293. White, D., Sandson, J. I., Rosenberg, L. and Schubert, M. (1963). *J. Clin. Invest.*, **42,** 992
294. Di Ferrante, N. (1964). *Science*, **143,** 250
295. Loewi, G. and Muir, H. (1965). *Immunology*, **9,** 119
296. Sandson, J. I., Rosenberg, L. and White, D. (1966). *J. Exp. Med.*, **123,** 817
297. Pankovich, A. M. and Korngold, L. (1967). *J. Immunol.*, **99,** 431
298. Brandt, K. D., Tsiganos, C. P. and Muir, H. (1973). *Biochim. Biophys. Acta*, **320,** 453
299. Boake, W. D. and Muir, H. (1955). *Lancet*, **2,** 1222

300. Quinn, R. W. and Cerroni, R. (1957). *Proc. Soc. Exp. Biol. Med.*, **96,** 268
301. Di Ferrante, N., Donnelly, P. V., Gregory, J. D. and Sajdera, S. W. (1970). *FEBS Lett.*, **9,** 149
302. Di Ferrante, N., Donnelly, P. V. and Sajdera, S. W. (1972). *J. Lab. Clin. Med.*, **80,** 364
303. Tsiganos, C. P. (1971). *Z. Klin. Chem. Klin. Biochem.*, **9,** 83
304. Sandson, J., Damon, H. and Mathews, M. B. (1970). In *Chemistry and Molecular Biology of the Intercellular Matrix*, Vol. 3, 1563 (E. A. Balazs, editor) (London & New York: Academic Press)
305. Baxter, E. and Muir, H. (1972). *Biochim. Biophys. Acta*, **279,** 276
306. Hardingham, T. E. and Muir, H. (1972). *Biochem. J.*, **126,** 791
307. Poole, A. R., Hembry, R. M. and Dingle, J. T. (1973). *Calc. Tissue Res.*, **12,** 313
308. Lowther, D. A., Toole, B. P. and Meyer, F. A. (1967). *Arch. Biochem. Biophys.*, **118,** 1
309. Meyer, F. A., Preston, B. N. and Lowther, D. A. (1969). *Biochem. J.*, **113,** 559
310. Lowther, D. A., Preston, B. N. and Meyer, F. A. (1970). *Biochem. J.*, **118,** 595
311. Buddecke, E. and Schubert, M. (1961). *Z. Physiol. Chem.*, **325,** 189
312. Toole, B. P. and Lowther, D. A. (1966). *Biochim. Biophys. Acta*, **121,** 315
313. Öbrink, B. (1972). *Biochim. Biophys. Acta*, **264,** 354
314. Jansson, L. and Lindahl, U. (1972). *Scand. J. Clin. Lab. Invest.*, **29,** Suppl. 123, 19
315. Preston, B. N. (1968). *Arch. Biochem. Biophys.*, **126,** 974
316. Kresse, E., Heidel, H. and Buddecke, E. (1971). *Eur. J. Biochem.*, **22,** 557
317. Buddecke, E. and Kresse, H. (1973). In *Connective Tissue and Ageing*, Vol. 1, 14 (H. G. Vogel, editor) (Amsterdam: Excerpta Medica)
318. Campo, R. D. and Tourtellotte, C. D. (1967). *Biochim. Biophys. Acta*, **141,** 614
319. Herring, G. M. (1968). *Biochem. J.*, **107,** 41
320. Radhakrishnamurthy, B., Srinivasan, S. R., Dalferes, E. R. and Berenson, C. S. (1970). *Comp. Biochem. Physiol.*, **36,** 107
321. Jacobs, S. and Muir, H. (1967). *Biochem. J.*, **87,** 38P
322. Jansson, L. and Lindahl, U. (1970). *Biochem. J.*, **117,** 699
323. Serafini-Fracassini, A., Durward, J. J. and Floreani, L. (1969). *Biochem. J.*, **112,** 167
324. Lindahl, U. (1970). *Biochem. J.*, **116,** 27
325. Horner, A. A. (1971). *J. Biol. Chem.*, **246,** 231
326. Olsson, I. (1968). *Biochim. Biophys. Acta*, **165,** 324
327. Olsson, I. (1969). *Exp. Cell. Res.*, **54,** 314
328. Olsson, I. (1969). *Exp. Cell Res.*, **54,** 318
329. Olsson, I. (1970). *Exp. Cell Res.*, **70,** 173
330. Mustard, J. F. and Packham, M. A. (1970). *Pharmacol. Rev.*, **22,** 67
331. Ridell, P. E. and Bier, A. M. (1965). *Nature*, **205,** 711
332. Revell, P. A., Dean, M. F., Vernon-Roberts, P. A., Muir, H. and Marshall, A. (1972). *Int. Arch. Allergy*, **43,** 813
333. Marshall, A. H. E., Swettenham, K. V., Vernon-Roberts, B. and Revell, P. A. (1971). *Int. Arch. Allergy*, **40,** 137
334. Dean, M. F., Muir, H., Marshall, A. H. E., Revell, P. A. and Vernon-Roberts, B. (1971). *FEBS Lett.*, **16,** 183
335. Toole, B. P., Jackson, G. and Gross, J. (1972). *Proc. Nat. Acad. Sci. USA*, **69,** 1384
336. Toole, B. P. and Gross, J. (1971). *Develop. Biol.*, **25,** 57
337. Toole, B. P. (1972). *Develop. Biol.*, **29,** 321
338. Toole, B. P. and Trelstad, R. L. (1971). *Develop. Biol.*, **26,** 28
339. Toole, B. P. (1973). *Amer. Zool.*, **13,** 1061
340. Wood, K. M., Curtis, C. G., Powell, G. M. and Wusteman, F. S. (1973). *Biochem. Soc. Trans.*, **1,** 840

# 5
# Regulation of Intermediary Carbohydrate Metabolism

**M. G. CLARK and H. A. LARDY**
University of Wisconsin

---

## 5.1 INTRODUCTION

From Gay-Lussac to Sutherland, the study of carbohydrate metabolism created biochemistry as well as several of its subsciences, and has led to our comprehension of the exquisite mechanism by which many hormones and other regulatory influences control cell function. The regulation of intermediary metabolism derives ultimately from the properties of the individual enzymes and of the carriers and other devices that control movements of materials across living membranes. Therefore this chapter will deal with the regulatory properties of the individual enzymes in sequence from glucose to pyruvate. An excellent review by Pitot and Yatvin[1] treats the influence of diet and hormonal status on the amounts of enzyme found in tissues. Specific consideration of the regulation of glycogen metabolism is omitted, as this is the subject of Chapter 7. Restrictions on space necessitated a limitation to the living forms that could be considered. We chose to emphasise non-ruminant, mammalian systems.

## 5.2 SOURCES OF CARBOHYDRATE

The principal carbohydrates in the mammalian diet are the sugars, starches, glycogen, and starch hydrolysis products such as the dextrins. The digestibility of cellulose and the many other plant structural polysaccharides varies greatly among animals, depending on the type of digestive system. Starches, dextrins and glycogen are hydrolysed to glucose and are absorbed primarily

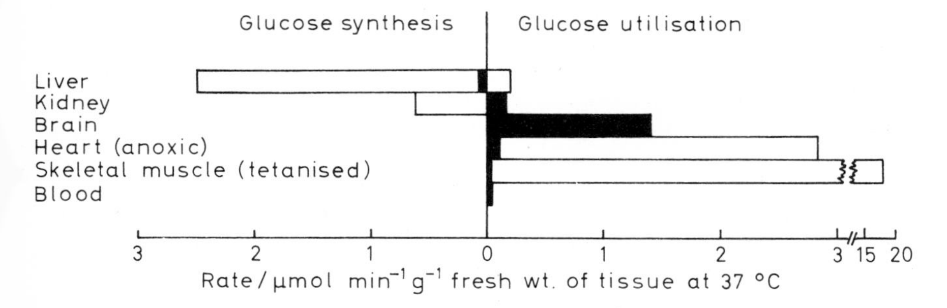


**Figure 5.1** Rates of mammalian tissue glucose metabolism. Average normal post-absorptive rates are shaded; maximum rates are unshaded. Data are taken from Refs. 57, 109, 351–355

by the portal blood. The glucose contributes to the blood sugar, to the glycogen of tissues and to many other carbohydrate-containing molecules. Fructose, which may appear in the diet as the free ketose or be liberated from sucrose, is converted into glucose phosphates by the liver and probably to some extent in the intestinal mucosa. Galactose, originating almost entirely from dietary lactose, is converted into glucose 1-phosphate (glucose 1-P) in the liver and other tissues. Although some pentoses are relatively good gluconeogenic substrates for liver, in non-herbivores many are largely excreted in the urine and most are not liberated by digestive enzymes from the dietary pentosans.

Carbohydrate metabolism in the mammal is thus largely concerned with the conversion of dietary carbohydrate, circulating amino acids and organic acids into glucose, and with the utilisation of this sugar in various tissues depending on their need. The approximate rates of utilisation and synthesis are shown in Figure 5.1.

## 5.3 REGULATORY PROPERTIES OF THE ENZYMES OF GLYCOLYSIS, GLUCONEOGENESIS AND THE PENTOSE PHOSPHATE PATHWAY*

### 5.3.1 Interconversion of glucose and glucose 6-phosphate

#### *5.3.1.1 Hexokinase (EC 2.7.1.1)*

In the vertebrate, glucose phosphorylation appears to be regulated at two levels: (*a*) constitutive isoenzymes and (*b*) tissue concentration of metabolite effectors.

(*a*) Tissues contain varying proportions of four different isoenzymes of hexokinase, distinct in their electrophoretic mobilities and in their catalytic properties. The two principal forms are (1) Types I, II and III, having relatively high affinities for their substrate with $K_m$ values of $10^{-5}$, $10^{-4}$ and $10^{-6}$ mol $l^{-1}$, respectively[1a], specifically inhibited by glucose 6-P; and (2) Type IV, often referred to as glucokinase (EC 2.7.1.2) or low affinity hexokinase ($K_m = 10^{-2}$ mol $l^{-1}$) and which is not affected by glucose 6-P [2]. The distribution of these four isoenzymes has great influences on the regulation of glucose catabolism in the various tissues. Brain and kidney contain predominently Type I, skeletal muscle Type II and adipose tissue, heart and intestine, Types I and II. The predominance of Type IV in liver[3] accounts for the requirement of high blood glucose concentrations for glycogen synthesis in this tissue; this isoenzyme of hexokinase is not inhibited by glucose 6-P[2] and glycogen synthetase is activated by this metabolite[4]. A further dimension is added to the regulation of glucose phosphorylation by changes in hexokinase isoenzymes induced by the dietary and hormonal status of the animal. Thus, although liver contains both the high $K_m$ and low $K_m$ enzymes[2,5,6], the high $K_m$ enzyme decreases preferentially during starvation or diabetes and increases to normal values following refeeding or insulin administration[2,6-8]. In adipose tissue[9,10], and in muscle[9], the Type II enzyme also appears to decrease with starvation or diabetes and then reappears with refeeding or insulin administration. Type II is characteristically found in insulin-sensitive tissues. More recent observations by Katzen *et al.*[11] disclosed that insulin administration to diabetic rats resulted in a rapid (less than 2 hours) net increase in the gastrocnemius hexokinase II activity, but similar treatment led to only a slow restoration of Type II activity in cardiac muscle. Further attention to decide whether hexokinase II removal is truly stimulated by the diabetic state, or if resynthesis is dependent on insulin, would seem merited.

For mammalian systems the substrate specificity of the four isoenzymes of hexokinase does not appear to differ widely (e.g. see Ref. 12) and therefore does not confer substrate selectivity on the tissues. However, in liver the low affinity of Type IV for all substrates renders its activity toward fructose negligible at physiological concentrations with the result that fructose is metabolised predominantly by fructokinase.

* See Figure 5.2.

(*b*) The fine control of tissue glucose phosphorylation may be achieved by the intracellular concentration of metabolite regulators of hexokinase. The unique feature of Type I, II and III hexokinases is the striking inhibition by glucose 6-P[13, 14]. ADP is also a possible regulator of hexokinase, although it is a weaker inhibitor than glucose 6-P[15]. Because inorganic phosphate ($P_i$) can attenuate the inhibitory capacity of glucose 6-P for hexokinase[16-18], elegant theories to explain the regulation of glycolysis have been proposed involving changes in the concentration of this metabolite[19]. This appeared attractive because $P_i$ could also reverse the inhibitory effects of ADP and other nucleotides. In situations requiring high glycolytic rates, an increase in $P_i$ concentration would not only deinhibit hexokinase but also activate phosphofructokinase[20].

Differences in relation to the $P_i$ effect have been detected by comparison of Types I and II hexokinases[21]; only Type I shows a $P_i$ effect on sensitivity to glucose 6-P inhibition, restricting this form of regulation to brain and kidney. Type II, on the other hand, shows the interesting phenomenon of a delayed onset of inhibition by glucose 6-P. Thus the different types of regulation may possibly relate to the different metabolic characteristics of the different tissues. Tumour and muscle tissues, which are rich in Type II enzyme, are thought to require adjustment to sudden changes of availability of intracellular glucose, whereas brain and erythrocytes, which are rich in Type I enzyme, are likely to have a constant supply of intracellular glucose. Indeed, for these latter tissues regulation of glucose utilisation may be achieved through alterations in the level of intracellular $P_i$.

One of the most exciting discoveries for the possible regulation of glucose phosphorylation comes from the initial finding by Crane and Sols[13, 22] that hexokinases in homogenates are associated with mitochondria[23-26]. Initial observations indicated that hexokinase could be dissociated by the addition of very low concentrations of glucose 6-P[25]. However, Purich and Fromm[15], studying tissue hexokinase–mitochondrial preparations, concluded that changes in the glucose 6-P level probably have little effect on the distribution of hexokinase under simulated intracellular conditions.

The role of the ATP:ADP ratio in controlling the distribution of brain hexokinase (i.e. soluble $\rightleftharpoons$ mitochondrial), as reported by Hochman[27], has recently been tested by Knull *et al.*[28]. An *in vivo* soluble $\rightleftharpoons$ mitochondrial equilibrium is indicated; it is responsive to the energy status of the brain, consistent with the involvement of such an equilibrium in regulation of carbohydrate metabolism[29]. For example, ischemia, which produced a decrease in ATP concentration, gave rise to more mitochondrial-bound enzyme and less of the soluble form. In the light that glucose 6-P inhibition alone may fall short of the control required[30], a further control exercised by the energy status of the cell would add considerably to a regulatory mechanism. Indeed Knull *et al.*[28] show an inverse correlation between the ATP concentration of brain regions and the fraction of the total hexokinase that is bound to mitochondria and in the active form. Finally, the observation that inorganic phosphate inhibits solubilisation[29] would fit the general pattern of a regulatory mechanism controlled by the ratio of ATP to ADP + $P_i$. As the cytosolic concentrations of ATP, ADP and $P_i$ are interrelated to one another, and linked to the $NAD^+$:NADH ratio through the equilibrium enzymes glyceraldehyde

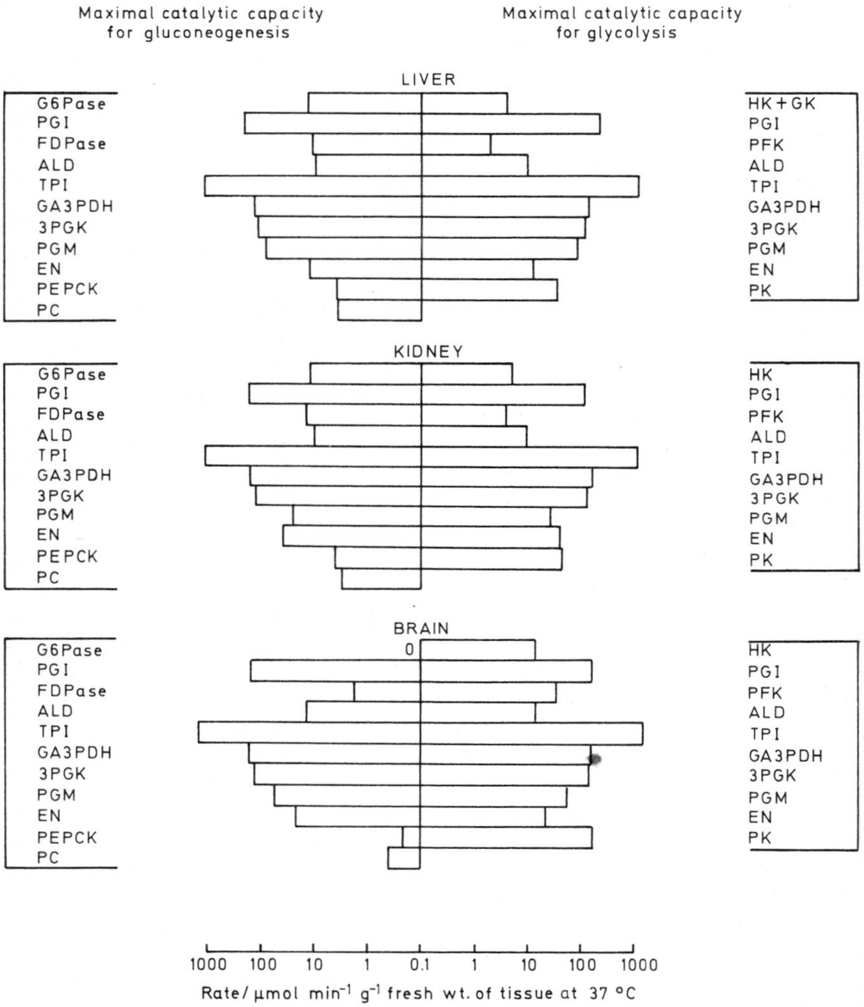
Maximal catalytic capacity for gluconeogenesis
Maximal catalytic capacity for glycolysis
LIVER
G6Pase
PGI
FDPase
ALD
TPI
GA3PDH
3PGK
PGM
EN
PEPCK
PC
HK+GK
PGI
PFK
ALD
TPI
GA3PDH
3PGK
PGM
EN
PK
KIDNEY
G6Pase
PGI
FDPase
ALD
TPI
GA3PDH
3PGK
PGM
EN
PEPCK
PC
HK
PGI
PFK
ALD
TPI
GA3PDH
3PGK
PGM
EN
PK
BRAIN
0
G6Pase
PGI
FDPase
ALD
TPI
GA3PDH
3PGK
PGM
EN
PEPCK
PC
HK
PGI
PFK
ALD
TPI
GA3PDH
3PGK
PGM
EN
PK
1000 100 10 1 0.1 1 10 100 1000
Rate/µmol min⁻¹ g⁻¹ fresh wt. of tissue at 37 °C

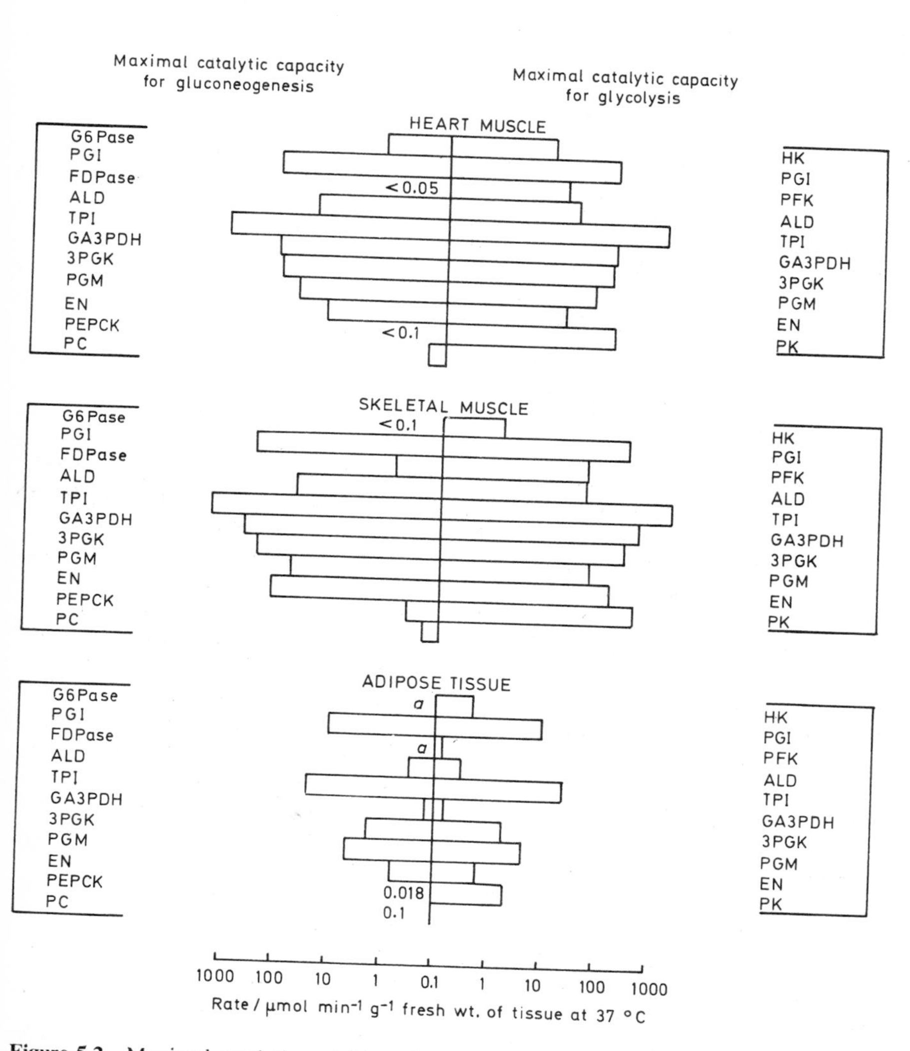


**Figure 5.2** Maximal catalytic activities of the enzymes of glycolysis and gluconeogenesis for tissues of the fed albino rat. Data are taken from Refs. 75, 132, 136, 293, 356–358 Abbreviations: G6Pase, glucose 6-phosphatase; HK, hexokinase; GK, glucokinase; PGI, glucose phosphate isomerase; FDPase, fructose 1,6-diphosphatase; PFK, phosphofructokinase; ALD, aldolase; TPI, triose phosphate isomerase; GA3PDH, glyceraldehyde 3-phosphate dehydrogenase; 3PGK, 3-phosphoglycerate kinase; PGM, phosphoglycerate mutase; EN, enolase; PC, pyruvate carboxylase; PEPCK, phosphoenolpyruvate carboxykinase; PK, pyruvate kinase; [a] Activity virtually absent

3-phosphate dehydrogenase and 3-phosphoglycerate kinase (as they are for liver[31]), decreases in ATP concentration would result in increased $P_i$ concentration. The combined effect would be both the attenuation of glucose 6-P inhibition and the formation of more mitochondrial-bound (active) enzyme.

#### *5.3.1.2 Glucose 6-phosphatase (EC 3.1.3.9)*

This enzyme is attached to the endoplasmic reticulum[32] and has been postulated to release its product, glucose, into the space that is continuous with the extracellular space. In addition to hydrolysing glucose 6-P [reaction (5.1)], this enzyme has the potential to catalyse the synthesis of glucose 6-P [reaction (5.2)] using a variety of phosphoryl donors[33].

$$\text{glucose 6-P} + H_2O \rightleftharpoons P_i + \text{glucose} \qquad (5.1)$$

$$\text{R—P} + \text{glucose} \rightleftharpoons \text{glucose 6-P} + \text{R} \qquad (5.2)$$

Whereas the activity of glucose 6-phosphatase is found in such diverse tissues as skeletal muscle, bone marrow, mammary gland, seminal vesicles and adipose tissue (e.g. see Ref. 32), its role until relatively recently has been assumed to be solely hydrolytic and thereby ultimately involved in reactions of either glycogenolysis or gluconeogenesis. The discovery that glucose 6-phosphatase is able, under appropriate conditions, to catalyse a synthetic reaction[33, 34] [reaction (5.2)] may to some degree confer a special significance of this enzyme to the synthesis of glucose 6-P from glucose. In particular, a role for the phosphorylation of glucose in liver of diabetic animals[35] and in bovine pancreatic β-cells[36] has been suggested. It must be stressed, however, that rather special circumstances may be required for glucose 6-phosphatase to function in a net glucose 6-phosphate synthesising capacity. These include: (*a*) conditions under which the hydrolytic activity of the enzyme is rendered inoperative (the simultaneous operation of both hydrolytic and phosphotransferring activities would constitute an energy dissipating process); (*b*) high concentration of glucose as the measured $K_m$ for glucose is *ca.* 80–90 mol $l^{-1}$; and (*c*) availability of a suitable phosphoryl donor (donors considered of possible physiological importance are $PP_i$ ($K_m$ = 1.8 mmol $l^{-1}$ [37]), carbamyl-P ($K_m$ = 1.6 mmol $l^{-1}$ [38]) and 1,3-diphosphoglycerate ($K_m$ = 1.4 mmol $l^{-1}$ [39]). As detailed by Nordlie[32], there are indeed several factors that serve to alter selectively the phosphohydrolase and phosphotransferase activities of the enzyme. These include pH (phosphotransferase activity is favoured by pH below 7), long-chain fatty acyl CoA esters (physiological concentrations appear to activate transferase activity and simultaneously act to decrease hydrolase activity[40]), cationic detergent, cetrimide with phlorizin (the combination of these agents selectively allows expression of the phosphohydrolase activity of the enzyme[41, 42]) and, finally, certain synthetic and natural detergents markedly potentiate inhibition by $P_i$, ATP and other nucleotides of the phosphohydrolysing activity of glucose 6-phosphatase[43, 44].

Regulation of glucose 6-phosphatase activity (both phosphotransferase and hydrolase activity) is also achieved by variation in the dietary and hormonal status of the animal[45]. Not surprisingly, the total activity of glucose

6-phosphatase changes in an inverse manner to the total activity of glucokinase. Conditions such as alloxan diabetes, acute fasting and high fructose diet[45] led to increases in the enzyme, whereas high maltose diet[46] and insulin administration to normal or diabetic animals[45] gave rise to decreases in the enzyme.

Recent studies indicate that the various phosphohydrolase and phosphotransferase activities of glucose 6-phosphatase are affected by a variety of metabolites (these are summarised in an excellent review by Nordlie[32]; see his Table X). The possible regulation of glucose 6-phosphatase by metabolites *in vivo* appears to be very important, particularly as this enzyme has relatively high $K_m$ values for its substrates. Thus, the $K_m$ for glucose 6-P is *ca.* 1–2 mmol $l^{-1}$, compared with hepatic glucose 6-P levels of 0.05 mmol $l^{-1}$ [43, 47, 48]; the $K_m$ for glucose is *ca.* 80–90 mmol $l^{-1}$, compared with normal physiological values of 5 mmol $l^{-1}$. Such a variance between $K_m$ values and normal physiological substrate levels suggests that the enzyme may be especially susceptible to competitive inhibition by such metabolites as ATP and other nucleotides[49].

### *5.3.1.3 The glucokinase–glucose 6-phosphatase substrate cycle*

Cahill and co-workers[50] in 1959 suggested that both glucokinase and glucose 6-phosphatase could operate simultaneously in liver to control blood glucose levels as well as the magnitude and direction of hepatic glucose metabolism. Since that time the glucose substrate cycle of liver has received surprisingly little attention, even though a mechanism of the kind suggested by Cahill *et al.*[50] would appear to be consistent with existing data for all but the pathological states. Thus gluconeogenesis *in vitro*[51] or *in vivo*[52] is diminished and glucose utilisation is enhanced at high ($>10$ mmol $l^{-1}$) extracellular glucose concentrations. Conversely, low glucose concentrations result in decreased rates of glucose utilisation and permit high rates of gluconeogenesis[51, 52]. At the enzyme level this inverse relationship between glucose phosphorylation and glucose 6-P dephosphorylation would also appear to be valid. Thus starvation, cortisol or glucagon administration[45] results in an increase in the activity of glucose 6-phosphatase and a decrease in glucokinase activity. In contrast, diabetes is characterised by elevated levels of blood glucose together with enhanced glucose 6-phosphatase activity[53] and a decreased glucokinase activity[6]. Thus the rates of glucose cycling in relation to the hepatic rates of glycolysis, gluconeogenesis and glycogen synthesis in normal and diabetic states should be investigated. A possible means of reciprocal control of glucose phosphorylation and glucose 6-P dephosphorylation, by either hormones or metabolites, would also merit investigation.

Preliminary estimations of the rate of substrate cycling of glucose in rat liver *in vivo*[54] and in isolated rat hepatocytes[55-57] indicate some interesting points: (i) the rate of hepatic glucose phosphorylation *in vivo* (0.13 μmol $min^{-1}$ $g^{-1}$) approximately equals the rate of fructose 6-P phosphorylation (0.12 μmol $min^{-1}$ $g^{-1}$) and is 35% of the expressed rate of glucose 6-phosphatase for the suckling 2-day-old rat[54]; (ii) for isolated hepatocytes from either starved or meal-fed rats the rates of phosphorylation of glucose are similar (15–25%

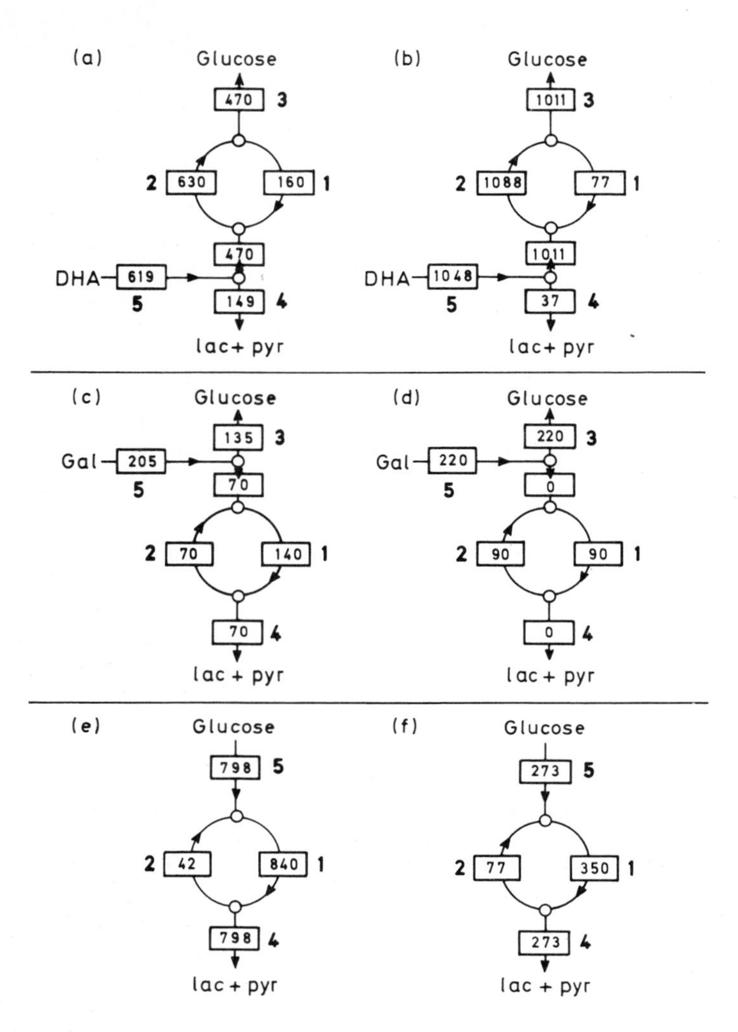


**Figure 5.3** Effect of glucagon on the rates of flux through reactions of gluconeogenesis and glycolysis in isolated rat-liver cells. Rates are expressed as nmol of glucose $min^{-1}$ $g^{-1}$ fresh weight of liver and were calculated from the estimated rates of net glucose formation, phosphorylation of fructose 6-phosphate and net lactate and pyruvate formation. 5.5 nM Glucose was used in experiments (a)–(d). Conditions were as follows: (a), 5 mM dihydroxyacetone (DHA); (b), 5 mM dihydroxyacetone plus 86 mM glucagon; (c), 2.5 mM galactose (GAL); (d), 2.5 mM galactose plus 86 mM glucagon; (e), 27.8 mM glucose; (f), 27.8 mM glucose plus 86 nM glucagon. Flux rates are represented: **1**, phosphorylation of fructose 6-phosphate; **2**, dephosphorylation of fructose 1,6-diphosphate; **3**, net glucose formation; **4**, net lactate and pyruvate formation; **5**, substrate utilisation. (From Clark *et al.*[57] by courtesy of the American Society of Biological Chemists.)

of the expressed rate of glucose 6-phosphatase) and independent of the presence of a gluconeogenic substrate (and hence the rate of gluconeogenesis)[55]; (iii) the rate of glucose phosphorylation is proportional to the extracellular concentration of glucose; the increase in rate was independent of the presence of lactate[55]; (iv) the rate of glucose phosphorylation at 5.5 mmol $l^{-1}$ glucose

**Table 5.1 Gluconeogenic and glycolytic rates for isolated rat hepatocytes†**

| *Substrate* | *Concentration* /mmol $l^{-1}$ | *Rate* /μmol glucose $min^{-1}$ $g^{-1}$ *Gluconeogenesis* | *Glycolysis* |
|---|---|---|---|
| Lactate | 2.5 | 0.06 | |
| Lactate | 5 | 0.08 | |
| Lactate | 10 | 0.20 | |
| Pyruvate | 1 | 0.08 | |
| Lactate + Pyruvate | 10 + 1 | 0.43 | |
| Dihydroxyacetone | 5 | 0.58 | |
| Galactose | 5 | 0.19 | |
| Glyceraldehyde | 5 | 0.74 | |
| Fructose | 5 | 1.48 | |
| Xylitol | 5 | 0.86 | |
| Dihydroxyacetone* | 5 | 0.47 | |
| Dihydroxyacetone* + glucagon | 5 | 1.01 | |
| Galactose* | 2.5 | 0.13 | |
| Galactose* + glucagon | 2.5 | 0.22 | |
| Fructose* | 2.5 | 0.69 | |
| Fructose* + glucagon | 2.5 | 1.10 | |
| Xylitol* | 2.5 | 0.48 | |
| Xylitol* + glucagon | 2.5 | 0.56 | |
| Glucose | 5.5 | | 0.06 |
| Glucose | 11.1 | | 0.30 |
| Glucose | 27.8 | | 0.74 |

* Rates of gluconeogenesis determined in the presence of 5.5 mM glucose

† Data taken from Ref. 57. Rates are calculated on the basis that 1 g fresh weight of liver is equivalent to $10^8$ cells

ranges between 0.16 and 0.2 μmol $min^{-1}$ $g^{-1}$ and is 60% and 25% of the rate of glucose 6-phosphatase, respectively, when 2.5 mmol $l^{-1}$ galactose and 5 mmol $l^{-1}$ dihydroxyacetone are used as gluconeogenic substrates (see Table 5.1 and Figure 5.3)[57]. These results, although preliminary, suggest that rates of cycling and 'futile' hydrolysis of ATP may be greatest when high rates of glucose formation and high blood glucose levels occur together.

### 5.3.2 Metabolism of phosphorylated glucose

#### *5.3.2.1 Glucose phosphate isomerase (EC 5.3.1.9)*

Despite the findings that several intermediates of the pentose phosphate pathway are potent competitive inhibitors of glucose phosphate isomerase (e.g. 6-phosphogluconate[58], apparent $K_i$ 5 μmol $l^{-1}$ [59]; erythrose 4-P[60], apparent $K_i$ 1–2 μmol $l^{-1}$ [60, 61]; and sedoheptulose 7-P [62]), there appears to be no experimental evidence to indicate that this enzyme fails to maintain equilibrium or near equilibrium conditions in the cell. This is particularly so for liver, where 6-phosphogluconate concentration is apparently sufficient (14 nmol per g wet weight of tissue[63]) to inhibit strongly both glycolysis and gluconeogenesis. Perhaps other effectors that nullify the inhibitors occur in tissues.

Interestingly, the inhibition of muscle glucose phosphate isomerase by 6-phosphogluconate is pH dependent[64] and the decrease in $K_i$ observed with decreasing pH has been suggested[65] as a possible means by which temporary decreases in the intracellular pH may exert a regulatory effect through increased inhibition of the glucose phosphate isomerase reaction. In tissues such as muscle, where glucose metabolism is essentially unidirectional, regulation of glucose phosphate isomerase may prove to be important; however, despite the potency of the inhibitors cited above and the claims that this enzyme was possibly regulatory[66, 67], no substantiating evidence has been forthcoming.

Of the many other reported inhibitors of glucose phosphate isomerase[68], ATP and $P_i$ have apparent $K_i$ values that would give them possible physiological significance[59].

Finally, yeast glucose phosphate isomerase has been demonstrated to possess anomerase activity[61] and can produce the open-chain form from either α- or β-glucose 6-P. Such anomerase activity possessed by glucose phosphate isomerase obviates the occurrence of an otherwise possible rate-limiting step and complies with the specificity of both glucose 6-P dehydrogenase (EC 1.1.1.49)[61] and phosphofructokinase (EC 2.7.1.11)[69] (each enzyme is specific for the β-anomer).

#### *5.3.2.2 Phosphoglucomutase (EC 2.7.5.1)*

As in the case of glucose phosphate isomerase, control at phosphoglucomutase would seem unlikely on the basis that (i) the enzyme's equilibrium constant is close to unity, (ii) it is present in relatively high concentration, and (iii) it has high catalytic efficiency. Nevertheless, recent evidence indicates that the divalent metal form of phosphoglcomutase may be important in regulating this enzyme's activity in the cell[70]. Indeed, a marked increase in the fraction of phosphoglucomutase in the active $Mg^{2+}$ form *in vivo* is produced in starved rabbits following insulin treatment[70]. In view of the work of Aikawa on the effect of insulin and alloxan diabetes on $Mg^{2+}$ flux in rabbit muscle[71, 72], it would appear likely that insulin acts to increase the intracellular availability of free $Mg^{2+}$. A resulting increase in free $Mg^{2+}$ within the cell may give rise to the observed increase in enzyme–Mg:enzyme–Zn ratio. This increase in phosphoglucomutase activity could be important in the increased deposition of glucose as glycogen which is observed on insulin treatment[73].

### 5.3.2.3 *Regulation of the pentose phosphate pathway*

The oxidative pentose phosphate pathway which proceeds via 6-phosphogluconate is essentially irreversible (e.g. see Ref. 74) and while the maximal catalytic activity of the rate-limiting enzyme of this pathway in most tissues is considerably less than that of glycolysis (e.g. compare the activities of glucose 6-P dehydrogenase with phosphofructokinase[75]), the pathway provides an important function for the production of NADPH for biosynthetic processes and ribose phosphate for nucleic acids. Regulation of the oxidative segment would seem to be implemented at three levels:

(*a*) Changes in maximal catalytic activity in response to hormonal or dietary status[76] (usually changes correspond to the demand for NADPH). Tissues possessing high activities of the oxidative enzymes show specific needs for NADPH (e.g. lactating mammary gland[77]).

(*b*) Possibly the most relevant regulation of the oxidative pathway is achieved indirectly by the controlled availability of cytosolic $NADP^+$. Calculations of the (free $NADP^+$):(free NADPH) ratio for the liver cytosol of the well-fed rat indicate[31] this value to be *ca.* 0.014, which is about five orders of magnitude less than that of the cytosolic (free $NAD^+$):(free NADH) ratio of the same tissue[31]. Such a low value would strongly suggest that $NADP^+$ is indeed a limiting factor for the operation of the oxidative pentose phosphate pathway enzymes, and that a tight coupling of NADPH utilising processes to NADPH-producing processes is maintained. Claims for this form of regulation of the pentose phosphate pathway have been further supported by the finding that phenazine methosulphate[63, 78] markedly enhances oxidation of [1-$^{14}$C]glucose to $^{14}CO_2$, indicating that the rate of oxidation depends on availability of coenzyme and that none of the enzymes of the oxidative system operates normally at maximal capacity.

(*c*) Regulation of the oxidative segment may also be achieved by metabolite effectors. Reported inhibitors of glucose 6-P dehydrogenase include stearyl CoA[79], nucleoside phosphates[80-82] and NADPH[83]. 6-Phosphogluconate dehydrogenase (EC 1.1.1.44) is inhibited by ATP[84], 2′-AMP[85] and potently by fructose 1,6-diP ($K_i$ 70 μmol $l^{-1}$ [86]). However, in view of the generality of the pronounced effect of phenazine methosulphate on oxidation via the pentose pathway[78], it would appear unlikely that regulation by means other than availability of $NADP^+$ is operative under these assay conditions.

Regulation of the non-oxidative segment of the pentose phosphate pathway has not been described. The non-oxidative segment is reversible and may function both as a parallel pathway[74] to the oxidative segment for the production of pentose 5-P and as a complementary component to the oxidative segment to complete a cyclic process for the oxidation of glucose 6-P[87, 88]. Operation in either direction may thus depend upon the cell's requirements for pentose phosphate and NADPH.

The reaction mechanism of the non-oxidative pentose phosphate pathway has been examined in detail recently and may involve new enzymes and new intermediates[89] (*cf.* the reaction scheme proposed by Horecker *et al.*[90]). Regulation of this new scheme is not defined, but D-arabinose 5-P, an intermediate of the proposed pathway[89], is a potent inhibitor of transaldolase ($K_i = 5.5 \times 10^{-5}$ mol $l^{-1}$ and competitive with glyceraldehyde 3-P[91]).

### 5.3.3 Interconversion of fructose 6-P and fructose 1,6-diP

#### *5.3.3.1 Phosphofructokinase*

Since the classical studies by Cori on the metabolism of working muscle and the report that the reaction catalysed by phosphofructokinase was rate-limiting[92], numerous investigations have concentrated on the regulatory aspects of this enzyme (e.g. see Refs. 69 and 93). As a result, phosphofructokinase has been recognised as having a major role for the regulation of glycolysis in many tissues. Furthermore, a possible role in the hormonal regulation of hepatic gluconeogenesis has been suggested recently[57] (Section 5.3.3.3b), and involvement in a substrate cycle with fructose 1,6-diP as a potential biochemical heat-producing mechanism has also been indicated[94,95] (Section 5.3.3.3c).

Recent studies on the electrophoretic and immunological properties of phosphofructokinase from several tissues of the rabbit indicate two distinct, and possibly a third, isoenzymes of phosphofructokinase[96]. Antisera to muscle enzyme showed only partial reaction with liver enzyme, and antisera to the liver enzyme showed very little reaction with the muscle enzyme. On the basis of these observations it was concluded that skeletal muscle and heart contain a single identical isoenzyme species (phosphofructokinase *A*) which is distinct from the major phosphofructokinase of liver and erythrocytes (phosphofructokinase *B*). Whereas most tissues besides liver, skeletal muscle and heart contain isoenzymes *A* and *B*, a third isoenzyme, not removed by the combined antisera *A* + *B*, was present in brain (designated phosphofructokinase *C*)[96]. The kinetic properties of the three isoenzymes remain to be examined in detail; however, Kemp[97] has compared the regulatory properties of rabbit phosphofructokinase from muscle and liver and found the liver enzyme is less inhibited by ATP, less inhibited by citrate, and more sensitive to 2,3-diphosphoglycerate inhibition. Furthermore, differences between rat-heart and rat-brain phosphofructokinases have also been noted[98,99]. While brain phosphofructokinase is inhibited by citrate, *cis*-aconitate, isocitrate, malate, succinate and α-ketoglutarate, rat-heart phosphofructokinase is inhibited by citrate but the other acids are without effect. These observations suggest that phosphofructokinase isoenzyme distribution may confer specific control properties on the individual tissues.

(a) *Regulation of phosphofructokinase at the metabolite level*—The inhibition of phosphofructokinase by ATP was first reported for rabbit muscle enzyme[100]. The importance of the energy status, i.e. the [ATP]/[ADP] [$P_i$] ratio or the 'energy charge' [ATP] + [$\frac{1}{2}$ADP]/[AMP] + [ADP] + [ATP][101], in the regulation of glycolytic flux gained attention when Passonneau and Lowry[99] discovered that metabolic degradation products of ATP (ADP, AMP and $P_i$) specifically reverse the inhibition of phosphofructokinase by ATP.

On teleological grounds, the ATP inhibition of phosphofructokinase would seem to be the most important as a probable regulatory mechanism for glycolysis *in vivo*. Increasing fructose 6-P concentrations decrease the inhibitory action of ATP[20, 103], demonstrating that the regulatory effect of either substrate is a function of the relative concentration of the other. ATP inhibition of phosphofructokinase is markedly dependent on pH (inhibition decreases as the pH increases from 6.5 to 8.0). An explanation for the Pasteur effect may derive from the inherent regulatory properties of phosphofructokinase as they relate to the cellular ATP concentration and energy status of the cell. Activation of glycolysis to compensate for the lack of energy production by oxidative phosphorylation is observed in tissues deprived of optimal oxygen tension. For perfused heart[104, 105], brain[106], kidney[107], adenocarcinoma[107] and Novikoff hepatoma[107], the onset of anaerobic conditions gives rise to distinct metabolite cross-over points at the level of phosphofructokinase. In all of these cells the increase in phosphofructokinase activity commensurate with increased glycolysis can be related to changes in the nucleotide concentrations. Thus in most cases under anaerobic conditions there is an increase in the concentrations of AMP and ADP paralleled by a decrease in ATP concentration. Since AMP and ADP are activators, whereas ATP is an inhibitor of phosphofructokinase *in vitro* (and phosphocreatine is an inhibitor of brain phosphofructokinase), it seems reasonable to propose that in anaerobiosis the altered concentration of these effectors would co-operate to increase the activity of phosphofructokinase.

Of the citric acid cycle intermediates reported to inhibit phosphofructokinase[98, 99], citrate is the most potent. It appears to be the most important factor in the regulation of glycolysis in cardiac muscle. Thus, a decrease in the rate of glycolyis in perfused rat heart coinciding with fatty acid addition does not appear to be related to changes in adenine nucleotide concentrations[98, 108-110], but may instead result from elevated citrate concentrations[108, 110]. Hearts from alloxan diabetic rats had a higher content of citrate than those from normal rats, and insulin administration to such rats resulted in a decrease in cardiac citrate concentration[110]. It was proposed by Randle *et al.*[110] that the increase in citrate concentration is sufficient to result in an inhibition of phosphofructokinase with a concomitant decrease in glycolysis; a fall in citrate will produce the opposite effect. Studies with fluoroacetate show that citrate can inhibit *in vivo*[111]. Parmeggiani and Bowman[108] have shown that under anaerobic conditions the ability of octanoate to raise the level of citrate is reduced and the fatty acid has no influence on the rate of glycolysis. A co-ordinated regulatory control of glycolysis may result from initial inhibition of phosphofructokinase by citrate with a secondary inhibition of hexokinase by elevated levels of glucose 6-P[13, 14].

Other inhibitors that may play an inhibitory role in the regulation of phosphofructokinase include phosphoenolpyruvate[97, 18, 112-116], phosphocreatine[116], 3-phosphoglycerate[97, 115, 116], 2-phosphoglycerate[97, 115, 116], 2,3-diphosphoglycerate[97, 116] and 6-phosphogluconate[115]. In the presence of inhibitory concentrations of either ATP or citrate, it is generally possible to activate the enzyme by adding agents such as AMP, cyclic AMP, ADP, $P_i$

and fructose 1,6-diP. Increasing concentrations of activators usually cause a decrease in the $K_m$ for fructose 6-P and a decrease in the co-operativity of fructose 6-P binding to the enzyme. Thus most inhibitors shift the hyperbolic plot of velocity as a function of fructose 6-P to the sigmoidal form; activators conversely change it to the hyperbolic form.

(b) *Phosphofructokinase and hormonal influence*—There has been no evidence to indicate that hormones or cyclic AMP are capable of influencing the activity of muscle phosphofructokinase by means other than their effects directly (i.e. cyclic AMP as an activator[117]) or indirectly through changes in the concentration of the metabolite regulators of phosphofructokinase. However, recent evidence obtained by the authors indicates that glucagon or exogenous cyclic AMP act to inhibit phosphofructokinase in isolated rat hepatocytes[57] (Section 5.3.3.3b); this effect appears to be independent of changes in the concentration of known effectors of liver phosphofructokinase.

Studies by Mansour and co-workers on rabbit muscle[118] and guinea-pig heart[119] confirm earlier findings[120, 121] and suggest that cyclic AMP or agents that act to increase the intracellular concentration of cyclic AMP achieve their effect either by direct activation of phosphofructokinase or by alteration of the metabolic profile of intracellular effectors of phosphofructokinase. Epinephrine administration to rabbits results in the conversion of muscle phosphofructokinase into a form that is more active under assay conditions that favour enzyme inhibition. Mansour[118] concludes that the activation is mediated through the combined concentration changes of hexose phosphates and adenylate nucleotides by epinephrine. For the isolated perfused guinea-pig heart[119], phosphofructokinase activity is inhibited following perfusion of the heart with pyruvate, β-hydroxybutyrate, octanoate or acetate, when assayed at pH 6.9. In addition, the inhibition of phosphofructokinase by perfusion with pyruvate was antagonised by the addition of isoproterenol (β-agonist, known to produce increase in intracellular concentration of cyclic AMP). None of the agents acted to alter the activity of phosphofructokinase at pH 8.5. The reduction in phosphofructokinase activity in hearts perfused with pyruvate was correlated with increased tissue citrate levels.

A direct effect of cyclic AMP to activate phosphofructokinase was deduced from studies on adipose tissue. Stimulation of glycolysis by epinephrine[122], although accompanied by a large increase in citrate concentration[123], resulted in increased glycolytic rates that were believed to result from an elevated concentration of cyclic AMP. Thus Denton and Randle[124] have shown that cyclic AMP is the most potent of all of the nucleotide activators of adipose tissue phosphofructokinase in reversing the inhibition by citrate. On the basis of the concentration of cyclic AMP in adipose tissue[125], it seems possible that cyclic AMP could activate phosphofructokinase.

Recent studies with phosphofructokinase from erythrocytes also indicate that cyclic AMP activates the enzyme[126]. Competition between cyclic AMP and ATP for the same binding site was suggested.

The proposed control of glycolysis by cyclic AMP as an activator of phosphofructokinase[127-129] should be regarded with some degree of caution. Although catecholamines stimulate glycolysis in muscle and produce increased levels of tissue cyclic AMP[130], cyclic AMP is no more effective than

AMP in activating muscle phosphofructokinase, and the concentration of the cyclic nucleotide in this tissue probably never reaches that required to overcome inhibition of phosphofructokinase by ATP[127-129].

### 5.3.3.2 *Fructose 1,6-diphosphatase (EC 3.1.3.11)*

Although considerable fructose 1,6-diphosphatase activity has been reported in rabbit skeletal muscle[131], brain[132] and, more recently, bumble-bee flight muscle[133], the highest activities of this enzyme occur predominantly in gluconeogenic tissues. Fructose 1,6-diphosphatase is essential in glucose-forming reactions of liver, for individuals lacking this enzyme show ketotic hypoglycemia, metabolic acidosis and hepatomegaly[134, 135]. The role of fructose 1,6-diphosphatase in muscle is less clear, as gluconeogenesis or glycogen synthesis from lactate is unlikely owing to an incomplete gluconeogenic pathway[136]. The reported activity of fructose 1,6-diphosphatase in brain has not been investigated further and a role for this enzyme in this tissue remains to be established. The high activity of fructose 1,6-diphosphatase in bumble-bee flight muscle has been proposed to facilitate a heat producing substrate cycle with phosphofructokinase[133] (Section 5.3.3.3c).

Immunological studies of liver, kidney and muscle fructose 1,6-diphosphatases from rabbit indicated that liver and kidney enzymes were identical and distinct from the muscle enzymes. The muscle enzyme has an isoelectric point of *ca.* 6.5, which is lower than that of the liver or kidney enzyme[137]. Kinetic differences between liver and muscle type fructose 1,6-diphosphatase also occur. Muscle enzyme is much more sensitive to inhibition by AMP than the liver enzyme, and is activated by both creatine phosphate and citrate (Section 5.3.3.2c).

(a) *Regulation of liver and kidney cortex fructose 1,6-diphosphatase*

(i) *Enzyme concentration.* Regulation of glucose synthesis in liver and kidney may be exercised at the enzyme level by adaptive changes in the concentration of fructose 1,6-diphosphatase protein[138]. Such changes are long-term effects and involve changes in the rate of protein turnover. Recent studies on the seasonal differences in the total activity of liver fructose 1,6-diphosphatase in the rabbit indicate a correlation between decreased fructose 1,6-diphosphatase activity and increased lysosomal activity as reflected by 'free' proteolytic activity[139, 140]. These results indicate an important relationship between the activity of an enzyme protein involved in gluconeogenesis and the levels and distribution of cellular proteases. Such changes in neutral (mol. wt. 36 000) to alkaline (mol. wt. 29 000) forms of fructose 1,6-diphosphatase with proteolytic activity result in decreased sensitivity to AMP[140]. It remains to be established whether a mechanism of this type operates *in vivo* for regulation.

(ii) *Changes in activity due to modification of the enzyme.* Since crystalline preparations of fructose 1,6-diphosphatase from rabbit liver have shown little activity at neutral pH, it has been proposed that the enzyme *in situ* may exist in a modified activated form. Pontremoli *et al.*[141] have suggested that the required conformational changes may involve a limited number of the 20 SH groups in the protein. Reacting two of these groups per mole of enzyme with *p*-hydroxymercuribenzoate doubled the activity at pH 7.5 [141]. Similar activation can be achieved by forming mixed disulphides of the enzyme with acyl carrier protein, coenzyme A[143, 144], homocystine or cystamine[145].

The conversion by cathepsins of hexosediphosphatase with a neutral pH optimum and high sensitivity to AMP to the less sensitive alkaline pH optimum enzyme may also play a role in regulation of this enzyme. An intermediate form of this enzyme, devoid of the tryptophan-bearing amino terminal segment, but still exhibiting optimum activity at pH 7.5, has been reported by Benkovic *et al.*[142]. Isolated bovine liver fructose 1,6-diphosphatase has a neutral pH optimum[146] and the above reasoning may not apply to it.

Modification of fructose 1,6-diphosphatase by phosphorylation as an additional type of regulatory mechanism has been suggested by Mendicino and co-workers[147]. They observed an inactivation of the enzyme in crude kidney extracts when incubated with ATP and cyclic AMP. These investigators suggested that this inactivation was associated with the phosphorylation of the protein. Subsequent studies from the same laboratory have not extended the initial observations.

(b) *Regulation of liver fructose 1,6-diphosphatase at the molecular level*—AMP is a specific and potent inhibitor of fructose 1,6-diphosphatase in every cell type examined except the slime mould[148] and bumble-bee flight muscle[133], and is an activator of phosphofructokinase[69]. Thus, on theoretical grounds, variations in the level of AMP would regulate both the magnitude and direction of carbon flux at the fructose 6-P level. In practice, however, the levels of hepatic AMP do not appear to fluctuate significantly in fed or fasted animals[149] and other factors must therefore contribute to the metabolic control by this substance. One such factor may be the concentration of fructose 1,6-diP, which itself inhibits fructose 1,6-diphosphatase, and which potentiates the inhibitory effect of AMP on this enzyme[150].

EDTA and other chelating agents enhance the activity of fructose 1,6-diphosphatase at neutral pH[151]; this effect is most striking with the enzyme from *Candida utilis*, which shows no activity at pH 7.5–8.0 in the absence of EDTA[152]. The physiological significance of such an activation of mammalian fructose 1,6-diphosphatases at neutral pH remains to be established.

The activation by EDTA and the anomalies associated with pH have prompted a search for a naturally occurring regulator of liver fructose 1,6-diphosphatase, capable of stimulating the activity of the enzyme at neutral pH. This substance was believed to be non-dialysable and present in homogenates of liver and Novikoff hepatoma[153]. Baxter *et al.*[154] further reported that a similar activator was present in the α-globulin fraction of rabbit serum, and more recently[155] this natural activator was reported to be free fatty acid. Fatty acids, of which oleate appeared to be the most potent, activate the enzyme at neutral pH and prevent inactivation of the enzyme by ATP. The first effect is common to a number of chelating agents, such as EDTA, certain thiol compounds (Section 5.3.3.2a) and histidine[153] and the second is shared with various compounds which include EDTA and 3-phosphoglycerate[156]. Fatty acids are unique among naturally occurring effectors of the enzyme in combining these two functions. However, it must be noted that oleate (1 mmol $l^{-1}$) does not enhance glucose production from dihydroxyacetone in the isolated perfused rat liver in the presence or absence of glucagon and under conditions where glucagon is an effective stimulant[157].

The second type of compound which may act as a natural activator of fructose 1,6-diphosphatase *in vivo* is phospholipid. Recent evidence is presented by Allen and Blair[158] that certain phospholipids are activators of rabbit-liver fructose 1,6-diphosphatase, and that it is the elimination of these phospholipids during purification that results in a loss of activity at neutral pH values. Analysis by thin-layer chromatography indicated the principal activator in rabbit-liver homogenate to be diphosphoinositide.

Guinea-pig-liver fructose 1,6-diphosphatase requires special mention, particularly as gluconeogenesis from lactate in the isolated perfused guinea-pig liver is not stimulated by the presence of fatty acids[159]. Thus the activation of fructose 1,6-diphosphatase in this tissue, if it is activated by fatty acids, may not bear importantly on the regulation of gluconeogenesis. The enzyme from guinea-pig liver is atypical in some respects; it is not stimulated by EDTA and is inhibited by ATP at concentrations exceeding 0.05 mmol $l^{-1}$, yet shows normal inhibition by fructose 1,6-diP and AMP[160]. In contrast to liver fructose 1,6-diphosphatase from other species, the guinea-pig-liver enzyme appears to be regulated by citrate; at low AMP concentrations, citrate increases fructose 1,6-diphosphatase activity by relieving substrate inhibition. At high concentrations of AMP, citrate increases AMP inhibition. Thus, changes in citrate concentration may result in large changes in the activity of guinea-pig-liver fructose 1,6-diphosphatase.

(c) *Regulation of muscle fructose 1,6-diphosphatase*—Fructose 1,6-diphosphatase from rabbit muscle is considerably more sensitive than either the liver or kidney enzyme to inhibition by AMP. At pH 7.5, 50% inhibition of fructose 1,6-diP hydrolysis was observed[131, 161] at an AMP concentration of $1\text{–}3 \times 10^{-7}$ mol $l^{-1}$ (with liver enzyme[162], $K_i = 8 \times 10^{-5}$ mol $l^{-1}$). In relation to the possible role of AMP in the regulation of fructose 1,6-diphosphatase in muscle, Opie and Newsholme[163] have estimated that the concentration of free AMP in muscle is sufficient to inhibit the enzymic activity. They found, however, that this inhibition was specifically relieved by $Mn^{2+}$ and suggested that this ion may participate in the regulation of fructose 1,6-diphosphatase activity in the intact cell.

Recently, Fu and Kemp[164] have shown that the activity at neutral pH of a highly purified preparation of rabbit skeletal muscle fructose 1,6-diphosphatase is increased by creatine phosphate and by citrate. The effects of the activators are additive and the activation is reversed by ADP. The concentrations of creatine phosphate and citrate that activate fructose 1,6-diphosphatase are within the physiological range and approximate to the same concentrations that result in inhibition of phosphofructokinase[97, 116]. Thus, fluctuations of creatine phosphate concentration may regulate muscle carbohydrate metabolism.

#### 5.3.3.3 *Phosphofructokinase–fructose 1,6-diphosphatase substrate cycle*

(a) *Regulation of glycolysis and gluconeogenesis at the phosphofructokinase—fructose 1,6-diphosphatase substrate cycle by effector concentration*—As already indicated, there is ample evidence to support regulatory roles for phosphofructokinase in the control of glycolysis and for fructose 1,6-diphosphatase in the control of gluconeogenesis. The fact that both phosphofructokinase and fructose 1,6-diphosphatase occur in tissues that are predominantly glycolytic, as well as those that are predominantly gluconeogenic, has led to a unified consideration of the regulatory properties of both enzymes for control of carbon flux at this level. Two possibilities have been formulated. In the first, modulation of the activity of each enzyme is proposed such that when one enzyme is active the other is inactive. The disadvantages of such a regulatory system are that a concentration range of regulator molecule would result within which neither enzyme was active, and the system would be relatively insensitive to changes in regulator concentration. Secondly, the opposite effects of regulators on each enzyme has led to a proposal that the activity of one enzyme would change in a converse manner to the other and there might be a region in which both enzymes are partially activated but no region in which both are inactive. Initially, fructose 1,6-diP was thought to be the key effector[165] such that changes in the concentration of this metabolite would amplify the actions of other regulators so that a switch in direction of glucose metabolism could occur without loss of sensitivity to the regulatory molecule. This particular mechanism is in doubt as inhibition of fructose diphosphatase by fructose 1,6-diP may not be of physiological significance[166] and correlation between changes in fructose 1,6-diP concentration and the rate of carbon flux apparently do not occur[167]. Subsequently, regulation of phosphofructokinase and fructose 1,6-diphosphatase by the adenine nucleotides has been the basis of proposals[101, 166, 168] for the control of carbon flux at the level of fructose 6-P. The most atractive mechanism that has been formulated is the 'amplification theory' of Newsholme and co-workers[166, 169]; the essential points for the regulation of glycolysis in muscle[169] and of glycolysis and gluconeogenesis in liver are as follows. (i) The regulatory properties of AMP assume key importance—to inhibit fructose 1,6-diphosphatase and de-inhibit ATP inhibition of phosphofructokinase. (ii) Adenylate kinase operates at a fixed ATP:ADP ratio (e.g. $x$) to produce amplified changes ($x$-fold) in the concentration of AMP in response to a change in the concentration of ATP. (iii) Special regulatory significance is accorded to the presence of fructose 1,6-diphosphatase in muscle. (iv) For both glycolytic and gluconeogenic systems, operation of the cycle is maximal when net carbon flux is at a minimum, and minimal when net carbon flux is at a maximum. For muscle the cycle is considered to operate only at rest, so that although the activity of phosphofructokinase is reduced to perhaps 2–10% of its maximum activity, the net rate of fructose 6-P phosphorylation is reduced to 1% or less of its maximum (see Ref. 169). Therefore, the effect of the cycle is to increase the sensitivity of the rate of fructose 6-P phosphorylation (and hence glycolytic flux) to changes in the concentration of AMP. In fact the cycle produces a threshold response of fructose 6-P phosphorylation to changes in the concentration of AMP. The inhibition of fructose 1,6-diphosphatase by AMP ensures that the cycle will operate only when the muscle is at rest, so that the maximum glycolytic flux will not be decreased in active muscle by the presence of the cycle. For liver, regulation of the magnitude and direction of net carbon flux at the level of fructose 6-P may be more complicated as this tissue can be either glycolytic or gluconeogenic. (v) It is implied that the wastage of energy during cycling of hexose phosphates may be compensated for by the increased versatility of control[166].

Indirect evidence for the control of hepatic gluconeogenesis at the level of fructose 6-P and involving the phosphofructokinase–fructose 1,6-diphosphatase substrate cycle comes from studies with the isolated perfused rat liver[170-173] In the majority of these studies there has been no experimental evidence in support of a reciprocal control of phosphofructokinase and fructose 1,6-diphosphatase by changes of adenine nucleotides. Indeed, only an alteration of the tissue citrate content appears to provide an adequate explanation for the changes in the rate of net glucose formation (i.e. agents leading to a decrease in the calculated concentration of citrate are assumed to increase flux through phosphofructokinase and thereby decrease the net rate of gluconeogenesis, and conversely agents which lead to an increase in cytosolic citrate are believed to produce decreases in phosphofructokinase and increased gluconeogenesis). As activation by citrate of rat-liver fructose 1,6-diphosphatase has not been demonstrated, situations of co-existing low citrate levels and high gluconeogenic rates would be favourable for high rates of substrate cycling and energy wastage. Furthermore, there appears to be no relationship between the intracellular liver citrate concentration and the rate of gluconeogenesis[174-178]. It would thus seem that regulatory processes other than those elicited by citrate, fructose 1,6-diP or the adenine nucleotides are operative at this level.

Until recently, quantitative data concerning the operation of the phosphofructokinase–fructose 1,6-diphosphatase substrate cycle had not been forthcoming and the operation of the cycle *in vivo* and its role in the regulation of carbohydrate metabolism were purely matters for speculation. In this laboratory a method has been developed for the estimation of the rate of substrate cycling between fructose 6-P and fructose 1,6-diP[179, 180]. The method depends upon the specific loss of $^{3}H$ from [2-$^{3}H$]glyceraldehyde 3-P at the reaction catalysed by triose phosphate isomerase when [5-$^{3}H$]glucose (or [5-$^{3}H$]-hexose 6-P) is the substrate for glycolysis. Thus [5-$^{3}H$,*U*-$^{14}C$]glucose is administered to the system and a decrease in the $^{3}H:^{14}C$ ratio in the hexose monophosphate is indicative of substrate cycling. Estimation of the rate of substrate cycling of fructose 6-P and its relationship to gluconeogenesis in rat liver *in vivo* has indicated that for a variety of situations the rate of phosphorylation of fructose 6-P is inversely proportional to the rate of dephosphorylation of fructose 1,6-diP[54]. In accord with this relationship, the rate of cycling of fructose 6-P is maximal (*ca.* 0.2 $\mu$mol min$^{-1}$ g$^{-1}$ wet wt. of liver) when carbon flux is low and decreases when appreciable net rates of either glycolysis or gluconeogenesis occur. A correlation was observed between decreases in the concentration of inhibitors of phosphofructokinase (i.e. ATP and cytosolic citrate) and increases in the rate of fructose 6-P phosphorylation. Although the conditions employed in these studies were extreme and essentially non-physiological, it would appear likely that phosphofructokinase and fructose 1,6-diphosphatase operate simultaneously in liver and that large changes in the concentrations of the effectors of these two enzymes could regulate the magnitude and direction of carbon flux in a manner consistent with the known properties of the enzymes. The existence of yet undiscovered factors controlling this cycle is highly likely.

(b) *Hormonal regulation of hepatic gluconeogenesis at the phosphofructokinase–fructose 1,6-diphosphatase substrate cycle*—Recent studies in the authors' laboratory with isolated rat hepatocytes indicate that both exogenous cyclic AMP and glucagon act to stimulate gluconeogenesis by inhibiting phosphofructokinase and activating fructose 1,6-diphosphatase[57]. The data from these studies indicate: (i) for galactose (entering the gluconeogenic pathway at glucose 6-P) both glucagon and cyclic AMP acted to increase glucose production, decrease lactate and pyruvate production, decrease the rate of fructose 6-P phosphorylation and increase the rate of fructose 1,6-diP dephosphorylation; (ii) for dihydroxyacetone and fructose, similar changes were observed as for galactose; (iii) for glycolysing liver cells (incubated with 27.8 mmol $l^{-1}$ glucose) both glucagon and cyclic AMP acted to decrease the rate of fructose 6-P phosphorylation and simultaneously to activate fructose 1,6-diphosphatase; (iv) no evidence was obtained for glucagon or cyclic AMP-mediated effects on the rate of glucose phosphorylation or glucose 6-P dephosphorylation; (v) changes in the rate of gluconeogenesis were always accompanied by reciprocal changes in the rate of phosphorylation of fructose 6-P and parallel changes in the rate of dephosphorylation of fructose 1,6-diP; (vi) there were no significant changes in the concentrations of known regulators of phosphofructokinase or fructose 1,6-diphosphatase (ATP, AMP or citrate).

These data support indications from the authors' laboratory[157, 181] and others[182-185] that a site for glucagon-mediated regulation of hepatic gluconeogenesis occurs between triose phosphate and glucose. As cyclic AMP and glucagon acted to inhibit fructose 6-P phosphorylation[57], these data are in contrast to the known effects of cyclic AMP on phosphofructokinase (i.e. as an activator). The data also explain in part the early observation by Haugaard and Haugaard[186] that the 'hyperglycemic factor' (later identified as glucagon) inhibited the incorporation of glucose carbon into fatty acid to a greater extent than it did the incorporation of fructose carbon (these experiments were later confirmed by Tepperman and Tepperman[187]). Although, at the present time, it is not possible to conclude whether cyclic AMP-mediated changes result from direct allosteric alteration, production of effectors, phosphorylation *via* protein kinase or some other unknown mechanism, no single experiment showed a separation between hormonally induced stimulation of fructose 1,6-diphosphatase and inhibition of phosphofructokinase. On this basis it is tempting to speculate that either of two possible regulatory mechanisms may be under hormonal control: (a) modification of each enzyme in a manner analogous to the cyclic AMP-mediated phosphorylase (EC 2.4.1.1) and glycogen synthetase (EC 2.4.1.11) enzymes; (b) cyclic AMP-mediated protein–protein interaction between fructose 1,6-diphosphatase and phosphofructokinase. Indeed, some of the existing data on the interaction between rabbit-liver fructose 1,6-diphosphatase and rabbit-muscle phosphofructokinase may be relevant to the observed hormone-induced changes. Pogell *et al.*[153] have demonstrated that liver fructose 1,6-diphosphatase is activated and desensitised towards AMP inhibition by purified muscle phosphofructokinase. El-Badry *et al.*[188] have observed that rabbit-liver fructose 1,6-diphosphatase acted to inhibit rabbit-muscle phosphofructokinase (possibly by the binding and destruction of the activator, fructose 1,6-diP). Although it must be stressed that there is no experimental evidence to support a cyclic AMP-mediated interaction between phosphofructokinase and fructose 1,6-diphosphatase of liver *in vivo* (to activate fructose 1,6-diphosphatase and inhibit phosphofructokinase), such a hormonally induced interaction would be consistent with the present data.

(c) *Phosphofructokinase–fructose 1,6-diphosphatase substrate cycle and heat production*—The simultaneous operation of phosphofructokinase and fructose 1,6-diphosphatase gives rise to cycling of fructose 6-P and the hydrolysis of ATP. Thus a major biological product of substrate cycling is heat. There is some evidence that substrate cycles may be adapted specifically for heat generation[94, 133].

In the flight muscles of bumble bees, the activities of fructose 1,6-diphosphatase are extremely high and comparable with those of phosphofructokinase in the same tissue[133]. The previously described method for the estimation of fructose 6-P substrate cycling (Section 5.3.3.3a) has been used in the authors' laboratory to investigate the effect of ambient temperature and flight on the rate of substrate cycling in the flight muscle of the bumble bee (*Bombus affinis*). Bees were injected with [5-$^{3}$H, *U*-$^{14}$C]glucose and after experimentation were freeze-clamped and glucose 6-P isolated from the flight muscles. The $^{3}$H:$^{14}$C ratio remains constant during flight, but is markedly reduced when the environmental temperature for the non-flying bee is reduced below 27 °C. It was calculated that at 5 °C the rate of substrate cycling in the non-flying bee was 10.4 μmol min$^{-1}$ g$^{-1}$ of fresh muscle. Whereas a significant correlation exists between elevated rates of substrate cycling and a need for heat production by this insect, this phenomenon may not occur in all heat-generating insects. For example, some insects known to exhibit pre-flight temperature increases in flight muscle contain very low activities of fructose 1,6-diphosphatase in this tissue and show visible signs of wing movement (e.g. wing whirring) and may thus utilise contractile processes for heat production. Most notable in this category of insects is the moth *Hyalophora cecropia*, which shows wing shivering during pre-flight heat production[189] and contains $<$0.1 units of fructose 1,6-diphosphatase per g wet weight of flight muscle[190].

Other evidence for the possible involvement of fructose 6-P substrate cycling in heat producing mechanisms comes from studies of muscle of normal and malignant hyperthermic pigs[95]. The anaesthetic halothane was found to trigger heat production and accelerate rates of substrate cycling of fructose 6-P in halothane-sensitive pigs but not in normal animals. In general, for the halothane-triggered hyperthermic pigs the final body temperature was proportional to the average rate of substrate cycling over 15 min[95].

### 5.3.4 Interconversion of fructose 1,6-diP and triose P

#### *5.3.4.1 Aldolase (EC 4.1.2.13)*

Three aldolase isoenzymes have been identified in animal tissues. The presence of a distinct aldolase in mammalian liver was first reported by Leuthardt and co-workers[191] and Hers and Kusaka[192], and the fact that this enzyme catalysed the cleavage of fructose 1,6-diP and fructose 1-P at nearly equal rates was established by Peanasky and Lardy[193]. Rutter and co-workers[194, 195] have identified three aldolase isoenzymes on the basis of their electrophoretic mobility, inhibition by specific antibodies and the ratio of cleavage of fructose 1,6-diP and fructose 1-P. Isoenzyme type *A* is found predominantly in muscle, has the lowest electrophoretic mobility and cleaves predominantly fructose 1,6-diP. Type *B* is found in liver and kidney, moves more rapidly than type *A* in the direction of the cathode and is able to hydrolyse fructose 1,6-diP and fructose 1-P with equal efficiency. Type *C*, as well as Type *A*, occurs in brain, moves towards the anode and can cleave fructose 1,6-diP at a rate twice that of fructose 1-P.

Despite the fact that there have been some observations of activation of aldolase by certain other proteins[196], the possible significance of this activation is not known and evidence for regulation at this site *in vivo* has not been obtained.

The recent observation that muscle aldolase is able to inhibit sheep-heart phosphofructokinase *in vitro*[188] may indicate specific properties of an interaction between these two enzymes *in vivo*; more likely, it involves depletion of fructose 1,6-diP, for the inhibition can be reversed by adding fructose 1,6-diP[188]. It remains to be established whether phosphofructokinase from all sources requires fructose 1,6-diP (which modifies the activity of phosphofructokinase by its reversal of inhibition at the allosteric site) and whether aldolase interacts in a similar manner with phosphofructokinase from the same tissue source.

#### 5.3.4.2 *Triose phosphate isomerase (EC 5.3.1.1)*

Triose phosphate isomerase is distinguished among the other glycolytic enzymes by its extraordinary catalytic capacity, *viz.* 7000–10 000 μmol of substrate conversion $min^{-1}$ $mg^{-1}$ of enzyme at 25 °C. From a quantitative viewpoint it is remarkable that in spite of its enormous turnover rate the enzyme does not maintain equilibrium between the two triose phosphates in rat liver[197, 198]. This disequilibrium has been explained as resulting from low activity of the enzyme relative to the metabolic flux, at the prevailing low physiological concentration (*ca.* 3 μmol $l^{-1}$) of unbound glyceraldehyde 3-P[197]. Regulation of carbon flux at this step thus appears unlikely and the rate is a function of substrate availability.

### 5.3.5 Interconversion of glyceraldehyde 3-P and phosphoenolpyruvate

#### 5.3.5.1 *Glyceraldehyde 3-P dehydrogenase (EC 1.2.1.12)*

A qualitative similarity between the catalytic properties of liver and muscle enzyme has been shown recently[199]. Although some quantitative differences that would favour the diversified metabolic requirements of liver were evident, no definitive structural differences have been identified. Regulation of the enzyme in muscle appears to be effected by ATP and phosphocreatine, as well as by the concentrations of substrates and products.

(a) *Glyceraldehyde 3-P dehydrogenase in glycolysis*—In several tissues, namely brain[30], heart[200] and tumor cells[201], the substrate and products of this enzyme are displaced from equilibrium during rapid glycolysis. Thus it has been proposed that glyceraldehyde 3-P dehydrogenase may under certain circumstances control the rate of glycolysis[30, 200].

Glyceraldehyde 3-P dehydrogenase occurs in high amounts in skeletal muscle[202, 203]. Cori *et al.*[202] proposed that excess enzyme is necessary because the catalysis occurs *in vivo* at a pH of 7.0, whereas the pH optimum *in vitro* is *ca.* 9.0 [202]. Recently, Oguchi *et al.*[204, 205] have shown that both ATP and phosphocreatine at physiological concentrations inhibit the muscle enzyme at neutral pH, which may be another reason cells have evolved with large amounts of this enzyme. The inhibition by ATP is enhanced at lower pH, is non-competitive with regard to glyceraldehyde 3-P and results from the obstruction of both $NAD^+$ and $P_i$ binding to the active centre[204].

The fact that the enzyme is sensitive to pH changes and is less active at lower pH would suggest that the glycolytic capacity of the cell is regulated during extremes of lactic acid production. Inorganic phosphate concentration, which increases during contraction, may also bear importantly on the regulation of glycolysis at glyceraldehyde 3-P dehydrogenase, as was originally postulated by Johnson[206] and Lynen[207]. Thus, the velocity of the dehydrogenase reaction may increase several-fold when phosphate levels increase to 5 mmol $l^{-1}$ during contraction because the apparent $K_m$ of $P_i$ is in the range of the varying physiological concentration. A co-ordinated regulation of glyceraldehyde 3-P dehydrogenase, phosphofructokinase ($P_i$ deinhibits ATP inhibition[127]) and hexokinase ($P_i$ activates[16-18]) may be facilitated by fluctuations in the concentration of $P_i$.

On telelogical grounds the regulation of muscle glycolysis at glyceraldehyde 3-P dehydrogenase would appear attractive. During the first few seconds of severe exercise, when no increase in glycolysis occurs, the levels of phosphocreatine decrease[208, 209] from 20 to 10 or even 5 mmol $l^{-1}$. Similarly, during cardiac anoxia, phosphocreatine breakdown and the resultant increased capacity to neutralise lactic acid may promote the activity of glyceraldehyde 3-P dehydrogenase, phosphofructokinase[18, 205, 210] and hexokinase[211]. With the accumulation of lactic acid and the decrease in pH[212, 213] the activities of phosphofructokinase and glyceraldehyde 3-P dehydrogenase would decrease. Such a decrease in activity of each of these enzymes would be accentuated by ATP, which remains at 6 mmol $l^{-1}$ during early contraction. The inhibition of these enzymes would slow glycolysis and protect the muscle against excessive acidity and depletion of glycogen stores. Thus the interrelated roles of phosphocreatine and ATP in glyceraldehyde 3-P dehydrogenase catalysis might represent an important coarse control of muscle glycolysis. In contrast, in brain the phosphocreatine concentration[106] is only 3 mmol $l^{-1}$; therefore glyceraldehyde 3-P dehydrogenase may not be inhibited unless the enzyme is unusually sensitive.

In contrast to the activating influence of cyclic AMP on phosphofructokinase, recent studies by Fife and Szabo[214] indicate that cyclic AMP acts directly to inhibit glyceraldehyde 3-P dehydrogenase. The inhibition by cyclic

AMP was competitive towards glyceraldehyde 3-P and non-competitive towards $NAD^+$. It is doubtful though whether this is of physiological significance as the concentration required to achieve a relevant inhibition is in excess of that normally found in muscle *in vivo*.

(b) *Glyceraldehyde 3-P dehydrogenase in gluconeogenesis*—Substrate analyses of freeze-clamped liver samples indicate that the coupled reactions catalysed by glyceraldehyde 3-P dehydrogenase and 3-phosphoglycerate kinase remain in equilibrium over a wide range of metabolic conditions[215]. Despite the adverse thermodynamics for phosphorylation of 3-phosphoglycerate and the reductive formation of glyceraldehyde 3-P, a number of factors tend to overcome this energy barrier in gluconeogenesis. The high $NAD^+$:NADH ratio in liver cytosol[31] is in part offset by the ATP:ADP ratio, and oxidative phosphorylation draws $P_i$ down to relatively low concentrations. The triose phosphate isomerase reaction removes 93% (at equilibrium[197]) of the glyceraldehyde 3-P formed by converting it into dihydroxyacetone phosphate. Aldolase withdraws both triose phosphates because the reaction it catalyses reaches equilibrium at 37 °C with 90% of the carbon in the form of hexose diphosphate. Finally, cleavage of the 1-phosphate from hexose diphosphate occurs[216] with a free energy change of $-3.8$ cal $mol^{-1}$.

Kinetically, the triose phosphate dehydrogenase of liver is elegantly controlled to facilitate catalysis in the direction dictated by the ATP:ADP ratio[199]. Both $NAD^+$ and NADH bind to the non-acylated catalytic sites of the tetrameric liver glyceraldehyde 3-P dehydrogenase and prevent access of 1,3-diphosphoglycerate at these sites. The bound $NAD^+$ decreases the $K_m$ of the unoccupied sites for acyl phosphate. This leads, in the presence of low concentrations of 1,3-diphosphoglycerate, to an increased rate of triose phosphate synthesis at low concentrations ($<100$ μmol $l^{-1}$) of $NAD^+$ and to an inhibition at higher concentrations. Higher concentrations of 1,3-diphosphoglycerate increase the rate of triose phosphate synthesis and only a weak inhibition by $NAD^+$ is observed. As the concentration of NADH is increased, it inhibits at low ($<6$ μmol $l^{-1}$) acyl phosphate concentrations but it becomes progressively less inhibitory as the concentration of acyl phosphate increases. Above 6 μmol $l^{-1}$ diphosphoglycerate the reaction rate is dependent on NADH concentration in a first-order relationship[199]. Thus by influencing the concentration of 1,3-diphosphoglycerate, the ATP:ADP ratio exerts a profound control of glyceraldehyde 3-phosphate dehydrogenase. An important factor in driving the reaction toward triose phosphate formation is the fact that the cytosolic $NAD^+$:NADH ratio drops from 1200–1800:1 in the fed rat's liver to *ca.* 500:1 in the fasted rat[31].

#### 5.3.5.2 3-Phosphoglycerate kinase (EC 2.7.2.3)

This enzyme catalyses the transfer of the energy-rich phosphoryl group from the acid anhydride bond of 1,3-diphosphoglyceric acid to the terminal phosphate of ADP. Divalent metal ions are required, for as with all phosphokinases the metal-ion complexes of the nucleotides are the true substrates[217-219]. The reaction equilibrium, essentially independent of pH favours ATP production ($K_{eq} = 3.1–3.4 \times 10^3$ at 25 °C [219, 220]). However in physiological conditions the reaction is tightly coupled to that of glyceraldehyde 3-P dehydrogenase[221] as described above (Section 5.3.5.1b). Thus the two enzymes can act concertedly in either direction, depending on ATP:ADP and NADH:$NAD^+$ ratios, $P_i$ concentration and pH. There is some evidence[222] that ADP may act specifically to inhibit this enzyme (at low $Mg^{2+}$, $ADP^{3-}$ has $K_i = 0.2$ mmol $l^{-1}$; at high $Mg^{2+}$, $MgADP^-$ has $K_i = 0.02$ mmol $l^{-1}$).

#### *5.3.5.3 Phosphoglycerate mutase and enolase*

Each of these enzymes catalyses a reaction with little free energy change and there is no indication of regulation of physiological significance at either site. The reaction catalysed by phosphoglycerate mutase is dependent, in most tissues, on 2,3-diphosphoglycerate as coenzyme.

The phosphate bond in 2-phosphoglycerate is converted into the high energy one of phosphoenolpyruvate by a dehydratase reaction catalysed by enolase. Studies with the yeast enzyme indicate that an enzyme–magnesium complex is required for active conformation[223]. The inhibition of glycolysis by fluoride is mediated at this enzyme. Fluoride forms a complex with Mg and orthophosphate, and the complex combines with the enzyme, presumably at the Mg binding site[223a].

### 5.3.6 Pyruvate metabolism

#### *5.3.6.1 Pyruvate kinase (EC 2.7.1.40)*

Pyruvate kinase catalyses the strongly exergonic conversion of ADP and phosphoenolpyruvate into ATP and pyruvate. Because the regulatory properties of the various forms of pyruvate kinase differ, it is relevant to consider the properties of each form and the forms present in each tissue. Recent kinetic studies of the pyruvate kinases of rat and human tissues have led to the identification of three classes of isoenzyme with qualitative differences in regulatory properties[224]. Class *L* is a major component in liver, minor in kidney, shows marked sigmoidal kinetics with phosphoenol pyruvate, allosteric inhibition by ATP and high concentrations of alanine, and activation by fructose 1,6-diP. Class *A* occurs predominantly in adipose tissue; it is the minor component in liver and major in kidney. This class shows slightly sigmoidal kinetics with phosphoenolpyruvate, is activated by fructose 1,6-diP, inhibited by alanine but not by ATP. Class *M* occurs predominantly in muscle and brain, and shows none of the regulatory properties described above for classes *L* and *A*. Classes *A* and *L* are markedly different in their specificity of inhibition by amino acids; class *L* is more sensitive to cysteine, and less sensitive to phenylalanine; class *A* is inhibited strongly by phenylalanine and only weakly by cysteine.

Two forms of pyruvate kinase occur in liver[225-230] and in adipose tissue[230a]. The form exhibiting co-operative interaction with phosphoenolpyruvate is converted by fructose 1,6-diP into that showing Michaelis–Menten kinetics with phosphoenolpyruvate, and the reverse conversion is mediated by ATP (liver enzyme[226]) and by EDTA, ATP or citrate (adipose tissue enzyme[230a]). Separation and isolation of two kinetically different forms of the enzyme can be achieved by extraction in the presence and absence of EDTA. Of the two forms isolated by isoelectric focusing[231], one has two moles of bound fructose 1,6-diP, but the other has none.

(a) *Regulation of cell content of pyruvate kinase*—High carbohydrate diets[232-235] increase the activity of the liver enzyme. Insulin can stimulate its synthesis in rat liver[236] and rat-liver-cell culture[237], and oestradiol in the rat uterus[238]. As the activities of pyruvate kinase in liver are sufficient to catabolise all phosphoenolpyruvate formed by the combined action of pyruvate carboxylase and phosphoenolpyruvate carboxykinase[132], a negative control of glycolysis at the level of pyruvate kinase would seem mandatory under conditions of increased gluconeogenesis. In starvation and the diabetic state this may be partially achieved by a lowering of the intracellular amounts of the enzyme.

(b) *Regulation of pyruvate kinase by metabolite effectors*—For pyruvate kinases other than muscle, high levels of ATP act as inhibitor. The regulation of muscle glycolysis in relation to phosphocreatine has been examined recently[239]. It is suggested that decreases in phosphocreatine levels may be a primary factor contributing to the increase in glycolytic flux that accompanies muscle contraction. The reported inhibition of muscle pyruvate kinase by creatine phosphate is competitive with phosphoenolpyruvate. This presents the attractive possibility that fructose 1,6-diphosphatase, phosphofructokinase and pyruvate kinase may be controlled by an interlinking mechanism so that as phosphocreatine increases, inhibition of pyruvate kinase[239] and phosphofructokinase[18] occurs and fructose 1,6-diphosphatase may be activated to allow glycogen synthesis from phosphorylated intermediates of the glycolytic pathway.

Seubert and Schoner have examined many citric acid cycle intermediates and some amino acids for possible effects on hepatic pyruvate kinase[240]. Alanine appeared to be the most effective inhibitor and the effect was specific for the L-isomer. However, significant inhibition is achieved only by concentrations (13 mmol $l^{-1}$) that far exceed those that occur naturally, even when concentrations of alanine are increased after treatment with glucocorticoids[241]. In the diabetic state or after starvation, the levels of alanine are in fact lowered[242]. As already pointed out, under the latter metabolic states, breakdown of newly formed phosphoenopyruvate is hindered to some degree by a decrease in the level of pyruvate kinase.

The allosteric activation of pyruvate kinase from several tissues (e.g. adipose tissue, erythrocytes, kidney and liver) by fructose 1,6-diP has been reported (see Table 1 of Ref. 240). In each instance the activating effect of fructose 1,6-diP was evident only at sub-optimal concentrations of phosphoenolpyruvate, indicating that the $K_m$ of the enzyme was decreased when fructose 1,6-diP was bound.

$Ca^{2+}$ ion efflux from mitochondria concomitant with increased rates of gluconeogenesis, and previous observations of $Ca^{2+}$ ion inhibition of muscle pyruvate kinase[243], provided the basis for the hypothesis[168] that $Ca^{2+}$ might act as a negative effector of liver pyruvate kinase. However, Gabrielli and Baldi[244] have recently shown that $Ca^{2+}$ is an inhibitor only when the liver enzyme (pigeon) is in the presence of a high concentration of phosphoenolpyruvate; at lower concentrations of this substrate, $Ca^{2+}$ is either an activator or inert. This finding diminishes the possibility that a flux of $Ca^{2+}$ into the cytosol could reduce the liver pyruvate kinase activity and result in the increase of phosphoenolpyruvate concentration which has been observed in the pigeon-liver homogenates when gluconeogenesis is stimulated[168]. $Ca^{2+}$ could inhibit the liver pyruvate kinase only when one or more negative effectors had already influenced the enzyme and resulted in an accumulation of phosphoenolpyruvate. Thus, the different sensitivity to calcium ions shown by the hepatic and muscle enzymes suggests that $Ca^{2+}$ may function differently in the control of the liver and muscle glycolytic pathways.

### *5.3.6.2 Pyruvate carboxylase (EC 6.4.1.1)*

Pyruvate carboxylase activity has been detected in most mammalian tissues, although highest maximal catalytic activities for this enzyme are confined to liver, kidney cortex, adrenal gland, lactating mammary gland and adipose tissue[245-247]. Metabolic roles for pyruvate carboxylase include participation in gluconeogenesis from three-carbon precursors[245, 248] in the urea cycle, glyceroneogenesis and lipogenesis[249], and other anaplerotic syntheses requiring oxalacetate[250]. In the tissues examined, pyruvate carboxylase activity appears to be wholly mitochondrial[246]; where previous reports had indicated some cytosolic activity, this would now appear to be the result of leaking mitochondria[246, 251].

Kinetic[252] and immunological[246] studies indicate that different forms of pyruvate carboxylase do not occur; the antibody prepared to pyruvate carboxylase purified from rat-liver mitochondria cross-reacts with the enzyme in rat-kidney cortex, brain and adipose tissue.

Agreement has not yet been reached concerning the behaviour of pyruvate carboxylase following physiological perturbation of the animal. Although some investigàtors have reported increased levels of activity in the livers of fasted, diabetic and glucocorticoid-treated rats[253-256], others have found no change in the activity under these conditions[257-260]. Taylor *et al.*[261] found that the level of pyruvate carboxylase activity in sheep liver increases several fold during diabetes and starvation.

Dietary availability of biotin or biotin precursors may also affect the maximal catalytic activity of pyruvate carboxylase by influencing the conversion of apoenzyme into holoenzyme. Thus biotin refeeding of biotin-deficient chickens restores normal rates of gluconeogenesis[262].

Pyruvate carboxylase has an absolute requirement for acetyl CoA, and early experiments on the stimulation of gluconeogenesis by fatty acids were frequently interpreted as evidence that the effect was mediated solely by enhancing the acetyl CoA concentration. At present it would appear that the regulation of pyruvate carboxylase is more complex and involves also, in some species, inhibition by acetoacetyl CoA, activation by increased ATP: ADP ratio, regulation of the rate of conversion of apoenzyme into holoenzyme, and possibly changes in the concentration of metal ions.

An important question for the regulation of pyruvate carboxylase by acetyl CoA concerns the intramitochondrial concentration of this effector, and whether fluctuations in concentration occur within a range that would enable regulation *in vivo*.

In chicken liver, the concentration of acetyl CoA greatly exceeds the apparent $K_a$ (2–3 μmol $l^{-1}$), but several other naturally occurring CoA esters including malonyl CoA[263], succinyl CoA[263] and acetoacetyl CoA[264] act as inhibitors. Thus in this species the combined effects of the variously acylated CoA compounds may under some circumstances serve to regulate pyruvate carboxylation. In rat liver, acetyl CoA concentration is usually 10 times higher than required for half-maximal activation of pyruvate carboxylase, and the acetoacetyl, succinyl and malonyl derivatives have no effect on the catalytic activity in the presence of limiting concentrations of acetyl CoA[252]. The β-hydroxybutyryl[264], propionyl and butyryl derivatives of CoA act as activators and the last-named may be of physiological importance in ruminants[265].

The chicken-liver enzyme is susceptible to inhibition by greater than physiological concentrations of sulphate and of phosphoenolpyruvate[263]. Neither sulphate nor any of a variety of dicarboxylic, organic or amino acids affect rat-liver pyruvate carboxylase[252]. Both the avian[266] and the rat-liver[252, 267] enzymes are regulated by the ATP:ADP ratio. The magnesium requirement of the enzyme is unusual[252]. It requires the Mg complex of ATP as substrate and, in addition, at pH 7.3 and 37 °C (but probably not at pH 8) requires free $Mg^{2+}$. Kinetic studies[252] suggest that an ordered addition of substrates is preferred; at 25 °C and pH 8, $Mg^{2+}$ adds before MgATP, whereas at 37 °C and pH 7.3, the order of addition is reversed. The mitochondrion contains at least 30 mmol $l^{-1}$ total Mg but it is possible that enough of it is bound by proteins, phospholipids, nucleotides and metabolites to bring the free $Mg^{2+}$ concentration into the range where small fluctuations in concentration may affect the activity of pyruvate carboxylase[252]. A monovalent cation is also required[268], but cellular concentrations of $K^+$ are always sufficient to support maximal rates. The most important factor controlling pyruvate carboxylase *in vivo* is the ambient concentration of pyruvate. The enzyme has a $K_m$ of 0.14 mmol $l^{-1}$ for pyruvate, at pH 7.3 [252], which is in the normal range of liver pyruvate concentration. A 'safety valve' is provided by the fact that at high concentrations of pyruvate (>0.5 mmol $l^{-1}$) a second linear relation is exhibited in Lineweaver–Burk plots with a $K_m$ approximately 8 times that in the low substrate region and a much higher $V_{max}$ than that obtained by extrapolating the 'low pyruvate' region[252]. It does not seem likely that this enzyme is controlled by an off–on mechanism analogous to phosphorylase, for it supplies oxalacetate for many different needs as described above.

Kimmich and Rasmussen[269] reported an inhibition of pyruvate carboxylase activity in rat-liver mitochondria in the presence of low concentrations of calcium (<100 μmol $l^{-1}$), and a regulation of gluconeogenesis by changes in the intracellular distribution of $Ca^{2+}$ was proposed[270, 271]. Studies with purified pyruvate carboxylase have shown that calcium is a potent inhibitor of this enzyme[252, 272], competing with $Mg^{2+}$ [252]. Recent evidence[273] suggests that with isolated mitochondria, 100 μmol $l^{-1}$ $Ca^{2+}$ inhibited pyruvate carboxylation only in media containing a high concentration of sucrose (240 mmol $l^{-1}$). Replacing sucrose with mannitol relieved the inhibition except when mitochondria were loaded with $Ca^{2+}$ before initiating the experiment. From this[273] and other[252] evidence, a role for $Ca^{2+}$ in regulating pyruvate carboxylation in liver appears unlikely.

#### *5.3.6.3 Phosphoenolpyruvate carboxykinase (EC 4.1.1.32)*

Phosphoenolpyruvate carboxykinase may occur in both the cytosol or the mitochondria. The distribution in liver varies from species to species and thus may give rise to specific regulatorydifferences. For the mouse, rat or hamster liver[274, 275] more than 90% of the total activity is cytosolic. In contrast, the enzyme is located mainly within the mitochondrion for birds[276-278] or the fed rabbit[274]. Roughly equal distribution between the compartments occurs for humans, cattle, sheep, guinea pigs and pigs[260, 261, 274, 279-284]. The mitochondrial enzyme appears to be immunologically[285] and physically[286] distinct from the cytosolic form. For many species the activity of the cytosolic enzyme increases in response to a demand for increased gluconeogenesis; *viz.* starvation[287-290], diabetes[287-289], immediately after birth[275] and following administration of glucocorticoids[287, 288, 291, 292,] or glucagon[287, 291]; in contrast, the mitochondrial activity of phosphoenolpyruvate carboxykinase remains unchanged. The location of the enzyme in rat liver changes from primarily mitochondrial in the foetal rat to primarily soluble in the adult[275].

For tissues other than liver the intracellular distribution of phosphoenolpyruvate carboxykinase is less well defined. At present it would appear that the enzyme of adipose tissue[293] is predominantly cytolsolic, regardless of species. For rat lung[294] and rat brain[295], the enzyme is predominantly intramitochondrial.

Adaptation of phosphoenolpyruvate carboxykinase to changes in hormonal and dietary status is pronounced and may result from alterations in either the rate of synthesis or degradation of the protein. Investigation of the effect of a starvation–re-feeding cycle on rats shows that, for both liver and adipose tissue, changes in the rate of synthesis of phosphoenolpyruvate carboxykinase predominate in influencing the enzyme's concentration[296]. Similarly, the effect of cortisol[287] or thyroxine[297] to increase rat-liver phosphoenolpyruvate carboxykinase levels also results from enhanced enzyme synthesis. Recent studies by Ballard and Hopgood[298] show that increases in the amount of hepatic phosphoenolpyruvate carboxykinase protein following the administration of L-tryptophan to starved rats is the result of increased synthesis as well as a retardation in the rate of enzyme degradation. Inhibition of protein degradation may also explain enhanced kidney-cortex levels of phosphoenolpyruvate carboxykinase following metabolic acidosis, although activation of an inactive form of the enzyme that is already present was not ruled out[299].

(a) *Relative contribution of cytosolic and mitochondrial phosphoenolpyruvate carboxykinase*—Hepatic gluconeogenesis in the rat from alanine, lactate or pyruvate may involve the translocation of either malate or aspartate, or both[279, 300, 301], across the mitochondrial membrane into the cytosol, where oxalacetate may be regenerated from these intermediates to form phosphoenolpyruvate[267, 302, 303]. Other species (rabbit, guinea pig, pigeon) that have a higher percentage of hepatic phosphoenolpyruvate carboxykinase in their mitochondria may produce varying amounts of phosphopyruvate in that organelle[281, 290, 304] from which it is transported to the cytosol by the tricarboxylate anion carrier[305]. In both the rabbit[288, 304] and guinea pig[279] the increased phosphoenolpyruvate carboxykinase activity, induced by conditions that enhance gluconeogenesis, occurs only in the cytosolic fraction. The high phosphoenolpyruvate carboxykinase activity in pigeon liver is not increased by fasting[306].

(b) *Regulation of phosphoenolpyruvate carboxykinase*—Several studies have suggested that regulation of mitochondrial phosphoenolpyruvate carboxykinase in guinea pig or rabbit liver is mediated by a form of physiological uncoupling by fatty acids. The basis of these suggestions has been the consistently increased rates of phosphoenolpyruvate formation by isolated mitochondria after partial uncoupling with dinitrophenol or oleic acid[290, 307-310]. Explanations of the uncoupler effect include: (*a*) an increase in the $NAD^+$ : NADH ratio induced by uncouplers[311, 312]; and (*b*) an increased rate of GTP formation linked to the oxidation of α-ketoglutarate[308, 313] (the dependence of phosphoenolpyruvate formation on the rate of oxidation of α-ketoglutarate appears to be particularly likely in rabbit-liver mitochondria, where nucleoside diphosphate kinase levels are low[314]).

Regulation of the mitochondrial $NAD^+$:NADH ratio merits special consideration as it bears importantly on controlling substrate availability for mitochondrial phosphoenolpyruvate carboxykinase. Oxidation of substrates that lead to a more reduced mitochondrion (e.g. succinate) substantially reduces the rate of mitochondrial phosphoenolpyruvate formation. In the rat, starvation decreases the liver mitochondrial $NAD^+$:NADH ratio, whereas in the guinea pig or rabbit the system becomes more oxidised; gluconeogenesis is enhanced in all three animals. On this basis, speculation that aspartate aminotransferase (EC 2.6.1.1), malate dehydrogenase (EC 1.1.1.37) and phosphoenolpyruvate carboxykinase compete for the same pool of oxalacetate seems justified[314]. Thus the relative rates of formation of the various gluconeogenic precursors may be determined largely by alteration of the intramitochondrial oxidation–reduction potential. For the rat, a decrease in the mitochondrial $NAD^+$ : NADH ratio favours malate formation and enhances gluconeogenesis. The $K_m$ for oxalacetate for phosphoenolpyruvate carboxykinase isolated from rat, chicken and sheep is 9–30 μmol $l^{-1}$ [315]. This is higher than the concentration usually found in liver and indicates that substrate availability may be the rate-determining factor. A new assay[315a] yields a lower $K_m$, but also results in apparently lower specific activity for the enzyme. Accurate determination of the compartmental concentration of oxalacetate should resolve this question.

Phosphoenolpyruvate carboxykinase has an intrinsic activity in the presence of MgGTP that is more than doubled in the presence of the divalent transition metals $Fe^{2+}$, $Mn^{2+}$, $Cd^{2+}$ or $Co^{2+}$ [316, 317]. The enzyme is inhibited by quinolinate (from tryptophan) *in vivo*[319] and in perfused liver[320]. Quinolinate inhibits *in vitro* when $Fe^{2+}$ is present but not when $Mn^{2+}$ is added[321] $Mn^{2+}$ also reverses quinolinate inhibition of gluconeogenesis in perfused rat liver[320]. These facts lead to the conclusion that $Fe^{2+}$ rather than $Mn^{2+}$ is the natural activator of this enzyme in the rat.

When purified to homogeneity, rat-liver phosphoenopyruvate carboxykinase is no longer stimulated by $Fe^{2+}$, $Cd^{2+}$ or $Co^{2+}$ [317]. The factor that renders the enzyme responsive to $Fe^{2+}$ in liver cytosol has been isolated in pure form[318] and has been characterised as a protein of mol. wt. *ca.* $7 \times 10^4$. The molar ratio of activator:enzyme required for full activation in the presence of 30 μmol $l^{-1}$ $Fe^{2+}$ is approximately 1:6. The findings clearly show that a specific protein entity is responsible for the effective activation of phosphoenolpyruvate carboxykinase by divalent transition metal ion *in vitro*, and direct attention to a possible regulatory mechanism to control this enzyme *in vivo*[318].

### *5.3.6.4 Pyruvate dehydrogenase (EC 1.2.4.1)*

The relevance of pyruvate metabolism to anabolic processes cannot be overstated, particularly as regulation of carbon flux through pyruvate dehydrogenase bears directly upon both the rates of acetyl CoA formation for lipogenesis and the 'sparing' of carbon for pyruvate carboxylation and gluconeogenesis. These considerations suggest that the pyruvate dehydrogenase complex is a likely site for metabolic regulation.

Regulation of pyruvate dehydrogenase appears to be mediated at two levels. First, the two principal products of pyruvate oxidation, acetyl CoA and NADH, exert feed-back inhibition on the enzyme ($K_i$ = 5–12 and 2 μmol $l^{-1}$, respectively). These product inhibitors are competitive with CoASH and $NAD^+$, respectively[322-324], and acetyl CoA appears to inhibit uncompetitively with respect to pyruvate[325]. The second and more complex mode of regulation originates from the work of Reed and co-workers (see the review[326]) and involves phosphorylation and dephosphorylation of the pyruvate dehydrogenase component of the mammalian pyruvate dehydrogenase complex. This discovery followed from the observation that purified preparations of the complex were inactivated by 1–10 μmol $l^{-1}$ ATP as a result of the transfer of the terminal phosphoryl moiety of ATP to the pyruvate dehydrogenase complex[327]. Incubation with $Mg^{2+}$ reactivated the enzyme and resulted in a simultaneous release of inorganic orthophosphate. These observations formed the basis of a proposed control mechanism involving pyruvate dehydrogenase kinase to inactivate, and pyruvate dehydrogenase phosphatase to activate, pyruvate oxidation. Similar properties were found for the pyruvate dehydrogenase complex from porcine liver, heart muscle and rat adipose tissue[324, 328, 329], but could not be demonstrated with preparations of α-ketoglutarate dehydrogenase.

Evidence for interconversion of active and inactive forms of pyruvate dehydrogenase *in vivo* is compelling. Wieland[330] found that the ratio of active to non-active pyruvate dehydrogenase complex decreases markedly in rat kidney and heart during starvation. Jungas[329] found that the ratio of active to inactive forms of the pyruvate dehydrogenase complex in rat adipose tissue may be modified by insulin (ratio increases) and epinephrine (ratio decreases). Söling and Bernhard[331] have provided evidence for substrate induced changes in pyruvate dehydrogenase activity *in vivo*. Intravenous injection of fructose resulted within a few minutes in a significant increase in the activity of pyruvate dehydrogenase and a marked decrease in the concentration of ATP in rat liver without affecting the total amount of the complex (measured after incubation of the homogenate with 15 mmol $l^{-1}$ $Mg^{2+}$ to convert inactive into active enzyme).

Evidence for interconversion of the two forms of pyruvate dehydrogenase in isolated perfused rat liver is also consistent with the proposal of Reed and co-workers[326]. Patzelt *et al.*[332] have shown that as the pyruvate concentration in the perfusion medium increased, the proportion of pyruvate dehydrogenase in the active form also increased. Substrates other than pyruvate also appeared to be involved in regulation of pyruvate dehydrogenase activity. Fructose was slightly less, and lactate considerably less, effective than pyruvate in activating the enzyme, whereas 2 mmol $l^{-1}$ oleate abolished the effects of fructose or pyruvate on the formation of active pyruvate dehydrogenase. The simultaneous addition of D-(+)-decanoyl carnitine inhibited the action of oleate, indicating that fatty acid oxidation may be required for the conversion of the active form of pyruvate dehydrogenase into the inactive form[332]. The importance of fatty acids in the control of pyruvate dehydrogenase active–inactive interconversion has been reported for liver[333], heart muscle and kidney[334]. Although treatment of rats with insulin gave rise to an increase in assayable pyruvate dehydrogenase activity[333], this effect could not be achieved in perfusion.

At present the regulatory properties of pyruvate dehydrogenase kinase and pyruvate dehydrogenase phosphatase are not totally defined. The following points summarise much of the current data. The true substrate for the kinase is $MgATP^{2-}$, and free $Mg^{2+}$ is required for pyruvate dehydrogenase phosphatase activity. Skeletal muscle, cyclic AMP-dependent protein kinase does not inactivate preparations of the bovine-kidney or -heart pyruvate dehydrogenase complexes in the presence of ATP and cyclic AMP. Pyruvate dehydrogenase kinase does not catalyse a phosphorylation of histone with or without cyclic AMP. Pyruvate protects the pyruvate dehydrogenase complex against inactivation by ATP in a non-competitive manner, and, finally, ADP has been found to inhibit the kinase competitively with respect to ATP. These observations underscore the importance of the ATP:ADP ratio for the regulation of pyruvate dehydrogenase activity. Reed and co-workers[326] visualise that a decrease in the ATP:ADP ratio inhibits the kinase and releases free $Mg^{2+}$ which in turn activates the phosphatase.

Controversy exists over the role of cyclic AMP in mediating pyruvate dehydrogenase activity. Reed and co-workers[326] as well as Jungas[329] and Randle and co-workers[335] report that cyclic AMP does not affect either pyruvate dehydrogenase kinase or the phosphatase. On the other hand, Wieland and Siess[336] reported that cyclic AMP resulted in the stimulation of pyruvate dehydrogenase phosphatase activity from porcine-heart muscle. These latter investigators attributed this effect to a hypothetical cyclic AMP-dependent phosphatase kinase. Schimmel and Goodman[337] reported dibutyryl cyclic AMP to activate pyruvate oxidation in adipose tissue.

An inhibition of pyruvate oxidation by glucagon may be inferred from data obtained with isolated hepatocytes[181, 338]. Whereas glucagon enhances gluconeogenesis from lactate, from mixtures of lactate and pyruvate, and from low concentrations of pyruvate, it strongly inhibits glucose formation from 5 or 10 mmol $l^{-1}$ pyruvate. These higher concentrations of pyruvate act as sinks for reducing equivalents generated during pyruvate oxidation and probably deprive the cytosol of NADH required for reduction of 3-phosphoglycerate to triose phosphate.

The inhibitory effect of glucagon on glucose synthesis from 10 mmol $l^{-1}$ pyruvate is abolished by 5 mmol $l^{-1}$ ethanol which supplies reducing equivalents in the cytosol. Glucagon decreased, and insulin increased, the rate of decarboxylation of [1-$^{14}$C]pyruvate[338]. Part of the stimulatory effect of insulin on pyruvate decarboxylation[338] could be mediated by increased calcium uptake with subsequent activation of pyruvate dehydrogenase phosphatase, as has been discussed by Martin *et al.*[339] for the adipose tissue enzyme.

In addition to the functions of $Mg^{2+}$ described above, $Ca^{2+}$ lowers the apparent $K_m$ of the phosphatase for binding to the transacetylase core of the pyruvate dehydrogenase complex[340]. The activity of the pyruvate dehydrogenase phosphatase is greatly enhanced when it is attached to the transacetylase and thus $Ca^{2+}$ may provide an important means of regulating the dephosphorylation–phosphorylation cycle. Further evidence for the regulation of pyruvate dehydrogenase activity by $Ca^{2+}$ is the finding[335] that the activity of the phosphatase on the phosphorylated pyruvate dehydrogenase of pig heart, pig-kidney cortex and rat epididymal adipose tissue is substantially decreased by EGTA in the presence of adequate $Mg^{2+}$. Use of CaEGTA buffers indicated that activation by $Ca^{2+}$ was detectable at concentrations of 10 nmol $l^{-1}$ and half maximal at 1 μmol $l^{-1}$, and that 4.8-fold activation was achieved by 0.1 mmol $l^{-1}$ $Ca^{2+}$. Furthermore, phosphorylation of pyruvate dehydrogenase by pyruvate dehydrogenase kinase was inhibited by $Ca^{2+}$ [335].

### *5.3.6.5 Malic enzyme (EC 1.1.1.40)*

Until the discovery of pyruvate carboxylase by Utter and Keech[248], it was assumed that malic enzyme and malate dehydrogenase provide the oxalacetate required for gluconeogenesis from pyruvate. However, a role for malic enzyme in lipogenesis instead of carbohydrate synthesis was clearly indicated by the finding that it was reduced to very low activitity in fasted or diabetic rats[287] and elevated far above normal after feeding a high carbohydrate diet[341]. Lipid synthesis from carbohydrate in liver involves carboxylation of pyruvate to oxalacetate, which condenses with acetyl CoA to form citrate. The latter is transported to the cytosol where it is cleaved[342] to acetyl CoA, the starting material for fatty acid synthesis, and oxalacetate, which is reduced to malate by the NADH generated during oxidation of triose phosphate. Malic enzyme forms NADPH for fat synthesis by reaction (5.3):

$$\text{malate} + \text{NADP}^+ \rightleftharpoons \text{pyruvate} + \text{CO}_2 + \text{NADPH} + \text{H}^+ \qquad (5.3)$$

While generation of NADPH may be the usual function for malic enzyme in cells that carry out lipogenesis, hydroxylation of steroids or other processes that use NADPH, the reversibility of reaction (5.3) indicates that it could be a source of $C_4$ acids for synthesis under proper circumstances.

No regulatory mechanisms have been described for this enzyme. It would seem likely that the reaction rate depends upon the availability of $NADP^+$ (in a similar manner to the oxidative pentose phosphate pathway enzymes).

### *5.3.6.6 The pyruvate carboxylase–phosphoenolpyruvate carboxykinase–pyruvate kinase substrate cycle*

The simultaneous operation of reactions leading to phosphoneolpyruvate synthesis and breakdown results not only in loss of cellular function, but also in the wasteful breakdown of ATP. Yet despite the abundance of apparent regulators of each of the three enzymes, simultaneous operation of all three reactions occurs in perfused liver (from starved or fed rats[344, 345]), kidney cortex slices[349] and possibly perfused guinea pig and pigeon livers[306]. While it is possible that operation of the cycle confers an added degree of versatility of control on both the reactions of glycolysis and gluconeogenesis, it is tempting to speculate that this cycle may prove to be under hormonal control in a manner analogous to that recently proposed for the phosphofructokinase–fructose 1,6-diphosphatase substrate cycle[57]. Experiments with perfused rat livers indicate that glucagon facilitates the formation of phosphoenolpyruvate from its precursors[346-350]. Obviously the pyruvate substrate cycle merits further attention in relation to the regulation of hepatic glycolysis, gluconeogenesis and lipogenesis.

## References


1. Pitot, H. C. and Yatvin, M. B. (1973). *Physiol. Rev.*, **53,** 228
1a. Katzen, H. M., Soderman, D. D. and Nitowsky, H. M. (1965). *Biochem. Biophys. Res. Commun.*, **19,** 377
2. Vinuela, E., Salas, M. and Sols, A. (1963). *J. Biol. Chem.*, **238,** PC1175
3. Dipietro, D. L., Sharma, C. and Weinhouse, S. (1962). *Biochemistry*, **1,** 455
4. Leloir, L. F., Olavarria, J. M., Goldemberg, S. H. and Carminatti, H. (1959). *Arch. Biochem. Biophys.*, **81,** 508
5. Walker, D. G. (1962). *Biochem. J.*, **84,** 118P
6. Sharma, C., Manjeshwar, R. and Weinhouse, S. (1963). *J. Biol. Chem.*, **238,** 3840
7. Niemeyer, H., Clark-Turri, L. and Rabajille, E. (1963). *Nature*, **198,** 1096
8. Salas, M., Vinuela, E. and Sols, A. (1963). *J. Biol. Chem.*, **238,** 3535.
9. Katzen, H. M. (1967). *Advan. Enzyme Regul.*, **5,** 335
10. Hansen, R., Pilkis, S. J. and Krahl, M. E. (1967). *Endocrinology*, **81,** 1397
11. Katzen, H. M., Soderman, D. D. and Wiley, C. E. (1970). *J. Biol. Chem.*, **245,** 4081
12. Grossbard, L. and Schimke, R. T. (1966). *J. Biol. Chem.*, **241,** 3546
13. Crane, R. K. and Sols, A. (1954). *J. Biol. Chem.*, **210,** 597
14. Weil-Malherbe, H. and Bone, A. D. (1951). *Biochem. J.*, **49,** 339
15. Purich, D. L. and Fromm, H. J. (1971). *J. Biol. Chem.*, **246,** 3456
16. Tiedemann, H. and Born, J. (1959). *Z. Physiol. Chem.*, **321,** 205.
17. Rose, I. A., Warms, J. V. B. and O'Connell, E. L. (1964). *Biochem. Biophys. Res. Commun.*, **15,** 33
18. Uyeda, K. and Racker, E. (1965). *J. Biol. Chem.*, **240,** 4682
19. Purich, D. L., Fromm, H. J. and Rudolph, F. B. (1973). *Advan. Enzymol.*, **39,** 249
20. Passonneau, J. V. and Lowry, O. H. (1962). *Biochem. Biophys. Res. Commun.*, **7,** 10
21. Kosow, D. P., Oski, F. A., Warms, J. V. B. and Rose, I. A. (1973). *Arch. Biochem. Biophys.*, **157,** 114
22. Crane, R. K. and Sols, A. (1955). *Methods in Enzymology*, Vol. 1, 277 (New York: Academic Press)

23. Lowry, O. H., Roberts, N. R., Schulz, D. W., Clow, J. E. and Clark, J. R. (1961). *J. Biol. Chem.*, **236,** 2813
24. Tanaka, R. and Abood, L. G. (1963). *J. Neurochem.*, **10,** 571
25. Rose, I. A. and Warms, J. V. B. (1967). *J. Biol. Chem.*, **242,** 1635
26. Craven, P. A. and Basford, R. E. (1969). *Biochemistry*, **8,** 3520
27. Hochman, M. S. (1972). *Fed. Proc.*, **31,** 463 abs.
28. Knull, H. R., Taylor, W. F. and Wells, W. W. (1973). *J. Biol. Chem.*, **248,** 5414
29. Wilson, J. E. (1968). *J. Biol. Chem.*, **243,** 3640
30. Lowry, O. H. and Passonneau, J. V. (1964). *J. Biol. Chem.*, **239,** 31
31. Krebs, H. A. and Veech, R. L. (1969). *The Energy Level and Metabolic Control in Mitochondria*, 329 (S. Papa, J. M. Tager, E. Quagliariello and E. C. Slater, editors) (Bari: Adriatica Editrice)
32. Nordlie, R. C. (1971). *The Enzymes*, Vol. IV, 543 (P. D. Boyer, editor) (New York: Academic Press)
33. Rafter, G. W. (1960). *J. Biol. Chem.*, **235,** 2475
34. Nordlie, R. C. and Lardy, H. A. (1961). *Biochim. Biophys. Acta*, **53,** 309
35. Nordlie, R. C. (1968). *Control of Glycogen Metabolism*, 153 (W. J. Whelan, editor) (New York: Academic Press)
36. Scott, D. B. M. and Jones, G. (1970). *Enzymes and Isoenzymes*, Abstr. 364 (D. Shugar, editor) (New York: Academic Press)
37. Nordlie, R. C. and Arion, W. J. (1965). *J. Biol. Chem.*, **240,** 2155
38. Lueck, J. D. and Nordlie, R. C. (1970). *Biochem. Biophys. Res. Commun.*, **39,** 190
39. Stettan, M. R. and Goldsmith, P. K. (1973). *Biochim. Biophys. Acta*, **327,** 82
40. Nordlie, R. C., Hanson, T. L. and Johns, P. T. (1967). *J. Biol. Chem.*, **242,** 4144
41. Soodsma, J. F., Legler, B. and Nordlie, R. C. (1967). *J. Biol. Chem.*, **242,** 1955
42. Lygre, D. G. and Nordlie, R. C. (1969). *Biochim. Biophys. Acta*, **185,** 360
43. Nordlie, R. C., Hanson, T. L., Johns, P. T. and Lygre, D. G. (1968). *Proc. Nat. Acad. Sci. USA*, **60,** 590
44. Vianna, A. L. and Nordlie, R. C. (1969). *J. Biol. Chem.*, **244,** 4027
45. Ashmore, J. and Weber, G. (1959). *Vitamins Hormones*, **17,** 91
46. Bot, G. and Vereb, G. (1966). *Acta Biochim. Biophys. Acad. Sci. Hung.*, **1,** 169
47. Hers. H. G. and de Wulf, H. (1968). *Control of Glycogen Metabolism*, 65 (W. J. Whelan, editor) (New York: Academic Press)
48. Young, D. A. (1966). *Arch. Biochem. Biophys.*, **114,** 309
49. Nordlie, R. C. (1969). *Ann. N.Y. Acad. Sci.*, **166,** 699
50. Cahill, G. F., Jr., Ashmore, J., Renold, A. E. and Hastings, A. B. (1959). *Amer. J. Med.*, **26,** 264
51. Herrera, M. G., Kamm, D., Ruderman, N. and Cahill, G. F., Jr. (1966). *Advan. Enzyme Regul.*, Vol. 4, 225 (G. Weber, editor) (New York: Pergamon)
52. Kahng, M. W. and Lardy, H. A. (1974). *Unpublished observations*
53. Ashmore, J., Hastings, A. B. and Nesbett, F. B. (1954). *Proc. Nat. Acad. Sci. USA*, **40,** 673
54. Clark, M. G., Bloxham, D. P., Holland, P. C. and Lardy, H. A. (1974). *J. Biol. Chem.*, **249,** 279
55. Clark, D. G., Rognstad, R. and Katz, J. (1973). *Biochem. Biophys. Res. Commun.*, **54,** 1141
56. Rognstad, R., Clark, D. G. and Katz, J. (1973). *Biochem. Biophys. Res. Commun.*, **54,** 1149
57. Clark, M. G., Kneer, N. M., Bosch, A. L. and Lardy, H. A. (1974). *J. Biol. Chem.*, **249,** 5695
58. Parr, C. W. (1957). *Biochem. J.*, **65,** 34P
59. Kahana, S. E., Lowry, O. H., Schulz, D. W., Passonneau, J. V. Crawford, E. J. (1960). *J. Biol. Chem.*, **235,** 2178
60. Grazi, E., De Flora, A. and Pontremoli, S. (1960). *Biochem. Biophys. Res. Commun.*, **2,** 121
61. Salas, M., Vinuela, E. and Sols, A. (1965). *J. Biol. Chem.*, **240,** 561
62. Venkataraman, R. and Racker, E. (1961). *J. Biol. Chem.*, **236,** 1876
63. Greenbaum, A. L., Gumaa, K. A. and McLean, P. (1971). *Arch. Biochem. Biophys.*, **143,** 617
64. Dyson, J. E. D. and Noltmann, E. A. (1968). *J. Biol. Chem.*, **243,** 1401

65. Geewater, D. M. J., Hanshaw, E. D., Martin, R. E. and Parr, C. W. (1965). *Biochem. J.*, **97**, 12P
66. Stadtman, E. R. (1966). *Advan. Enzymol.*, **28**, 41
67. Racker, E. (1965). *Mechanisms in Bioenergetics*, 207 (New York: Academic Press)
68. Noltmann, E. A. (1972). *The Enzymes*, Vol. VI, 271 (P. D. Boyer, editor) (New York: Academic Press)
69. Bloxham, D. P. and Lardy, H. A. (1973). *The Enzymes*, Vol. VIII, 239 (P. D. Boyer, editor) (New York: Academic Press)
70. Peck, E. J. Jr. and Ray, W. J., Jr. (1971). *J. Biol. Chem.*, **246**, 1160
71. Aikawa, J. K. (1960). *Proc. Soc. Exp. Biol. Med.*, **103**, 363
72. Aikawa, J. K. (1960). *Amer. J. Physiol.*, **199**, 1084
73. Villar-Palasi, C. and Larner, J. (1970). *Ann. Rev. Biochem.*, **39**, 651
74. Horecker, B. L. (1968). *Carbohydrate Metabolism and Its Disorders*, Vol. 1, 139 (F. Dickens, P. J. Randle and W. J. Whelan, editors) (New York: Academic Press)
75. Shonk, C. E. and Boxer, G. E. (1964). *Cancer Res.*, **24**, 709
76. Novello, F., Gumaa, K. A. and McLean, P. (1969). *Biochem. J.*, **111**, 713
77. Abraham, S., Hirsch, P. F. and Chaikoff, I. L. (1954). *J. Biol. Chem.*, **211**, 31
78. Katz, J. and Wals, P. A. (1971). *Arch. Biochem. Biophys.*, **147**, 405
79. Eger-Neufeldt, I., Teinzer, A., Weiss, L. and Wieland, O. (1965). *Biochem. Biophys. Res. Commun.*, **19**, 43
80. Avigad, G. (1966). *Proc. Nat. Acad. Sci. USA*, **56**, 1543
81. Passonneau, J. V., Schulz, D. W. and Lowry, O. H. (1966). *Fed. Proc.*, **25**, 219
82. Horne, R. N., Anderson, W. B. and Nordlie, R. C. (1970). *Biochemistry*, **9**, 610
83. Glaser, L. and Brown, D. H. (1955). *J. Biol. Chem.*, **216**, 67
84. Glock, G. E. and McLean, P. (1953). *Biochem. J.*, **55**, 400
85. Rippa, M., Picco, C. and Pontremoli, S. (1970). *J. Biol. Chem.*, **245**, 4977
86. Dyson, J. E. D. and D'Orazio, R. E. (1973). *J. Biol. Chem.*, **248**, 5428
87. Wood, H. G. and Katz, J. (1958). *J. Biol. Chem.*, **233**, 1279
88. Katz, J. and Wood, H. G. (1960). *J. Biol. Chem.*, **235**, 2165
89. Williams, J. F. and Clark, M. G. (1971). *Search*, **2**, 80
90. Horecker, B. L., Gibbs, M., Klenow, H. and Smyrniotis, P. Z. (1954). *J. Biol. Chem.*, **207**, 393
91. Clark, M. G. and Williams, J. F. (1971). *Proc. Aust. Biochem. Soc.*, **4**, 37
92. Cori, C. F. (1942). *A Symposium on Respiratory Enzymes*, 175 (Madison: Univ. of Wisconsin Press)
93. Mansour, T. E. (1972). *Current Topics in Cellular Regulation*, Vol. 5, 1 (B. L. Horecker and E. R. Stadtman, editors) (New York: Academic Press)
94. Clark, M. G., Bloxham, D. P., Holland, P. C. and Lardy, H. A. (1973). *Biochem. J.*, **134**, 589
95. Clark, M. G., Williams, C. H., Pfeifer, W. F., Bloxham, D. P., Holland, P. C., Taylor, C. A. and Lardy, H. A. (1973). *Nature*, **245**, 99
96. Tsai, M. Y. and Kemp, R. G. (1973). *J. Biol. Chem.*, **248**, 785
97. Kemp, R. G. (1971). *J. Biol. Chem.*, **246**, 245
98. Garland, P. B., Randle, P. J. and Newsholme, E. A. (1963). *Nature*, **200**, 169
99. Passonneau, J. V. and Lowry, O. H. (1963). *Biochem. Biophys. Res. Commun.*, **13**, 372
100. Lardy, H. A. and Parks, R. E. (1956). *Enzymes: Units of Biological Structure and Function*, 584 (O. H. Gaebler, editor) (New York: Academic Press)
101. Atkinson, D. E. and Walton, G. M. (1967). *J. Biol. Chem.*, **242**, 3239
103. Mansour, T. E. and Ahlfors, C. E. (1968). *J. Biol. Chem.*, **243**, 2523
104. Newsholme, E. A. and Randle, P. J. (1961). *Biochem. J.*, **80**, 655
105. Williamson, J. R. (1966). *J. Biol. Chem.*, **241**, 5026
106. Lowry, O. H., Passonneau, J. V., Hasselberger, F. X. and Schultz, D. W. (1964). *J. Biol. Chem.*, **239**, 18
107. Wu, R. (1964). *Biochem. Biophys. Res. Commun.*, **14**, 79
108. Parmeggiani, A. and Bowman, R. H. (1963). *Biochem. Biophys. Res. Commun.*, **12**, 268
109. Underwood, A. H. and Newsholme, E. A. (1967). *Biochem. J.*, **104**, 300
110. Randle, P. J., Garland, P. B., Hales, C. N., Newsholme, E. A., Denton, R. M. and Pogson, C. I. (1966). *Recent Progr. Hormone Res.*, **22**, 1
111. Williamson, J. R., Jones, E. A. and Azzone, G. F. (1964). *Biochem. Biophys. Res. Commun.*, **17**, 696

112. Yoshida, M., Oshima, T. and Imahori, K. (1971). *Biochem. Biophys. Res. Commun.*, **43,** 36
113. Kelly, G. J. and Turner, J. F. (1968). *Biochem. Biophys. Res. Commun.*, **30,** 195
114. Kelly, G. J. and Turner, J. F. (1969). *Biochem. J.*, **115,** 481
115. Kelly, G. J. and Turner, J. F. (1970). *Biochim. Biophys. Acta*, **208,** 360
116. Krzanowski, J. and Matschinsky, F. M. (1969). *Biochem. Biophys. Res. Commun.*, **34** 816
117. Mansour, T. E. (1970). *Advan. Enzyme Regul.*, Vol. 8, 37 (G. Weber, editor) (New York: Pergamon Press)
118. Mansour, T. E. (1972). *J. Biol. Chem.*, **247,** 6059
119. Nakatsu, K. and Mansour, T. E. (1973). *Mol. Pharmacol.*, **9,** 405
120. Mansour, T. E. (1962). *J. Pharmacol. Exp. Ther.*, **135,** 94
121. Mansour, T. E. and Mansour, J. M. (1962). *J. Biol. Chem.*, **237,** 629
122. Flatt, J. P. and Ball, E. G. (1964). *J. Biol. Chem.*, **239,** 675
123. Denton, R. M., Yorke, R. E. and Randle, P. J. (1966). *Biochem. J.*, **100,** 407
124. Denton, R. M. and Randle, P. J. (1966). *Biochem. J.*, **100,** 420
125. Butcher, R. W., Ho, R. J., Meng, H. C. and Sutherland, E. W. (1965). *J. Biol. Chem.*, **240,** 4515
126. Lee, L. M. Y., Krupka, R. M. and Cook, R. A. (1973). *Biochemistry*, **12,** 3503
127. Lowry, O. H. and Passonneau, J. V. (1966). *J. Biol. Chem.*, **241,** 2268
128. Stone, D. B. and Mansour, T. E. (1967). *Mol. Pharmacol.*, **3,** 177
129. Mansour, T. E. (1966). *Pharmacol. Rev.*, **18,** 173
130. Robison, G. A., Butcher, R. W. and Sutherland, E. W. (1971). *Cyclic AMP*, 159 (New York: Academic Press)
131. Black, W. J., Van Tol, A., Fernando, J. and Horecker, B. L. (1972). *Arch. Biochem. Biophys.*, **151,** 576
132. Scrutton, M. C. and Utter, M. F. (1968). *Ann. Rev. Biochem.*, **37,** 249
133. Newsholme, E. A., Crabtree, B., Higgins, S. J., Thornton, S. D. and Start, C. (1972). *Biochem. J.*, **128,** 89
134. Pagliara, A. S., Karl, I. E., Keating, J., Brown, B. and Kipnis, D. M. (1970). *J. Lab. Clin. Med.*, **76,** 1020
135. Melancon, S. B., Khachadurian, A. K., Nadler, H. L. and Brown, B. I. (1973). *J. Pediatrics*, **82,** 650
136. Crabtree, B., Higgins, S. J. and Newsholme, E. A. (1972). *Biochem. J.*, **130,** 391
137. Esner, M., Shapiro, S. and Horecker, B. L. (1969). *Arch. Biochem. Biophys.*, **129,** 377
138. Weber, G., Singhal, R. L. and Srivastava, S. K. (1965). *Advan. Enzyme Regul.*, Vol. 3, 43 (G. Weber, editor) (New York:Pergamon)
139. Pontremoli, S., Melloni, E., Salamino, F., Franzi, A. T., De Flora, A. and Horecker, B. L. (1973). *Proc. Nat. Acad. Sci. USA*, **70,** 3674
140. Pontremoli, S. (1972). *Biochem. J.*, **130,** 1P
141. Pontremoli, S., Luppis, B., Traniello, S., Rippa, M. and Horecker, B. L. (1965). *Arch. Biochem. Biophys.*, **112,** 7
142. Benkovic, S. J., Frey, W. A., Libby, C. B. and Villafranca, J. J. (1974). *Biochem. Biophys. Res. Commun.*, **57,** 196
143. Nakashima, K., Horecker, B. L., Traniello, S. and Pontremoli, S. (1970). *Arch. Biochem. Biophys.*, **139,** 190
144. Nakashima, K., Pontremoli, S. and Horecker, B. L. (1969). *Proc. Nat. Acad. Sci. USA*, **64,** 947
145. Pontremoli, S., Traniello, S., Enser, M., Shapiro, S. and Horecker, B. L. (1967). *Proc. Nat. Acad. Sci. USA*, **58,** 286
146. Marcus, C. J., Geller, A. M. and Byrne, W. L. (1973). *J. Biol. Chem.*, **248,** 8567
147. Mendicino, J., Beaudreau, C. and Bhattacharyya, R. N. (1966). *Arch. Biochem. Biophys.*, **116,** 436
148. Rosen, O. M. (1966). *Arch. Biochem. Biophys.*, **114,** 31
149. Start, C. and Newsholme, E. A. (1968). *Biochem. J.*, **107,** 411
150. Pontremoli, S., Grazi, E. and Accorsi, A. (1968). *Biochemistry*, **7,** 3628
151. Pontremoli, S. and Horecker, B. L. (1971). *The Enzymes*, Vol. IV, 611 (P. D. Boyer, editor) (New York: Academic Press)
152. Rosen, O. M., Rosen, S. M. and Horecker, B. L. (1965). *Arch. Biochem. Biophys.*, **112,** 411

153. Pogell, B. M., Tanaka, A. and Siddons, R. C. (1968). *J. Biol. Chem.*, **243,** 1356
154. Baxter, R. C., Carlson, C. W. and Pogell, B. M. (1972). *Fed. Proc.*, **31,** 837
155. Carlson, C. W., Baxter, R. C., Ulm, E. H. and Pogell, B. M. (1973). *J. Biol. Chem.*, **248,** 5555
156. Taketa, K., Sarngadharan, M. G., Watanabe, A., Aoe, H. and Pogell, B. M. (1971). *J. Biol. Chem.*, **246,** 5676
157. Blair, J. B., Cook, D. E. and Lardy, H. A. (1973). *J. Biol. Chem.*, **248,** 3601
158. Allen, M. B. and Blair, J. McD. (1972). *Biochem. J.*, **130,** 1167
159. Söling, H. D., Willms, B. and Kleineke, J. (1971). *Regulation of Gluconeogenesis*, 210 (H. D. Söling and B. Willms, editors) (New York: Academic Press)
160. Jones, C. T. (1972). *Biochem. J.*, **130,** 23P
161. Fernando, J., Enser, M., Pontremoli, S. and Horecker, B. L. (1968). *Arch. Biochem. Biophys.*, **126,** 599
162. Taketa, K. and Pogell, B. M. (1965). *J. Biol. Chem.*, **240,** 651
163. Opie, L. H. and Newsholme, E. A. (1967). *Biochem. J.*, **104,** 353
164. Fu, J. Y. and Kemp, R. G. (1973). *J. Biol. Chem.*, **248,** 1124
165. Underwood, A. H. and Newsholme, E. A. (1965). *Biochem. J.*, **95,** 868
166. Newsholme, E. A. and Gevers, W. (1967). *Vitamins Hormones*, **25,** 1
167. Newsholme, E. A. and Underwood, A. H. (1966). *Biochem. J.*, **99,** 24C
168. Gevers, W. and Krebs, H. A. (1966). *Biochem. J.*, **98,** 720
169. Newsholme, E. A. and Crabtree, B. (1973). *Symp. Soc. Exp. Biol.*, **27,** 429
170. Williamson, J. R., Browning, E. T. and Scholz, R. (1969). *J. Biol. Chem.*, **244,** 4607
171. Williamson, J. R., Scholz, R. and Browning, E. T. (1969). *J. Biol. Chem.*, **244,** 4617
172. Williamson, J. R., Scholz, R., Browning, E. T., Thurman, R. G. and Fukami, M. (1969). *J. Biol. Chem.*, **244,** 5044
173. Williamson, J. R., Anderson, J. and Browning, E. T. (1970). *J. Biol. Chem.*, **245,** 1717
174. Lynen, F., Matsuhashi, M., Numa, S. and Schweizer, E. (1963). *The Control of Lipid Metabolism*, 43 (J. K. Grant, editor) (New York: Academic Press)
175. Tubbs, P. K. and Garland, P. B. (1964). *Biochem. J.*, **93,** 550
176. Williamson, J. R., Kreisberg, R. A. and Felts, P. W. (1966). *Proc. Nat. Acad. Sci. USA*, **56,** 247
177. Lowenstein, J. M. (1966). *Control of Energy Metabolism*, 261 (B. Chance, R. W. Estabrook and J. R. Williamson, editors) (New York: Academic Press)
178. Start, C. and Newsholme, E. A. (1967). *Biochem. J.*, **104,** 46P
179. Bloxham, D. P., Clark, M. G., Holland, P. C. and Lardy, H. A. (1973). *Biochem. J.*, **134,** 581
180. Bloxham, D. P., Clark, M. G., Goldberg, D. M., Holland, P. C. and Lardy, H. A. (1973). *Biochem. J.*, **134,** 586
181. Lardy, H. A., Zahlten, R. N., Stratman, F. W. and Cook, D. E. (1973). *Regulation of Hepatic Metabolism*, 19 (F. Lundquist, editor) (Alfred Benzon Symposium VI, Munksgaard, Copenhagen)
182. Veneziale, C. M. (1971). *Biochemistry*, **10,** 3443
183. Veneziale, C. M. (1972). *Biochemistry*, **11,** 3286
184. Veneziale, C. M. and Lohmar, P. H. (1973). *J. Biol. Chem.*, **248,** 7786
185. Tolbert, M. E. M., and Fain, J. N. (1974). *J. Biol. Chem.*, **249,** 1162
186. Haugaard, E. S. and Haugaard, N. (1954). *J. Biol. Chem.*, **206,** 641
187. Tepperman, H. M. and Tepperman, J. (1972). *Insulin Action*, 543 (I. B. Fritz, editor) (New York: Academic Press)
188. El-Badry, A. M., Otani, A. and Mansour, T. E. (1973). *J. Biol. Chem.*, **248,** 557
189. Hanegan, J. L. and Heath, J. E. (1970). *J. Exp. Biol.*, **53,** 349
190. Clark, M. G., Huang, M. and Lardy, H. A. (1974). Unpublished observations
191. Leuthardt, F., Testa, E. and Wolf, H. P. (1952). *Helv. Physiol. Pharmacol. Acta*, **10,** C57
192. Hers, H. G. and Kusaka, T. (1953). *Biochim. Biophys. Acta*, **11,** 427
193. Peanasky, R. J. and Lardy, H. A. (1958). *J. Biol. Chem.*, **233,** 365
194. Penhoet, E., Rajkumar, T. and Rutter, W. J. (1966). *Proc. Nat. Acad. Sci. USA*, **56,** 1275
195. Lebherz, H. G. and Rutter, W. J. (1969). *Biochemistry*, **8,** 109
196. Kwon, T.-W. and Olcott, H. S. (1965). *Biochem. Biophys. Res. Commun.*, **19,** 300
197. Veech, R. L., Raijman, L., Dalziel, K. and Krebs, H. A. (1969). *Biochem. J.*, **115,** 837

198. Rose, I. A., Kellermeyer, R., Stjernholm, R. and Wood, H. G. (1962). *J. Biol. Chem.*, **237**, 3325
199. Smith, C. M. and Velick, S. F. (1972). *J. Biol. Chem.*, **247**, 273
200. Williamson, J. R. (1965). *J. Biol. Chem.*, **240**, 2308
201. Kosow, D. P. and Rose, I. (1972). *Fed. Proc.*, **31**, 434
202. Cori, G. T., Slein, M. W. and Cori, C. F. (1948). *J. Biol. Chem.*, **173**, 605
203. Klingenberg, M. and Bücher, T. (1960). *Ann. Rev. Biochem.*, **29**, 669
204. Oguchi, M., Meriwether, B. P. and Park, J. H. (1973). *J. Biol. Chem.*, **248**, 5562
205. Oguchi, M., Gerth, E., Fitzgerald, B. and Park, J. H. (1973). *J. Biol. Chem.*, **248**, 5571
206. Johnson, M. J. (1941). *Science*, **94**, 200
207. Lynen, F. (1942). *Naturwiss.*, **30**, 398
208. Hohorst, H. J., Reim, M. and Bartels, H. (1962). *Biochem. Biophys. Res. Commun.*, **7**, 142
209. Danforth, W. (1965). *Control of Energy Metabolism*, 287 (B. Chance, R. W. Estabrook, and J. R. Williamson, editors) (New York: Academic Press)
210. Trivedi, B. and Danforth, W. H. (1966). *J. Biol. Chem.*, **241**, 4110
211. Kosow, D. P. and Rose, I. A. (1971). *J. Biol. Chem.*, **246**, 2618
212. Caldwell, P. (1956). *Int. Rev. Cytol.*, **5**, 229
213. Hill, A. V. (1955). *Proc. Roy. Soc. Ser. Biol. Sci.*, **144**, 11
214. Fife, T. H. and Szabo, A. (1973). *Arch. Biochem. Biophys.*, **157**, 100
215. Veech, R. L., Raijman, L. and Krebs, H. A. (1970). *Biochem. J.*, **117**, 499
216. Hanson, R. L., Rudolph, F. B. and Lardy, H. A. (1973). *J. Biol. Chem.*, **248**, 7852
217. Lardy, H. A. (1951). *Phosphorus Metabolism*, Vol. 1, 477 (W. D. McElroy and B. Glass, editors) (Baltimore: The Johns Hopkins Press)
218. Larsson-Raźnikiewicz, M. and Malmström, B. G. (1961). *Arch. Biochem. Biophys.*, **92**, 94
219. Larsson-Raźinikiewicz, M. (1967). *Biochim. Biophys. Acta*, **132**, 33
220. Krietsch, W. K. G. and Bücher, T. (1970). *Eur. J. Biochem.*, **17**, 568
221. Scopes, R. K. (1973). *The Enzymes*, Vol. VIII, 335 (P. D. Boyer, editor) (New York: Academic Press)
222. Larsson-Raźnikiewicz, M. and Arvidsson, L. (1971). *Eur. J. Biochem.*, **22**, 506
223. Brewer, J. M. and Weber, G. (1966). *J. Biol. Chem.*, **241**, 2550
223a. Warburg, O. and Christian, W. (1942). *Biochem. Z.*, **310**, 384
224. Carbonell, J., Feliu, J. E., Marco, R. and Sols, A. (1973). *Eur. J. Biochem.*, **37**, 148
225. Tanaka, T., Harano, Y., Morimura, H. and Mori, R. (1965). *Biochem. Biophys. Res. Commun.*, **21**, 55
226. Tanaka, T., Sue, F. and Morimura, H. (1967). *Biochem. Biophys. Res. Commun.*, **29**, 444
227. Taylor, C. B. and Bailey, E. (1967). *Biochem. J.*, **102**, 32C
228. Passeron, S., Jiménez de Asúa, L. and Carminatti, H. (1967). *Biochem. Biophys. Res. Commun.*, **27**, 33
229. Seubert, W., Henning, H. V., Schoner, W. and L'Age, M. (1968). *Advan. Enzyme Regul.*, Vol. 6., 153 (G. Weber, editor) (New York: Pergamon Press)
230. Llorente, P., Marco, R. and Sols, A. (1970). *Eur. J. Biochem.*, **13**, 45
230a. Pogson, C. I. (1968). *Biochem. J.*, **110**, 67
231. Hess, B. and Kutzbach, C. (1971). *Z. Physiol. Chem.*, **352**, 453
232. Krebs, H. A. and Eggleston, L. V. (1965). *Biochem. J.*, **94**, 3C
233. Tanaka, T., Harano, Y., Sue, F. and Morimura, H. (1967). *J. Biochem. (Tokyo)*, **62**, 71
234. Bailey, E., Stirpe, F. and Taylor, C. B. (1968). *Biochem. J.*, **108**, 427
235. Szepesi, B. and Freedland, R. (1969). *Proc. Soc. Exp. Biol. Med.*, **132**, 489
236. Weber, G., Stamm, N. B. and Fisher, E. A. (1965). *Science*, **149**, 65
237. Gerschenson, L. and Andersson, M. (1971). *Biochem. Biophys. Res. Commun.*, **43**, 1211
238. Jiménez de Asúa, L., Rozengurt, E. and Carminatti, H. (1968). *Biochim. Biophys. Acta*, **170**, 254
239. Kemp, R. G. (1973). *J. Biol. Chem.*, **248**, 3963
240. Seubert, W. and Schoner, W. (1971). *Curr. Top. Cell. Regul.*, Vol. 3, 237 (B. L. Horecker and E. R. Stadtman, editors) (New York: Academic Press)
241. Betheil, J. J., Feigelson, M. and Feigelson, P. (1965). *Biochim. Biophys. Acta*, **104**, 92
242. Williamson, D. H., Lopes-Vieiro, O. and Walker, B. (1967). *Biochem. J.*, **104**, 497
243. Boyer, P. D., Lardy, H. A., and Phillips, P. H. (1943). *J. Biol. Chem.*, **149**, 529

244. Gabrielli, F. and Baldi, S. (1972). *Eur. J. Biochem.*, **31**, 209
245. Utter, M. F. and Keech, D. B. (1963). *J. Biol. Chem.*, **238**, 2603
246. Ballard, F. J., Hanson, R. W. and Reshef, L. (1970). *Biochem. J.*, **119**, 735
247. Anderson, J. W. (1970). *Biochim. Biophys. Acta*, **208**, 165
248. Utter, M. F. and Keech, D. B. (1960). *J. Biol. Chem.*, **235**, PC17
249. Hanson, R. W. and Ballard, F. J. (1967). *Biochem. J.*, **105**, 529
250. Kornberg, H. L. (1966). *Essays Biochem.*, **2**, 1
251. Walter, P. and Anabitarte, M. (1971). *FEBS Lett.*, **12**, 289
252. McClure, W. R. and Lardy, H. A. (1971). *J. Biol. Chem.*, **246**, 3591
253. Freedman, A. D. and Kohn, L. (1964). *Science*, **145**, 58
254. Henning, H. V., Stumpf, B., Ohly, B. and Seubert, W. (1966). *Biochem. Z.*, **344**, 274
255. Prinz, W. and Seubert, W. (1964). *Biochem. Biophys. Res. Commun.*, **16**, 582
256. Wagle, S. R. (1964). *Biochem. Biophys. Res. Commun.*, **14**, 533
257. Shrago, E. and Lardy, H. A. (1966). *J. Biol. Chem.*, **241**, 663
258. Struck, E., Ashmore, J. and Wieland, O. (1966). *Enzymol. Biol. Clin.*, **7**, 38
259. Krebs, H. A. (1966). *Advan. Enzyme Regul.*, **4**, 339
260. Brech, W., Shrago, E. and Wilken, D. (1970). *Biochim. Biophys. Acta*, **201**, 145
261. Taylor, P. H., Wallace, J. C. and Keech, D. B. (1971). *Biochim. Biophys. Acta*, **237**, 179
262. Madapally, M. and Mistry, S. P. (1970). *Biochim. Biophys. Acta*, **215**, 316
263. Utter, M. F. and Scrutton, M. C. (1969). *Curr. Top. Cell. Reg.*, Vol. 1, 253 (B. L. Horecker and E. R. Stadtman, editors) (New York: Academic Press)
264. Fung, C. H. and Utter, M. F. (1970). *Fed. Proc.*, **29**, 542
265. Ballard, F. J., Hanson, R. W. and Kronfeld, D. S. (1969). *Fed. Proc.*, **28**, 218
266. Keech, D. B. and Utter, M. F. (1963). *J. Biol. Chem.*, **238**, 2609
267. Walter, P., Paetkau, V. and Lardy, H. A. (1966). *J. Biol. Chem.*, **241**, 2523
268. McClure, W. R., Lardy, H. A. and Kneifel, H. P. (1971). *J. Biol. Chem.*, **246**, 3569
269. Kimmich, G. A. and Rasmussen, H. (1969). *J. Biol. Chem.*, **244**, 190
270. Friedmann, N. and Rasmussen, H. (1970). *Biochim. Biophys. Acta*, **222**, 41
271. Rasmussen, H. (1970). *Science*, **170**, 404
272. Wimhurst, J. M. and Manchester, K. L. (1970). *Biochem. J.*, **120**, 79
273. Mörikofer-Zwez, S., Kunin, A. S. and Walter, P. (1973). *J. Biol. Chem.*, **248**, 7588
274. Nordlie, R. C. and Lardy, H. A. (1963). *J. Biol. Chem.*, **238**, 2259
275. Ballard, F. J. and Hanson, R. W. (1967). *Biochem. J.*, **104**, 866
276. Utter, M. F. (1959). *Ann. N.Y. Acad. Sci.*, **72**, 451
277. Gevers, W. (1967). *Biochem. J.*, **103**, 141
278. Felicioli, R. A., Gabrielli, F. and Rossi, C. A. (1967). *Eur. J. Biochem.*, **3**, 19
279. Lardy, H. A. (1965). *Harvey Lect.*, **60**, 261
280. Ballard, F. J., Hanson, R. W. and Kronfeld, D. S. (1968). *Biochem. Biophys. Res. Commun.*, **30**, 100
281. Söling, H. D., Willms, B., Kleineke, J. and Gehlhoff, M. (1970). *Eur. J. Biochem.*, **16**, 289
282. Garber, A. J. and Hanson, R. W. (1971). *J. Biol. Chem.*, **246**, 589
283. Smith, R. M. and Osborne-White, W. S. (1971). *Biochem. J.*, **124**, 867
284. Swiatek, K. R., Chao, K. L., Chao, H. L., Cornblath, M. and Tildon, T. (1970). *Biochim. Biophys. Acta*, **206**, 316
285. Ballard, F. J. and Hanson, R. W. (1969). *J. Biol. Chem.*, **244**, 5625
286. Diesterhaft, M., Shrago, E. and Sallach, H. J. (1971). *Biochem. Med.*, **5**, 297
287. Shrago, E., Lardy, H. A., Nordlie, R. C. and Foster, D. O. (1963). *J. Biol. Chem.*, **238**, 3188
288. Ilyin, V. S., Usatenko, M. S. and Evstratora, L. A. (1966). *Zh. Evol. Biokhim. Fiziol.*, **2**, 185
289. Johnson, D. C., Ebert, K. A. and Ray, P. D. (1970). *Biochem. Biophys. Res. Commun.* **39**, 750
290. Garber, A. J. and Hanson, R. W. (1971). *J. Biol. Chem.*, **246**, 5555
291. Lardy, H. A., Foster, D. O., Shrago, E. and Ray, P. D. (1964). *Advan. Enzyme Regul.* Vol. 2, 39 (G. Weber, editor) (New York: Pergamon Press)
292. Nordlie, R. C., Varricchio, F. E. and Holten, D. D. (1965). *Biochim. Biophys. Acta*, **97** 214
293. Ballard, F. J., Hanson, R. W. and Leveille, G. A. (1967). *J. Biol. Chem.*, **242**, 2746
294. Evans, R. M. and Scholz, R. W. (1973). *Biochim. Biophys. Acta*, **321**, 671

295. Cheng, S.-C. and Cheng, R. H. C. (1972). *Arch. Biochem. Biophys.*, **151**, 501
296. Hopgood, M. F., Ballard, F. J., Reshef, L. and Hanson, R. W. (1973). *Biochem. J.*, **134**, 445
297. Nagai, K. and Nakagawa, H. (1972). *J. Biochem.*, **71**, 125
298. Ballard, F. J. and Hopgood, M. F. (1973). *Biochem. J.*, **136**, 259.
299. Flores, H. and Alleyene, G. A. O. (1971). *Biochem. J.*, **123**, 35
300. Anderson, J. H., Nicklas, W. J., Blank, B., Refino, C. and Williamson, J. R. (1971). *Regulation of Gluconeogenesis*, 293 (H. D. Söling, and B. Willms, editors) (New York: Academic Press)
301. Zahlten, R. N., Kneer, N. M., Stratman, F. W. and Lardy, H. A. (1974). *Arch. Biochem Biophys.*, **161**, 528
302. Lardy, H. A., Paetkau, V. and Walter, P. (1965). *Proc. Nat. Acad. Sci. USA*, **53**, 1410
303. Krebs, H. A., Gascoyne, T. and Notton, B. M. (1967). *Biochem. J.*, **102**, 275
304. Johnson, D. C., Brunsvold, R. A., Ebert, K. A. and Ray, P. D. (1973). *J. Biol. Chem.*, **248**, 763
305. Robinson, B. H. (1971). *FEBS Lett.*, **14**, 309
306. Söling, H.-D., Kleineke, J., Willms, B., Janson, G. and Kuhn, A. (1973). *Eur. J. Biochem.*, **37**, 233
307. Stanbury, S. W. and Mudge, G. H. (1954). *J. Biol. Chem.*, **210**, 949
308. Nordlie, R. C. and Lardy, H. A. (1963). *Biochem. Z.*, **338**, 356
309. Ishihara, N. and Kikuchi, G. (1968). *Biochim. Biophys. Acta*, **153**, 733
310. Davis, E. J. and Gibson, D. M. (1969). *J. Biol. Chem.*, **244**, 161
311. Granger, M. and Harris, E. J. (1971). *Bioenergetics*, **2**, 151
312. Krebs, H. A. (1970). *Advan. Enzyme Regul.*, **8**, 335
313. Wilson, M. B. (1973). *Biochem. J.*, **132**, 553
314. Garber, A. J., Ballard, F. J. and Hanson, R. W. (1972). *Energy Metabolism and the Regulation of Metabolic Processes in Mitochondria*, 109 (M. A. Mehlman and R. W. Hanson, editors) (New York: Academic Press).
315. Ballard, F. J. (1970). *Biochem. J.*, **120**, 809
315a. Walsh, D. A. and Chen, L-J. (1971). *Biochem. Biophys. Res. Commun.*, **45**, 669
316. Foster, D. O., Lardy, H. A., Ray, P. D. and Johnston, J. B. (1967). *Biochemistry*, **6**, 2120
317. Bentle, L. A., Snoke, R. E. and Lardy, H. A. (1973). *ACS Div. Biol. Chem.*, **166*th ACS Nat. Meeting*, Abstr. No. 113
318. Bentle, L. A., and Snoke, R. E. (1974). *Fed. Proc.*, **33**, 1272
319. Ray, P. D. Foster, D. O. and Lardy, H. A. (1966). *J. Biol. Chem.*, **241**, 3904
320. Veneziale, C. M., Walter, P., Kneer, N. and Lardy, H. A. (1967). *Biochemistry*, **6**, 2129
321. Snoke, R. E., Johnston, J. B. and Lardy, H. A. (1971). *Eur. J. Biochem.*, **24**, 342
322. Garland, P. B. and Randle, P. J. (1964). *Biochem. J.*, **91**, 6C
323. Bremer, J. (1969). *Eur. J. Biochem.*, **8**, 535
324. Wieland, O. (1969). *Z. Physiol. Chem.*, **350**, 329
325. Hucho, F., Burgett, M. W. and Reed, L. J. Unpublished observations referred to by Reed, L. J., Linn, T. C., Pettit, F. H., Oliver, R. M., Hucho, F., Pelley, J. W., Randall, D. D. and Roche, T. E. (1972). *Energy Metabolism and the Regulation of Metabolic Processes in Mitochondria*, 253 (M. A. Mehlman and R. W. Hanson, editors) (New York: Academic Press)
326. Reed, L. J., Linn, T. C., Pettit, F. H., Oliver, R. M., Hucho, F., Pelley, J. W., Randall, D. D. and Roche, T. E. (1972). *Energy Metabolism and the Regulation of Metabolic Processes in Mitochondria*, 253 (M. A. Mehlman and R. W. Hanson, editors) (New York: Academic Press)
327. Linn, T. C., Pettit, F. H. and Reed, L. J. (1969). *Proc. Nat. Acad. Sci. USA*, **62**, 234
328. Wieland, O. and Von Jagow-Westermann, B. (1969). *FEBS Lett.*, **3**, 271
329. Jungas, R. L. (1971). *Metabolism*, **20**, 43
330. Wieland, O. (1970). *8th International Congress of Biochemistry Lucerne, Abstr.*, 238
331. Söling, H. D. and Bernhard, G. (1971). *FEBS Lett.*, **13**, 201
332. Patzelt, C., Löffler, G. and Wieland, O. H. (1973). *Eur. J. Biochem.*, **33**, 117
333. Wieland, O. H., Patzelt, C. and Löffler, G. (1972). *Eur. J. Biochem.*, **26**, 426
334. Wieland, O. H., Siess, E., Schulze-Wethmar, F. H., Funcke, H. G. V. and Winton, B. (1971). *Arch. Biochem. Biophys.*, **143**, 593

335. Randle, P. J., Denton, R. M. and Pask, H. T. (1973). *Birmingham Meeting of the Biochemical Society, Abstracts*, 30
336. Wieland, O. and Siess, E. (1970). *Proc. Nat. Acad. Sci. USA*, **65,** 947
337. Schimmel, R. J. and Goodman, H. (1972). *Biochim. Biophys. Acta*, **260,** 153
338. Zahlten, R. N., Stratman, F. W. and Lardy, H. A. (1973). *Proc. Nat. Acad. Sci. USA*, **70,** 3213
339. Martin, B. R., Denton, R. M., Pask, H. T. and Randle, P. J. (1972). *Biochem. J.*, **129,** 763
340. Pettit, F. H., Roche, T. E. and Reed, L. J. (1972). *Biochem. Biophys. Res. Commun.*, **49,** 563
341. Young, J. W., Shrago, E. and Lardy, H. A. (1964). *Biochemistry*, **3,** 1687
342. Srere, P. A. and Lipmann, F. (1953). *J. Amer. Chem. Soc.*, **75,** 4874
343. Ochoa, S., Mehler, A. H. and Kornberg, A. (1948). *J. Biol. Chem.*, **174,** 979
344. Friedmann, B., Goodman, E. H. Jr., Saunders, H. L., Kostos, V. and Weinhouse, S. (1971). *Metabolism*, **20,** 2
345. Friedmann, B., Goodman, E. H. Jr., Saunders, H. L. Kostos, V. and Weinhouse, S. (1971). *Arch. Biochem. Biophys.*, **143,** 566
346. Exton, J. H. and Park, C. R. (1969). *J. Biol. Chem.*, **244,** 1424
347. Exton, J. H., Mallette, L. E., Jefferson, L. S., Wong, E. H. A., Friedmann, N., Miller, T. B. Jr. and Park, C. R. (1970). *Recent Progr. Hormone Res.*, **26,** 411
348. Exton, J. H., Ui, M., Lewis, S. B. and Park, C. R. (1971). in *Regulation of Gluconeogenesis*, 160 (H. D. Söling and B. Willms, editors) (New York: Academic Press)
349. Rognstad, R. and Katz, J. (1972). *J. Biol. Chem.*, **247,** 6047
350. Blair, J. B., Cook, D. E. and Lardy, H. A. (1973). *J. Biol. Chem.*, **248,** 3608
351. Underwood, A. H. and Newsholme, E. A. (1967). *Biochem. J.*, **104,** 300
352. Randle, P. J. (1964). *Symp. Soc. Exp. Biol.*, **18,** 129
353. Hohorst, H. J., Reim, M. and Bartels, H. (1962). *Biochem. Biophys. Res. Commun.*, **7,** 137
354. Wu, R. (1965). *J. Biol. Chem.*, **240,** 2373
355. Cahill, G. F. Jr. and Owen, O. E. (1968). *Carbohydrate Metabolism and Its Disorders*, Vol. 1, 497 (F. Dickens, P. J. Randle and W. J. Whelan, editors) (New York: Academic Press)
356. Opie, L. H. and Newsholme, E. A. (1967). *Biochem. J.*, **103,** 391
357. Weber, G., Banerjee, G. and Ashmore, J. (1960). *Biochem. Biophys. Res. Commun.*, **3,** 182
358. Wise, E. M. Jr. and Ball, E. G. (1964). *Proc. Nat. Acad. Sci. USA*, **52,** 1255

# 6
# Chemistry and Biochemistry of Starch

D. FRENCH
Iowa State University

---

---

## Symbols and abbreviations

$G_1$ = glucose; $G_2$, $G_3$, etc; maltose, maltotriose, etc., $B_2$, $B_3$, etc., isomaltose and higher homologues containing a single 1,6-α-glucosidic linkage.

Ø = glucose; O—Ø = maltose; O—O—Ø = maltotriose, etc. The circle is a shorthand notation for an α-D-glucopyranosyl ring; the diagonal slash denotes the reducing group. A horizontal bar indicates an 1,4-α-glucosidic bond; an arrow (usually downward-pointing) indicates a 1,6-α-glucosidic bond. An indefinite extension of a 1,4-α-linked chain is indicated by three dots (· · ·). Enzymic cleavage is indicated by a curved arrow (↷) at a point attacked by the enzyme. Long starch chains are sometimes indicated simply by a line, which may be terminated at the reducing end by Ø or by attachment to another symbol. Substitution of a position in an oligosaccharide is indicated by a large number, indicating the position on a given glucose unit, with a superscript, designating the particular glucose unit, counting from the reducing end. For example, $6^2$-α-maltosyl maltotriose is:

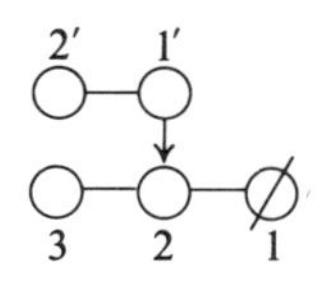


The most widely accepted abbreviations for oligosaccharides or polysaccharides involve the use of Glc (for glucose), *p* for pyranose, and parenthetically enclosed modes of linkage, e.g. (1→4), to indicate a 1→4 glycosidic bond. Thus panose, O↓O—Ø, would be α-D-Glc*p*-(1→6)-α-D-Glc*p*-(1→4)-D-Glc

and fast $B_5$, 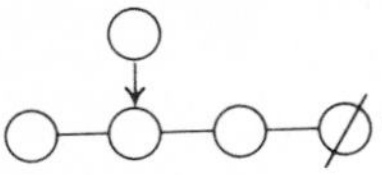 , would be [α-D-Glc*p*-(1→4)-, α-D-Glc*p*-(1→6)-]-α-D-Glc*p*-(1→4)-α-D-Glc*p*-(1→4)-D-Glc or bis-[α-D-Glc*p*]-(1→4)-, (1→6)-]-α-D-Glc*p*-(1→4)-α-D-Glc*p*-(1→4)-D-Glc. The cumbersome nature of such expressions illustrates the starch scientists' need for simple trivial names and symbolic representations.

Other abbreviations: G-1-P, α-D-glucose 1-phosphate; ADPG, adenosine diphosphate glucose; UDPG, uridine diphosphate glucose; DMSO, dimethyl sulphoxide; DP, degree of polymerisation.

The designations *A*, *B* and *C* distinguish between the types of chains in starch: an *A*-chain is one which is unsubstituted at position 6 of any of the constituent glucose units, and is linked at its reducing end to a *B* or *C* chain. A *B*-chain is substituted at position 6 of one or more of its glucose units by *A* or other *B* chains, and it is linked at its reducing end to a *B* or *C* chain. A *C*-chain is substituted at position 6 of one or more of its glucose units by *A* or *B* chains, and has a free reducing group. In Figure 6.6 (p. 275) the chain designated (*C*) would be a *C*-chain only if it is unsubstituted at the reducing end; otherwise it would be a *B*-chain.

---

## 6.1 BRIEF HISTORICAL OVERVIEW*

Knowledge of starch and its domestic application to cooking and laundering antedates recorded history. The word 'starch' is of Teutonic origin and has the connotation of strength or stiffness, as applied to a fabric or to paper. The Greco-Latin term *amylon* or *amylum* connotes a flour-like material prepared without milling from tuberous or succulent vegetable sources.

Man's transition from a predominantly hunting and gathering nomadic society to a more stationary civilisation would not have been possible without the domestication of plants and the establishment of agriculture. Those plants which we presently know as our important foods or field crops were selected not only on the basis of their ease of culture, but also for their intrinsic food value. Starch comprises the main digestible or nutrient material of most plant foods, for example the cereal grains (rice, maize, wheat), various tubers (potatoes, sweet potatoes, yams, cassava), fruits (bananas and plantains), and other seeds (especially the pulses such as beans, peas and lentils).

Many of the starchy foods require processing to optimise their useful

* Unsupported statements, except those of a propositional or speculative nature, are part of the general body of starch knowledge. For detailed references to the original literature, consult one of the compendia or references cited in Section 6.14. The author wishes to acknowledge the valued discussions and personal communications in advance of publication of his colleagues in starch science, many of whom have contributed the ideals presented herein.

An enormous amount of information is contained in the patent literature. Most of this information relates to large-scale industrial practice, but in fact only a very small proportion of the patents represent actual industrial use, and no attempt has been made to cover this literature.

properties. Corn and sorghum have a tough pericarp which can be removed after soaking, and the softened endosperm can be ground into flour for various culinary purposes (tortillas, etc.) or for fermentation. Barley becomes much more fermentable by malting (germination). Wheat becomes much more digestible by making it into flour and baking it in bread. The starch-bearing cassava root contains a poisonous glycoside, which is readily removed by pounding the roots into a pulp and washing the starch with water (see below). Other fleshy tubers such as the potato and sweet potato tend to spoil on storage, but the starch can be obtained readily by pulping the potatoes, removing the coarse fibre, allowing the starch to settle and finally drying it in the sun.

Using such primitive methods, starch has been prepared since prehistoric times. In many parts of the world, particularly in the tropics, starch is still produced essentially in this way. However, with the industrial revolution, it was possible to mechanise these processes. Starch is now manufactured on a very large scale, and has become a cheap, versatile raw material for use in food products, paper and textile manufacture and in countless other applications.

Development of the sciences made possible the chemical and biochemical characterisation of starch. In 1811 Kirchoff discovered that starch could be hydrolysed by dilute acids to give glucose. This single discovery is the basis for one of the major industrial applications of starch. Many of the early scientific studies of starch were in the botanical laboratories of universities. With the application of technology to brewing, particularly in Europe and the British Isles, there arose laboratories for the analysis of malt and grain. These industrial laboratories furnished much of our early scientific knowledge of starch. This tradition has expanded enormously so that now starch is being studied in universities, government research institutes and industrial laboratories around the world.

Discovery of maltose (malt sugar) as a crystalline 'starch' sugar, distinct from glucose, was announced by Kirchoff in 1815. This sugar is formed by the action of cereal and malt amylases on starch. Though not as sweet as cane sugar, and more difficult to purify and crystallise, it could be manufactured cheaply on a very large scale, if there existed a suitable market for it.

Many of the food and industrial uses of starch depend on its colloidal properties. When starch is heated with water, the granules swell and burst, giving a viscous paste or gel. Gelatinised starch is then used as a thickener, as an adhesive and for the sizing of textiles and paper. Scores of chemical or physical treatments are used to improve the pasting properties of starch and to increase its usefulness. Many of these modifications were discovered accidentally; for example, the conversion of starch into dextrin by roasting (discovered after a fire in a textile factory) or the oxidation of starch by chlorine (discovered in an attempt to 'bleach' or 'chlorinate' starch by using chlorine water or hypochlorite).

Although Fischer had an excellent understanding of glucose and other low molecular weight sugars by the beginning of the 20th century, a real understanding of the chemical nature of starch was impossible until the development of the concept of high polymers in the 1920s and 1930s. Thus the natural materials starch, cellulose and rubber, along with the synthetic

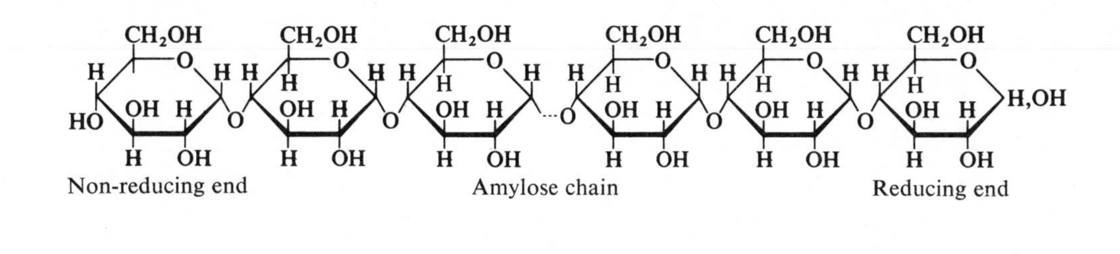


**Figure 6.1** Amylose: structural and symbolic representation. The actual size of native amylose may range from a few hundred to several thousand glucose units

polymer nylon, were among the first materials to be recognised as giant molecules.

Prior to 1930, it was supposed by the organic chemists that the high molecular weight 'particles' (in reality, molecules) of polysaccharides, proteins, etc., were held together by mysterious 'association forces.' The true nature of polymers, or macromolecules, was first realised by physical chemists through the study of the viscosity of polymers and the x-ray diffraction of polymer fibres. The principle emerged that the natural high polymers consist of monomers, containing at least two points of reaction, linked together in an almost endless chain. Although Fischer had synthesised a pentadecapeptide as early as 1900, the discovery of nylon by Carrothers in 1938 was the first practical application of this simple principle.

The mode of linkage of glucose units into starch chains was established in 1928 by methylation analysis (Figure 6.1)[1]. This technique showed that 96% of the glucose units in starch are linked 1→4, and that approximately 4% of the units are linked only at the 1-position and constitute end groups. Haworth initially interpreted these results as indicating that starch consists of unit chains of approximately 25 glucose units, held together by 'associative forces.' However, in reality the methylation analysis was the first chemical evidence for branching in starch (Figure 6.2), a concept incorporated by Haworth into his famous, but obsolete, laminated formula (Figure 6.3)[2]. Staudinger, on the basis of viscosity measurements, interpreted the methylation and enzymic results in terms of a comb formula (Figure 6.4)[3]. In retrospect, it is obvious that the Haworth and Staudinger formulae are essentially identical if the mode of linkage at the branching point is unspecified (Figure 6.5).

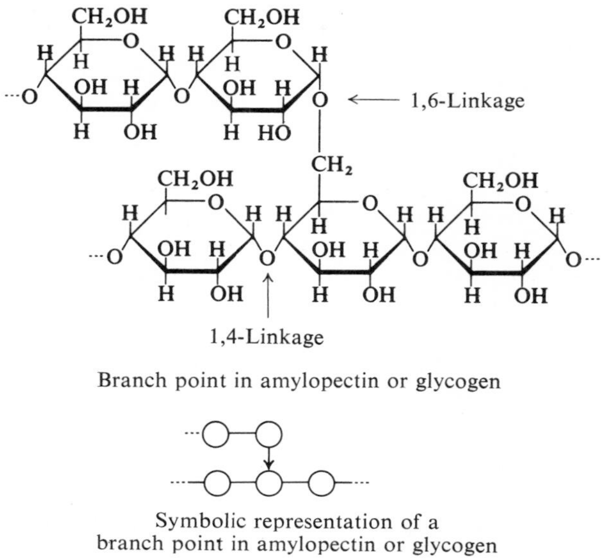

1,6-Linkage

1,4-Linkage

Branch point in amylopectin or glycogen

Symbolic representation of a branch point in amylopectin or glycogen

**Figure 6.2** Branching point in starch: structural formula and symbolic representation

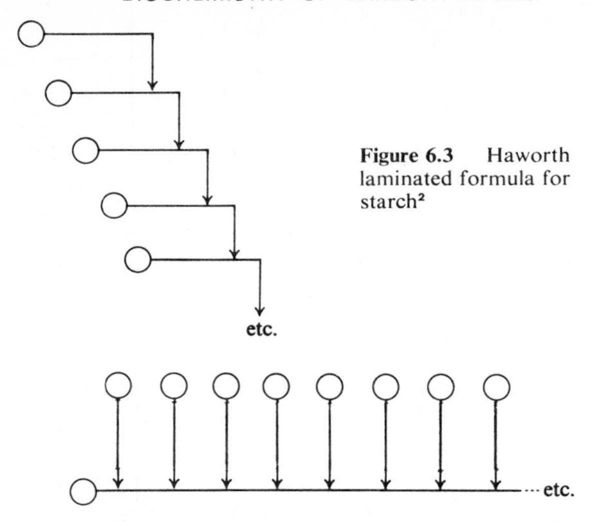


**Figure 6.3** Haworth laminated formula for starch[2]

**Figure 6.4** Staudinger comb formula for starch[3]

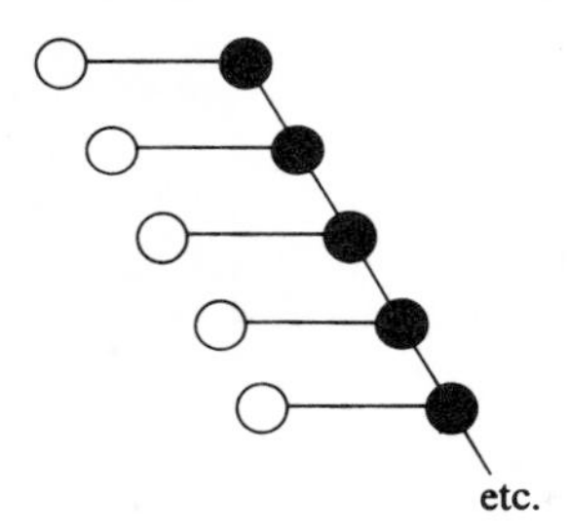


**Figure 6.5** Scheme showing the similarity of the Haworth and Staudinger formulae, if the mode of linkage at the *filled circles* is not specified

Nearly simultaneously, K. H. Meyer in Switzerland (1940)[4] and T. J Schoch in the USA (1941)[5] worked out practical laboratory procedures fo separating starch into its component fractions, amylose and amylopectin Structural analysis of the fractions showed that amylose is a linear polymer whereas amylopectin is branched. These findings constitute the sound basi for all modern starch chemistry and biochemistry. On the basis of methyla tion analysis and enzymic studies, Meyer proposed the tree formula[4] depicte in Figure 6.6.

The biochemistry of starch originated on the one hand in the brewin laboratories through a study of malt amylase action, and on the other han in the physiological laboratories of medical schools through the study o starch digestion by salivary and pancreatic amylases. These studies indicate maltose as a chief product of the enzymic breakdown of starch. During th middle half of the 20th century, many additional amylases have been dis covered from plant and microbial sources, and some of these enzymes hav assumed high industrial importance.

The enzymic polymerisation of glucose, by the enzyme phosphorylas (EC 2.4.1.1) acting on α-glucose 1-phosphate, was discovered in 1939 b Cori *et al.*[6] (mammalian muscle) and by Hanes[7] in 1940 (peas and potatoes)

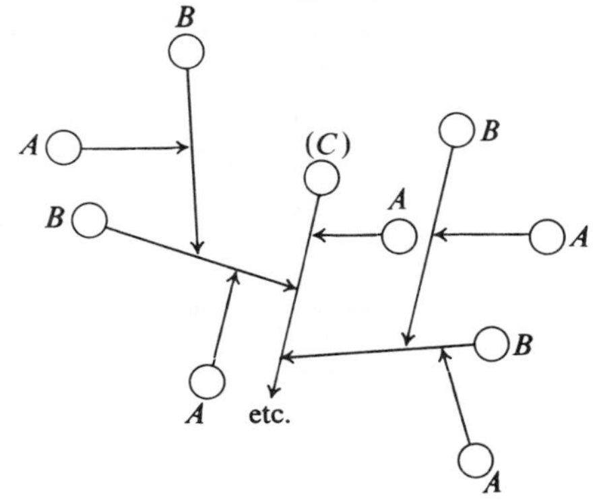


**Figure 6.6** Meyer tree formula for amylopectin[4]. The chain labelled '(*C*)' would be a *C* chain only if it terminates in a reducing end group

Originally it was thought that phosphorylase is the enzyme which synthesises starch in Nature, and in fact it is possible to synthesise a branched amylopectin-like polymer by the joint action of phosphorylase and branching enzyme[8] ('Q-enzyme') (EC 2.4.1.18). However, most starch scientists now think that natural starch is produced by a starch synthase (EC 2.4.1.21) discovered in 1961 by Leloir[9], which uses ADPG as a glucose donor. The biological role of phosphorylase, found universally in starch-synthesising plants, remains an enigma.

## 6.2 OCCURRENCE AND PREPARATION

Starch occurs in most green plants, in practically every type of tissue: leaves, stems, roots, seeds, fruits, even pollen grains. The most important starches of commerce are from seeds of cereal grains (maize, rice, wheat, etc.) and from underground storage organs (potato, sweet potato, cassava).

### 6.2.1 Leaf starch

Starch is produced in the chloroplasts of the majority of higher plants, where it constitutes a temporary storage form of photosynthetic carbohydrates[10]. The amount of starch increases during active photosynthesis, and decreases as the starch is enzymically reconverted into sugars for metabolism or translocation to other parts of the plant[11]. Leaf starch is formed as tiny granules, only about 1 μm in diameter, which are scarcely visible by optical microscopy. Starch in green leaves is most readily visualised by treating the leaves with ethanol to extract the pigment, then staining the starch with dilute iodine solution. Electron microscopy of thin sections through chloroplasts often shows one or more starch granules imbedded between the chlorophyll-containing membrane layers.

Green leaves, especially those of tobacco or bean, are often used for the preparation of radioactive starch by photosynthesis in $^{14}CO_2$. After extracting the leaf pigments with boiling ethanol, the starch is extracted with hot water or alkali and precipitated by adding iodine–potassium iodide. Leaf starch can also be obtained by homogenising the leaf tissue, straining through cloth to remove fibres and centrifuging out the starch[12]. Because the method of preparation is somewhat tedious, and the yield of starch is small, there have been few studies on its constitution.

Starch can be synthesised *in vitro* in isolated chloroplasts, but such starch has received very little attention. Starch granules are also found in green algae.

### 6.2.2 Stem, root and tuber starches

Storage of starch in the vegetative tissues of plants is used by the plant to permit a rapid return of activity, after a period of quiescence, or to permit rapid development of fruit or seed. Many of the practical starches of commerce represent such vegetative storage. Starch granules deposited in storage tissues are not subject to the daily increase and decrease characteristic of leaf starch, and they may attain a much larger size. For example, with potato, arrowroot, or canna, it is not unusual for the length of some of the granules to exceed 100 μm.

Starch manufacture from fleshy vegetative tissues consists primarily of breaking the cells and separating the starch from the cellular debris. The plant tissues are ruptured by crushing, grating, rasping, etc. The starch is separated from the fibrous material by mixing with water and straining, and the resulting starch 'milk' is purified by allowing the starch to settle out. To obtain a pure product the starch needs to be washed several times. In modern starch factories, potatoes and sweet potatoes are mechanically processed using large rasps and graters to break the plant tissues, and centrifuges and filters to purify and de-water the starch. But, especially in the tropics, much starch is still produced without use of sophisticated machinery.

Manufacture of starch from potatoes, sweet potatoes, cassava and various edible yams was originally possible when there was a surplus of these vegetables not needed as such for food or fodder. Later, cultivated varieties of potatoes and sweet potatoes were developed specifically for use in starch production. During growth of these underground storage organs the plant cells become very full of starch granules. Large yields of potatoes are possible in cool, moist climates, and, particularly in northern Europe, potatoes are an important source of industrial starch. Sweet potatoes are well adapted to a warmer climate and longer growing season. Sweet potatoes are an important starch source in Japan, for example.

The edible or sweet cassava (manioc, manihot, tapioca) has a large starchy root. However, the bitter cassava is more important for starch production. The roots attain a size of 20 cm or more in diameter and 1 m in length, and one plant may bear several roots. Bitter cassava contains the cyanogenetic glycoside linamarin[13], which must be removed or destroyed during starch processing for food use.

The trunk of the sago palm becomes filled with a starch-bearing pith during its growth and maturation and thus constitutes a rich source of starch. After 15 years it sends up an inflorescence and the starch is rapidly mobilised, eventually leaving only a hollow trunk. For starch production

It is necessary to cut down the trees before the inflorescence appears, usually after about 6 years of growth. The trunk is split open, the starchy pith mixed with water is run through sieves, and the starch is settled out, washed and dried. Only the most primitive equipment is required.

During development of the pineapple plant, a large amount of starch is stored in the fleshy stem (very little in the developing fruit). Then as the fruit matures there is a massive conversion of the starch into sugar, which is translocated to the ripening fruit.

Small starch granules are often deposited in stems or shoots of rapidly growing plants, for example in bean seedlings or potato sprouts.

### 6.2.3 Seed and fruit starches

Seeds of many wild plants contain starch as a reserve nutrient (energy source) for the eventual germination of the seed. Some seeds (for example, acorns, chestnuts, wild cereals and grasses) have far more starch than could reasonably be used for the germination process. It seems likely that the surplus starch has made these seeds attractive to various birds and mammals and thereby played a part in the evolutionary and survival process. The selection and domestication of plants by man has also been influenced in large part by the starch content of the seeds. Thus the cereals rice, wheat, maize, etc. contain 70% or more by weight of starch, which is the principal nutrient. During domestication of the cereals, and more recently through scientific plant breeding and genetic manipulation, new races of cereals have been developed which are highly superior to their wild ancestors not only in agronomic characteristics but also in nutritive value. Mutants of maize, rice, sorghum, barley, etc. are now cultivated for their special starch characteristics. For example, the waxy* or glutinous cereals (e.g. waxy maize, waxy or glutinous rice) contain starch with no amylose[14, 15]. Similarly, plant breeders have developed varieties of maize, 'amylomaize', with 70% or more amylose in the starch[16].

Most of the cereals grown in the world are used directly or after processing for human food or animal feed. Only a few per cent is channelled into industrial starch manufacture and processing. Yet this minor fraction forms the basis for a giant starch manufacturing and processing industry. In the United States alone, of the $5.6 \times 10^9$ bushels ($1.4 \times 10^8$ metric tons) of maize harvested in 1973, approximately $1.8 \times 10^8$ bushels (over 4 million metric tons) were processed for starch and starch syrup production.

Starch has been produced from cereals for untold centuries. Pliny and Cato describe starch preparation by soaking the ground cereal in water in a cloth bag, and kneading to permit the starch to come through the cloth, leaving the sticky gluten and fibrous material within the bag. Modern starch manufacture from maize, wheat, sorghum, rice, etc. involves the steeping of the grain to soften the kernel, cracking to separate the germ and hull from the starch-bearing endosperm, grinding the endosperm to break up the starch–gluten association, and separating the starch from the

* The term 'waxy' relates to the translucent appearance of the endosperm; glutinous refers to the sticky character of the cooked starch or grain, not to the protein gluten.

gluten by screening, settling or centrifugation. The process is almost entirely a physical separation. The only chemical treatment is in the steeping, where $SO_2$ (or bisulphite) is added, in part to inhibit microbial growth, but also to help break up the starch–protein association. After washing the starch it can be dried or used wet in innumerable ways.

Starch is produced in a similar way from the dry seeds of pulses (peas, beans, lentils, grams).

Starch occurs as a component of many fruits, for example, unripe apples and other pomaceous fruits[17], mangoes, green tomatoes and potato fruit (seed ball). The major high starch fruits of which the author is aware are the various bananas and plantains (genus *Musa*). Starch can be readily prepared in quantity from green bananas or plantains, but such starch has never achieved any commercial significance.

## 6.3 STARCH FRACTIONS

Starch consists of two distinct polysaccharide types: amylose, ideally a linear polymer of α-D-glucopyranose units linked 1→4; and amylopectin, a branched polymer containing short amylose chains (18–25 glucose units) linked by α-1,6-branching points. Natural starch is a mixture of various proportions of amylose and amylopectin, each having a wide range of molecular sizes. Particularly in the root and tuber starches, starch is slightly esterified at position-6 with phosphate[18]. For example, potato starch contains 0.04–0.10% P as ester-linked phosphate, and this anionic group markedly affects the colloidal properties of the starch. The phosphate is linked mainly or exclusively to the amylopectin, which Schoch has separated into phosphate-rich and phosphate-poor fractions[15]. Many starches, particularly those derived from the cereal grains, contain up to about 0.6% lipids which complex with the amylose component and thereby substantially alter the starch colloidal behaviour[19]. Minor amounts of proteins, nucleotides, and various inorganic substances usually found within the starch granule are generally regarded as impurities since they are non-carbohydrate and not covalently linked to the polysaccharide. Starch fractionation is now understood to mean the separation of the various component polysaccharides, primarily amylose and amylopectin.

In a series of papers which revolutionised starch science, K. H. Meyer demonstrated separation of starch into two fractions which differed chemically[20]. Fractionation was achieved by partly swelling a dilute suspension of starch granules in hot water, so that a part of the linear component was gradually leached into solution, leaving the branched amylopectin in the undissolved residue. The dissolved amylose was shown to be linear by methylation analysis, which gave primarily 2,3,6-trimethyl glucose and only 0.4% of 2,3,4,6-tetramethyl glucose (from the non-reducing end group). Both the yield of 2,3,4,6-tetramethyl glucose and the reducing value of the polymer were compatible with a linear chain of approximately 250 glucose units. The insoluble material gave a much higher yield of tetramethyl glucose (*ca.* 4–5%) and had a very low reducing value, as appropriate for a high molecular weight branched polysaccharide.

A second critical indication of the linearity of the amylose fraction was its essentially quantitative conversion by β-amylase into maltose. By contrast, the amylopectin fraction gave only about 50% conversion, owing to the inability of β-amylase to by-pass branch points.

At the same time Meyer was studying the chemical and biochemical differences of his fractions, Schoch[5] in the U.S.A. was examining the colloidal interaction between starch and various aliphatic alcohols, e.g. butanol. He discovered that if an autoclaved dilute starch sol is slowly cooled in the presence of excess butanol, part of the starch crystallises in the form of beautiful needles or rosettes. These rosettes were sufficiently large that they could be isolated by centrifugation; the remaining non-crystallising polysaccharide could be recovered by adding a de-watering agent such as methanol. Schoch showed that, in contrast to the non-crystallising fraction, the crystalline fraction had high alkali lability (an index of the number of reducing end groups), remarkable gel-forming ability, and an exaggerated tendency to retrograde. Schoch generously made these fractions available to other researchers, and work at Iowa State University showed that the crystalline fraction gave a strong, unique x-ray diffraction pattern characteristic of amylose helices[21]. Moreover, dilute solutions of the butanol complex had a high iodine-binding capacity, as measured by potentiometric titration[14]. The various physical, chemical and biochemical differences indicated clearly that the crystalline fraction was linear* (amylose) and that the non-crystalline fraction was branched (amylopectin).

Study of starch fractions isolated from natural maize or potato starch received further impetus by the availability of waxy-maize starch, from a maize mutant discovered in China, which contains no amylose[22]. Waxy-maize starch has been produced commercially since the 1940s and is now used widely because of its colloidal texture and paste stability. It can be regarded as a type of amylopectin, although it probably differs in average chain length, molecular size and fine structure from the amylopectin fraction of ordinary maize starch.

A second factor of significance in the study of starch fractions was the development of synthetic amylose[7], obtained by action of phosphorylase on glucose 1-phosphate plus a suitable primer (see Section 6.12.1 on phosphorylase). By enzymic synthesis it is possible to prepare in quantity amyloses of any desired chain length, with a purity dependent only on the quality of the enzyme and substrates[23].

Starch can be fractionated very successfully on a large scale by precipitation from hot concentrated salt solutions[24]. Whole starch is dissolved in hot, strong, aqueous magnesium sulphate. On cooling, the amylose component first precipitates. This component can be removed by centrifugation and washed with water to remove salt. The amylopectin component is then removed by further cooling the solution.

Various specific complex-forming reagents have been suggested for the precipitation of amylose. Schoch initially used butanol, but later recommended n-amyl alcohol or a commercial mixture of amyl alcohols ('Pentasol')[25].

* The term 'linear' is used in a topological sense, as opposed to 'branched'. The actual three-dimensional conformation in the butanol complex is helical, and the conformation in solution is probably a loose, flexible helical coil.

Nitroparaffins, cyclic alcohols and phenols have also been suggested. In the author's experience, the cleanest fractions have been obtained using hydrophobic complexing agents, especially 1,1,2,2-tetrachloroethane or cyclohexane[26].

Fractionation of high amylose starch requires special methods. This starch cannot be dissolved in water except under high pressure at a very high temperature (~150 °C). If the starch is dissolved by using alkali, it immediately retrogrades on neutralisation. The problem is complicated by the unusual composition of its fractions[27]. The amylose fraction consists of both a normal high molecular weight component and a unique low molecular weight component. Moreover, the 'amylopectin' fraction either has unusually long outer chains, or is contaminated with low molecular weight amylose. One method of fractionation involves the initial hot water extraction of low molecular weight amylose. The residual starch is then dissolved in DMSO, diluted with water and fractionated by using butanol or other specific amylose precipitant. It has also been recommended to disaggregate the starch in DMSO and precipitate it with ethanol before attempting fractionation.

The low molecular weight amylose can be separated from the amylopectin by careful fractional precipitation using iodine–potassium iodide[27].

Starches, particularly the cereal starches, which contain fatty materials are more difficult to disperse and fractionate. It is common laboratory practice[19] to remove such fatty material, prior to fractionation, by repeated extraction with boiling 85% methanol or with a methanol–DMSO mixture (40:60, by volume).

### 6.3.1 Subfractionation

Amylose sub-fractions have been obtained in various ways. The classical method of linear polymer fractionation is to add a slight excess of a poor solvent to a solution of the polymer in a good solvent and to separate the phases[28]. Usually the high molecular weight component or fraction is in the polymer-rich phase. By using exceedingly dilute polymer solutions, and only a slight excess of poor solvent, it is possible in principle to achieve an excellent fractionation. The best solvents for unsubstituted amylose appear to be DMSO, DMF, chloral hydrate, etc. Suitable non-solvents are the lower aliphatic alcohols, particularly methanol or ethanol.

Amylose can be successfully subfractionated by first converting it into a derivative, such as the triacetate or tricarbanilate[29].

An interesting amylose fraction, so-called 'crystalline amylose', was obtained by Kerr by using the Meyer hot-water extraction method, followed by the Schoch butanol-complex precipitation[30]. This amylose component is of relatively low molecular weight (DP ≈ 250) and it is highly linear as judged by its high iodine binding capacity (*ca.* 200 mg $I_2$ $g^{-1}$ polysaccharide) and by its complete conversion into maltose by β-amylase.

Amylose subfractionation has been attempted by adding increments of a specific complex-forming precipitant (e.g. butanol or iodine[31]). In general, results by this method have not been as useful as by adding a non-solvent to a solution of the polymer in a good solvent.

## 6.4 CHEMICAL STRUCTURE OF STARCH

### 6.4.1 Amylose

Ideally, and by definition, amylose consists of a long unbranched chain of 1,4-linked α-D-glucopyranose units. The reducing end is either free, or readily removed by alkali. Crude β-amylase, containing a trace of α-amylase ('Z-enzyme'), converts natural amylose quantitatively into maltose[32]. However, with pure β-amylase the limit is usually *ca.* 90% maltose, indicating a minor degree of branching. Although other structural irregularities would interfere with complete conversion by β-amylase, the β-amylase limits of carefully fractionated amylose can be raised from *ca.* 90% to 100% by the joint action of β-amylase and the debranching enzyme pullulanase[33].

### 6.4.2 Amylopectin

Amylopectin consists of short amylose chains (DP in the range 12–50 or more glucose units, with an average of *ca.* 20) linked into a branched structure. There is no single definite structure, as with the globular proteins or the nucleic acids. Rather, the structure of amylopectin can only be described in the statistics of its structural features and details[34]. The component chains of amylopectin are conveniently divided into three categories: the *A*-chains, short amylose chains unsubstituted except at the reducing end; *B*-chains, which are substituted at one or more C-6 OH groups by *A*-chains or other *B*-chains, and which are also substituted at the reducing end; and *C*-chains, which are substituted at one or more C-6 OH groups, but unsubstituted at the reducing end, there being only one such *C*-chain per molecule. As with amylose, it is not known with certainty whether the reducing end of amylopectin is free or substituted.

#### *6.4.2.1 Average chain length by methylation*

The average chain length of amylopectin was originally measured by determining the ratio of the total number of glucose units to the number of non-reducing termini. Non-reducing glucose residues give 2,3,4,6-tetramethyl glucose on methylation, whereas the chain-forming units give 2,3,6-trimethyl glucose and the branch points give 2,3-dimethyl glucose[35]. Tetramethyl glucose can be accurately determined by gas or paper chromatography[36]. In principle, the molecular proportion of dimethyl glucose should be stoichiometrically equal to the tetramethyl glucose (except for the extra tetramethyl glucose originating from the *C*-chain). But in case of a slightly incomplete methylation, calculations based on the proportion of dimethyl glucose would give an unreliable average chain length.

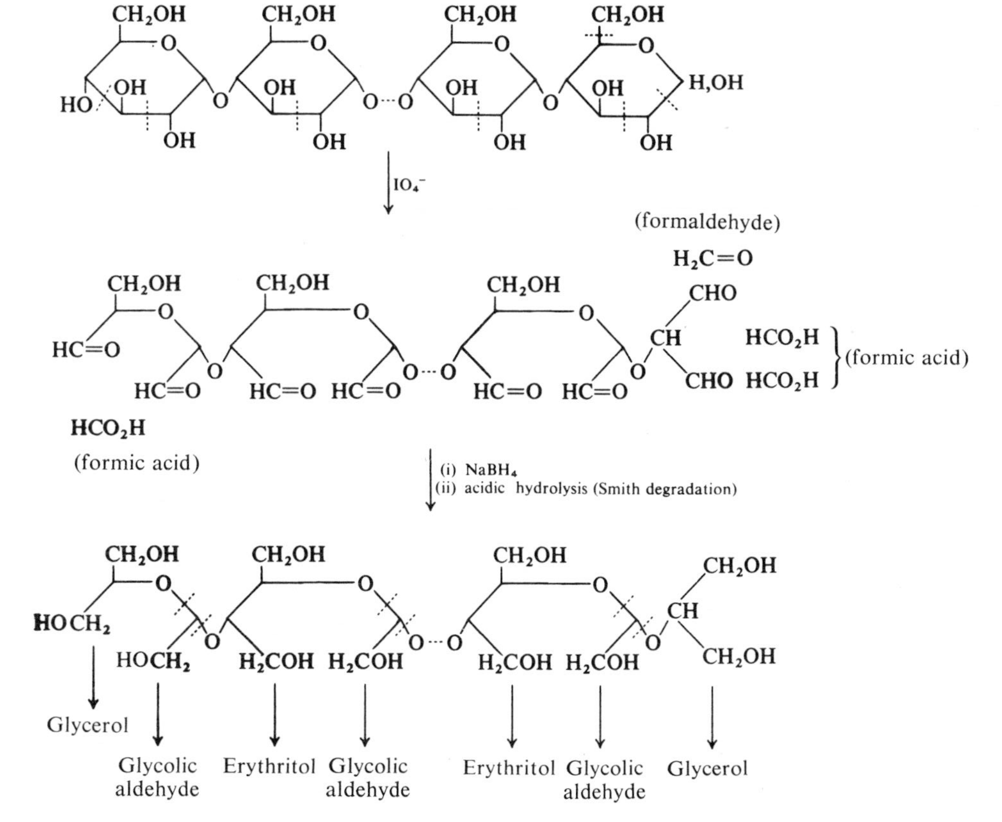


**Figure 6.7** Periodate oxidation of amylose chain elements and end-groups; Smith degradation

### *6.4.2.2 End-group analysis by periodate oxidation*[37]

Periodate oxidation is also an extremely useful method for investigating the structure of amylopectin. The basic reaction involves glycol cleavage such that each non-reducing terminal glucose unit generates one molecule of formic acid (Figure 6.7). The single reducing glucose unit gives two molecules of formic acid and one molecule of formaldehyde, but this contribution from the reducing glucose unit in an enormous molecule of amylopectin is usually negligible. The chain-forming internal glucose units and the branch points are also oxidised, but produce no formic acid or formaldehyde. The formic acid is analysed by titration.

In a further refinement, known as the Smith degradation[38], the carbonyl groups produced by periodate oxidation are reduced (e.g. by sodium borohydride). The reduced polymer is then acid-hydrolysed. Characteristic products are glycerol (one molecule from each terminal non-reducing glucose unit) and erythritol (one molecule from each interior glucose unit), which are analysed by paper or gas chromatography.

### *6.4.2.3 Structure at the reducing end by periodate over-oxidation*

Periodate over-oxidation (Figure 6.8) involves more extensive periodate treatment. Specifically, oxidation of the reducing glucose unit produces a

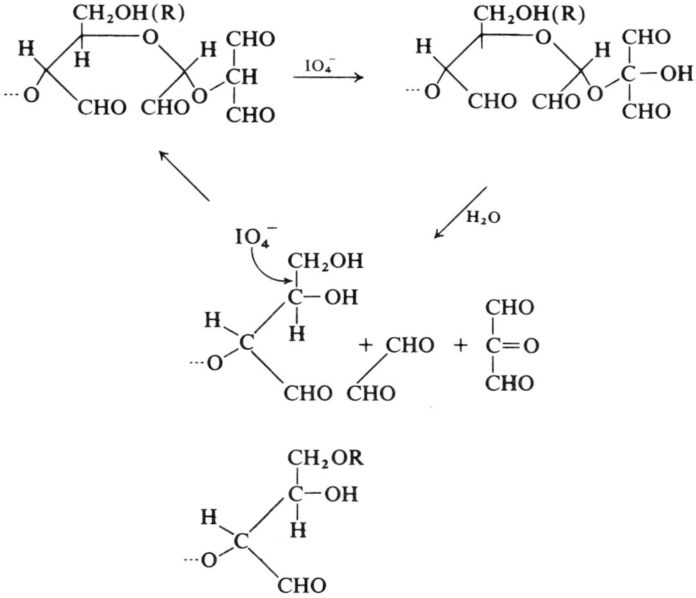

(stable to periodate)

**Figure 6.8** Periodate over-oxidation. After 'normal' oxidation, over-oxidation begins at the reducing end of a starch chain and is repeated until further over-oxidation is blocked by substitution on O-6

substituted malonic dialdehyde, which is only slowly oxidised under the usual conditions of the periodate analysis. However, with more rigorous treatment, oxidation occurs with concomitant cleavage of the adjacent acetal linkage. This results in further susceptibility to periodate oxidation, so that eventually the *C*-chain is destroyed, up to the first branch point. *A*- and *B*-chains attached to the *C*-chain do not over-oxidise, since to form the crucial malonic dialdehyde it is essential to start with a free reducing group. The reaction is followed by measuring the formaldehyde produced.

### 6.4.2.4 *Identification of the branch linkage*

'Fine structure' of amylopectin was originally intended to describe the mode of branching. Identification of the α-1,6-linkage was first performed by methylation analysis (isolation of 2,3-dimethyl glucose)[35]. Later Wolfrom identified isomaltose (α-D-glucopyranosyl-1→6-D-glucose) and panose (α-D-glucopyranosyl-1→6-α-D-glucopyranosyl-1→4-D-glucose) as products of acidic hydrolysis[40].

Isolation after hydrolysis or acetolysis of amylopectin of very small amounts of nigerose (α-D-glucopyranosyl-1→3-D-glucose) suggested the possibility of α-1,3-links in the amylopectin structure[41]. However, a careful re-examination has shown that such α-1,3-links were probably artifacts produced by acid-catalysed transglycosylation[42]. At present it is generally agreed that amylopectin contains only α-1,4- and α-1,6-links.

### 6.4.2.5 *Outer and inner chain lengths*

The average lengths of outer chains have been established by enzymatic analysis, primarily by the action of β-amylase. It is clear that the branch linkage imposes a total barrier to β-amylase. After extensive β-amylase action, the outer *A*-chains are either two or three glucose units in length (depending on whether the original chain contains an even or odd number of glucose units), and the outer *B*- or *C*-chains are also two or three glucose units long (Figure 6.9). (See Section 6.10). If β-amylase action is somewhat

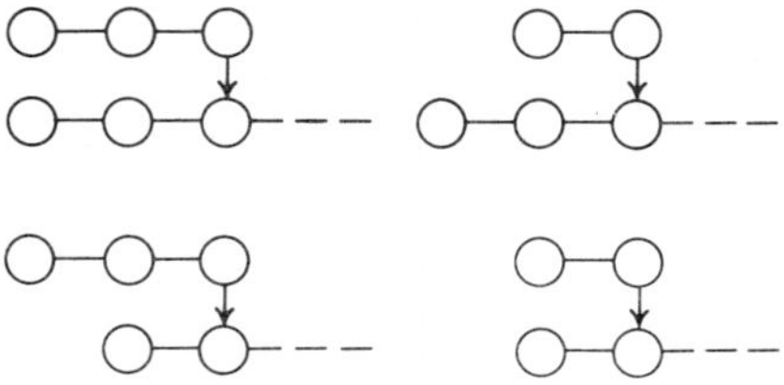

**Figure 6.9** Branching configuration resistant to sweet potato β-amylase

less extensive, fewer 2-unit outer chains are formed, leading to a residual polysaccharide ('limit dextrin') containing 3- and 4-unit outer chains[43].

On this basis the average outer chain length of most amylopectins is *ca.* 12 glucose units. Since the total chain length is *ca.* 20, and each chain on the average is branched just once*, the average inner chain length is *ca.* 8 glucose units. These statistical values are those incorporated into the classical models of amylopectin (Haworth, Staudinger and Meyer models).

### 6.4.2.6 *Tier structure for amylopectin*

The 'tier' concept of amylopectin is the suggestion that all outer chains form a tier which can be removed by an endwise acting enzyme (an exoase such as β-amylase or phosphorylase) until the enzyme is blocked by the outermost branch point. Removal of the outermost branch points exposes another tier of chains to further exoase action until the next branching point is encountered. Successive alternate action by an exoase and a suitable debranching enzyme (e.g. a glucosidase) would eventually completely break down the polysaccharide, and by measuring the yield of products at each stage one can infer details of the original polysaccharide structure. Using this experimental approach, Meyer found that several cycles of enzyme action were required, and that at each cycle approximately one-half of the remaining polysaccharide was converted into low molecular weight products. This led Meyer to propose his famous tree formula. The tier concept has been rigorously confirmed by Larner *et al.*[44].

### 6.4.2.7 *Proportion of A and B chains*

In a randomly rebranched tree formula, it is equally likely that any outer chain might be an *A*-chain or a *B*-chain, and in the absence of any experimental evidence, Meyer assumed this to be the case. An experimental approach involves conversion of amylopectin into its β-amylase limit dextrin. The *A*-chains are now only two or three units in length, whereas the *B*- and *C*-chains are substantially longer. After hydrolysis with a specific debranching enzyme, the yield of $G_2$ plus $G_3$ indicates that the *A*-chains are approximately equal in number to the *B*-chains, as required by the tree formula[45-47].

### 6.4.2.8 *Details of branching*

If the branches in amylopectin are sufficiently isolated or spaced from each other, action of α-amylase converts each branch point into a singly branched oligosaccharide (Figure 6.10). However, clusters or regions of dense branching lead to multiply branched oligosaccharides[48] whose structures reflect the branching pattern of the parent polysaccharide. With glycogen, which has a much higher degree of branching than amylopectin, α-amylase leads to multiply branched oligosaccharides and macrodextrin[49], an α-amylase-resistant cluster of branches with *ca.* one branch per four glucose units. Some glycogens, particularly shellfish glycogen, give yields of 10% or more macrodextrin. However, with amylopectin only insignificant amounts of

* From a theorem of polymer chemistry, for $(n + 1)$ chains there are $n$ branches. Hence as $n$ increases, the ratio $(n + 1)/n$ approaches unity.

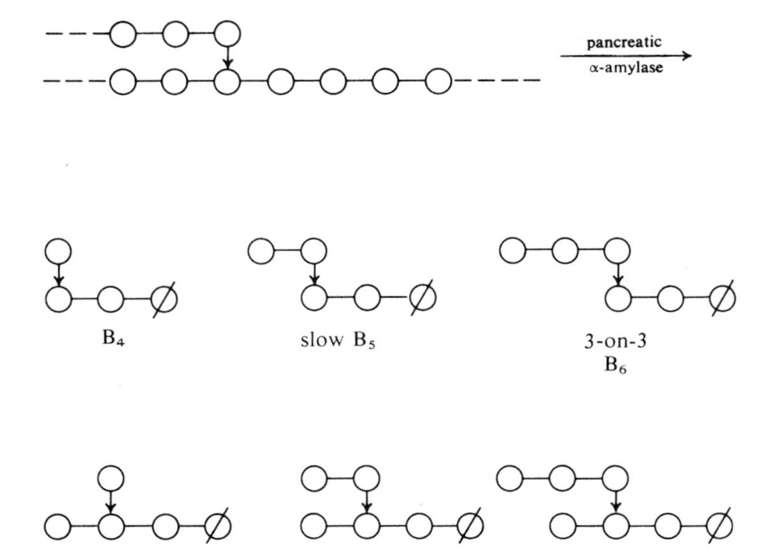


**Figure 6.10** Singly branched oligosaccharides formed by action of pancreatic amylase on isolated branch points in amylopectin

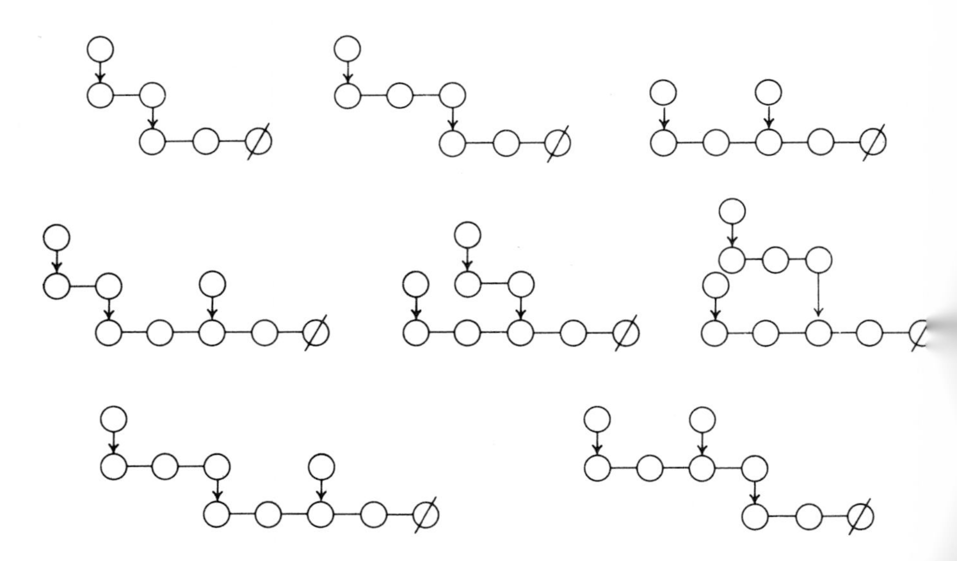

**Figure 6.11** Doubly and triply branched oligosaccharides formed by action of pancreatic amylase on amylopectin

macrodextrin are formed. Approximately 70% of the branches are isolated and are converted into singly branched oligosaccharides. The remaining 30% of the branches are converted into doubly and triply branched oligosaccharides. Structural analysis of these oligosaccharides shows that they have the structures shown in Figure 6.11.

Thus a very significant proportion of the branches are separated by no more than a single glucose unit. Actually, analysis of the specificity of

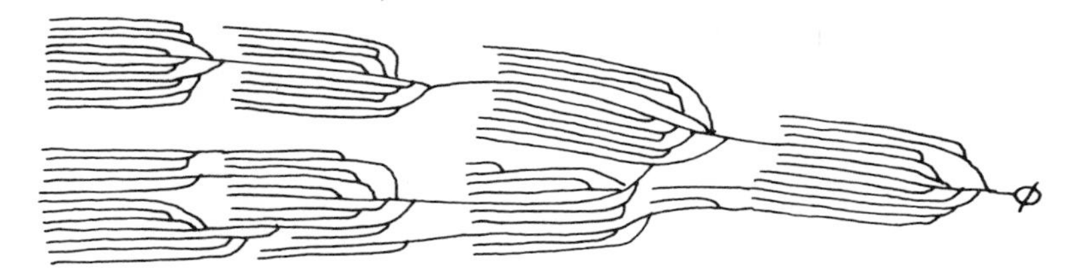

**Figure 6.12** Cluster model of amylopectin[50]. This model is proposed as an alternative to the Meyer 'tree' formula to account for the high viscosity of amylopectin, which requires a very asymmetric structure, and the high crystallinity, which indicates that many of the branches must be able to run parallel to each other

pancreatic α-amylase shows that where the branches are spaced by two or more glucose units, the interbranch regions are cleaved by pancreatic amylase, so that no such configurations would accumulate in the amylase-resistant oligosaccharides[50]. The existence of regions of dense branching in amylopectin has been one piece of experimental evidence leading to the concept of the cluster formula for amylopectin (Figure 6.12)[50].

## 6.5 CHEMICAL REACTIONS OF STARCH

The significant features of starch are the secondary OH groups on positions-2 and -3, the primary OH group on position-6, and the glycosidic linkage. The OH groups undergo a multitude of typical reactions such as oxidation, ether and ester formation. The glycosidic bond is susceptible to hydrolysis, alcoholysis, transglycosylation, chlorinolysis, etc. Many of these reactions are of great industrial significance and form the basis for the commercial chemical modification of starch.

### 6.5.1 Low degree of substitution

Reactions of the OH groups are significant in modifying the solubility and colloidal properties of starch (Figure 6.13). Specifically, substitution of only a few per cent of the OH groups, for example by reaction with acetic anhydride, markedly retards starch retrogradation. A low degree (e.g. D.S. = 0.2) of carboxymethylation by reaction with chloroacetate imparts a negative charge to the starch and greatly enhances solution stability. Similarly, reaction with diethylaminoethyl chloride gives a cationic starch which has a high affinity

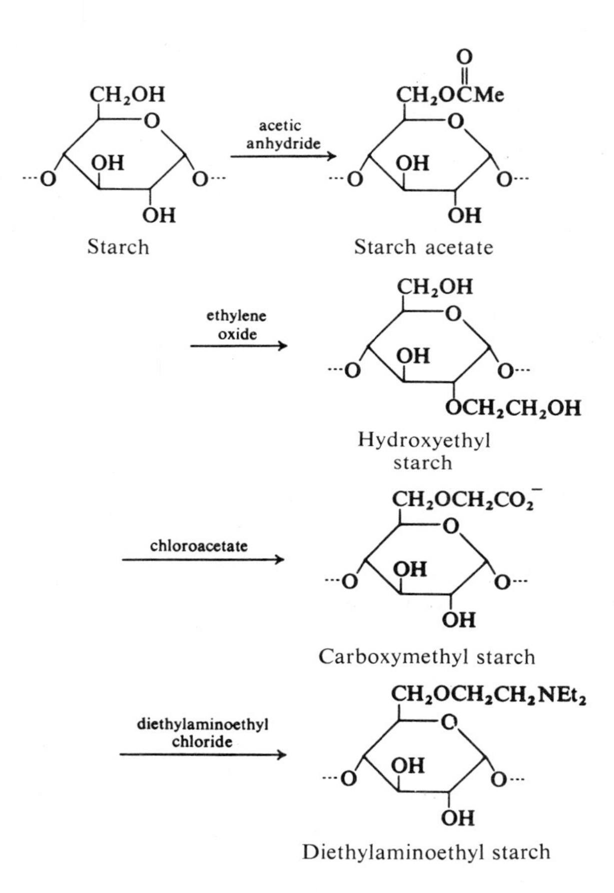


**Figure 6.13** Some examples of low D.S. (degree of substitution) starch derivatives

for cellulose, clays and other negatively charged colloids. Grafting starch with polystyrene gives a polymer of potential utility as a rubber compounding agent. A polyacrylic acid graft polymer has ability to absorb up to 300 times

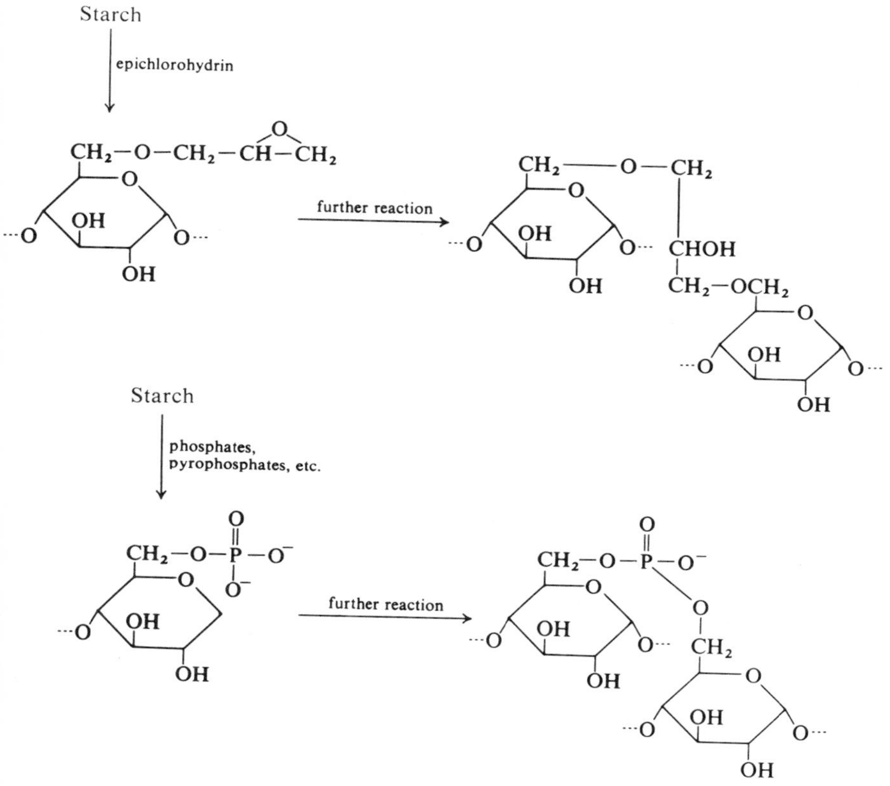

**Figure 6.14** Cross-linking reactions of starch by epichlorohydrin and phosphate

its weight of water, with formation of a gel. Reaction with ethylene oxide or propene oxide gives hydroxyethyl or hydroxypropyl ethers. Low degrees of substitution (e.g. 0.1) give improved water solubility and decreased retrogradation. Reaction with epichlorohydrin gives cross-linked glycerol diethers (Figure 6.14). Less than 0.1% cross-linking substantially restricts swelling, and higher amounts give non-swelling starch granules.

### 6.5.2 Oxidation

Reaction with chlorine or hypochlorite gives oxidised starches of much importance as sizing agents for paper and fabric manufacture. Although the chemistry is not totally understood, it probably involves the initial oxidation of a few secondary OH groups to give keto groups (Figure 6.15). Such

**Figure 6.15** Hypochlorite oxidation of a secondary hydroxyl group in a glucose unit of starch

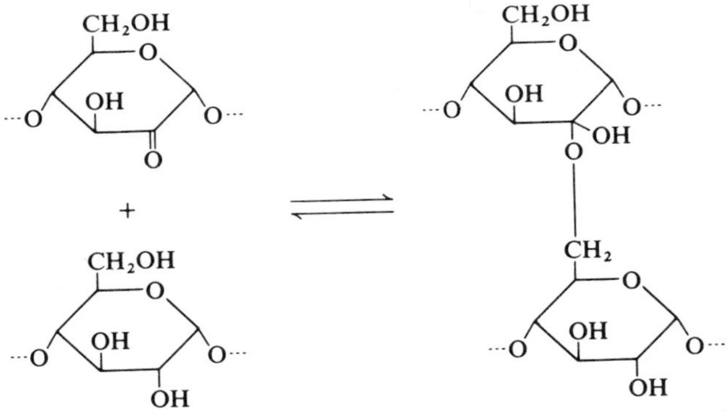

**Figure 6.16** Cross-linking of starch chains through carbonyl groups

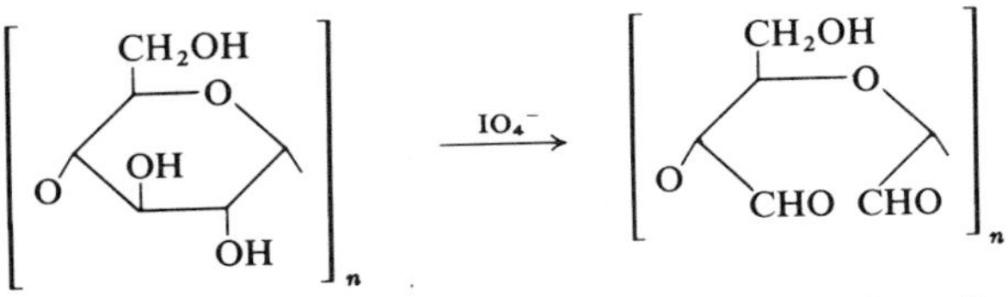

**Figure 6.17** Conversion of starch by periodate into dialdehyde starch

carbonyl groups disrupt the regular structure of starch and also facilitate glycosidic bond cleavage, thus leading to improved dispersibility of oxidised starch. However, as shown in Figure 6.16, carbonyl groups are also capable of forming cross-linking hemiacetals with either starch or cellulose, and this probably improves the adhesive characteristics of oxidised starch.

Periodate-oxidised starch ('dialdehyde starch'; see Sections 6.4.2.2 and 6.4.2.3 and Figure 6.17) shows promise as an adhesive or sizing agent in paper manufacture.

### 6.5.3 High degree of substitution

Starch can be made soluble in organic solvents, such as acetone or chlorinated hydrocarbons, by replacing a sufficient number of hydroxyl groups with non-polar substituents. Starch triacetate, trimethyl starch and starch tricarbanilate, for example, are soluble in organic solvents. Although not widely used for industrial purposes, these derivatives have often been used for research where it is desired to have materials soluble in non-polar organic solvents.

Reaction of starch with alkali and carbon disulphide gives starch xanthate, a derivative of promise for removing heavy metals from water.

### 6.5.4 Acid hydrolysis

The acidic cleavage of glycosidic bonds in starch is of inestimable significance, both from a theoretical and practical point of view. The literature on this one subject alone is enormous, and no attempt will be made to cover it. Basically, the reaction involved (Figure 6.18) is the simple acid-catalysed cleavage of an acetal. The initial attack by the proton is on the glycosidic bond oxygen, giving a protonated acetal. This cleaves by an elimination reaction, giving the $C_4OH$ sugar as the right-hand fragment, and on the left a resonance-stabilised oxycarbonium ion as a fleeting intermediate. Hydration

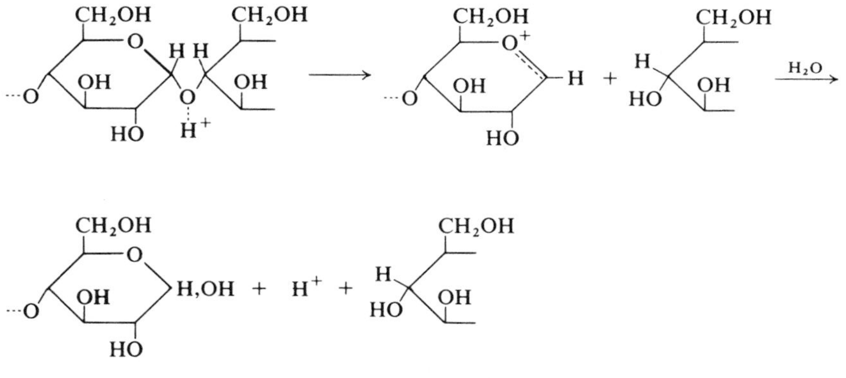

**Figure 6.18** Acid catalysis of glycosidic linkage through an oxy-carbonium ion intermediate

of the oxycarbonium ion gives a mixture of anomeric products which immediately release a proton to become a new free 'reducing end.' The kinetics of the process depend mainly on acid concentration, the kind of acid, and temperature. The activation energy[51] is *ca.* 31 kcal $mol^{-1}$.

To a first approximation, the hydrolysis of dissolved starch (homogeneous reaction) is random, i.e. all of the glycosidic bonds in starch are equally susceptible to cleavage. However the α-1,6-links are more resistant by a factor of 4–10, depending on the conditions used, and non-reducing terminal bonds are hydrolysed about 1.8 times as fast as interior bonds[51]. The exact effect of a branch point has never been adequately analysed.

Hydrolysis of granular or retrograded starch is extremely slow, by comparison with dispersed starch. The glycosidic linkages in starch crystallites are sterically protected. However, the amorphous part of starch granules is readily attacked by aqueous acid (Figure 6.19). Heterogenous acid

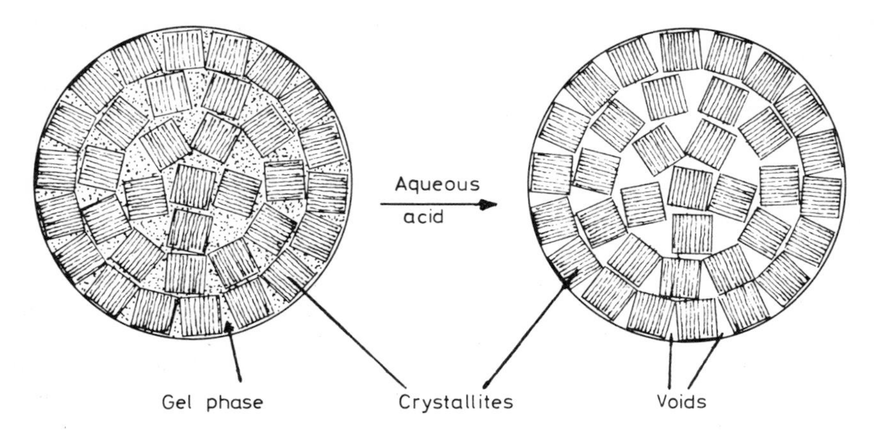


**Figure 6.19** Acid erosion of the gel phase of a starch granule, leaving a crystalline 'amylodextrin' residue (From Kainuma and French[52], by courtesy of Interscience)

hydrolysis is widely used in the starch industry to produce 'soluble starch' or 'thin-boiling starch.' If heterogeneous hydrolysis is continued for a long time, different types of starch erode at different rates, with potato starch and high-amylose maize starch being notably resistant[52]. There is on record a case in which potato starch was left in contact with 10% hydrochloric acid for a period of 6 years, without the starch becoming fully hydrolysed. Granular starches which have been extensively degraded are called 'Nägeli amylodextrins' (see Section 6.8.4).

Reaction with moist acid and heat not only hydrolyses glycosidic bonds, but also gives some transglycosylation (Figure 6.20) with a big increase in branching. The resulting dextrins have greatly improved water solubility, and are suitable for use as adhesives.

## 6.6 COMPLEX FORMATION

Starch, particularly amylose, and the cyclic Schardinger dextrins form inclusion complexes with a variety of *guest* molecules. The complexing agents which have been studied most are iodine and polar organic compounds (e.g. butanol). However, amylose[26] and the Schardinger dextrins[53] form strong, crystalline complexes with such hydrophobic materials as hydrocarbons and halogenated hydrocarbons.

### 6.6.1 Amylose–organic complexes

Amylose forms crystalline, helical complexes which are essentially insoluble in water and which have been used to fractionate starch (see Section 6.3). Complexes with butanol and other primary linear alcohols, linear fatty acids, etc. form helices with six glucose residues per turn[54]. Both electron microscopy

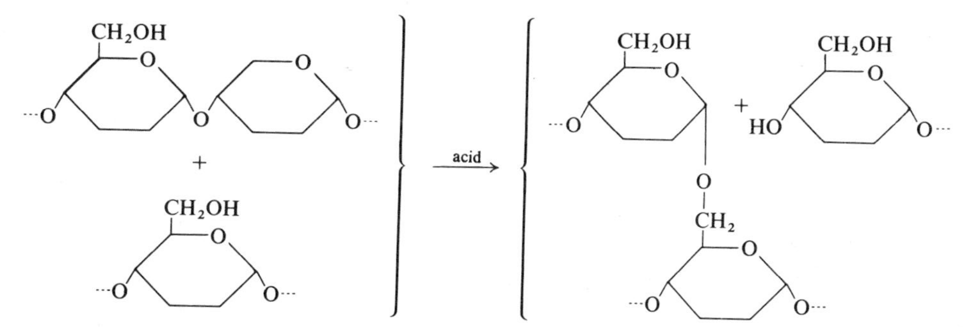


**Figure 6.20** Cross-linking of starch chains by acid-catalysed transglycosylation

and x-ray diffraction indicate that these helices pack together in a pseudo-hexagonal array. Complexes with larger guests such as tertiary alcohols or tetrachloroethane may form seven-unit helices[55].

Examination of models shows that the interior of the helix is lined with C—H groups and thus has a hydrophobic character. The guest is not necessarily in any exact arrangement, but may simply fill the available space. Thus the bonding between amylose and guest is essentially a kind of space-filling hydrophobic bonding. For non-halogenated organic molecules, such as alcohols or fatty acids, amylose complexes contain about 5–10% by weight of the complexing agent.

Amylopectin chains are too short to form good complexes, and because of the branching it is impossible to obtain crystalline complexes.

Amphipathic complexing agents (e.g. monoglycerides or detergents) may induce short helices, but having a hydrophilic 'head' cannot give long helices or crystalline complexes. Such reagents are often used in modifying the rheological and colloidal behaviour of starch[56].

### 6.6.2 Schardinger dextrin complexes

The cyclic α-, β- and γ-Schardinger dextrins (Section 6.13.2.2) form numerous inclusion complexes with iodine (with or without cationic iodides), and with a host of hydrophobic or amphipathic organic complexing agents[53]. These complexes can be regarded as the cyclic analogues of the helical amylose complexes. The iodine complexes have been useful for the identification of Schardinger dextrins and for distinctive colour reactions on paper chromatograms. Because of the differences in geometry of the α-, β- and γ-dextrins, various organic reagents form different, distinctive complexes which can be used to separate dextrin mixtures.

The hydrophobic cavity of the cyclodextrins has provided a useful model of enzyme substrate binding[57].

### 6.6.3 Starch–iodine reaction

The element iodine was discovered in 1811, and soon afterwards it was realised that it gives a blue complex with starch[58]. The starch–iodine reaction has been very useful both in analysis and as a probe into the structure of starch. Iodine reacts with linear components or segments of starch. For a blue iodine colour, a chain length in excess of 40 glucose units is normally required. Shorter chains give red, brown or yellow complexes[59].

The starch–iodine reaction can be quantitated by measuring the free iodine concentration (potentiometric or amperometric method) or by spectrophotometric measurement of the complex. Countless variations have been proposed.

Analysis by some kind of iodine complex formation has been the most valuable tool in differentiating between linear and branched starch materials. Native amylose adsorbs 19–20% of its weight of iodine from a dilute solution of $I_2$ in KI, whereas at low iodine concentrations 'normal' amylopectin adsorbs none, or very little (i.e. *ca.* 0.5%)[31]. Short amylose chains, or amylopectins with long outer chains, also adsorb iodine, but at a higher iodine concentration. Since iodine complex formation is temperature dependent, such short-chain amyloses may be differentiated from normal amylose by running the analysis at two different temperatures[60].

The mechanism of the starch–iodine reaction involves the inclusion of a linear polyiodine-iodide array within an amylose helix (Figure 6.21). Iodide ions are required[61], but the exact nature of the polyiodine array is not known. At low iodide levels, the approximate composition[62] is $I_8^{2-}$.

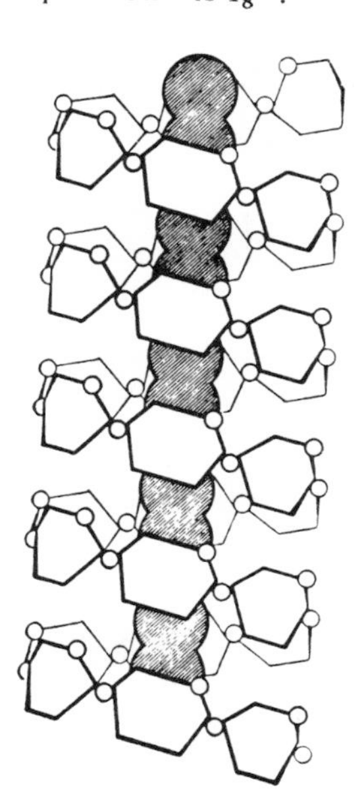

**Figure 6.21** Amylose–iodine complex

**Figure 6.22** Maltohexaose–iodine complex photographed in polarised light

Reaction of iodine with starch oligosaccharides as small as maltotetraose can be measured by potentiometric titration. The binding constants fall in a regular series with increasing chain length except that there is a break between maltohexaose and maltoheptaose, indicative of a 'loop' to 'helix' transition in the configuration of the oligosaccharide chain[63]. As with the Schardinger α- and β-dextrins, maltotetraose and the higher oligosaccharides form beautiful dichroic crystalline complexes[64] with iodine–potassium iodide (Figure 6.22).

## 6.7 PHYSICAL PROPERTIES OF STARCH

Various physical measurements have been used widely in industrial research to gain insight into the performance of starch in practical applications. The rheology (viscosity, flow properties, resistance to shear, thixotropy, etc.) of starch 'cooks' is generally the most significant single factor in determining the suitability of a particular starch or modification. Starch films can be cast from solution in water or other volatile solvent, and measurement of film properties often gives an index of the performance of a starch when used for a coating, sizing or adhesive.

### 6.7.1 Solution behaviour

The theory of viscosity has been worked out on a rigorous mathematical basis for solutions of molecules that are hard, rigid, spherical or ellipsoidal and large compared with the solvent molecules. Molecules, such as those of starch, that are soft, flexible (deformable), hydrated, non-ellipsoidal or having dimensions not much larger than the solvent molecules, require adjustable parameters and empirical modifications of the theory[65].

#### *6.7.1.1 Amylose*

Studies on amylose have used solutions of the native or subfractionated polymer, primarily in aqueous salt, alkali or DMSO, or solutions of amylose triesters in organic solvents. Initially studies were directed to the relationship between intrinsic viscosity of polymers and molecular weight. Viscosity measurements have also been exceedingly useful in detecting degradation of polymers. With modern theory it is possible to draw strong inferences about the shape of the polymer in solution, its interaction with the solvent, the local molecular conformation (secondary structure) and the flexibility of the polymer chain. Complementary information comes from light scattering, ultracentrifugation, nuclear magnetic resonance and other physical techniques. However, viscosity measurements require no complicated equipment and the accuracy and precision are excellent.

The current picture[69] of amylose in dilute aqueous solution (or in salt solutions such as 0.5 M KCl) is that the chain exists as a 'loose' (as opposed to 'tight' or 'compact') segmented, flexible helix (Figure 6.23)[66]. The helix

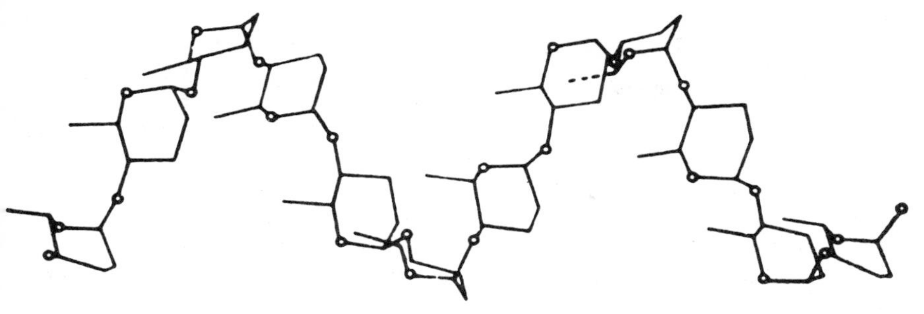

**Figure 6.23** 1,4-α-Glucan conformation. (From Rees and Scott[69], by courtesy of the Chemical Society.)

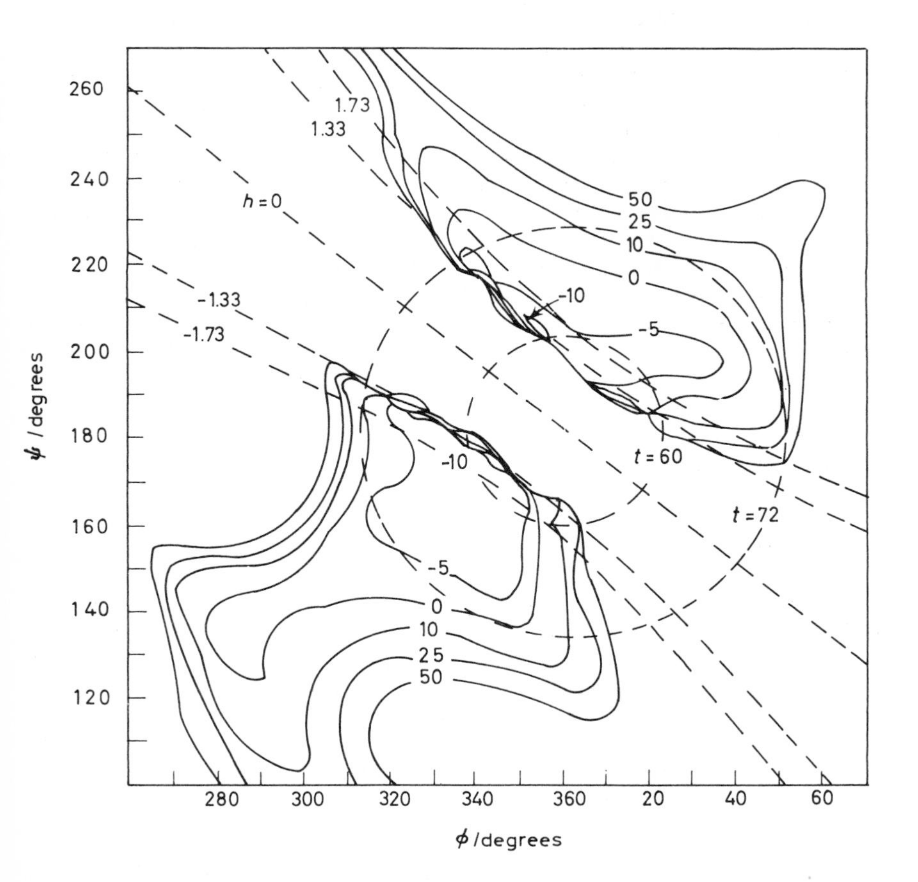


**Figure 6.24** Conformational energy map for helical amylose. Dashed lines show contours of the rise in Å per glucose residue ($h$) and the axial rotation per residue ($t$). Contour levels in kcal mol$^{-1}$ of glucose residues. (From Goebel et al.[70], by courtesy of the American Chemical Society.)

turns are far enough apart that there are few, if any, inter-turn hydrogen bonds, the amylose OH groups being simply hydrogen-bonded to water. Statistical mechanical considerations require that a small proportion of the glucose units exist in high-energy forms, perhaps in the flexible skew or boat forms[67]. To the extent that such conformations occur, there are irregularities in the helix conformation and as a consequence the chain becomes substantially more flexible than it would otherwise be.

Conformation maps of amylose[68-70] show the energy calculated for any particular geometry about a given glycosidic bond (Figure 6.24). Such maps give quantitative information which supplements molecular or computer models[71]. On a statistical basis the most frequent geometries are those of the lowest energy. The degree of flexibility of a polymer such as amylose depends primarily on the range of low-energy conformations available to the glycosidic bond*. If, for example, there was only a single low-energy conformation, the molecule would be a stiff helix. At the other extreme, if the glycosidic bond could adopt any conformation whatsoever, with equal probability, the molecule would exist as a truly random coil. The reality of the situation lies between these extremes, but there exists no agreement among investigators regarding the 'stiffness' of the molecule. Bulky, charged, or hydrogen-bond-blocking substituents also alter the range of suitable conformations.

The term 'worm-like' has been used to express the conformation of polymers with limited flexibility. The description is probably appropriate for amylose, if the 'worm' is considered to have a considerable degree of helical character.

Conformation maps and computer models suggest that left-handed helices are more stable than right-handed helices. Thus an extended helix is far more likely to be left-handed. Any transition between left- and right-handed helices must pass through some kind of extended conformation of higher energy[72].

Iodine-complex formation[66] leads to an appreciable shrinking of the polymer and an increase in rigidity owing to dehydration of the amylose (through hydrophobic interaction with iodine) and concomitant inter-turn hydrogen bond formation.

### *6.7.1.2 Amylopectin*

There have been few studies of the hydrodynamic properties of amylopectin. One problem is that amylopectin consists of a wide range of molecular sizes, and some amylopectin molecules are so large that it has not been certain whether they are truly dispersed or molecular aggregates. Molecular weight determinations for amylopectin by light scattering gave values 100 times or more as large as by osmotic pressure. At first this was thought to be due to molecular association. However, statistical theory for branched polymers shows that such a ratio is to be expected[34]. A second problem is the lack of a suitable hydrodynamic theory for branched molecules, especially when the exact branching pattern is poorly understood[65].

Glycogen[73] has a low intrinsic viscosity ($[\eta] = 6\text{–}13.5\ cm^3\ g^{-1}$) essentially independent of molecular size, as appropriate for a rigid spherical particle

* Amylose conformation maps are usually based on the approximation that the glucose ring is rigid, and that all the molecular flexibility comes from twisting the glycosidic bonds.

containing 66–88% water. Amylopectin by contrast has a very high intrinsic viscosity ($[\eta] \approx 90$–$150\ cm^3\ g^{-1}$), dependent on molecular size, in a range similar to that of amylose[74]. This indicates that amylopectin has an entirely different form from glycogen. Presumably amylopectin is highly elongated, or possibly rope-like, and heavily hydrated. The molecular weight of amylopectin is much greater than that of amylose with an equivalent viscosity.

### 6.7.2 Swelling, gelatinisation and retrogradation

Heating starch granules in water initiates a series of irreversible changes in which the native starch granule first swells (Figure 6.25). Then at a critical temperature it undergoes a rapid gelatinisation with a many-fold increase in volume and total loss of the native granule organisation. As heating is continued to 100 °C or above, particularly when the starch suspension is subjected to shear or agitation, the swollen starch granule gradually disintegrates, hydrates and forms a colloidal sol. Essentially complete disaggregation of starch in water can be achieved by heating under pressure (e.g. at 120–130 °C) for 2–4 h.

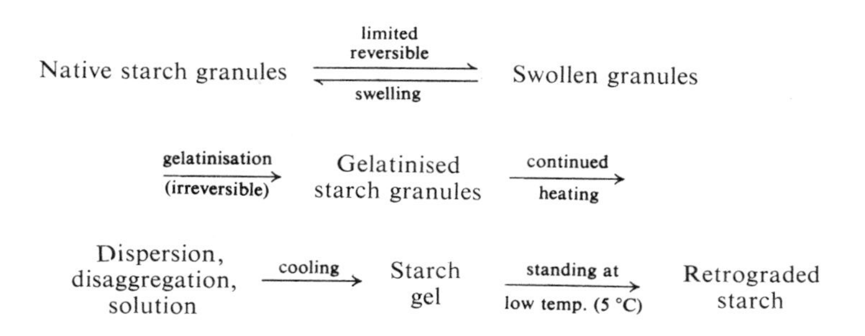


**Figure 6.25** Stages in the gelatinisation and dispersion of starch granules

During gelatinisation of suspensions containing 3% or more starch, the rapidly swelling starch granule may absorb all the available water in the system. The result is a viscous or semi-solid starch paste, which on cooling may set to a gel. This 'viscosity' or thickening capacity of cooked starch is of very great practical utility in many starch applications, particularly in the food industry. So important is the measurement of starch pasting properties that instruments have been developed exclusively for this purpose. The Brabender 'amylograph' is the most widely known. This instrument permits the operator to pass a starch suspension through a carefully controlled cycle of heating, holding and cooling, and provides a continuous record of the paste viscosity during this process.

During gelatinisation, starch granules lose their strong, positive birefringence and become optically isotropic or very weakly negative. The loss of birefringence is often used to indicate the temperature or extent of gelatinisation. There is also a total loss of crystallinity as measured by x-ray diffraction.

#### 6.7.2.1 *Mechanism of swelling*

The mechanism of swelling is only partly understood. In the initial, limited swelling, water is imbibed into the amorphous gel phase of the starch granule, without significantly altering the crystalline phase[75]. This limited swelling is largely reversible. With continued heating, there appears to be a melting of the starch crystallites with simultaneous water uptake and lateral or tangential swelling of the granule[76]. This tangential swelling may be so strong and rapid that a vacuum bubble is created in the starch granule interior[77]. With further swelling, the granule structure becomes weaker and the bubble collapses, but the swelling continues often in a bizarre asymmetric pattern.

The complexity and intricacy of starch gelatinisation can only be appreciated by microscopic observation of starch granules during gelatinisation, e.g. by heating on a hot stage, or better by studying motion pictures of the process.

#### 6.7.2.2 *Retrogradation*

The term retrogradation, as applied to starch, means a return from a solvated, dispersed, amorphous state to an insoluble, aggregated or crystalline condition. Retrogradation of starch dispersions is enhanced by a low temperature (i.e. 0–5°C) and by a high starch concentration. Linear starch chains (amylose) by themselves have a greatly exaggerated tendency to retrograde, and this is enhanced if the amylose is degraded to an average chain length of *ca.* 150–200 glucose units. Conditions for retrogradation are met during the freezing of starch dispersions, especially if the freezing is slow so that there is ample opportunity for the starch molecular chains to undergo association. Slowly freezing a starch paste, then thawing, results in a starch sponge in which the starch is largely insolubilised, and the voids created by large ice crystals have the ability to soak up water by capillary absorption.

Starch retrogradation accompanies the aging of cooked starch foods such as bread, potatoes, cooked cereals and puddings, sauces or gravies. With bread, the staling process involves hardening of the starch by retrogradation and redistribution of water from the crumb (thereby reducing the water content) to the crust (initially low in moisture). This transfer accelerates retrogradation in both the crumb and crust.

Retrogradation is partly reversible simply by heating the retrograded material to destroy the crystalline aggregates. However, repeated cycles of retrogradation usually lead to irreversibly retrograded materials.

Although retrogradation is promoted by increasing the starch content of a dispersion, the process requires moisture, and dry starch materials (e.g. soda crackers, prepared cereals, popped corn, potato chips) retain their 'freshness' as long as they are kept dry. However, absorption of a small amount of water hydrates the starch chains and permits them to associate slowly and crystallise.

## 6.8 STRUCTURE AND ORGANISATION OF STARCH GRANULES

The architecture of the starch granule has been studied by optical microscopy, particularly by using polarised light, by transmission and scanning electron microscopy (Figures 6.26–6.28), by x-ray diffraction and by optical interference patterns[78]. Although the approaches are substantially different, the

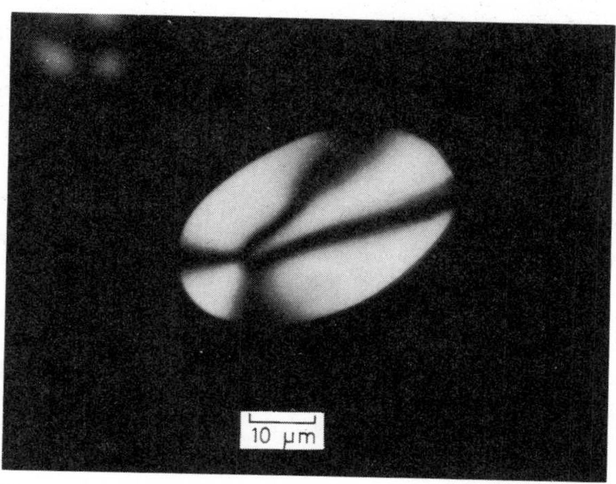


**Figure 6.26** Potato starch granule photographed between crossed polarisers. The dark cross is characteristic of a spherocrystalline arrangement. The plane of the polarisers is vertical and horizontal. Note the asymmetric form of the polarisation cross, indicating a non-radial arrangement of starch chains. (Photograph by courtesy of Dr. D. Outka.)

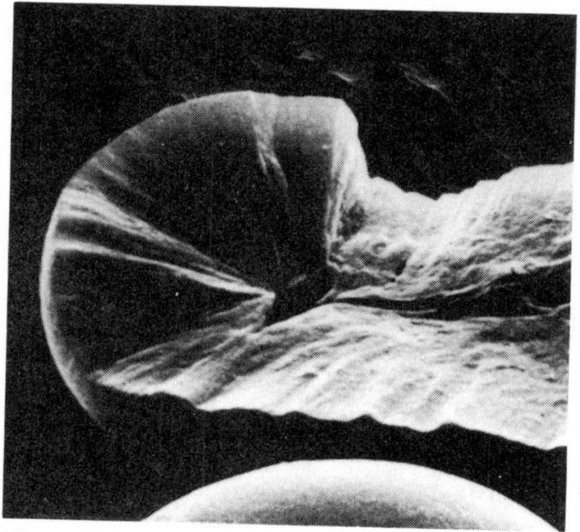

**Figure 6.27** Scanning electron micrograph of crushed potato starch granule, showing radial arrangement of 'crystalline' structure. (Photograph by courtesy of D. E. Hall.)

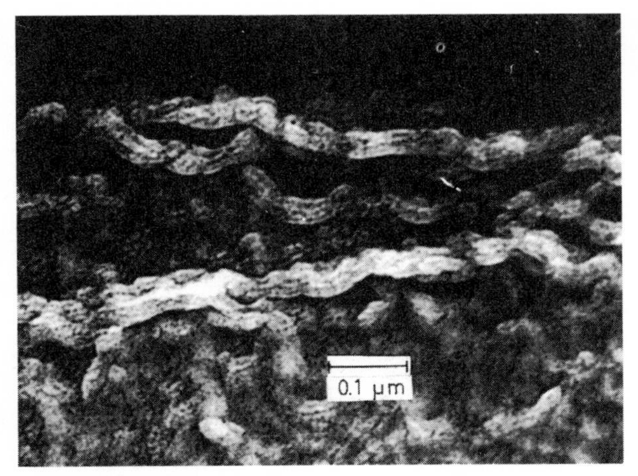


**Figure 6.28** Fracture surface of potato starch granule showing ropy texture. (Photograph by courtesy of C. E. Sterling.)

results of such studies correlate well and indicate that the structural elements or 'crystallites' initially radiate from the hilum. In the peripheral parts of large asymmetric granules, the crystallites are perpendicular to the surface, not necessarily in a 'radial' orientation.

### 6.8.1 Development of starch granules[10]

Growth of starch granules occurs in membrane-bound subcellular organelles called 'plastids' or 'amyloplasts'. These are thought to be similar to chloroplasts, except that they lack the internal membrane organisation and photosynthetic machinery. Actually, some plastids, such as those in *Pellionia* stems, develop chlorophyll and become chloroplasts when exposed to light. Conversely, normally green plant tissues fail to develop chlorophyll when kept in darkness.

The development of external form and internal organisation of starch granules is characteristic of the particular kind of plant, even of the particular tissue and location within the tissue. An experienced microscopist can often identify starches from various sources by examining their microscopic appearance in ordinary light and between crossed polarisers with or without a first-order red-retardation plate. The appearance of iodine-stained granules is also helpful in identification.

Development of starch within a plastid commences with the accumulation of poorly organised, amorphous material of unknown chemical composition. At a certain moment there is the deposition of a minute 'brush-heap' of insoluble polysaccharide, which acts as a nucleus for further starch deposition. Deposition of starch can continue with one nucleus, leading to a single starch granule, or with several nuclei, leading to a cluster of small granules which may grow together giving a compound granule. Initial growth gives nearly spherical granules. However, as the granules are enlarged they often become elongated or flattened. Large granules may have a conspicuous hilum, the original growing point of the granule. The granule usually has a very irregular texture at and near the hilum, and often cracks or dislocations of the granule radiate from it. With some starches, e.g. rice and maize, the granules become so crowded that they are forced to adopt a polygonal form.

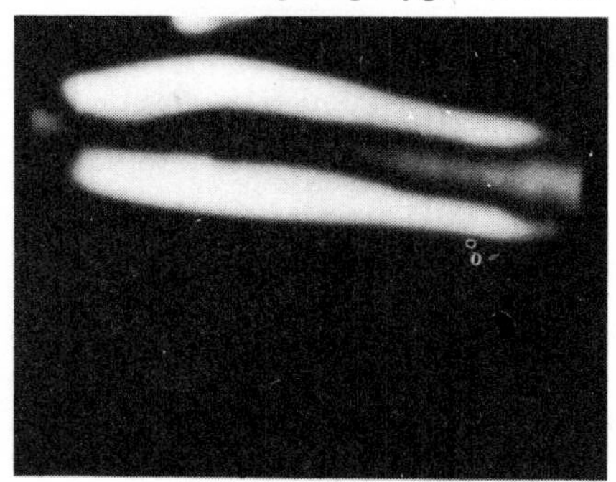

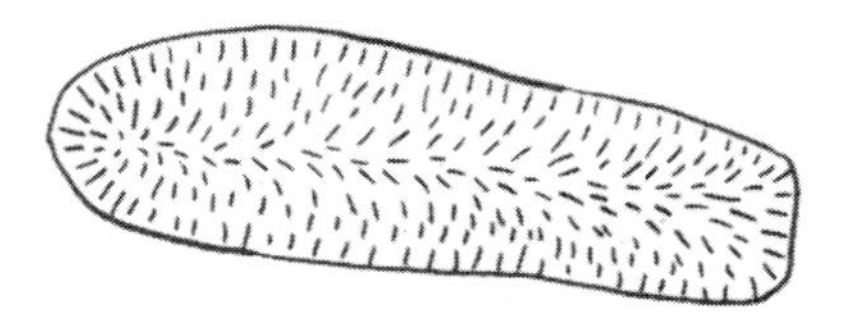

**Figure 6.29** Birefringence map of the orientation of starch chains of *Dieffenbachia sp*. Above: view in light polarised at 45° to horizontal; below: optical map

Orientation of starch crystallites perpendicular to the granule surface implies that the molecular axes of the starch molecules are also arranged in this fashion. The sign of birefringence of starch granules is positive (highest refractive index perpendicular to the granule surface). Most polymers, including cellulose, and presumably starch, have the highest refractive index along the molecular axis. By using the polarisation patterns of starch granules, it is possible to prepare optical maps[79] of the molecular orientation within

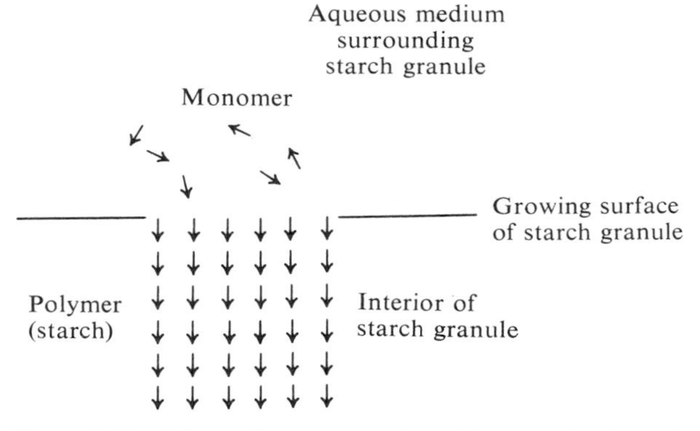


**Figure 6.30** Schematic representation of the growth of starch chains at the surface of the starch granule. As the monomer units are linked to the growing polymer, they simultaneously crystallise so that the orientation of molecular chains is perpendicular to the starch granule surface

individual granules (Figure 6.29). Also, x-ray diffraction studies on single large starch granules show that the molecular axis of starch is perpendicular to the growth rings and to the surface of the granule[80]. These observations taken together indicate that the starch molecular chains grow in an orientation perpendicular to the (growing) surface of the starch granule (Figure 6.30), a mode of synthesis referred to by Stöckmann[81] as simultaneous polymerisation and crystallisation.

Crystallisation of preformed polymers from solution (or from a melt) usually occurs by apposition of successive molecules, or molecular segments, parallel to the growing crystal surface. Such a mode of growth would give negative birefringence, as is observed for artificial spherocrystals of short amylose chains.

## 6.8.2 X-ray diffraction of starch

### *6.8.2.1 A–B starch*

Native starch granules give an x-ray diffraction pattern indicative of a high degree of crystalline order. Patterns of various starches are classed as *A* (Figure 6.31), characteristic of the cereal starches, *B* (Figure 6.32), characteristic of the tuber starches and retrograded starch, and *C*, patterns intermediate

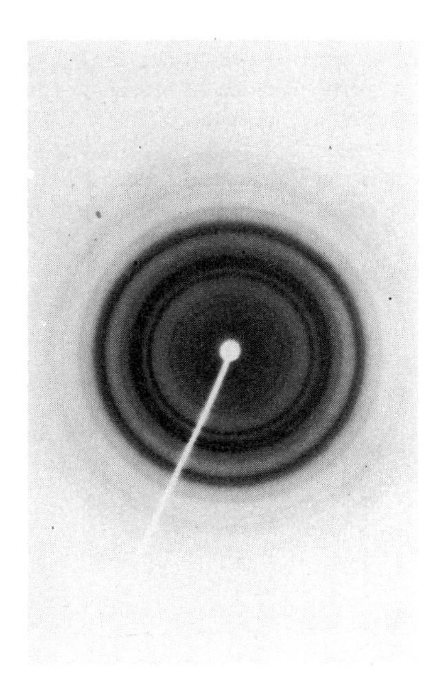

**Figure 6.31** A-type x-ray diffraction pattern, characteristic of the cereal starches and recrystallised amylodextrins

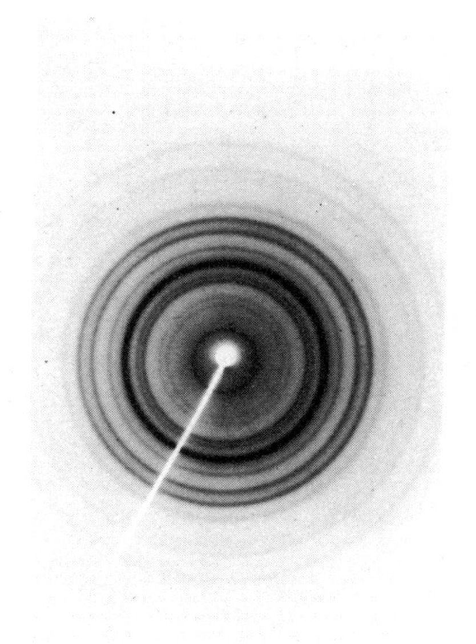

**Figure 6.32** B-type x-ray diffraction pattern, characteristic of the root starches and retrograded starch

between *A* and *B*. The crystallographic interpretation of x-ray patterns depends heavily on the availability of fibres or other preparations which give good oriented patterns. Stretched fibres of amylose acetate give an excellent x-ray pattern[82], and, on de-acetylation of these fibres in various media, various complexes are formed (e.g. with alkali, KBr). Exposure to moisture converts alkali amylose fibres into *A* and *B* fibres[83]. Both *A* and *B* fibres show that there is a periodicity along the fibre axis of *ca.* 10.5 Å. Fibre data, other than the fibre period, have never been published for *A*-starch, and it has not been possible to deduce the crystallographic unit cell from powder data alone[90]. For *B*-starch, good fibre patterns have been published[84] and two basically different unit cells have been proposed. The first is the Rundle pseudohexagonal cell[85]. To fit this cell an extended amylose chain conformation with two glucose units per 10.5 Å was originally proposed, but this structure requires grossly distorted bond angles and ring conformations. A hydrated helix was also suggested[84], but the amount of water required does not agree with actual *B*-starch, which is essentially anhydrous[75]. Moreover, to fit such a helix into the Rundle cell it is necessary to double one of the cell dimensions. A second, basically different unit cell was proposed by Schieltz[86]. This cell agrees slightly better with the x-ray data than the Rundle cell. It easily accommodates amylose if the molecular conformation is that of a parallel double helix (Figure 6.33) with three glucose residues per 10.5 Å. Many facets of starch behaviour are harmonised

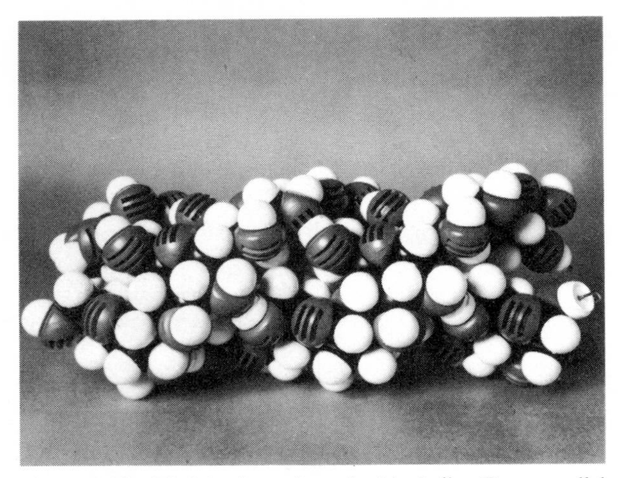

**Figure 6.33** Model of amylose double helix. Two parallel amylose molecules are twisted about each other and locked together with hydrogen bonds. (From Kainuma and French[75], by courtesy of Interscience.)

by assuming a double helix conformation, and the author is strongly attracted to this structure[75].

### *6.8.2.2 V-patterns*

Various preparations of helical starch give a V-pattern, distinctly different from that of granular or retrograded starch. Analysis of the x-ray patterns, particularly using fibres, show that the fibre or molecular axis spacing is *ca.* 8 Å, in agreement with the measurements obtained from molecular models[87]. The helical conformation is stabilised by complexing agents such as aliphatic alcohols (butanol, pentanol), iodine–potassium iodide, hydrophobic agents (cyclohexane, tetrachloroethane), DMSO, glycerol, ethylenediamine, and a host of other organic compounds. With butanol, amylose forms a helix with six glucose units per turn. However, bulkier complexing agents such as t-butanol give a seven-unit helix[55].

## 6.8.3 Growth rings, crystal and gel phase

Optical microscopy of large starch granules (e.g. potato, canna, arrowroot) shows a high degree of internal organisation as evidenced by strong birefringence of the spherocrystal type. Although starch grains may be grossly distorted from the ideal spherocrystal form, well-formed granules show a 'Maltese cross' polarisation pattern, the arms of which intersect at the hilum. The arms of the polarisation cross also intersect the growth rings perpendicularly. Although the growth rings look like surface irregularities, it can readily be seen by micromanipulation of individual large granules

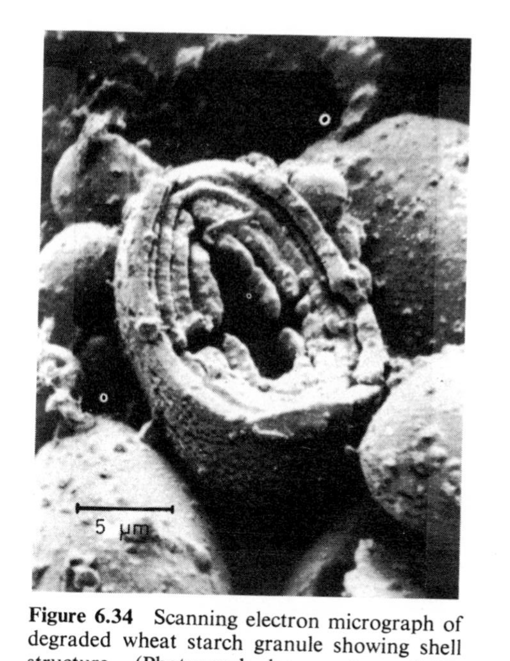


**Figure 6.34** Scanning electron micrograph of degraded wheat starch granule showing shell structure. (Photograph by courtesy of W. Bushuk[89] and the American Association of Cereal Chemists.)

**Figure 6.35** Acid-modified waxy maize starch. (From Musselman and Wagoner[88], by courtesy of the American Association of Cereal Chemists.)

that in reality the growth rings are concentric shells of alternating refractive index. These shells are also readily seen by transmission[88] and scanning electron microscopy (Figure 6.34)[89]. During protracted acid treatment of starch granules a large proportion of the starch is degraded and solubilised leaving a residue (Naegeli amylodextrin) which has enhanced x-ray crystallinity over that of the native starch[52,90]. Simultaneously there is a pronounced separation of the layers, as indicated by electron microscopy (Figure 6.35)[88].

X-Ray *d*-values for dry and fully hydrated starch are indistinguishable. Hence the crystallites and the unit cells themselves do not swell during moisture sorption below the gelatinisation temperature. The pronounced swelling which occurs during starch granule hydration[91] must then be a property of a gel or amorphous phase, presumably a phase of lower refractive index which alternates with the crystalline phase of higher refractive index. The refractive index of starch is approximately 1.53 whereas that of water is 1.33. Therefore a hydrated gel phase would have an intermediate refractive index depending on its water content.

### 6.8.4 Acid hydrolysis of starch granules; Naegeli amylodextrin[52]

The amorphous gel phase of starch granules readily absorbs aqueous chemical reagents and is degraded by aqueous acid (e.g. 15% $H_2SO_4$ or 7–10% HCl). The crystalline phase is relatively resistant to weeks or months of such treatment. The product of brief treatment (e.g. 2 days at 20 °C with 7.5% HCl) is 'soluble starch', so called because it readily dissolves in boiling water without gelatinisation or paste formation. If acidic treatment is more extensive (e.g. 3 months with 16% $H_2SO_4$) the residual material is called 'Naegeli amylodextrin'. With the highly crystalline starches, such as potato, the yield of Naegeli amylodextrin is *ca.* 50% of the weight of the original starch. With poorly crystalline starches, such as waxy rice, the yield of

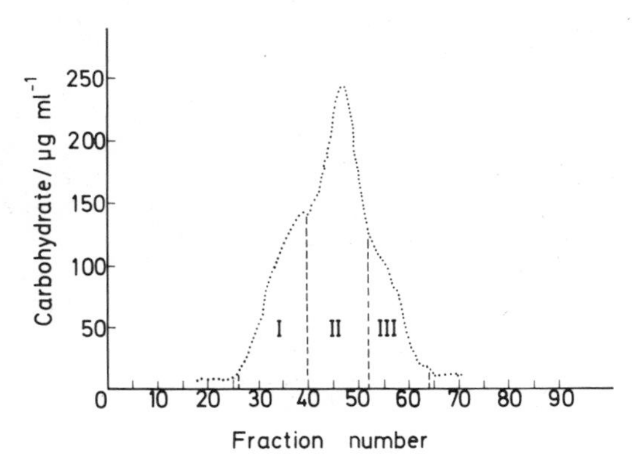


**Figure 6.36** Elution diagram of waxy maize amylodextrin on Sephadex G-50[93,94]. I, highest molecular weight, complexly branched; II, DP 20–30, singly branched; III, DP 10–20, linear

crystalline residue is small and decreases rapidly. The acidic hydrolysis can be accelerated by raising the temperature keeping under the gelatinisation point, especially if alcohol is added to decrease the tendency to gelatinise[92]. Treatment with anhydrous alcoholic acid gives alcoholysis, with formation of non-reducing starch glycoside and alcohol-soluble glycosides.

Acidic treatment of waxy-maize starch granules gives a highly crystalline residue which consists initially of three populations of molecules: fraction I, multiply branched, fraction II, singly branched, and fraction III, linear (Figure 6.36)[93,94]. Further treatment preferentially destroys fraction I. The crystalline residue has an average DP of *ca.* 25 and is 90% degradable by β-amylase. Study of the molecular structure and physical properties of these amylodextrins has contributed greatly to our understanding of the molecular structure of amylopectin and the detailed architecture of the starch granule.

## 6.9 α-AMYLASES

α-Amylases[95] are produced by all types of life: animals, plants and micro-organisms. Most of the study of α-amylases has been focused on the amylases of mammalian digestion (e.g. pancreatic and salivary amylase), the α-amylases of germinating cereals (malt α-amylase) and amylases of starch-digesting micro-organisms (molds, bacteria). These enzymes are of great importance, not only for the physiology of the organism that synthesises the amylase, but also in technology.

α-Amylases are designated alpha (as opposed to beta) because the newly formed reducing groups have an α-anomeric configuration (downward mutarotation). It is found as a corollary that all α-amylases are also endoases, i.e. they are able to attack the interior of starch chains, as opposed to β-amylase, an exoase, which is an endwise attacking enzyme.

### 6.9.1 Mammalian α-amylases: pancreatic and salivary

These enzymes have been very extensively studied owing to their accessibility and significance in digestion. Human saliva can readily be collected in quantities of several litres from cooperative colleagues or students. Saliva contains almost no other starch converting enzyme. The preparation of amylase in pure, crystalline form[96] involves only a few steps. The crude enzyme is fractionally precipitated with acetone and ammonium sulphate to give a concentrated, purified enzyme which can be readily crystallised from aqueous acetone. Even without such purification, saliva can be used directly or with minimal clean-up for many purposes, as an enzyme source.

Pancreas glands can be obtained in quantity from the slaughter of swine. Preparation of amylase involves extracting defatted pancreas powder with buffer, fractionally precipitating the extract with acetone and ammonium sulphate, removing sulphate ions by ion-exchange resin, removing inactive material with chloroform–isoamyl alcohol and crystallising[97].

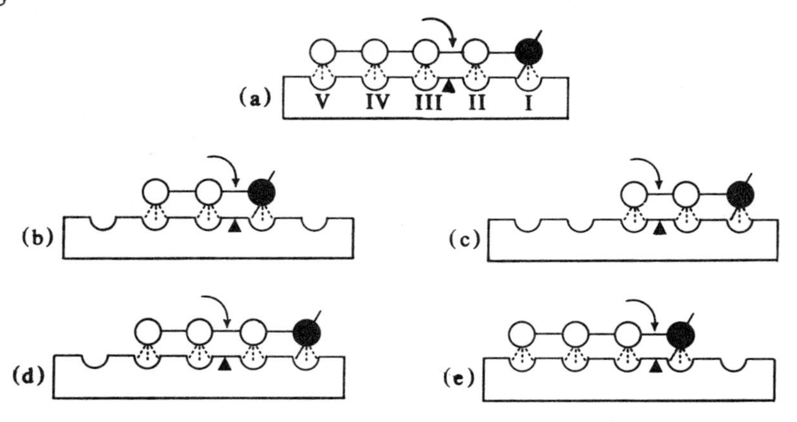


**Figure 6.37** Productive hydrolytic enzyme–substrate complexes of porcine pancreatic α-amylase. (a), maltopentaose complex; (b), major maltotriose complex; (c) minor maltotriose complex; (d) major maltotetraose complex; (e) minor maltotetraose complex. The indentations in the block and the roman numerals I–V represent glucosyl binding sites of the enzyme; ○, a non-labelled glucosyl unit; ∅, a reducing end $^{14}C$-labelled glucosyl unit; ——, an α-1 → 4 glycosidic bond; ▲ and ↷, catalytic groups of the enzyme. The substrate chain is oriented so that its non-reducing end is on its left and its reducing end is at its right. (From Robyt and French[99], by courtesy of the American Society of Biological Chemists.)

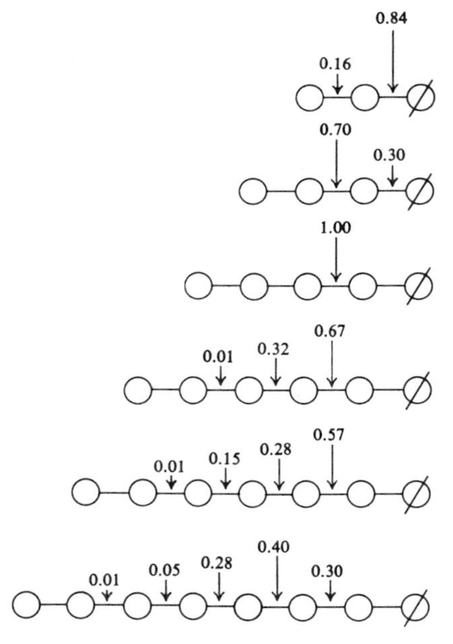


**Figure 6.38** The frequency distribution of bond cleavage during initial action of porcine pancreatic α-amylase on reducing end-labelled oligosaccharides. (From Robyt and French[99], by courtesy of the American Society of Biological Chemists.)

The crystalline enzyme is available commercially. It has a molecular weight of *ca.* 50 000 and is composed of two subunits. Each subunit is enzymically active. Little is known about the detalied structure of pancreatic amylase. There are five disulphide links which maintain its tertiary and quaternary structure[108].

The animal amylases are absolutely dependent on a monovalent anion which presumably is an allosteric activator[98]. Chloride, about 0.01 mol $l^{-1}$, is optimal; bromide and other ions are less effective. Concentrations of chloride as low as $10^{-4}$ mol $l^{-1}$ have an appreciable activating effect.

### 6.9.1.1 *Substrate binding site*

The substrate binding site of pancreatic[99] or salivary[100] amylase binds a chain of five glucose units (Figure 6.37). This was established by the pattern of action on reducing end-labelled oligosaccharides (quantitative values in Figure 6.38 are given for porcine pancreatic amylase[99]).

### 6.9.1.2 *Action pattern*

The action pattern of porcine pancreatic amylase is unique in displaying a very high degree of multiple attack. This action pattern can be demonstrated by comparing the blue value–reducing value plots for various amylases[101, 102].

For an enzyme which has a very high degree of multiple attack, there should be essentially a linear relationship between the decrease in blue value and the increase in reducing value, such that by extrapolation of this plot to zero blue value one would arrive at the final reducing value when all the substrate has been converted into products. Such plots are obtained in the first part of the digest with porcine pancreatic α-amylase, and show that for a given decrease in iodine colour there is a large production of reducing sugar. Enzymes with non-repetitive, random attack, bacterial and fungal α-amylases give plots which descend much more steeply initially, indicative of a dextrinising activity.

The *rationale* of multiple attack by α-amylase is as follows (Figure 6.39)[103]. The enzyme initially encounters the substrate molecule at a random position and forms an enzyme–substrate complex of a conformation suitable for catalysis. Hydrolysis occurs. The right-hand fragment, being held to the enzyme less strongly than the left-hand fragment, readily dissociates, leaving the right-hand binding site unoccupied. The left-hand fragment then rearranges in its association with the enzyme to occupy the entire binding site, then giving an enzyme–substrate complex with a geometry suitable for further catalysis. Eventually the left-hand fragment dissociates from the enzyme (this is the $k_{-1}$ process in the Michaelis enzyme–substrate equilibrium).

The degree of multiple attack is the ratio $k_2/k_{-1}$; the multiple attack parameter $f$ is $k_2/(k_{-1} + k_2)$. For porcine pancreatic amylase, we have obtained multiple attacks in the range of 8–10 per effective encounter. For salivary amylase the value is much lower, about 3.

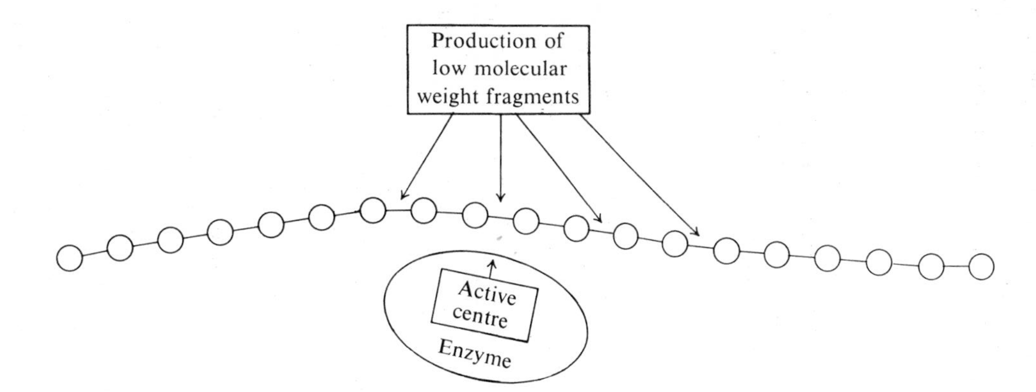


**Figure 6.39** Action of α-type amylases on long-chain starch molecules. The initial attack cleaves the substrate into large fragments. Subsequent local (repetitive) attack produces small oligosaccharides

### .9.1.3 *Mechanism of action*[104]

The mechanism of action of pancreatic amylase probably involves proton-ation of the glycosidic bond by an imidazolium ion, attack on the anomeric

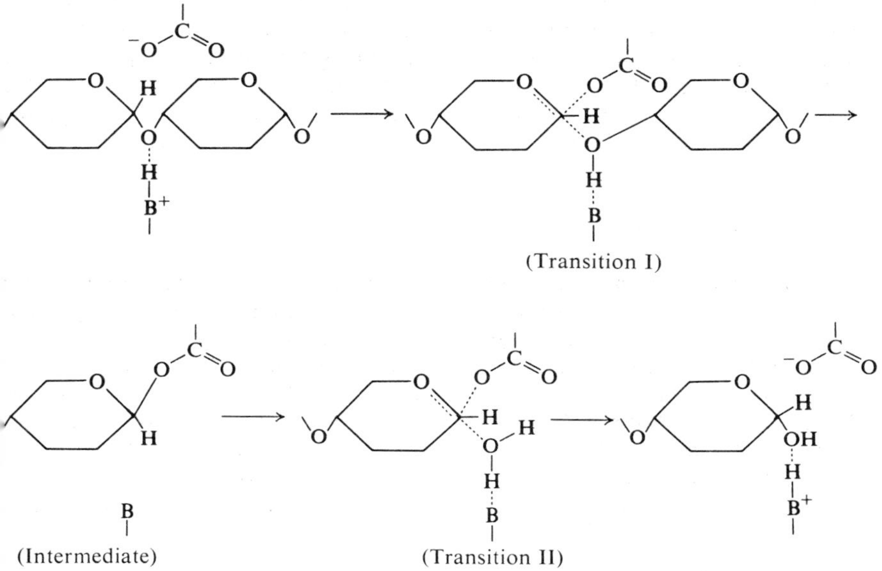
(Transition I)
(Intermediate)
(Transition II)

**Figure 6.40** Proposed mechanism of action of α-amylase[104]

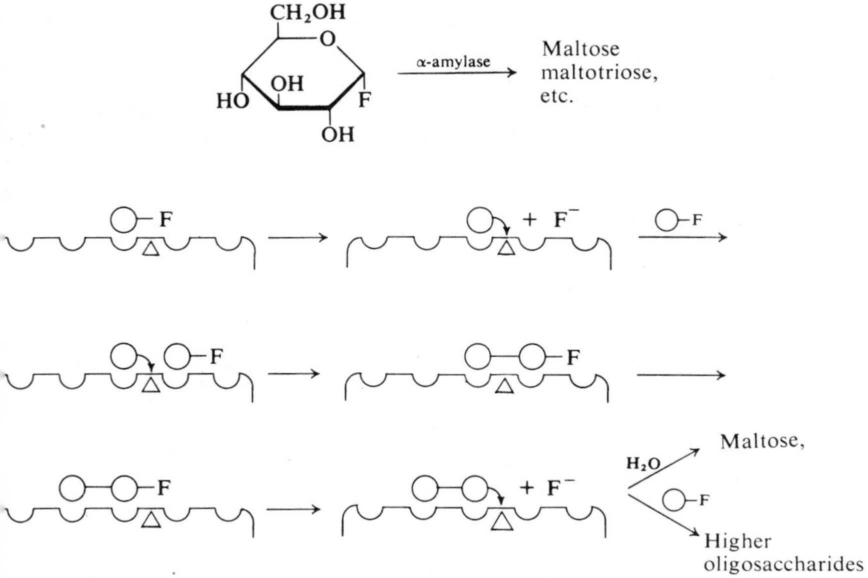
α-amylase
Maltose
maltotriose,
etc.
Maltose,
$H_2O$
Higher
oligosaccharides

**Figure 6.41** Oligosaccharide synthesis using α-D-glucosyl fluoride as a glucosyl donor[106]

C atom by a $CO_2^-$ group of the enzyme with formation of a glycosyl–enzyme ester intermediate, and hydrolysis of the ester by water (attack on the anomeric C atom) (Figure 6.40). This mechanism accounts for the known pH–activity curve (optimal around pH 6.5–7), the glycosidic bond cleavage with retention of anomeric configuration (double displacement), the polarity of multiple attack requiring preferential binding on the left-hand fragment[103], and inhibition by maltobionic acid lactone[105]. Such a mechanism is in harmony with the unusual polymerisation of α-D-glucosyl fluoride catalysed by this enzyme (Figure 6.41)[106]. However, at present no one has performed the crucial experiment of isolating and identifying the postulated glycosyl enzyme ester intermediate.

#### 6.9.1.4 *Action on branched substrates*

Branch linkages are not cleaved by α-amylases, and the branch point cannot be accommodated at all positions of the substrate binding site. Consequently, action on amylopectin or glycogen is limited, and produces α-limit dextrins or oligosaccharides containing all the branch linkages. Both salivary and pancreatic amylase give the singly branched oligosaccharides shown in Figure 6.10 (p. 286)[107]. More complex branched configurations in the substrate (particularly with glycogen) lead to doubly branched, triply branched and more highly branched dextrins. Structural analysis of these dextrins has given important clues regarding the detailed structure of starch and glycogen[48].

#### 6.9.1.5 *Conformation of the substrate at the binding site*

Starch chains, though generally drawn in an extended conformation (e.g. the Haworth depiction) must in reality have a twisted or helical arrangement (see Section 6.7.1.1). The most extended, undistorted conformation is like that in amylose acetate (Figure 6.42) or in the KBr complex. In such extended conformations the glycosidic bonds are relatively accessible, and probably more reactive, than in compact helical or double helical arrangements. At the same time, the substrate, in the E–S complex, must exist in a strained conformation, especially at the point where catalysis is to occur.

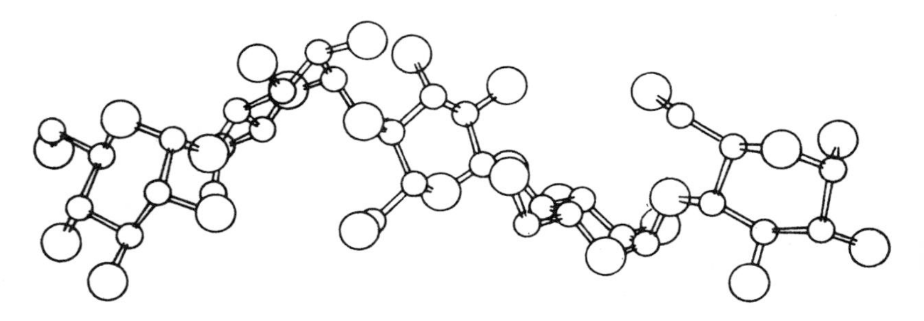

**Figure 6.42** Conformation of amylose chain in amylose triacetate. (From Sarko and Marchessault[82], by courtesy of the American Chemical Society.)

By substituting various positions in the amylose molecule, it is possible to find which positions are critical in formation of the E–S complex and catalysis. For example, substitution at position 6, or branching, can occur at subsites I or V, and to a less degree at subsite II, without blocking enzyme action. However, 6-substitution at subsites III or IV totally blocks enzyme action. Position 2 can be substituted in subsites I and IV, and position 3 can be substituted in subsites II or V. No substitution is tolerated at subsite III[105].

### 6.9.2 Bacterial amylases

Amylases from the Bacilli, especially *Bacillus subtilis* var. *amyloliquefaciens*, are of major industrial importance as starch liquefying or solubilising agents. They are particularly useful because they are able to act at high temperatures, above the gelatinising temperature of starch granules. Thus they are important not only in the liquefaction of starch pastes, but also in various industrial processes such as the removal of starch sizing from fabrics.

Bacterial amylase is produced commercially on a large scale, and sold with a minimum of refinement or purification. Some of the bacterial enzymes used in laundry detergents have a high amylase activity. For scientific studies, *B. subtilis* amylase can be readily purified and crystallised. The zinc complex has a molecular weight of 96 000 and consists of 2 subunits held together by a $Zn^{2+}$ ion. The enzyme, like most other α-amylases, requires $Ca^{2+}$ for its stability[109].

The substrate binding site, originally proposed by Robyt and French[110], has been further detailed by Thoma (Figure 6.43)[111]. He shows that the critical subsite VI is actually an anti-binding site, i.e. it requires energy to force the substrate to occupy this subsite. Clearly this means that the substrate

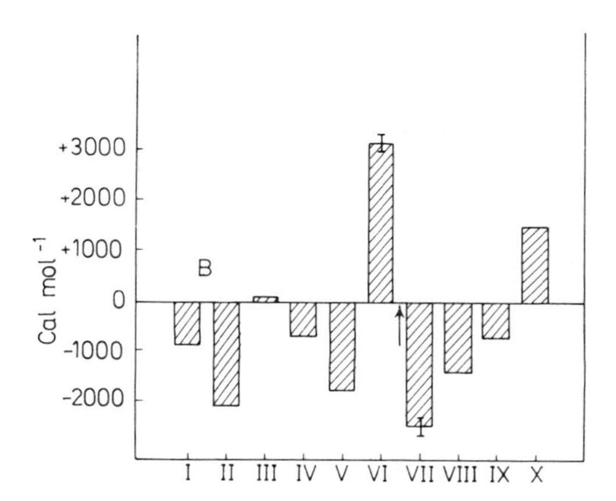


**Figure 6.43** Histogram showing subsite-monomer interaction energies for *Bacillus subtilis* α-amylase. The vertical arrow represents the position of cleavage. (From Thoma *et al.*[111], by courtesy of the American Society of Biological Chemists.)

is highly distorted, and thereby rendered much more reactive, when it is bound to the enzyme.

This enzyme shows a dual product specificity. With amylose as substrate there are two product distributions: a higher oligosaccharide component, centering on maltoheptaose ($G_6$, $G_7$, $G_8$), and a lower component, primarily maltose and maltotriose. Only small amounts of glucose, maltotetraose or maltopentaose are produced. The minimum substrate size for rapid enzyme action is eight or nine glucose units. Maltoheptaose is attacked only slowly and the lower oligosaccharides are essentially inert.

The explanation of the product distribution is probably that the enzyme acts best when the binding site is fully occupied. Thus the left-hand fragment would have a minimum of six glucose units, and the right-hand fragment a minimum of three. As a high molecular weight substrate is degraded, the end-effects become more and more important in determining the eventual product distribution.

With amylopectin, the bimodal product distribution is more pronounced than with amylose. Glycogen gives primarily $G_6$, while amylopectin β-amylase limit dextrin gives only the $G_2 + G_3$ component[110]. With these branched substrates the enzyme also forms branched limit dextrins such as $6^2$-α-maltosyl maltotriose[112]:

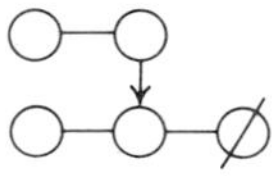

### 6.9.3 Malt α-amylase

During germination of the cereals, particularly barley, there is a rapid synthesis of α-amylase in the aleurone layer of cells. This amylase has as its physiological function the solubilisation and mobilisation of starch in the endosperm for use by the growing seedling. To produce malt, the seed is germinated until development of the amylase is at a maximum. Further growth of the seed is then arrested by kilning (heating and drying). Malt is used mainly to convert starch and starchy cereals into sugars for alcoholic fermentation. Also, the cheapness and availability of malt has made it a useful source of amylase for other purposes.

During germination of barley, at the same time as α-amylase is being synthesised, there is a concomitant increase in the amount of β-amylase. In the diastatic activity of malt, the α-amylase has primarily a liquefying, dextrinising, solubilising action, whereas the β-amylase has a saccharifying action. Both activities are necessary to get good yields of fermentable sugars. α-Amylase has been separated from the other carbohydrates in malt by taking advantage of the heat stability of malt α-amylase, especially in the presence of $Ca^{2+}$ ions. Besides, it is readily absorbed on starch, either on starch granules or on a suspension of dispersed starch in aqueous ethanol[113].

Malt α-amylase acts rapidly on starch until *ca.* 15% of the linkages have been hydrolysed, when action becomes exceedingly slow. The products at this stage consist of linear and branched oligosaccharides like those produced by *B. subtilis* (*amyloliquefaciens*) α-amylase. The product distribution has never been studied very carefully.

### 6.9.4 Saccharifying α-amylases

*B. subtilis* var. *saccharitictus* and certain strains of the molds *Aspergillus niger* and especially *A. oryzae* produce large amounts of saccharifying α-amylase, the crude amylase of the latter being called Takadiastase. These enzymes are of considerable importance, both in the koji process of rice fermentation and as commercial amylases for ethanol production, textile finishing, etc. Takadiastase has also been sold as a digestive supplement. The enzyme is similar in many respects to the mammalian α-amylases in that it hydrolyses the linear parts of starch to maltose, maltotriose, etc. The branch points are left in resistant branched oligosaccharides in the range $B_4$–$B_7$, presumably similar in structure to the pancreatic amylase limit dextrins.

The fungal amylases are accompanied by other carbohydrases, for example maltase, transglucosylase, glucoamylase and 'limit dextrinase.' The possibility of side reactions through these other enzymes has somewhat limited the utility of the fungal amylases, both from the industrial and the scientific standpoint.

## 6.10 β-AMYLASE

β-Amylase occurs widely in plant tissues where it functions in the breakdown of starch to maltose. It is particularly abundant in certain seeds, e.g. soybeans, wheat and barley, and it rises to very high levels during germination or malting. Sweet potatoes are a very rich source of β-amylase. This amylase is of culinary interest, since the sweet taste of sweet potatoes is due to maltose, produced by the action of β-amylase on starch during the cooking process. The enzyme is readily purified[114] and crystalline sweet potato β-amylase is commercially available. Sweet potato β-amylase has a molecular weight of *ca.* 200 000[115]. It contains four sub-units of 50 000 dalton each. The pH optimum is in the range 4.5–6, and being an 'SH' enzyme it is very susceptible to inactivation by heavy metals and denaturing agents. Dilute solutions of the enzyme must be protected by adding SH compounds and protective colloids such as serum albumin.

Action of β-amylase attacks the next-to-last glycosidic bond at the non-reducing end of a starch chain, liberating β-maltose. Repeated action leads to the total destruction of amylose, but the enzyme is blocked by a branching point or other irregularity in the chain. Thus amylopectin and glycogen are degraded only to the extent of about 55% and 45%, respectively. The residual high molecular weight limit dextrin contains all the branching of the parent molecule, and such limit dextrins have been of singular importance in elucidating starch and glycogen structure.

The specificity of β-amylase with respect to the branch point has been worked out by use of model oligosaccharides and by specific debranching enzymes[116]. The configurations in Figure 6.44 represent the outer chains

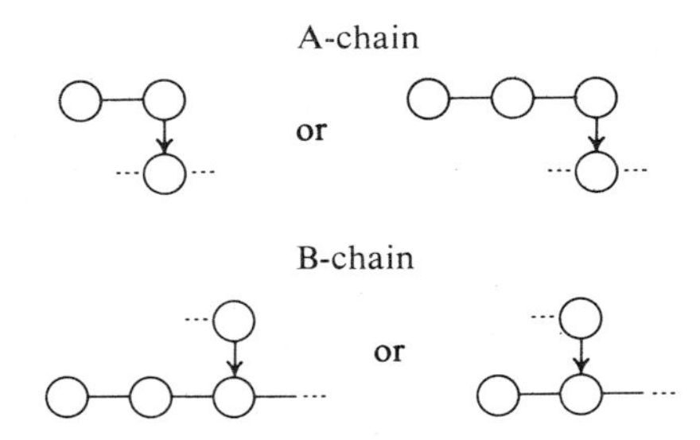


**Figure 6.44** Outer A and B chain configurations of amylopectin and glycogen resistant to β-amylase

remaining after extensive β-amylase action. By limiting the degree of β-amylase attack it is possible to stop degradation when the A chains contain primarily 3 or 4 glucose units[43].

### 6.10.1 Action pattern

β-Amylase shows a high degree of multiple attack, i.e. the ability of the enzyme to cleave repetitively a given substrate chain, without intervening dissociation of enzyme and partly degraded substrate[117]. Thus during a single encounter between enzyme and substrate the enzyme may act several times, removing the terminal maltose unit at each action, before the enzyme and substrate chain dissociate. For low molecular weight amyloses the average number of such cleavages is *ca.* four[117, 118]. With high molecular weight amyloses the degree of multiple attack appears to be much higher.

### 6.10.2 Mechanism of action

The action mechanism of β-amylase involves a carboxyl group and an imidazole group of the enzyme[119]. Consideration of the effects of inhibitors, and the fact that the product is β-maltose, have led the author to propose the mechanism in Figure 6.45. The novel facet of this proposal is the existence of an intermediate enzyme glycosyl ester. This ester can be decomposed by attack of water on the ester carbonyl group, liberating maltose in the β-configuration and regenerating the enzyme carboxyl function. Enzyme esters are recognised intermediates in the 'serine enzymes' (e.g. chymotrypsin) where the carboxyl function is contributed by the substrate rather than by the enzyme.

The critical SH group of β-amylase has no known function in the catalytic step. Rather, it has been suggested[120] that the SH group is the point of attachment of an inactivating or blocking peptide in the inactive or pro-enzyme form.

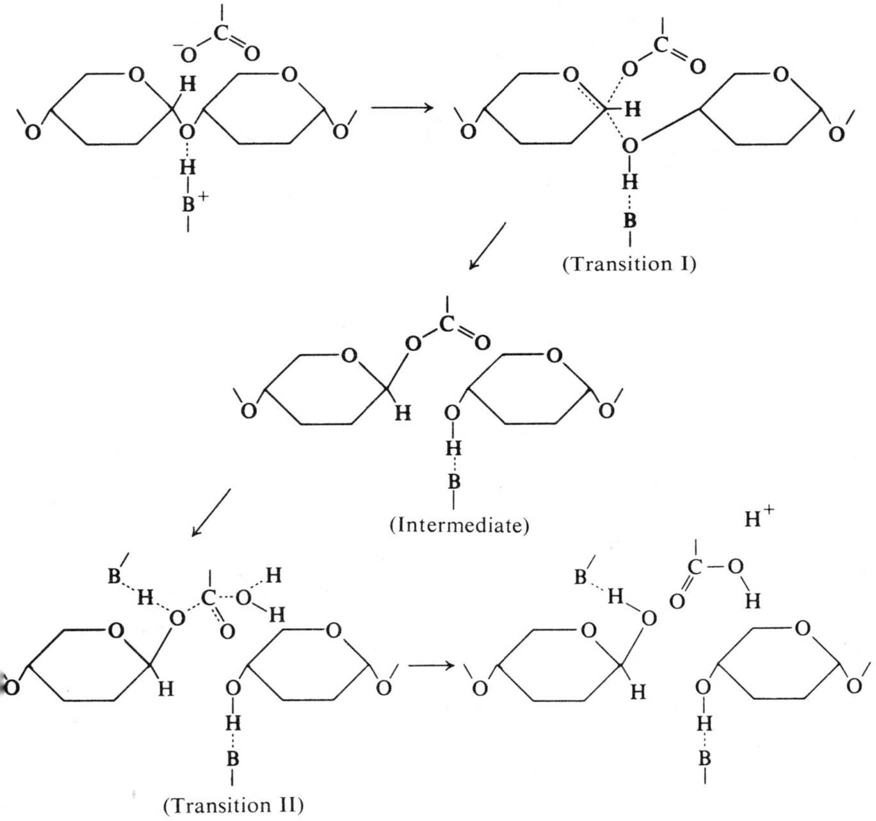

(Transition I)

(Intermediate)

(Transition II)

Figure 6.45 Proposed mechanism of β-amylase action

### 6.10.3 Substrate binding site

The substrate binding site of β-amylase has been probed by using linear oligosaccharides as substrates. The substrate affinities and maximal velocities increase very rapidly with increasing chain length in the region of 3–6 glucose units, reaching a maximum at 6 glucose units. We conclude that the binding site spans 6 glucose units of a polymer chain. For multiple attack to occur it is necessary that the portion of the binding site to the right of the catalytic site must bind the substrate much more firmly than the portion to the left. Maltose[118] and maltotetraose[121] greatly reduce the rate of action on amylose, but do not alter the degree of multiple attack, and are probably bound on the right-hand side (Figure 6.46).

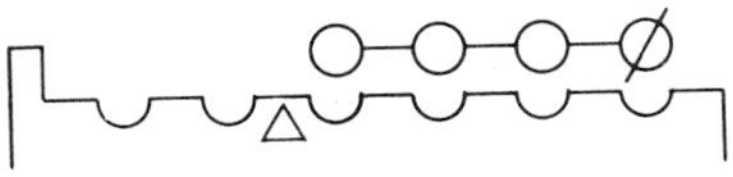

**Figure 6.46** Binding of $G_4$ as a competitive inhibitor to β-amylase

## 6.11 GLUCOAMYLASE

Glucoamylases occur in many micro-organisms, particularly in the starch-degrading molds (*Aspergillus*[122], *Rhizopus*[123]), certain bacteria, and in the brush border of the mammalian digestive tract[124]. Mold enzymes are commercially important in the conversion of starch into glucose, and the mold enzymes have been more extensively studied than other glucoamylases. The action is similar to that of β-amylase in that the enzyme acts only on non-reducing chain ends with the cleavage of α-1,4-bonds and the liberation of β-glucose[125]. Terminal α-1,6-bonds are also cleaved, but much more slowly. Maltose is only slowly attacked; increasing the chain length up to 5 or 6 glucose units gives faster attack[126]. In principle, action of glucoamylase could give complete degradation of amylose or amylopectin. However, with enzyme rigorously purified and free of α-amylase, it is usually found that amylose gives incomplete hydrolysis, presumably because of structural irregularities[127].

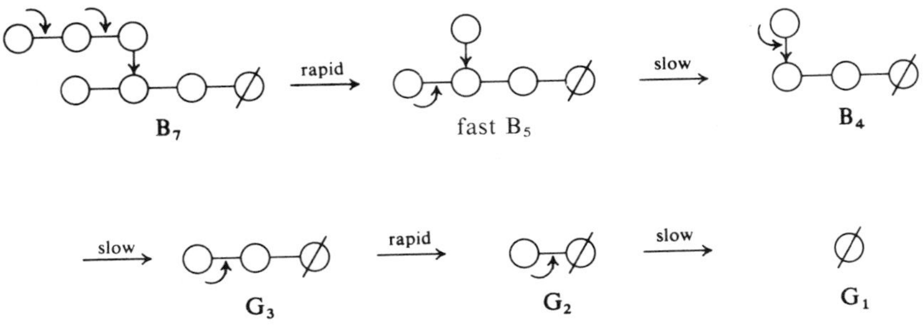

**Figure 6.47** Action of glucoamylase (curved arrows) on singly branched oligosaccharides

With the branched oligosaccharides resulting from action of α-amylase on glycogen or amylopectin, action of glucoamylase 'trims' the more accessible glucose units from the non-reducing chain end, leading to trimmed oligosaccharides (Figure 6.47)[48]. Glucoamylase is incapable of removing the single glucose side chain of fast $B_5$ and analogous compounds; fast $B_5$ is instead degraded to $B_4$. This pattern is just the opposite of that shown by the amylo-1,6-glucosidase of glycogen metabolism; this enzyme requires a single glucose unit linked α-1,6 to one of the interior glucose units of a chain or cyclic molecule (Figure 6.48)[128, 129].

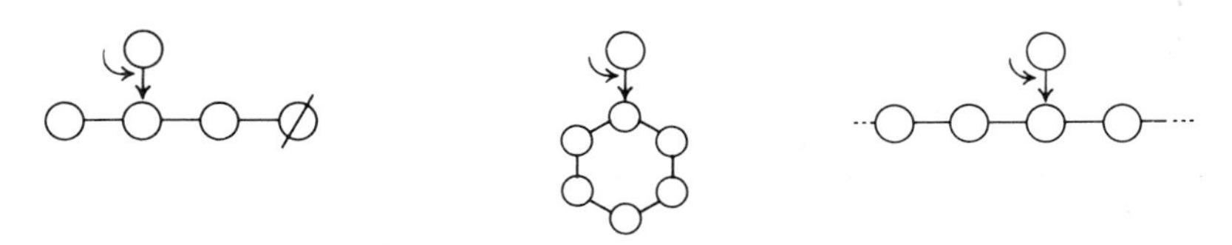

**Figure 6.48** Action of amylo-1,6-glucosidase on branched oligosaccharides

Glucoamylase has been attached to various solid supports, especially glass beads, to give an immobilised enzyme[130]. Such immobilised enzymes have great potential utility in the industrial conversion of starch into glucose.

## 6.12 BIOSYNTHESIS OF STARCH

Until 1961 it appeared to be firmly established that biosynthesis of starch involved two enzymes: a phosphorylase ('P-enzyme') which has the capacity to lengthen existing chains, and a branching enzyme ('Q-enzyme'). However in 1961 a new enzyme was discovered: starch synthase, using UDPG[131] or ADPG[9] as the glucosyl donor. Now most starch scientists think that this is the true chain-lengthening enzyme in normal starch biosynthesis. How amylose escapes branching, in the obvious presence of Q-enzyme, is still a puzzle.

### 6.12.1 Phosphorylase

Starch phosphorylase has been prepared from potatoes, peas and related seeds, and the cereals; its presence has been indicated in countless other plants. It has been crystallised from potato juice[132]. In homogeneous solution, starch phosphorylase catalyses the reversible reaction (6.1).

$$\underset{\text{donor}}{\text{G-1-P}} + \underset{\text{primer}}{G_n} \rightleftharpoons \underset{\text{elongated primer}}{G_{(n+1)}} + \underset{\text{inorganic phosphate}}{P_i} \qquad (6.1)$$

The minimum effective primer for phosphorylase is maltotetraose $(G_4)$[133]. Repetition of the above reaction can lead to the eventual synthesis of extremely long chains, e.g. containing 10 000–20 000 glucose units[134, 135]. Under normal laboratory conditions, however, it is difficult to avoid effects of traces of α-amylase, which greatly reduces the average chain length, or Q-enzyme, which gives a partially branched product. Unless the primer and α-glucose

1-phosphate are sufficiently dilute, and unless the phosphorylase concentration is sufficiently high, reaction is usually arrested by the precipitation (retrogradation) of the synthesised amylose when it is in the range of 150–200 glucose units. Also, accumulation of inorganic phosphate gradually slows down the reaction. The equilibrium constant for the process is a function of pH, temperature and ionic strength. At pH 6, and 30 °C, the equilibrium ratio[136] of orthophosphate to glucose phosphate is *ca.* 7.5. This ratio is not unfavourable for the laboratory synthesis of polysaccharide. However, in green plants the cytoplasmic ratio is much higher, and therefore the reaction would proceed from right to left, giving polysaccharide degradation rather than synthesis.

### 6.12.2 Branching enzyme ('Q-enzyme')

Action of branching enzyme has been studied both on isolated polysaccharide and on polysaccharide which is being elongated by phosphorylase plus glucose 1-phosphate. Very little, if any, branching action occurs on amyloses of chain length less than *ca.* 35–40 glucose units[137]. However, with a branched substrate such as amylopectin, branching enzyme is capable of introducing additional branches to give a glycogen-like polymer.

Recent studies[138] on branching enzyme indicate that the enzyme catalyses an inter-chain transfer [reaction (6.2)], as opposed to an intra-chain transfer

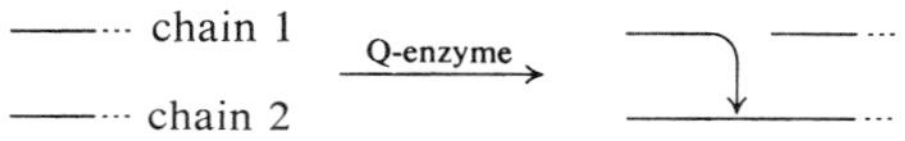

——··· chain 1
——··· chain 2
Q-enzyme →
(6.2)

[reaction (6.3)]. An attractive mechanism for branching enzyme involves

(6.3)

interturn transfer between the chains of a double helix [reaction (6.4)][75, 138].

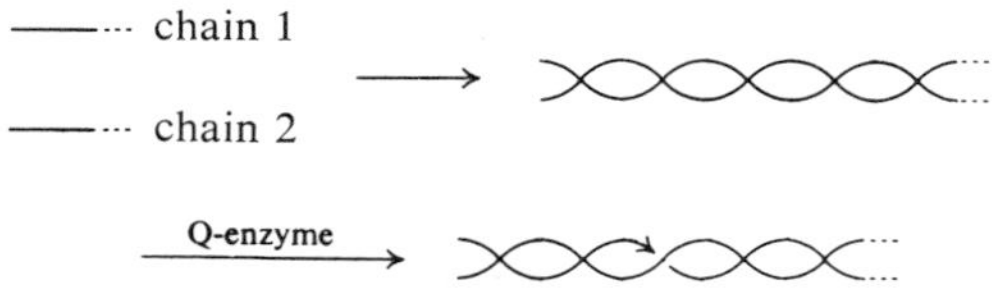

——··· chain 1
——··· chain 2
Q-enzyme →
(6.4)

Such a mechanism would account for the requirement for long chains when amylose is the substrate, as short amylose chains could not be expected to form a stable double helix. However, if the substrate is already a branched molecule, a branching point could lock the chains together and thereby initiate double helix formation, and hence the chains would not need to be so long [reaction (6.5)]. Manipulation of double helix models shows clearly

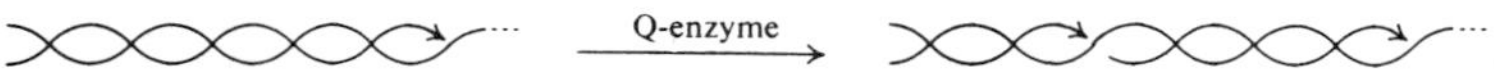

Q-enzyme →
(6.5)

that a branch point can initiate double helix formation, and that the inter-chain transfer is sterically feasible.

With a temperature at or near 30 °C (where most such experiments have been performed), Q-enzyme requires a minimum chain length of *ca.* 40 glucose units. However, at low temperatures (0–10 °C) much shorter chains (15–20 glucose units) can serve as substrates[166]. These temperature effects strongly suggest that it is the association of two chains in a particular conformation, rather than the chain length itself, which is critical for enzyme action. Increasing the temperature tends to promote 'melting' (loss of secondary structure) and dissociation, effects which can be opposed by increased length of the polysaccharide chains.

By the joint action of phosphorylase, to give chain elongation, with branching enzyme, it is possible *in vitro* to produce a branched polysaccharide resembling natural amylopectin with regard to iodine colour, frequency of branching (average chain length) and degree of susceptibility to β-amylase.

### 6.12.3 Starch synthase

The main objection to phosphorylase as the chain-lengthening enzyme is that the *in vivo* ratio of inorganic phosphate to glucose 1-phosphate is too high to permit chain elongation. Actually it is difficult to measure this ratio, and it would only be the ratio within the plastids which would be relevant. This problem disappears if a nucleoside diphosphate sugar (ADPG or UDPG) is the glucosyl donor [reaction (6.6)].

$$\underset{\text{donor}}{\text{ADPG (or UDPG)}} + \underset{\text{acceptor}}{G_n} \rightarrow \underset{\substack{\text{elongated}\\\text{acceptor}}}{G_{(n+1)}} + \text{ADP (or UDP)} \qquad (6.6)$$

The equilibrium constant for the above reaction is very favourable for reaction from left to right as written ($K \approx 260$).

Granule-bound starch synthase, as initially reported[131], could utilise either UDPG or ADPG as the glycosyl donor. However, at present, ADPG is generally thought to be the normal donor[9, 139, 140].

Most studies of starch synthase have used $^{14}C$-labelled substrates. Up to now it has been difficult to synthesise a significant mass of polysaccharide, and some researchers feel that this is evidence against the role of synthase in normal starch synthesis.

### 6.12.4 Side-by-side synthesis of amylose and amylopectin

The problem of the side-by-side synthesis of amylose with amylopectin is not solved simply by the substitution of phosphorylase by starch synthase. However, recently it has been demonstrated that extraction of the soluble enzymes of corn endosperm gives multiple forms of starch synthase, which can be readily visualised by disc gel electrophoresis (a trace of primer being present in the gel), incubation with ADPG and staining with iodine[141]. With sweet corn ($su_1$ maize) as many as 17 bands of activity were seen. With another mutant, *ae* maize, at least 8 bands were seen, but they were substantially different in intensity distribution from the sweet corn bands. Most interestingly, the slow-moving bands gave a pure blue colour, whereas some of the faster bands stained more violet, suggesting that they synthesised a branched polysaccharide. The results indicated that the 'violet' bands are complexes containing both synthase and branching enzyme, whereas branching enzyme is absent from the 'blue' bands. Thus, the amount of synthesis of amylose or amylopectin might depend simply on the ratio of activity of amylose-synthesising particles (enzymes or multi-enzyme complexes) to amylopectin-synthesising particles. Conceivably the polysaccharide-synthesising particles may well be immobilised or adsorbed at or near the surface of the growing starch granule. If so, branching might occur only during polymer elongation by a synthase-branching enzyme complex.

Although the above-mentioned experiments are very illuminating, it still remains a possibility that the soluble enzymes are only involved in the synthesis of branched polysaccharides, and that the true mechanism for biosynthesis of amylose remains unknown. The disc gel results with waxy-maize endosperm extracts were said to be similar to those with sweet corn, yet waxy maize normally does not synthesise amylose. Also, a substantial amount of starch synthase remains firmly attached to the starch granule. Conceivably, only the granule-bound synthase synthesises amylose. Finally, it has never been established that amylose is synthesised by the same type of chain elongation as is characteristic of amylopectin. It is also possible that amylose may be synthesised by elongation at the reducing end, through an insertion mechanism[79]. Until the above problems are satisfactorily solved, the mode of biosynthesis of amylose will not be entirely clear.

## 6.13 OTHER ENZYMES OF STARCH METABOLISM

### 6.13.1 Debranching enzymes: R-enzyme (EC 3.2.1.41), pullulanase (EC 3.2.1.41), isoamylase (EC 3.2.1.68)

These enzymes are specific for the hydrolysis of α-1,6-branch points in otherwise 1,4-linked α-glucans. The original preparation of the plant debranching enzyme, R-enzyme, antedated modern methods of enzyme purification, but the crude preparations were very useful, particularly for the study of branched oligosaccharides. The specificity of R-enzyme was never studied extensively, but in general it showed the capacity to remove α-1,6-linked side branches containing two or more 1,4-linked α-glucose units. Single α-1,6-linked 'stubs' are immune to attack by R-enzyme, pullulanase and isoamylase, but are cleaved by the glycogen-metabolising enzyme amylo-1,6-glucosidase[129].

#### *6.13.1.1 Pullulanase*

Pullulanase (bacterial R-enzyme) occurs as the main extracellular glucan-metabolising enzyme of certain strains of *Aerobacter aerogenes* (*Klebsiella aerogenes*), and it probably occurs widely among glucan-metabolising bacteria. The discovery of pullulanase by Wallenfels[142] resulted from the chance infection and degradation of a pullulan solution by an airborne organism. Pullulanase acts readily on the glucan pullulan, a 1→6-linked olymer of α-maltotriose, cleaving the 1,6-links and liberating maltotriose. )ligomers of maltotriose are intermediate products (see Figure 6.49). )riginally, pullulanase was thought to be inactive on starch since it did ot appreciably increase the reducing value or decrease the iodine stain. Iowever, it is now recognised that pullulanase acts specifically on the -1,6-links of starch, starch dextrins, pullulan and oligosaccharides[143]. The ninimum chain which pullulanase will remove is maltose. A three-unit hain (maltotriose) is optimal; increasing the chain length tends to retard ullulanase action.

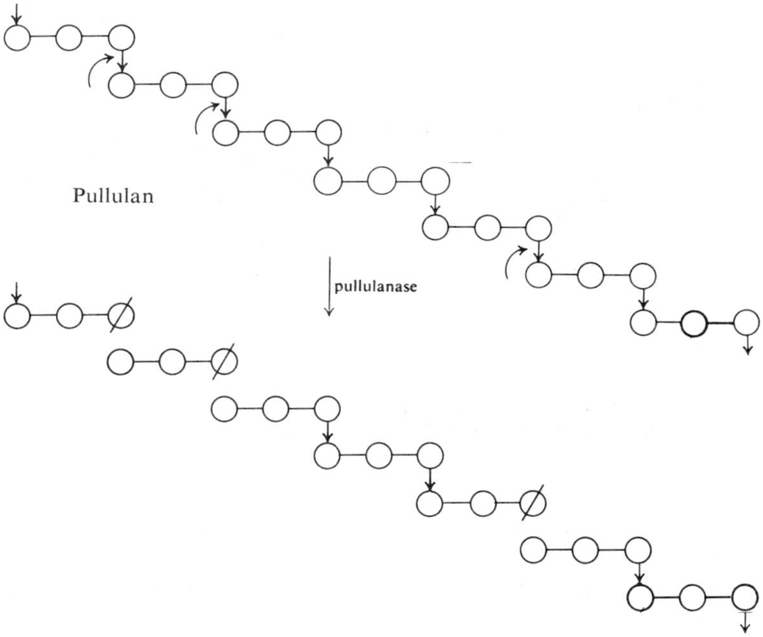

Pullulan

pullulanase

**Figure 6.49** Partial cleavage of pullulan by pullulanase produces $G_3$ and its α-1,6-linked oligomers

Use of pullulanase has been of the utmost value in structure analysis of ranched oligosaccharides and polymers[48]. The enzyme is available in rystalline form. The bacterial culture filtrates have good enzyme activity nd for many purposes there is no necessity of purifying the enzyme. However, ome crude pullulanase preparations are contaminated with $G_6$-enzyme Section 6.13.3.3), an exoase which converts linear starch chains into malto-exaose. Presence of the $G_6$-enzyme may lead to anomalous high levels of naltohexaose in enzyme digests of polysaccharides.

Action of pullulanase on amylopectin requires thorough disruption of the econdary structure (polymer chain association). If native amylopectin is nerely dissolved in DMSO, pullulanase has little action on it. However, if he amylopectin is treated with alkali and neutralised, it is much more readily ttacked[144].

#### *6.13.1.2 Isoamylase*

This enzyme attacks the α-1,6-links ('iso-links') in starch, glycogen and dextrins. It occurs in yeast[145] and it is being produced commercially from *Cytophaga*[146] and *Pseudomonas*[147]. It differs from R-enzyme and pullulanase in that it acts more readily on the native polymers and can completely debranch amylopectin and glycogen. It requires a minimum of three glucose units on the A-chain[146], whereas pullulanase requires only two. It fails to

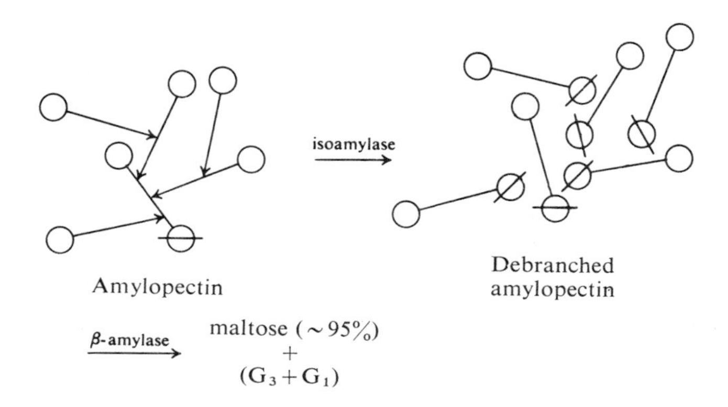


**Figure 6.50** Conversion of amylopectin into maltose by the combined or successive actions of isoamylase and β-amylase

depolymerise pullulan and it does not attack the smaller branched oligosaccharides.

In addition to its laboratory use in polysaccharide structure analysis, isoamylase is being used commercially to convert starch into a mixture of linear polymers. This also forms the basis of a commercial process to convert starch totally to maltose (see Figure 6.50).

### 6.13.2 Transglucanosylases

#### *6.13.2.1 Amylomaltase, D-enzyme*

Certain strains of *E. coli* produce amylomaltase[148], an enzyme which in effect disproportionates maltose to give glucose plus higher oligosaccharides[149]. If the glucose is continuously removed, as by use of glucose oxidase, the polymer chains can become sufficiently long to give a blue iodine stain. Amylomaltase has a remarkable specificity in that it fails to act on maltose unless a catalytic amount of maltotriose (or higher oligosaccharide) is present. Presumably it acts by transfer of a glucose or oligosaccharide residue.

Maltotriose is the smallest oligosaccharide which can function as a maltosyl donor and reaction (6.7) could ensue. An alternative interpretation[150] is that

●—●—∅ + ○—∅* ⟶ ●—●—○—∅* + ∅ (6.7)

maltose is a glucosyl donor, and maltotriose is the minimum acceptor [reaction (6.8)]. As yet the action pattern of amylomaltase has not been

●—∅ + ○—○—∅ ⟶ ∅ + ●—○—○—∅ (6.8)

critically tested by using radioactively labelled substrates.

The plant enzyme counterpart of amylomaltase is called D-enzyme (disproportionating enzyme). The action pattern of D-enzyme[151] is depicted in reaction (6.9). where the heavy bars represent linkages that are inert to

○━○—○—···○━○—∅ + ∅ ($G_1$) or ○—···○—∅ ($G_3$ or higher) (6.9)

donor ($G_3$ minimum) acceptor

the enzyme. Maltose does not function as a donor or acceptor and it does not appear among the products. The physiological function of these enzymes is unknown.

### 6.13.2.2 **Bacillus macerans** ***'amylase' (transglycosylase)*** (EC 2.4.1.19) ***and the Schardinger dextrins***[53]

*Bacillus macerans* produces an extracellular transglycosylase which has the ability to transform linear starch chains into cyclic molecules, specifically the α- and β-Schardinger dextrins (cyclodextrins, cycloamyloses) [reaction (6.10)]. These compounds have long held a fascination for the starch scientist

$$G_n \xrightarrow[\text{transglycosylase}]{B.macerans} \begin{cases} \alpha\text{ (ring of 6)} + G_{n-6} \\ \beta\text{ (ring of 7)} + G_{n-7} \end{cases} \quad (6.10)$$

because they form crystalline inclusion complexes (see Section 6.6.2). The α-dextrin consists of a ring of six α-D-glucopyranose units linked 1→4. The higher homologues are named according to the Greek alphabet, the β-dextrin having seven glucose units, the γ-dextrin eight, and so on. The δ-dextrin and higher dextrins are formed in only trivial amounts and have received little study[152].

Cyclodextrin formation is only one characteristic feature of the enzyme. The reaction is reversible; thus it is possible to open the ring and couple the open chain to a suitable acceptor, which may be glucose, substituted glucose or a molecule like glucose. The enzyme (E) also catalyses transglucanosylase reactions [reactions (6.11) and (6.12)], where the transferred fragment ($G_x$)

$$G_n + E \rightleftharpoons G_{n-x} + E{-}G_x \qquad (6.11)$$

$$G_m + E{-}G_x \rightleftharpoons G_{m+x} + E \qquad (6.12)$$

can be one or more glucose units. These coupling and homologising reactions are of great value in the synthesis of reducing-end labelled oligosaccharides, by using radioactive glucose as the acceptor [reactions (6.13) and (6.14)].

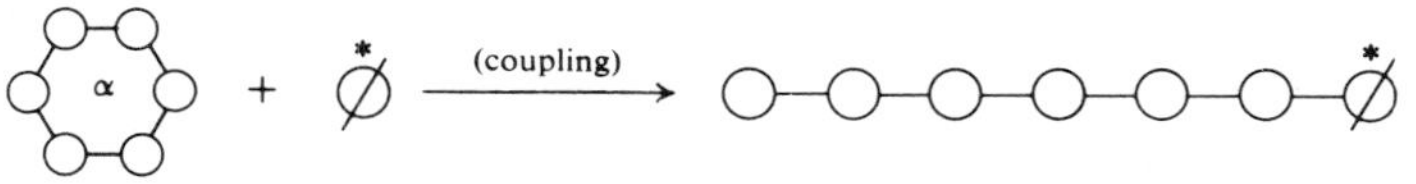

(coupling) (6.13)

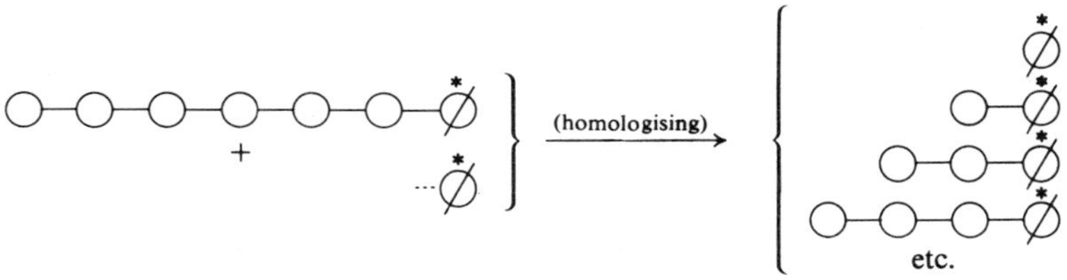

(homologising) etc. (6.14)

Coupling reactions with an excess of cyclodextrin and only a small amount of acceptor can lead to the synthesis of amylose chains sufficiently long to stain blue with iodine and retrograde from solution.

The substrate specificity for cyclic dextrin formation or transglucanosylase action includes linear portions of starch chains or linear starch oligosaccharides. Branch links if present may be incorporated into the products, giving rise, for example, to branched cyclic molecules[48].

Glucose is the best acceptor, but glucosyl oligosaccharides or glycosides also accept. Both glucose units of isomaltose, and the second and third glucose units of panose, can accept starch chains. However, almost any modification of the glucose structure reduces its capacity to react as an acceptor.

Enzymes similar to that of *Bacillus macerans* have been reported[153] from other microbes, but so far they have received little study.

Mechanism of action probably involves formation of a glucanosyl enzyme[105]. If the glucanosyl chain is of the right length (e.g. six or seven glucose units), a cyclic molecule is formed and released from the enzyme. Alternatively, a glucanosyl chain of any length can react with a suitable acceptor. If the acceptor concentration is too low for reaction, the glucanosyl enzyme probably undergoes slow hydrolysis with the formation of a new reducing group.

The Schardinger dextrins have been produced on a semi-commercial basis. They could be manufactured on a very large scale if there were a suitable market for them.

### 6.13.3 Newer microbial enzymes

#### *6.13.3.1* **Bacillus polymyxa** *amylase system*

*Bacillus polymyxa* produces a system of carbohydrases capable of converting starch primarily into maltose[154]. The principal component is a β-type amylase, in that it produces β-maltose in very high yields. The enzyme system has the ability to degrade the cyclic Schardinger dextrins, presumably by action of an α-amylase. With the β-cyclodextrin (cyclomaltoheptaose) it forms maltose and glucose in a 3:1 ratio. Paper chromatography gives no evidence for formation of branched oligosaccharides from amylopectin, thus suggesting the presence of a debranching enzyme[155]. The enzyme system acts on pullulan, but the products have not been adequately characterised.

#### *6.13.3.2* **Pseudomonas stutzeri** *amylase*

*Pseudomonas stutzeri* produces a unique amylase[156] which converts starch in high yield into maltotetraose. The enzyme acts, like β-amylase, from the non-reducing ends of the starch chains and produces maltotetraose in the β-configuration. Acting on amylopectin, the amylase produces maltotetraose plus a high molecular weight limit dextrin, thus indicating its inability to bypass the branch points of the substrate.

#### *6.13.3.3 $G_6$ enzyme*

$G_6$ enzyme is produced from some strains of *Aerobacter aerogenes* (*Klebsiella aerogenes*). During a study of pullulanase production from various strains of *Aerobacter*, it was discovered that some of these have the ability to convert amylose into maltohexaose[157]. Initial study of the enzyme indicates that it is an exoase, but it has the ability to by-pass branch points so that branching is incorporated into the oligosaccharides. Remarkably, the six-unit basic chain can include the fragmentary A chain, the 1→6-linkage, and the reducing end of the B chain [reaction (6.15)]. Taking advantage of this enzyme for

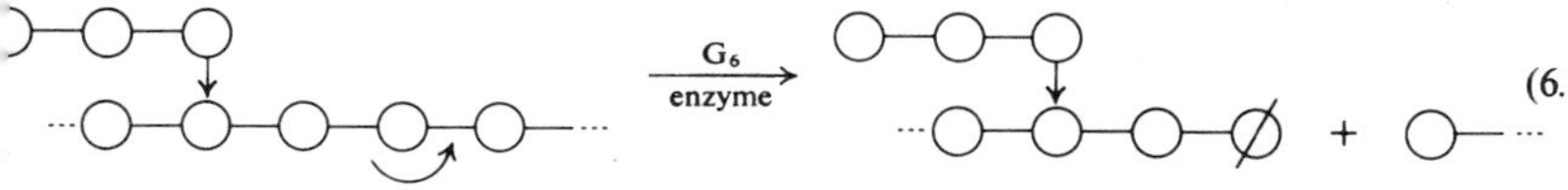

$$\xrightarrow[\text{enzyme}]{G_6} \qquad (6.15)$$

formation of large amounts of pure maltohexaose, Kainuma has succeeded in crystallising maltohexaose from an acetone–water mixture.

#### *6.13.3.4* ***Isopullulanase*** (EC 3.2.1.57)

Isopullulanase[158] is produced by some strains of *Aspergillus niger*. It has the capacity to cleave pullulan, producing isopanose [reaction (6.16)]. It also cleaves certain oligosaccharides which contain an α-1,6-link, provided that the B or C chain does not extend past the α-1,6-link (Figure 6.51).

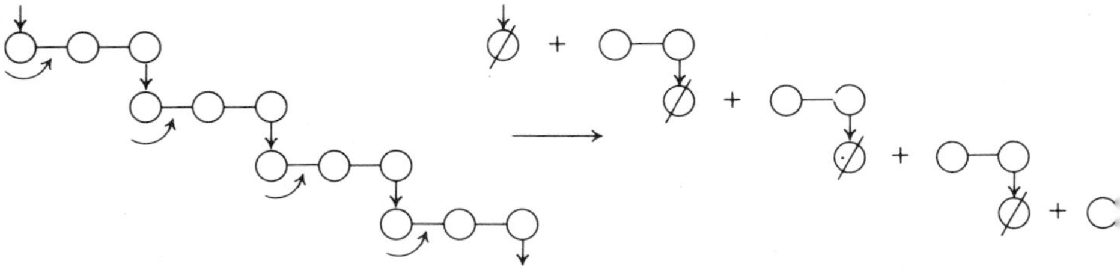

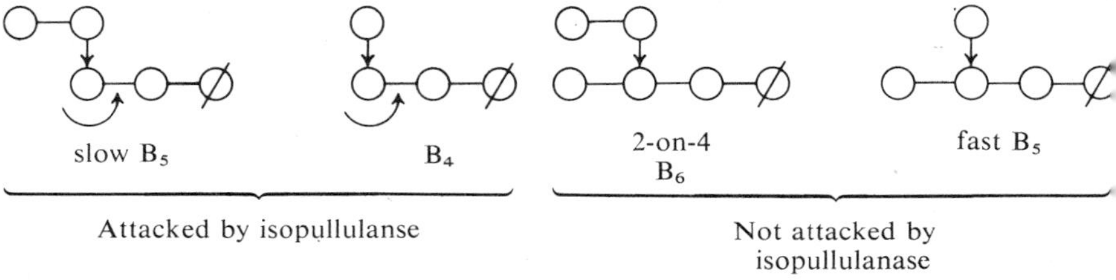

slow $B_5$ | $B_4$ | 2-on-4 $B_6$ | fast $B_5$

Attacked by isopullulanse | Not attacked by isopullulanase

**Figure 6.51** Substrate specificity of isopullulanase

## 6.14 LITERATURE AND COMMUNICATION IN STARCH SCIENCE

### 6.14.1 Reference volumes and compendia

A Comprehensive Survey of Starch Chemistry[159], Vol. 1. A planned second volume dealing with the patent literature was never published. The first part of the book consists of a collection of papers from a 1927 starch symposium. The papers are mainly of historical interest and reflect the state of knowledge at that time. The second part of the volume consists of titles and brief abstracts of over 3000 papers dealing with the chemistry and technology of starch and amylases, up until 1925.

Chemistry and Industry of Starch[160]. This reference book stresses the industrial aspects and application of starch.

Starch and its Derivatives[161]. This reference consists of 17 chapters dealing with starch, chemistry, physics, biochemistry and biology, both at the fundamental and applied level. Most of the chapters were contributed by specialists in the appropriate areas.

Starch: Chemistry and Technology[162]. A reference work in two volumes: Vol. I, Fundamental Aspects (20 contributers), Vol. II, Industrial Aspects (35 contributors). Vol. I covers the chemistry, biochemistry, genetics, botany, physical chemistry and physics of starch. Vol. II covers the commercial production of starch from various sources, starch products and starch modification, applications of starch, and starch analysis and identification. Vol. II contains over 30 optical micrographs of starch granules from different sources.

### 6.14.2 Specialised monographs

The Biogenesis of Starch Granules in Higher Plants[10]. Covers starch synthesis in leaves (chloroplasts) as well as storage tissues. Illustrated extensively with original electron micrographs.

The Differentiation and Specificity of Starches in Relation to Genera, Species, etc.[163]. Mammoth work, in two volumes. Contains hundreds of microphotographs in ordinary light and between crossed polarisers. The chemical part is largely of historical interest.

Die Mikroscopie der Stärkekörner (Microscopy of starch granules)[164]. The author classifies starch granules of approximately 900 plant species into over 30 structural types, on the basis of optical microscopy, swelling behaviour, iodine reaction, etc.

Methods in Carbohydrate Chemistry, Vol. IV, Starch[36]. A collection of authoritative procedures for preparation, analysis, characterisation and chemical modification of starch.

### 6.14.3 Periodicals devoted to starch and related subjects

*Die Stärke/Starch*. An international journal publishing original research and reviews. Articles are in German, English or French. A most valuable section contains abstracts (in German) of current papers and patents dealing with starch, amylases, etc. Also contains news and notices of meetings of interest to starch scientists.

*Denpun Kagaku* (Journal of the Japanese Society of Starch Science). Published quarterly by the Japanese Society of Starch Science. Original research, articles, reviews and reports of meetings. Articles and reviews are in Japanese with English summaries. Captions of Figures are usually in English.

*Cereal Chemistry*. Published bi-monthly by the American Association of Cereal Chemists. Articles, in English, describe original research on starch and other cereal products.

### 6.14.4 Meetings dealing with starch science

*Stärke-Tägung*, Detmold, Germany. Annual meeting, generally in March. Language is primarily German with some English and French. Sponsored by the Arbeitsgemeinschaft Getreideforschung e. V., Detmold, Germany.

*Starch Round Table.* Previously held annually at various places in the USA. Sponsored by the Corn Industries Research Foundation, now the Corn Refiners Association. Currently inactive.
*Japanese Starch Round Table.* Held annually for four days at various locations in the vicinity of Tokyo, usually during May. Sponsored by the Japanese Society of Starch Science (Nihon Denpun Gakkai).
*Japanese Society of Starch Science* has annual meetings and sponsors special meetings and lectures.
*Indian Starch Round Table.* A new venture designed as a forum for facets of starch science and technology relevant to India.
*American Association of Cereal Chemists.* Annual meeting, usually in November. Covers all aspects of cereal science and technology, generally including special sessions on starch or related carbohydrates.
*International Carbohydrate Symposium.* Held on alternate years. The programme may include papers on starch, etc.

## References


1. Irvine, J. C. (1926). *J. Chem. Soc.*, 1502
2. Haworth, W. N., Hirst, E. L. and Isherwood, F. A. (1937). *J. Chem. Soc.*, 577
3. Staudinger, H. and Husemann, E. (1937). *Annalen*, **527**, 195
4. Meyer, K. H. and Bernfeld, P. (1940). *Helv. Chim. Acta*, **23**, 875
5. Schoch, T. J. (1941). *Cereal Chem.*, **18**, 121
6. Cori, C. F., Schmidt, G. and Cori, G. T. (1939). *Science*, **89**, 464
7. Hanes, C. S. (1940). *Nature*, **145**, 348
8. Haworth, W. N., Peat, S. and Bourne, E. J. (1944). *Nature*, **154**, 236
9. Recondo, E. F. and Leloir, L. F. (1961). *Biochem. Biophys. Res. Commun.*, **6**, 85
10. Badenhuizen, N. P. (1969). *The Biogenesis of Starch Granules in Higher Plants* (New York: Appleton-Century-Crofts)
11. Alexander, A. G. (1973). *Ann. N.Y. Acad. Sci.*, **210**, 64
12. Badenhuizen, N. P. (1964). *Methods in Carbohydrate Chemistry*, Vol. 4, 14 (R. Whistler, editor) (New York and London: Academic Press)
13. Clapp, R. C., Bissett, F. H., Coburn, R. A. and Long, L. Jr. (1966). *Phytochemistry*, **5**, 1323
14. Bates, F. L., French, D. and Rundle, R. E. (1943). *J. Amer. Chem. Soc.*, **65**, 142
15. Schoch, T. J. (1942). *J. Amer. Chem. Soc.*, **64**, 2957
16. Alexander, D. E., Dudley, J. W. and Creech, R. G. (1970). *Corn: Culture, Processing, Products*, 6 (Westport, Conn.: Avi Publishing Co.)
17. Hanes, C. S. (1936). *Biochem. J.*, **30**, 168
18. Posternak, T. (1935). *Helv. Chim. Acta*, **18**, 1351
19. Schoch, T. J. (1942). *J. Amer. Chem. Soc.*, **64**, 2954
20. Meyer, K. H. (1952). *Experientia*, **8**, 405
21. Rundle, R. E. and Edwards, F. C. (1943). *J. Amer. Chem. Soc.*, **65**, 2200
22. Brimhall, B. and Hixon, R. M. (1939). *Ind. Eng. Chem.*, **11**, 358
23. Husemann, E. (1960). *Die Makromolekulare Chem.*, **35**, 239
24. Muetgeert, J. (1961). *Advances in Carbohydrate Chemistry*, Vol. 16, 299 (M. L. Wolfrom, editor) (New York and London: Academic Press)
25. Schoch, T. J. (1945). *Advances in Carbohydrate Chemistry*, Vol. 1, 247 (W. W. Pigman and M. L. Wolfrom, editors) (New York: Academic Press)
26. French, D., Pulley, A. O. and Whelan, W. J. (1963). *Stärke*, **15**, 349
27. Adkins, G. K. and Greenwood, C. T. (1969). *Carbohyd. Res.*, **11**, 217
28. Banks, W. and Greenwood, C. T. (1968). *Carbohyd. Res.*, **7**, 349
29. Husemann, E. and Pfannemüller, B. (1961). *Makromol. Chem.*, **49**, 214
30. Kerr, R. W. and Severson, G. M. (1943). *J. Amer. Chem. Soc.*, **65**, 193
31. Lansky, S., Kooi, M. and Schoch, T. J. (1949). *J. Amer. Chem. Soc.*, **71**, 4066

32. Peat, S., Pirt, S. J. and Whelan, W. J. (1952). *J. Chem. Soc.*, 714
33. Banks, W. and Greenwood, C. T. (1967). *Stärke*, **7,** 197
34. Erlander, S. and French, D. (1956). *J. Polymer Sci.*, **20,** 7
35. Freudenberg, K. and Boppel, H. (1940). *Naturwissenschaften*, **28,** 264
36. Ingle, T. R. and Whistler, R. L. (1964). *Methods in Carbohydrate Chemistry*, Vol. 4 (R. L. Whistler, editor) (New York and London: Academic Press)
37. Shasha, B. and Whistler, R. (1964). *Methods in Carbohydrate Chemistry*, Vol. 4, 86 (R. L. Whistler, editor) (New York and London: Academic Press)
38. Hay, G. W., Lewis, B. A. and Smith, F. (1965). *Methods in Carbohydrate Chemistry*, Vol. 5, 357 (R. L. Whistler, editor) (New York and London: Academic Press)
39. Hough, L. (1965). *Methods in Carbohydrate Chemistry*, Vol. 5, 370 (R. L. Whistler, editor) (New York and London: Academic Press)
40. Wolfrom, M. L. and Thompson, A. (1956). *J. Amer. Chem. Soc.*, **78,** 4116
41. Wolfrom, M. L. and Thompson, A. (1955). *J. Amer. Chem. Soc.*, **77,** 6403
42. Manners, D. J., Mercer, G. A. and Rowe, J. J. M. (1965). *J. Chem. Soc.*, 2150
43. Lee, E. Y. C. (1971). *Arch. Biochem. Biophys.*, **146,** 488
44. Larner, J., Illingworth, B., Cori, G. T. and Cori, C. F. (1952). *J. Biol. Chem.*, **199,** 641
45. Peat, S., Whelan, W. J. and Thomas, G. J. (1956). *J. Chem. Soc.*, 3025
46. Bathgate, G. N. and Manners, D. J. (1966). *Biochem. J.*, **101,** 3C
47. Lee, E. Y. C., Mercier, C. and Whelan, W. J. (1968). *Arch. Biochem. Biophys.*, **125,** 1028
48. Abdullah, M. and French, D. (1970). *Arch. Biochem. Biophys.*, **137,** 483
49. Heller, J. and Schramm, M. (1964). *Biochim. Biophys. Acta*, **81,** 96
50. French, D. (1973). *J. Animal Sci.*, **37,** 1048
51. Weintraub, M. and French, D. (1970). *Carbohyd. Res.*, **15,** 251
52. Kainuma, K. and French, D. (1971). *Biopolymers*, **10,** 1673
53. French, D. (1957). *Advances in Carbohydrate Chemistry*, Vol. 12, 189 (M. L. Wolfrom, editor) (New York: Academic Press)
54. Rundle, R. E. and Edwards, F. C. (1943). *J. Amer. Chem. Soc.*, **65,** 2200
55. Zaslow, B. (1963). *Biopolymers*, **1,** 165
56. Osman, E., Leith, S. J. and Fles, M. (1961). *Cereal Chem.*, **38,** 449
57. Van Etten, R. L., Sebastian, J. F., Clowes, G. A. and Bender, M. L. (1967). *J. Amer. Chem. Soc.*, **89,** 3242
58. Colin and Gaultier, de C. (1814). *Ann. Phys.*, **48,** 297
59. Bailey, J. M. and Whelan, W. J. (1961). *J. Biol. Chem.*, **236,** 969
60. Adkins, G. K. and Greenwood, C. T. (1966). *Carbohyd. Res.*, **3,** 152
61. Thoma, J. A. and French, D. (1960). *J. Amer. Chem. Soc.*, **82,** 4144
62. Gilbert, G. A. and Marriott, J. V. R. (1948). *Trans. Faraday Soc.*, **44,** 84
63. Thoma, J. A. and French, D. (1961). *J. Phys. Chem.*, **65,** 1825
64. French, D. and Youngquist, R. W. (1960). *Abstr. Papers Amer. Assoc. Cereal Chem., 45th Meeting*, Abstr. 42
65. Banks, W. and Greenwood, C. T. (1963). *Advances in Carbohydrate Chemistry*, Vol. 18, 357, (M. Wolfrom, editor) (New York and London: Academic Press)
66. Senior, M. B. and Hamori, E. (1973). *Biopolymers*, **12,** 65
67. Brant, D. A. and Dimpfl, Wm. L. (1970). *Macromolecules*, **3,** 655
68. Rao, V. S. R., Sundararajan, P. R., Ramakrishnan, C. and Ramachandran, G. N. (1967). *Conformation of Biopolymers*, Vol. 2, 721
69. Rees, D. A. and Scott, W. E. (1971). *J. Chem. Soc. B*, 469
70. Goebel, C. V., Dimpfl, Wm. L. and Brant, D. A. (1970). *Macromolecules*, **3,** 644
71. Flory, P. J. (1973). *J. Polymer Sci.*, **2,** 621
72. Brant, D. A. and Dubin, P. L. (1973). *Polymer Preprints*, **14,** 169
73. Bell, D. J., Gutfreund, H., Cecil, R. and Ogston, A. G. (1948). *Biochem. J.*, **42,** 405
74. Greenwood, C. T. (1956). *Advances in Carbohydrate Chemistry*, Vol. 11, 335 (M. L. Wolfrom, editor) (New York: Academic Press)
75. Kainuma, K. and French, D. (1972). *Biopolymers*, **11,** 2241
76. Bear, R. S. and Samsa, E. G. (1943). *Ind. Eng. Chem.*, **35,** 721
77. Sandstedt, R. M. (1965). *Cereal Sci. Today*, **10,** 305
78. Finkelstein, R. S. and Sarko, A. (1972). *Biopolymers*, **11,** 881
79. French, D. (1972). *J. Jap. Soc. Starch Sci.*, **19,** 8
80. Kreger, D. R. (1951). *Biochim. Biophys. Acta*, **6,** 406

81. Stöckmann, V. E. (1972). *Biopolymers*, **11**, 251
82. Sarko, A. and Marchessault, R. H. (1967). *J. Amer. Chem. Soc.*, **89**, 6454
83. Senti, F. R. and Witnauer, L. P. (1946). *J. Amer. Chem. Soc.*, **68**, 2407
84. Blackwell, J., Sarko, A. and Marchessault, R. H. (1969). *J. Mol. Biol.*, **42**, 379
85. Rundle, R. E., Daasch, L. and French, D. (1944). *J. Amer. Chem. Soc.*, **66**, 130
86. Schieltz, N. C. (1965). *Quart. Colorado School Mines*, **60** (4), 84
87. French, A. D. and Murphy, V. G. (1973). *Carbohydr. Res.*, **27**, 391
88. Mussulman, W. C. and Wagoner, J. A. (1968). *Cereal Chem.*, **45**, 162
89. Dronzek, B. L., Hwang, P. and Bushuk, W. (1972). *Cereal Chem.*, **49**, 232
90. Bear, R. S. and French, D. (1941). *J. Amer. Chem. Soc.*, **63**, 2305
91. Hellman, N. N., Boesch, T. F. and Melvin, E. H. (1952). *J. Amer. Chem. Soc.*, **74**, 348
92. Hizukuri, S., Takeda, Y. and Imamura, S. (1972). *J. Agric. Chem. Soc. Jap.*, **46**, 119
93. Robin, J. P., Mercier, C., Charbonniere, R. and Guilbot, A. (1974). *Cereal Chem.*, **51**, 389
94. French, D., Kainuma, K. and Watanabe, T. (1971). *Cereal Sci. Today*, **16**, 299
95. Thoma, J. A., Spradlin, J. E. and Dygert, S. (1971). *The Enzymes*, Chap. 6 (P. Boyer, editor) (New York and London: Academic Press)
96. Fischer, E. H. and Stein, E. A. (1961). *Biochem. Preprints*, **8**, 27
97. Fischer, E. H. and Bernfeld, P. (1948). *Helv. Chim. Acta*, **31**, 1831
98. Levitzki, A. and Steer, M. L. (1974). *Eur. J. Biochem.*, **41**, 171
99. Robyt, J. F. and French, D. (1970). *J. Biol. Chem.*, **245**, 3917
100. French, D. (1957). *Brewers Digest*, **32**, 50
101. Kung, J. T., Hanrahan, V. M. and Caldwell, M. L. (1953). *J. Amer. Chem. Soc.*, **75**, 5548
102. Robyt, J. F. and French, D. (1967). *Arch. Biochem. Biophys.*, **122**, 8
103. Robyt, J. F. and French, D. (1970). *Arch. Biochem. Biophys.*, **138**, 662
104. Wakim, J., Robinson, M. and Thoma, J. A. (1969). *Carbohyd. Res.*, **10**, 487
105. French, D., Chan, Y.-C. and England, B. (1974). *Fed. Proc.*, **33**, 1313 (Abstr. 510)
106. Hehre, E. J. and Genghof, D. S. (1971). *Arch. Biochem. Biophys.*, **142**, 382
107. Nordin, P. and French, D. (1958). *J. Amer. Chem. Soc.*, **80**, 1445
108. Robyt, J. F., Chittenden, C. G. and Lee, C. T. (1971). *Arch. Biochem. Biophys.*, **144**, 160
109. Robyt, J. F. and Ackerman, R. J. (1973). *Arch Biochem. Biophys.*, **155**, 445
110. Robyt, J. and French, D. (1963). *Arch. Biochem. Biophys.*, **100**, 451
111. Thoma, J. A., Rao, G. V. K., Brothers, C. and Spradlin, J. (1971). *J. Biol. Chem.*, **246**, 5621
112. French, D., Smith, E. E. and Whelan, W. J. (1972). *Carbohyd. Res.*, **22**, 123
113. Schwimmer, S. and Balls, A. K. (1949). *J. Biol. Chem.*, **180**, 883
114. Takeda, Y. and Hizukuri, S. (1969). *Biochim. Biophys. Acta*, **185**, 469
115. Colman, P. M. and Matthews, B. W. (1971). *J. Mol. Biol.*, **60**, 163
116. Summer, R. and French, D. (1956). *J. Biol. Chem.*, **232**, 469
117. Bailey, J. M. and French, D. (1956). *J. Biol. Chem.*, **266**, 1
118. French, D. and Youngquist, R. W. (1963). *Stärke*, **12**, 425
119. Thoma, J. A. (1968). *J. Theoret. Biol.*, **19**, 297
120. Spradlin, J. and Thoma, J. A. (1969). *J. Biol. Chem.*, **245**, 117
121. Bailey, J. M. and Whelan, W. J. (1957). *Biochem. J.*, **67**, 540
122. Pazur, J. H. (1972). *Methods in Enzymology*, Vol. 28, 931 (V. Ginsburg, editor) (New York and London: Academic Press)
123. Pazur, J. H. and Okada, S. (1967). *Carbohyd. Res.*, **4**, 371
124. Kelly, J. J. and Alpers, D. H. (1973). *Biochim. Biophys. Acta*, **315**, 113
125. Reese, E. T., Maguire, A. H. and Parrish, F. W. (1967). *Can. J. Biochem.*, **46**, 25
126. Abdullah, M., Fleming, I. D., Taylor, P. M. and Whelan, W. J. (1963). *Biochem. J.*, **89**, 35
127. Marshall, J. J. and Whelan, W. J. (1970). *FEBS Lett.*, **9**, 85
128. Taylor, P. M. and Whelan, W. J. (1966). *Arch. Biochem. Biophys.*, **113**, 500
129. Illingworth, B. and Brown, D. H. (1962). *Proc. Nat. Acad. Sci. USA*, **48**, 1619, 1783
130. Smiley, K. L. (1971). *Biotechnol. Bioeng.*, **13**, 309
131. Leloir, L. F., De Fekete, M. A. R. and Cardini, C. E. (1961). *J. Biol. Chem.*, **236**, 636
132. Kamogawa, A., Fukui, T. and Nikuni, Z. (1967). *J. Biochem.*, **63**, 361
133. French, D. and Wild, G. M. (1953). *J. Amer. Chem. Soc.*, **75**, 4490

134. Pfannemüller, B. (1968). *Stärke*, **20,** 351
135. Bittiger, H., Husemann, E. and Pfannemüller, B. (1971). *Stärke*, **23,** 113
136. Trevelyan, W. E., Mann, P. F. E. and Harrison, J. S. (1952). *Arch. Biochem. Biophys.*, **39,** 419
137. Whelan, W. J. (1971). *Biochem. J.*, **122,** 609
138. Borovsky, D. and Whelan, W. J. (1972). *Fed. Proc.*, **31,** 477
139. Murata, T., Sugiyama, T. and Akazawa, T. (1964). *Arch. Biochem. Biophys.*, **107,** 92
140. Hawker, J. S., Ozbun, J. L., Ozaki, H., Greenberg, E. and Preiss, J. (1974). *Arch. Biochem. Biophys.*, **160,** 530
141. Schiefer, S., Lee, E. Y. C. and Whelan, W. J. (1973). *FEBS Lett.*, **30,** 129
142. Bender, H. and Wallenfels, K. (1961). *Biochem. Z.*, **334,** 79
143. Abdullah, M., Catley, B. J., Lee, E. Y. C., Robyt, J., Wallenfels, K. and Whelan, W. J. (1966). *Cereal Chem.*, **43,** 111
144. Whelan, W. J. (1973). Personal communication
145. Gunja, Z. H., Manners, D. J. and Maung, K. (1961). *Biochem. J.*, **81,** 392
146. Gunja-Smith, Z., Marshall, J. J., Smith, E. E. and Whelan, W. J. (1970). *FEBS Lett.*, **12,** 96
147. Harada, T., Yokobayashi, K. and Misaki, A. (1968). *Appl. Microbiol.*, **16,** 1439
148. Wiesmeyer, H. (1962). *Methods in Enzymology*, Vol. 5, 141 (S. P. Colowick and N. O. Kaplan, editors) (New York and London: Academic Press)
149. Palmer, T. N., Ryman, B. E. and Whelan, W. J. (1968). *FEBS Lett.*, **1,** 1
150. Häselbarth, V., Schulz, G. V. and Schwinn, H. (1971). *Biochim. Biophys. Acta*, **227,** 296
151. Jones, G. and Whelan, W. J. (1969). *Carbohyd. Res.*, **9,** 483
152. French, D., Pulley, A. O., Effenberger, J. A., Rougvie, M. A. and Abdullah, M. (1965). *Arch. Biochem. Biophys.*, **111,** 153
153. Kitahata, S., Tsuyama, N. and Okada, S. (1974). *Agric. Biol. Chem.*, **38,** 387
154. Robyt, J. and French, D. (1964). *Arch. Biochem. Biophys.*, **104,** 338
155. Griffin, P. J. and Fogarty, W. M. (1973). *Biochem. Soc. Trans.*, **1,** 397
156. Robyt, J. F. and Ackerman, R. J. (1971). *Arch. Biochem. Biophys.*, **145,** 105
157. Kainuma, K., Kobayashi, S., Ito, T. and Suzuki, S. (1972). *FEBS Lett.*, **26,** 281
158. Sakano, Y., Higuchi, M. and Kobayashi, T. (1972). *Arch. Biochem. Biophys.*, **153,** 180
159. Walton, R. P. (1928). *A Comprehensive Survey of Starch Chemistry*, Vol. 1 (New York: Chemical Catalog Co.)
160. Kerr, R. W. (1950). *Chemistry and Industry of Starch* (New York: Academic Press)
161. Radley, J. A. (1968). *Starch and its Derivatives* (London: Chapman and Hall)
162. Whistler, R. L. and Paschall, E. F. (1965). *Starch: Chemistry and Technology*, 2 Vols. (New York and London: Academic Press)
163. Reichert, E. T. (1913). *The Differentiation and Specificity of Starches in Relation to Genera, Species, etc.* (Washington, D.C.: Carnegie Institute)
164. Czaja, A. T. (1969). *Die Mikroscopie der Stärkekörner* (Berlin and Hamburg: Paul Parey)
165. Hall, D. E. and Sayre, J. G. (1970). *Textile Res. J.*, **40,** 147
166. Borovsky, D. and Smith, E. E. (1974). *Fed. Proc.*, **33,** 1559

# 7
# Regulatory Mechanisms in Glycogen Metabolism

E. G. KREBS and J. PREISS
University of California School of Medicine

---

## 7.1 SCOPE

One part of this article will concern itself with the regulation of glycogen metabolism in higher animals, with special reference to skeletal muscle. For this portion the reader is also referred to the elegant chapter on the same subject by Newsholme and Start[1]. Other noteworthy reviews on related topics include those by E. H. Fischer and his collaborators on glycogen phosphorylase[2,3] as well as that by Graves and Wang[4], the chapter by Larner and Villar-Palasi on glycogen synthase[5] and that by Stalmans and Hers on glycogen synthesis from UDPG[6]. Intimately related to the regulation of glycogen metabolism in animals is the subject of protein phorphorylation reactions and protein kinases which has been reviewed recently[7,8], as well as the all encompassing topic of cyclic AMP[9]. Although a general treatment of cyclic AMP is obviously beyond the scope of the present chapter, it is inevitable that many aspects of this subject will be discussed in the present review. Cyclic AMP[9] was originally discovered by Sutherland and co-workers as a result of their work on the stimulation of glycogenolysis in liver. Moreover, the mechanism of action of cyclic AMP as a cofactor for protein kinases[10] grew out of work in the regulation of glycogenolysis and glycogenesis[11-16] and the best characterised protein phosphorylation reactions are those involved in glycogen metabolism.

The second portion of the review deals mainly with the regulation of glycogen metabolism in bacteria and to a small extent with the regulation of leaf and algal starch synthesis. Starch metabolism is comprehensively covered in Chapter 6 of this volume by French. The regulatory modes observed for bacterial glycogen and starch synthesis appear to be very similar. Moreover, the pathways involved in the debranching and subsequent degradation of the plant and bacterial glucans to the monosaccharide level have been postulated to be the same, and they differ[17] from what has been described for mammalian glycogen catabolism[18]. Thus, the similarities in the regulation of metabolism of the 1,4-$\alpha$-glucans in bacteria and plants will be pointed out where appropriate.

A review in 1969 dealt with the regulation of the biosynthesis of 4,1-$\alpha$-glucans in bacteria and plants[19]. Recently a review describing the detailed properties of bacterial and plant ADPglucose pyrophosphorylases has appeared[20]. A review on the role and regulation of the energy reserve polymers glycogen, polyphosphate and poly-β-hydroxybutyrate in micro-organisms has recently appeared[21] and a number of earlier reviews dealing with earlier work on the endogenous metabolism of energy reserve compounds are also relevant to the material presented on regulation of bacterial glycogen metabolism[22-24].

## 7.2 FUNCTIONS OF GLYCOGEN

### 7.2.1 Animal cells

It is generally accepted that glycogen serves as a reserve source of energy in animal cells and is present in essentially all types of cell in the body. Glycogen accumulates under conditions in which exogenous supplies of energy and glucose are abundant and is utilised when these supplies are diminished. It is recognised, however, that vast quantitative differences exist between the various tissues and organs with respect to the importance of glycogen metabolism. Moreover, in liver and muscle the metabolism of glycogen must also be viewed in relation to its impact on the total body economy and not solely in reference to its cellular function. In this connection it is known that liver glycogen serves the body differently than muscle glycogen.

Does glycogen serve any function other than being a reserve source of energy in animal cells? Does it, for example, serve as an essential precursor for the biosynthesis of important metabolites or cellular components other than the recognised intermediates of energy metabolism? Obviously, the carbon of glycogen is utilised for many biosynthetic purposes, but no examples come readily to mind in which glycogen is *required* for this function. If a specific precursor role for glycogen were to be discovered, it would be anticipated that the regulation of glycogenolysis would be geared to the biosynthetic pathway.

Another possible function of glycogen is its potential regulatory role in metabolism. Glycogen is known to interact with the enzymes glycogen synthase and phosphorylase, for which it is substrate, but it also interacts with other enzymes. For example, glycogen binds to phosphorylase kinase, affecting not only its activity in the phosphorylase *b* to *a* reaction, but also its activation by phosphorylation[13,16]. Glycogen also stimulates the slow phosphorylation of casein catalysed by phosphorylase kinase[25]. The dephosphorylation of glycogen synthase D is inhibited by glycogen[26], but this action is probably due to the interaction of glycogen with its substrate rather than with the synthase phosphatase (see below). In studies on the phosphorylation and dephosphorylation of phosphorylase in the muscle 'glycogen particle', E. H. Fischer and his collaborators[27,28] found that these processes were strongly affected by glycogen. Destruction of glycogen with α-amylase increased the affinity of phosphorylase kinase for $Ca^{2+}$ 10-fold (but see later) and abolished the $Ca^{2+}$-dependent inhibition of phosphorylase phosphatase.

### 7.2.2 Bacterial cells

Table 7.1 lists those bacteria in which glycogen has been reported to be present. Glycogen is not restricted to any class of bacteria; many Gram-positive as well as gram-negative bacteria synthesise glycogen. Glycogen accumulation in bacteria usually occurs as a result of limited growth conditions in the presence of an excess carbon source[21-23]. For many bacterial species glycogen accumulation occurs in the stationary phase when growth ceases because of either nitrogen depletion[29,31,34,36,43,44,58], low pH[34], sulphur depletion[34,58] or phosphate depletion[34], but in the presence of an excess carbon source. Exceptions where glycogen accumulation has been shown to occur at optimal rates during exponential growth have been in the growth of *Streptococcus mitis*[52] and *Rhodopseudomonas capsulata*[49].

In bacteria, the function of glycogen is not as clear-cut as in animal cells. It is believed, however, that bacterial glycogen is utilised for energy required for the preservation of cell integrity especially during conditions of starvation, i.e. when an exogenous source of carbon is not available. Experiments carried out in the early 1960s suggested that bacteria required, in addition to energy for growth, energy for maintenance under non-growing conditions[59-63]. This 'energy of maintenance' has been defined as that energy required for processes such as turnover of RNA and protein, for the maintenance of motility, for osmotic regulation, for maintenance of intracellular pH, and for the chemotactic response.

**Table 7.1 Occurrence of glycogen in bacteria***

| *Organism* | *Ref.* |
|---|---|
| *Aerobacter aerogenes* | 29 |
| *Agrobacterium tumefaciens* | 30, 31 |
| *Arthrobacter crystallopoietes* | 32 |
| *Arthrobacter viscosus* | 33–37 |
| *Bacillus cereus* | 36 |
| *Bacillus megaterium* | 38 |
| *Bacillus stearothermophilus* | 39 |
| *Clostridium pasteurianum* | 40, 41 |
| *Escherichia coli* | 42–44 |
| *Mycobacterium phlei* | 45 |
| *Mycobacterium smegmatis* | 45, 46 |
| *Mycobacterium tuberculosis* | 45, 47, 48 |
| *Rhodopseudomonas capsulata* | 49 |
| *Rhodospirillum rubrum* | 50 |
| *Salmonella montevideo* | 42 |
| *Salmonella typhimurium* | 51 |
| *Shigella* | 42 |
| *Streptococcus mitis* | 52–54 |
| *Streptococcus mutans* | 54 |
| *Streptococcus salivarius* | 55, 56 |
| *Streptococcus sanguis* | 54, 57 |

* This list is not presumed to be complete either in the listing of all bacteria accumulating glycogen or in the references documenting the occurrence of glycogen in each organism

In support of this a number of experiments have shown that incubation of *E. coli* cells containing glycogen in media devoid of a carbon source do not degrade the nitrogen-containing constituents of the cell[64]. Ammonia is released from RNA and protein only after the glycogen reserve is depleted. This is in contrast to *E. coli* cells containing no glycogen. Ammonia is immediately secreted by glycogen-less cells upon incubation in media containing no carbon source. Similar experiments have been done with *Aerobacter aerogenes*[29]. There is no complete suppression of the release of ammonia in the catabolism of nitrogenous materials by glycogen in this species, but cells that do contain glycogen release ammonia at a much lower rate than cells that do not contain glycogen. Thus these experiments suggest that endogenous glycogen accumulation aids in the preservation of cellular constituents undergoing turnover.

Consistent with this is the evidence suggesting that glycogen-containing *Aerobacter aerogenes*[29] or *E. coli*[65] survive better than the same organisms containing no glycogen. Other experiments, however, have shown that in media containing 0.5–1.0 mM $MgCl_2$ the survival rate of glycogen-containing and glycogen-deficient cells were the same[65, 66]. It is quite possible that the $MgCl_2$ in the media caused greater stabilisation of the ribosomes present in the cell, thereby avoiding their degradation by latent ribonucleases and proteases. This stabilisation would then decrease the necessity for protein and RNA turnover and possibly obviate much of the energy requirement for maintenance. Marr *et al.*[60] have shown that the resynthesis of proteins in *E. coli* during turnover appears to be a major portion of the maintenance energy requirement. Therefore, the observation that glycogen has no effect on prolonging viability of bacteria in $Mg^{2+}$-sufficient media does not rule out the role of glycogen as an energy reserve in media having low levels of $Mg^{2+}$ and in other probable normal physiological situations.

The survival characteristics of *Streptococcus mitis* containing glycogen has been compared with cells containing no glycogen[53]. Cells which possessed high levels of glycogen survived well in 20 mM phosphate buffer, pH 6.8 containing 0.13 M NaCl at 37 °C, while cells containing little or no glycogen died rapidly.

Another function for glycogen or a 1,4-α-glucan similar to glycogen, named granulose, has been alluded to in the studies of the obligate anaerobes *Clostridium pasteurianum*[67] and *Clostridium botulinum* type E[68, 69]. *Cl. pasteurianum* accumulates little granulose during exponential growth, but, just prior to or during initiation of sporulation, accumulates up to 60% dry weight as granulose. Strains of *Cl. pasteurianum* which were poor spore formers only accumulate 15% of their dry weight as granulose prior to sporulation. In *Cl. botulinum* type E it has been demonstrated that granulose is rapidly accumulated and then lost during growth and spore formation in glucose–trypticase medium. Non-proliferating cells containing granulose rapidly degraded the polysaccharide when placed in carbon-free media and transformed into mature spores. These observations and others suggested to Strasdine[68] that granulose served as an endogenous source of carbon and energy for spore maturation. Thus granulose (glycogen) in *Clostridia* may play a role in the formation of spores and in the subsequent survival of these anaerobes.

In summary, a number of experiments suggest that glycogen may play a role in the survival of the bacterial cell. The precise function of glycogen in this role, however, remains to be elucidated. Moreover, a number of studies seem to obfuscate the issue. As mentioned above, the survival rates of magnesium-rich *E. coli* and *A. aerobacter* cells are not affected whether they contain glycogen or not. Moreover, glycogen-rich *Sarcina lutea* dies at a faster rate than cells containing no polysaccharide when starved in phosphate buffer[70]. It is clear that additional studies are required to clarify the function of glycogen in bacteria.

## 7.3 PATHWAYS AND SITES OF REGULATION IN GLYCOGEN METABOLISM

### 7.3.1 Animal cells

No definition of what is meant by 'glycogen metabolism' is entirely satisfactory, because it is not possible to segregate a given set of reactions and say that they constitute the only pathways of importance with respect to any specific area of metabolism. Nevertheless, in an attempt to narrow the range of topics to be discussed, an arbitrary section of the metabolic map, i.e. the reactions between glucose 6-P and glycogen (Figure 7.1), will be considered with respect to regulation. Also to be considered are the reactions catalysed by the branching enzyme and the debranching enzyme[71] or 'transferase-glucosidase' (EC 3.2.1.33), which are not illustrated in Figure 7.1. Because

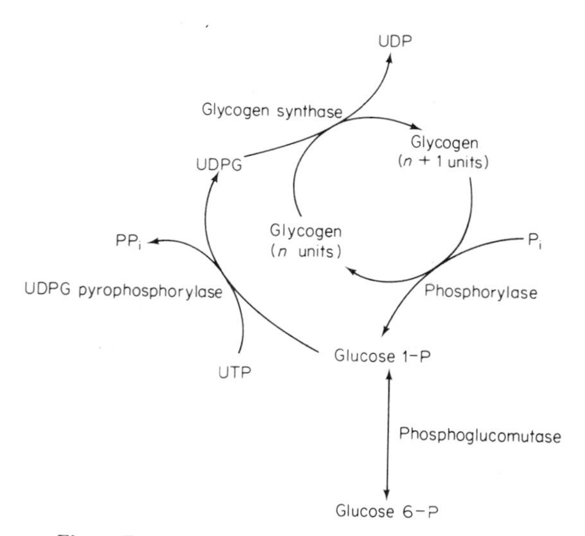


**Figure 7.1** Pathways of glycogen metabolism

there is so much uncertainty as to whether or not hydrolytic enzymes other than the transferase-glucosidase have a dynamic role in glycogen metabolism, the reactions catalysed by these enzymes will not be taken up.

What are the regulatory enzymes of glycogen metabolism, i.e. which steps in the pathway of Figure 7.1 should be considered with respect to control? In animal species it would appear that only two reactions are of major importance. These are the glycogen synthase (EC 2.4.1.11) reaction in the biosynthetic pathway and the phosphorylase (EC 2.4.1.1) reaction in the degradative pathway. The synthase reaction is the only 'committed step' in glycogen synthesis from glucose 6-P, i.e. the phosphoglucomutase (EC 2.7.5.1) and UDPG pyrophosphorylase (EC 2.7.7.9) reactions serve other purposes. Similarly, the phosphorylase reaction is the only step between glycogen and glucose 6-P which serves the glycogenolytic process exclusively. Furthermore, investigation of the properties of glycogen synthase and phosphorylase has revealed numerous potential control devices, in contrast to what has been seen with the other enzymes. With reference to experimental work serving to delineate points of control, Sutherland and Cori[72, 73] showed that phosphorylase is rate-limiting in the conversion of glycogen into glucose in liver slices, as had been anticipated even earlier[74]. These observations and conclusions were especially noteworthy in that they were made at a time when it was believed that phosphorylase catalysed glycogen synthesis as well as the phosphorolysis of glycogen. Shortly after the discovery of glycogen synthesis from UDPG[75], attention was focused on glycogen synthase when two of its important regulatory properties, namely activation by glucose 6-P[76] and its existence in more than one form[77, 78], became known. In keeping with their importance as regulatory sites, the glycogen synthase and phosphorylase reactions are essentially irreversible under physiological conditions, the marked displacement of the otherwise reversible phosphorylase reaction being brought about by the high intracellular concentration of inorganic phosphate and the relatively low level of glucose 1-P.

Is there reason to believe that enzymes other than glycogen synthase and phosphorylase (Figure 7.1) are involved in control mechanisms important with respect to glycogen metabolism? Phosphoglucomutase is readily inhibited by heavy metal ions[79, 80] that bind to the enzyme and compete with $Mg^{2+}$, an essential activator of the enzyme[81]. Peck and Ray[82] have reported that insulin administered to fasting animals results in the displacement of $Zn^{2+}$ by $Mg^{2+}$ *in vivo*, a phenomenon which might indicate that phosphoglucomutase is subject to hormonal control. The UDPG pyrophosphorylase of animals cells does not appear to be involved in dynamic aspects of metabolic regulation[83]. Regulation of the branching or debranching enzymes, which would be made manifest by changes in the degree of branching or the average chain length of glycogen, is not known to exist.

### 7.3.2 Bacterial cells

#### *7.3.2.1 Enzymatic reactions leading to 1,4-α-glucan synthesis*

Previous to 1957, enzymatic synthesis of bacterial 1,4-α-glucan was believed to occur either from sucrose via the amylosucrase (EC 2.4.1.4) reaction [equation (7.1)] or from α-D-glucose 1-P via the phosphorylase reaction or from maltose via the amylomaltase (EC 2.4.1.25) reaction [equation (7.2)][84].

$$\text{Sucrose} + (1,4\text{-}\alpha\text{-D-glucosyl})_n \rightleftharpoons \text{D-fructose} + (1,4\text{-}\alpha\text{-D-glucosyl})_{n+1} \qquad (7.1)$$

$$\text{Maltose} + (1,4\text{-}\alpha\text{-D-glucosyl})_n \rightleftharpoons \text{D-glucose} + (1,4\text{-}\alpha\text{-D-glucosyl})_{n+1} \qquad (7.2)$$

Amylosucrase activity has been detected in a few bacteria. The enzyme is constitutive in *Neisseria* strains, which accumulate large quantities of a water-soluble glycogen-type polysaccharide when grown on sucrose[84-89].

In *Corynebacteria*, certain *Streptococci* and in *Neisseria perflava*, synthesis of amylose and amylopectin-type polymers occurs *in vivo* or *in vitro* from glucose 1-P and is believed to be due to phosphorylase action[86, 90].

When *Streptococcus mutans*[17], *Aerobacter aerogenes*[17] or many strains of *E. coli*[17,91], *Streptococcus mitis*[92], *Diplococcus pneumoniae*[93] or *Pseudomonas stutzeri*[94] are grown on maltose or on maltodextrins, a number of enzymes, including amylomaltase, involved in utilisation of these oligosaccharides for growth are induced.

The occurrence of amylosucrase in bacteria appears to be limited to a few species and these bacteria are capable of catalysing 1,4-α-glucan synthesis in large quantities when sucrose is supplied in the media. Presently there are no reports indicating that *Neisseria* or other bacteria are capable of synthesising sucrose *in vitro* or *in vivo*. Thus, amylosucrase activity could not account for the observed accumulation of glycogen in micro-organisms grown on carbon sources besides sucrose. Similarly, amylomaltase synthesis is only induced in the presence of maltose or maltodextrins and is repressed by glucose[95]. Its activity, therefore, appears not to be responsible for the synthesis of glycogen in organisms grown on glucose as a carbon source.

Glycogen phosphorylase[86,89,96-98] and maltodextrin phosphorylase[17,93,94,97,99-101] have been reported to be present in a number of bacteria. However, the maltodextrin phosphorylase activity is only induced in the presence of maltose, and in many bacteria the glycogen phosphorylase activity is too low to account sufficiently for the rate of glycogen accumulation observed[96-98]. Furthermore, mutants of *E. coli* containing defective maltodextrin phosphorylase cannot utilise maltodextrins, suggesting that the phosphorylase is involved in phosphorolysis of the oligosaccharides rather than in their synthesis[102,103].

After the discovery of the UDPglucose pathway for glycogen synthesis in mammalian tissues by Leloir and his colleagues[75,76], Madsen in 1961 reported the presence of a UDPglucose:1,4-α-D-glucan 4-α-glucosyl transferase (glycogen synthase) in extracts of *Agrobacterium tumefaciens* (A6)[104]. A subsequent report by Sigal *et al.* in 1964 showed, however, that several UDPglucose pyrophosphorylase-negative mutants of *E. coli* still accumulated normal amounts of glycogen during the stationary phase of growth on limited nitrogen media[44]. This strongly suggested that some precursor other than UDPglucose served as a glycosyl donor for the *E. coli* glycogen synthesis. At the same time it was reported that extracts of several bacteria, including *Agrobacterium tumefaciens* (A6), contained potent activities of an ADPglucose:1,4-α-D-glucan 4-α-glucosyl transferase [starch (bacterial glycogen) synthase][105,106]. These same bacterial extracts also contained ADPglucose pyrophosphorylase (EC 2.7.7.27) activity. The unique reactions in the ADPglucose pathway of bacterial glycogen synthesis may then be shown in equations (7.3) and (7.4).

$$\text{ATP} + \text{glucose 1-P} \rightleftharpoons \text{ADPglucose} + \text{PP}_i \quad \text{(ADPglucose pyrophosphorylase)} \qquad (7.3)$$

$$\text{ADPglucose} + (1,4\text{-}\alpha\text{-D-glucosyl})_n \rightleftharpoons \text{ADP} + (1,4\text{-}\alpha\text{-D-glucosyl})_{n+1} \quad \text{(glycogen synthase)} \qquad (7.4)$$

The formation of the 1,6-α-branch linkages in glycogen is presumably catalysed by the 1,4-α-glucan branching enzyme (EC 2.4.1.18). Branching enzyme activity has been detected in *Arthrobacter globiformis*[107] and in *E.*

*coli.*[108, 109]. The isolation of glycogen-deficient mutants of *E. coli*[19, 20, 110-11] and *Salmonella typhimurium*[51] containing either defective glycogen synthases or defective ADPglucose pyrophosphorylases indicates that at least in these organisms the ADPglucose pathway is the predominant if not the sole pathway for glycogen synthesis. Table 7.2 lists those bacteria wherein glycogen synthase and/or ADPglucose pyrophosphorylase have been found.

**Table 7.2 The occurrence of ADPglucose pyrophosphorylase and ADPglucose-specific glycogen synthase in bacteria***

| *Organism* | *Glycogen synthase* | *ADPglucose pyrophosphorylase* |
|---|---|---|
| *Aerobacter aerogenes* | 105, 116, 117, 118 | 116 |
| *Aerobacter cloacae* | 116 | 116 |
| *Aeromonas formicans* | NR | 20 |
| *Aeromonas hydrophila* | 119 | 119 |
| *Agrobacterium tumefaciens* | 105, 116 | 120, 121 |
| *Arthrobacter viscosus* | 106, 122 | 123–125 |
| *Bacillus stearothermophilus* | 126 | NR |
| *Chromatium vinosum* | 127 | 127 |
| *Citrobacter freundii* | 116 | 116 |
| *Clostridium pasteurianum* | 41 | 41 |
| *Escherichia aurescens* | 116 | 116 |
| *Escherichia coli* B | 105, 128 | 114, 129–133 |
| *Escherichia coli* K-12 | 110, 134 | 110, 134 |
| *Micrococcus lysodeiktikus* | 105 | 135 |
| *Mycobacterium smegmatis* | 136 | 137 |
| *Pasteurella pseudotuberculosis* | 138 | NR |
| *Rhodomicrobium vanniellii* | 127 | 127 |
| *Rhodopseudomonas capsulata* | 139 | 120 |
| *Rhodopseudomonas gelatinosa* | 127 | 127 |
| *Rhodopseudomonas palustris* | 127 | 127 |
| *Rhodospirillum rubrum* | 105 | 139, 140 |
| *Salmonella typhimurium* | 116 | 116 |
| *Serratia marcescens* | 116 | 20, 116 |
| *Shigella dysenteriae* | 119 | 119 |
| *Streptococcus mitis* | 141 | NR |
| *Streptococcus salivarius* | 142 | 142 |

* Numbers listed are reference numbers. NR signifies not reported

### 7.3.2.2 *Regulation of the ADPglucose pathway*

As indicated previously, the site of regulation of glycogen synthesis in mammals is at the glycogen synthase level. In contrast, no regulatory phenomena are apparent with the bacterial glycogen synthase. Of these bacterial glycogen synthases tested, none was activated by glycolytic intermediates,

NADPH or pyridoxal phosphate[122,128,138,141]. However, the bacterial ADPglucose pyrophosphorylases appear to be allosterically regulated and an important feature of the enzyme catalysing the synthesis of ADPglucose is its activation by glycolytic intermediates and inhibition by 5′-adenylate, ADP or orthophosphate[19,20]. Modulation of the activity of ADPglucose pyrophosphorylase by the above metabolites has been the main emphasis in studies of the ADPglucose pyrophosphorylase from many sources[113-116,120,121,123-125,129-133,137,139,140]. These studies strongly suggest that regulation of bacterial glycogen synthesis occurs at the level of ADPglucose synthesis and that both ADPglucose and 1,4-α-glucan synthesis are regulated by the level of glycolytic intermediates and the energy state in the cell[19,20,113,114,143].

The differences in sites of regulation between those systems utilising the ADPglucose pathway can be easily rationalised. UDPglucose is utilised for the synthesis of other sugar nucleotides, mainly UDPgalactose and UDP-glucuronate, which are required for the synthesis of several cellular constituents. The first unique reaction for glycogen synthesis in the UDPglucose pathway is the UDPglucose: 1,4-α-D-glucan 4-α-glucosyl transferase step. As will be described, both allosteric control and hormonal mediated control is exerted on the activity of this enzyme. However, in bacteria and leaf tissues, ADPglucose probably has no other physiological function besides being the sole glucosyl nucleotide precursor of the 1,4-α-glucosyl linkages. It would be advantageous therefore for the prokaryote and plant cell to regulate glycogen synthesis at the level of ADPglucose formation so as to conserve the ATP utilised for synthesis of the sugar nucleotide. One would consider the ADP-glucose pyrophosphorylase step as the first unique step in 1,4-α-glucan synthesis. The finding of regulation occurring at this step is consistent with the concept that regulation of a biosynthetic pathway occurs at the first unique step of the pathway.

## 7.4 THE REGULATION OF GLYCOGEN METABOLISM IN SKELETAL MUSCLE

### 7.4.1 General considerations

#### *7.4.1.1 Skeletal muscle glycogen and the total body economy*

The total amount of muscle glycogen in a 70 kg man is *ca.* 0.15 kg and represents approximately two-thirds of all the glycogen present in the body[144]. In comparison with fat stores, this glycogen cannot be looked upon as constituting a major 'fuel depot'. Nevertheless, it is available for use on a short-term or emergency basis and constitutes an important source of readily available energy. Glycogenolysis and glycolysis in muscle provide for the synthesis of ATP within that tissue and also give rise to blood lactate that is utilisable as a fuel by the heart and other tissues. The blood lactate serves as a source of carbon for glucose synthesis in the liver. Because muscle glycogen has functions which go beyond the cellular level, it is not surprising that its metabolism is subject to regulation by the endocrine system, as well as by processes centred in muscle *per se*.

#### *7.4.1.2 Different types of skeletal muscle*

As is well recognised, skeletal muscle can be subdivided into various types. Early classification schemes recognised only red and white muscle, but more recently workers have spoken of three types of muscle fibres. In a recent study[145] the three types of fibres were referred to as fast-twitch-glycolytic, fast-twitch-oxidative-glycolytic and slow-twitch-oxidative. The fast-twitch-glycolytic fibres are primarily anaerobic fibres and are characterised by having a high glycogen level, high levels of certain glycolytic enzymes, including phosphorylase, and low levels of mitochondrial enzymes and myoglobin. The slow-twitch-oxidative fibres are adapted for aerobic metabolism and have less glycogen and lower concentrations of glycolytic enzymes. The fast-twitch-oxidative-glycolytic fibres appear to be richly endowed with the capacity for anaerobic as well as aerobic metabolism.

It is difficult to make precise statements as to the relationship between the metabolite or enzyme content of tissues and the importance of the metabolic pathways in which they appear. With respect to the finding of high glycogen levels in those muscles in which glycogen metabolism is believed to be prominent, one can only conclude with certainty that this represents a rich source of potential energy. The high levels of glycogen may simply indicate an imbalance between biosynthetic and degradative rates, and it should be recalled that the highest glycogen levels ever seen are those in certain glycogen storage diseases in which glycogen turnover is diminished. Under these conditions high glycogen levels are associated with a decreased importance of this substance to the economy of the cell. With respect to the significance of glycogenolytic and glycolytic enzyme levels as being indicative of the extent to which a given type of muscle fibre depends upon glycogen metabolism, a note of caution should again be introduced. A great many enzymes are present in great excess and variations in the amounts of such enzymes would appear to be of secondary importance. It is of interest that glycogen phosphorylase, the level of which is often used as a marker for the placement of a muscle on the anaerobic–aerobic scale, is an enzyme generally considered to be present in excess in terms of total enzyme concentration[146]. Despite these problems of interpretation, it is clear that glycogen does play a greater or a lesser role in the various types of fibres. This concept is supported by a number of pertinent physiological measurements and various biological observations in addition to the measurement of glycogen and enzyme levels. One would expect, therefore, to find that differences in regulatory mechanisms between the type of fibres will also be revealed. The latter may, in fact, be more striking than differences in the concentrations of muscle components.

### 7.4.2 Potential regulatory mechanisms

There is a distinct difference between the elucidation of likely regulatory mechanisms as revealed by the work of enzymologists and those which actually operate *in vivo*. Moreover, there is a tendency for investigators to treat all *in vivo* situations as being identical and to accept one set of data obtained under a specific set of experimental conditions as being indicative of a universal pattern. For these reasons the treatment of *potential* regulatory mechanisms in muscle will be handled separately from the operation of these mechanisms in specific physiological situations.

### 7.4.2.1 *Allosteric control*

The two key enzymes involved in the regulation of glycogenesis and glycogenolysis, glycogen synthase and phosphorylase, respectively, are both richly endowed with allosteric properties, suggesting that their activities can be readily controlled through their interaction with metabolites. Phosphorylase, particularly in its non-phosphorylated *b* form, is strongly activated by AMP[147, 148] and is inhibited by ATP[149, 150] and glucose 6-P. An analysis of the effects of these components in terms of the allosteric interactions involved is presented in Table 7.3 from the review by Graves and Wang[4].

**Table 7.3 Allosteric interactions in rabbit muscle phosphorylase *b* from many sources**

| | *Effect* of other ligands (heterotropic interactions)* | | | |
|---|---|---|---|---|
| | *Positive effectors* | | *Negative effectors* | |
| *Homotropic interaction ligand†* | *Affecting affinity* | *Affecting homotropic interaction* | *Affecting affinity* | *Affecting homotropic interaction* |
| AMP | $P_i$, G 1-P glycogen | $P_i$ | G 6-P ATP‡ | ATP G 6-P |
| G 1-P | AMP | AMP | ATP | ATP |
| $P_i$ | AMP | AMP | ATP G 6-P | ATP G 6-P |

* The effect is directed toward the ligands in the first column
† No homotropic interactions have been demonstrated for glycogen binding. Other small molecules which show homotropic interactions are ATP and glucose 6-P
‡ This effect arises from a direct competition of ATP and AMP for the nucleotide site

The sensitivity of phosphorylase *b* to AMP can be affected not only by heterotropic (or competitive) interactions involving low molecular weight metabolites, but can also be influenced by protein–protein interactions. For example, the interactions of the enzymes with protamine or polylysine increases its affinity for AMP[152] whereas its interaction with proteins in the glycogen particle decreases its affinity for this nucleotide[153].

Glycogen synthase is activated by glucose 6-P[76]. In this instance the phosphorylated form is affected much more than the non-phosphorylated form. The effect of the ligand on substrate affinities and $V_{max}$ values has been discussed by Stalmans and Hers[6]. In addition to the effects of glucose 6-P, inorganic phosphate, ATP and $Mg^{2+}$ have been implicated as potential effectors of glycogen synthase. The possible role of the various effectors in the physiological control of phosphorylase and glycogen synthase will be discussed later.

### *7.4.2.2 Interconversion between enzyme forms*

(a) *Glycogen phosphorylase*—Glycogen phosphorylase has the distinction of being the first enzyme known to exist in interconvertible forms[154]; these forms are the phosphorylated and non-phosphorylated species of the enzyme[155-157]. The interconversion reactions involving skeletal muscle phosphorylase are shown in equations (7.5) and (7.6):

$$\underset{\text{(dimer)}}{\text{2Phosphorylase } b} + \text{4ATP} \xrightarrow[\text{kinase}]{\text{phosphorylase } b} \underset{\text{(tetramer)}}{\text{Phosphorylase } a} + \text{4ADP} \quad (7.5)$$

$$\text{Phosphorylase } a + 4H_2O \xrightarrow[\text{phosphatase}]{\text{phosphorylase } a} \text{2Phosphorylase } b + 4P_i \quad (7.6)$$

These equations are usually presented as shown above, but it is now known that although muscle phosphorylase *a* occurs as a tetramer under laboratory conditions, the form that is most active catalytically is a dimer[158]. The non-phosphorylated form of the enzyme, phosphorylase *b*, is essentially inactive at finite substrate concentrations unless AMP is present, whereas phosphorylase *a* possesses nearly full activity without any cofactor. Thus the conversion of phosphorylase *b* into phosphorylase *a* is looked upon as an activation of the enzyme. The introduction of the covalently bound 'ligand', i.e. the phosphoryl group, favours the formation of an active conformation of the enzyme, a transition that is also brought about by the binding of AMP to phosphorylase. E. H. Fischer and his collaborators[159] have shown that the conversion of phosphorylase *b* into phosphorylase *a* takes place in a stepwise manner with the formation of partially phosphorylated intermediates. The latter may manifest almost full activity, especially in the presence of traces of AMP, but the activity of the phospho–dephospho hybrid molecules is strongly repressed in the presence of glucose 6-P. It is probable that the existence of the hybrids may be of importance with respect to fine aspects of control.

It is apparent that with any interconvertible enzyme system of the type represented by equations (7.5) and (7.6), the amount of each form of the enzyme present at a given time will be determined by the relative rates of the conversion reactions. These rates in turn will be governed by the respective activities of the interconverting enzymes. With respect to the interconverting enzymes involved in phosphorylation–dephosphorylation systems, much more is known about the protein kinases and the mechanism of their control than is known about the phosphoprotein phosphatases. Furthermore, Danforth *et al.*[146] have presented evidence supporting the concept that phosphorylase kinase plays a more dominant role than phosphorylase phosphatase in regulating the interconversion of phosphorylase *b* and phosphorylase *a*.

(b) *Phosphorylase kinase* (EC 2.7.1.38)—This enzyme, which represents the first protein kinase to have been recognised as an entity[155, 157], has been obtained in the pure form and a number of its physiochemical properties determined[160, 161]. Its molecular weight is $1.33 \times 10^6$, and it is made up of three types of subunits, designated as A, B and C, having molecular weights of 118 000, 108 000 and 41 000, respectively. As described in a different laboratory, the subunits have been designated $\alpha$, $\beta$ and $\gamma$ and assigned molecular weights of 145 000, 128 000 and 45 000[161]. The empirical formula of the enzyme would appear to be either $A_4B_4C_8$[160] or $A_4B_4C_4$[161]. No definitive information is available regarding the function of the individual subunits, but it is of interest to speculate as to why this enzyme has such a complicated quaternary structure. It is possible that all of the phosphoryl transferase activity resides in only one of the subunits, and in this connection it is known that the A and B subunits can be degraded extensively by trypsin without loss of enzyme activity, suggesting that subunit C is the catalytic subunit[162]. Subunits A and B may be regulatory in nature or they may be involved in determining the specificity of the kinase. As will be discussed later, the A and B subunits undergo enzymatic phosphorylation.

Phosphorylase kinase is a specific protein kinase that catalyses the phosphorylation of a single serine residue in phosphorylase. The sequence of amino acids at the phosphorylated site in rabbit skeletal muscle phosphorylase *a* is Lys-Glu-Ile-Ser(P)-Val-Arg[163]. Casein can be phosphorylated by phosphorylase kinase but at an extremely slow rate[16]. Similarly, the enzyme undergoes an autophosphorylation reaction that is also very slow. No sequence determinations have been made of the phosphorylated sites in these latter substrates. More recently[164] it was found that TN-I, one of the subunits of skeletal muscle troponin, serves as a relatively good substrate for phosphorylase kinase. It is of interest that the sequence of amino acids at the phosphorylated site in TN-I, Arg-Ala-Ile-Thr(P)-Ala-Arg, bears certain similarities to that in phosphorylase *a*. This TN-I sequence was determined independently in the laboratory of Dr. S. V. Perry (personal communication) and by Dr. T. S. Huang in the laboratory of E. G. Krebs.

From the viewpoint of enzyme regulation, several properties of muscle phosphorylase kinase are of potential importance. One of these is the requirement of the enzyme for $Ca^{2+}$, which was first revealed by the work of Meyer *et al.*[165], who found that the enzyme could be inhibited by EGTA and reactivated by $Ca^{2+}$. Later, Ozawa *et al.*[166] determined that the concentration of $Ca^{2+}$ required for half-maximal activation of the enzyme was *ca.* $1 \times 10^{-7}$ mol $l^{-1}$, somewhat lower than that ($\sim 3 \times 10^{-6}$ mol $l^{-1}$) required for the development of maximal tension in muscle. Another properties of phosphorylase kinase which may be important with respect to regulation is the strong inhibition of the enzyme that occurs when ATP concentration exceeds that of $Mg^{2+}$ (or $Mn^{2+}$)[13,57,167]. This may be due to the fact that free $Mg^{2+}$ activates the kinase, or it may be due to the presence of an inhibitory site for ATP on the enzyme. If the latter exists, however, it would not appear to be analogous to the site on phosphofructokinase since ATP inhibition is not overcome in the presence of 5′-AMP[168]. Finally, a property of phosphorylase kinase that is of major importance with respect to regulation is that this enzyme, like phosphorylase itself, undergoes activation and inactivation as a result of phosphorylation and dephosphorylation [equations (7.7) and (7.8)][16,169]:

$$\text{Non-activated phosphorylase kinase} + n\text{ATP} \xrightarrow[\text{dependent protein kinase)}]{\text{Phosphorylase } b \text{ kinase kinase (cyclic AMP-}} \text{Activated phosphorylase kinase} + n\text{ADP} \quad (7.7)$$

$$\text{Activated phosphorylase kinase} + n\text{H}_2\text{O} \xrightarrow[\text{kinase phosphatase}]{\text{Phosphorylase } b} \text{Non-activated phosphorylase kinase} + n\text{P}_i \tag{7.8}$$

Both forms of phosphorylase kinase require $Ca^{2+}$, but the activated form is more sensitive to the metal ion than the non-activated form.[170] The activities of non-activated and activated phosphorylase kinase as a function of pH are shown in Figure 7.2.

The relationship between phosphorylation and the activation of phosphorylase kinase is complex. More than one site becomes phosphorylated when the purified enzyme is incubated with MgATP[169]. Moreover, phosphate uptake does not correlate precisely with the increase in activity that occurs during the activation reaction. The situation is further complicated by the fact that

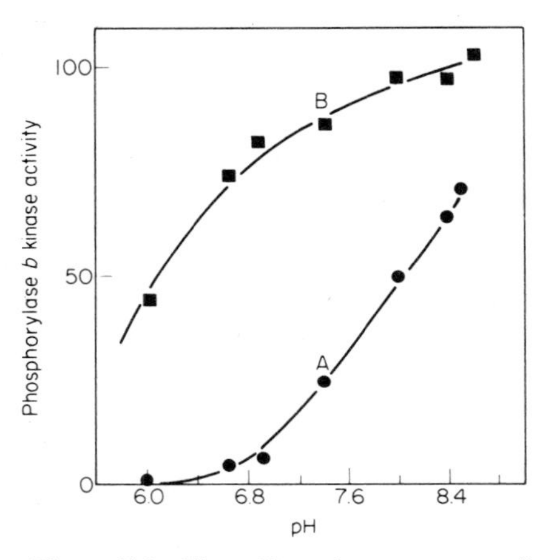


**Figure 7.2** The pH optimum curves of non-activated (A) and activated (B) rabbit skeletal muscle phosphorylase *b* kinase

activation is not only catalysed by the cyclic AMP-dependent protein kinase (see below) but is also autocatalytic[16, 171]. When activation is catalysed by the protein kinase, the B subunit is the first to become phosphorylated[160, 161, 172] and the increase in activity of the enzyme more or less parallels phosphate uptake, although this depends somewhat on how one measures activity. With longer phosphorylation times the A subunit also becomes phosphorylated. Evidence has been presented that this latter phase of phosphorylation may influence the susceptibility of the kinase to dephosphorylation by phosphorylase *b* kinase phosphatase[173]. The phosphatase was originally described by Riley *et al.*[169], who found that when it acts on activated phosphorylase kinase, various phosphorylated sites are attacked at different rates.

(c) *Glycogen synthase*—The two interconvertible forms of glycogen synthase, recognised[77, 78] by the different extent of their requirement for glucose 6-P, are referred to as form I (glucose 6-P independent) and form D (glucose 6-P dependent); these are non-phosphorylated and phosphorylated forms of the enzyme, respectively. Thus the pattern differs from that existing for the two forms of phosphorylase in which the non-phosphorylated form of phosphorylase is more dependent on a regulatory cofactor than the phosphorylated form. The interconversion reactions for glycogen synthase are shown in equations (7.9) and (7.10):

$$\text{Glycogen synthase I} + n\text{ATP} \xrightarrow[\text{protein kinase)}]{\text{Synthase I kinase (cyclic AMP-dependent}} \text{Glycogen synthase D} + n\text{ADP} \quad (7.9)$$

$$\text{Glycogen synthase D} + n\text{H}_2\text{O} \xrightarrow{\text{Synthase I kinase phosphatase}} \text{Glycogen synthase I} + n\text{P}_i \quad (7.10)$$

As with phosphorylase kinase, discussed above, the precise relationship between phosphorylation of glycogen synthase and the ensuing change in enzyme activity is poorly understood. Soderling *et al.*[174] found a fairly good correlation between the loss of glucose 6-P independent activity of the I form and phosphate uptake, amounting to 1 mole of phosphate per mole of synthase subunit (90 000–100 000 g protein). On the other hand, Brown and Larner[175] found that complete conversion of muscle glycogen synthase I into synthase D resulted in the uptake of 6 moles of phosphate per mole of enzyme subunit. The immediate site of phosphorylation in muscle glycogen synthase has been reported to have the same sequence of amino acids as is found in phosphorylase[176], a surprising finding in view of the fact that glycogen synthase kinase and phosphorylase kinase are different enzymes. If all six of the phosphate binding sites in glycogen synthase had the same structure it would suggest the presence of a repeating amino acid sequence in the enzyme.

Glycogen synthase kinase has been shown to be identical to phosphorylase *b* kinase kinase, i.e. to the cyclic AMP-dependent protein kinase (EC 2.7.1.37)[174, 177]. This arrangement provides a mechanism whereby an effector molecule, cyclic AMP, can decelerate glycogen synthesis and accelerate glycogenolysis by combining with a single receptor protein (Figure 7.3). A

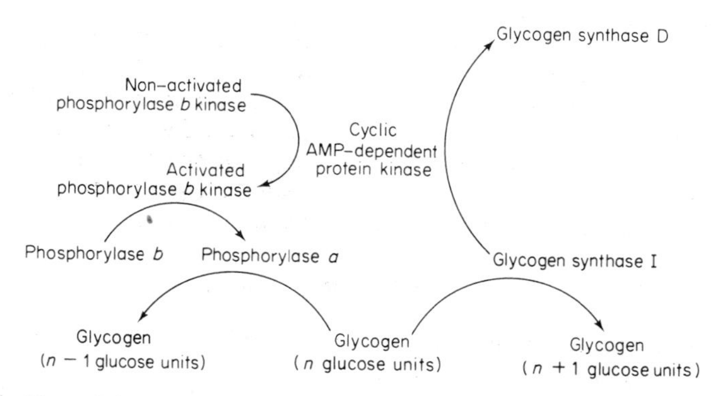


**Figure 7.3** The regulation of glycogenolytic enzymes and glycogen synthase by cyclic AMP

system of this type requires that in one instance protein phosphorylation results in the formation of a more active form of an enzyme whereas in the other instance it causes inactivation. It will be of interest to see whether this type of a reciprocal arrangement exists in other biosynthetic-degradative pathways outside the area of glycogen metabolism.

(d) *Cyclic AMP-dependent protein kinase*—Shortly after the discovery of cyclic AMP as the mediator of hormonal stimulation of phosphorylase activation in liver[178], it was determined that in the skeletal muscle system this compound probably acts on phosphorylase *b* kinase rather than phosphorylase phosphatase[11]. The effect of cyclic AMP on the kinase[11, 13] required MgATP and was shown to be due to an acceleration of the phosphorylation of that enzyme[179]. These findings subsequently led to the discovery of a cyclic AMP-dependent phosphorylase *b* kinase kinase which was named 'cyclic AMP-dependent protein kinase' when it was found that the enzyme had a substrate specificity broader than that which was implied by the more restrictive name[10]. Originally shown to phosphorylate protamine and casein, in addition to phosphorylase kinase, the protein kinase was soon found to be identical to the liver histone kinase studied by Langan[180]. Subsequently the enzyme was shown to catalyse the activation and presumably the phosphorylation of the hormone-sensitive lipase of fat cells[181, 182]. As noted earlier, the enzyme is identical to glycogen synthase kinase which had been known as a cyclic AMP-stimulated enzyme[183]. It was postulated that still other actions of cyclic AMP might be mediated by the kinase[184]. Kuo and Greengard[185], noting that cyclic AMP-dependent protein kinases are widely distributed in Nature, postulated that *all* effects of the cyclic nucleotide are due to its stimulation of protein kinases, a hypothesis which holds for animal systems, but is apparently not valid in bacteria.

The cyclic AMP-dependent protein kinase of skeletal muscle, like the enzyme from heart[186], the adrenal cortex[187], reticulocytes[188] and liver[189], is made up of two types of subunits, R and C[190]. R is a regulatory subunit which binds cyclic AMP and C is a catalytic subunit. The intact enzyme, i.e. the RC complex, is inactive, but dissociates to yield an active C subunit in the presence of cyclic AMP. The molecular weight of the protein kinase has been determined to be approximately 160 000 based on sedimentation equilibrium data using pure rabbit skeletal muscle enzyme[191]. Erlichman *et al.*[192] have reported a value of 174 700 for bovine heart cyclic AMP-dependent protein kinase. In terms of subunit structure the enzyme from either of these sources can be represented by the formula $R_2C_2$. The monomeric molecular weights of R and C are 48 000 and 38 000, respectively, for the skeletal muscle enzyme[191], or 49 000 and 38 000, respectively, for the heart muscle protein kinase[192], as determined by sodium dodecyl sulphate gel electrophoresis. The equation for activation of rabbit skeletal muscle protein kinase can be written as follows:

$$R_2C_2 + 2 \text{ cyclic AMP} \rightleftharpoons R_2(\text{cyclic AMP})_2 + 2C \qquad (7.11)$$

Activation of the heart muscle enzyme is similar except that the dimeric regulatory subunit binds only one mole of cyclic AMP instead of two[192].

It is probable that the mechanism of activation of the cyclic AMP-dependent protein kinase is a more complicated process than that shown in equation

(7.11). It is known, for example, that MgATP binds strongly to the enzyme[193]. MgATP also binds to C, since it serves as a substrate for the catalytic subunits, but this binding is much weaker than that observed originally by Haddox *et al.*[193] for the protein kinase itself. Work in this laboratory[191, 194] has revealed that the only component of equation (7.11) that binds MgATP with high affinity is $R_2C_2$. Thus, a more complete representation of the activation process would be:

$$R_2C_2(MgATP)_2 \underset{+2MgATP}{\overset{-2MgATP}{\rightleftharpoons}} R_2C_2 \underset{-2\text{ cyclic AMP}}{\overset{+2\text{ cyclic AMP}}{\rightleftharpoons}} R_2(\text{cyclic AMP})_2 + 2C$$

(a) (b)

(7.12a, b)

The high affinity binding of MgATP to $R_2C_2$ serves to prevent the metal–nucleotide complex from dissociating $R_2C_2$, i.e. as a result of its binding to C. Another factor of significance in relation to the activation of the protein kinase by cyclic AMP is that the protein kinase concentration in muscle and other tissues is of the same order of magnitude as the basal cyclic AMP concentration[195]. This has the effect of making the apparent $K_a$ for cyclic AMP activation higher than that usually measured in the laboratory at very low enzyme concentrations. A third point of interest is the possibility that it may not be essential for the $R_2C_2$ complex to dissociate in order for cyclic AMP to activate the enzyme; for example, a $R_2C_2$(cyclic AMP)$_2$ complex may be formed as an intermediate in the reaction of equation (7.11) and could constitute an active species of enzyme. Regarding this last point, it is significant that reports from a number of different laboratories have shown overlapping peaks of cyclic AMP binding protein and protein kinase activity in chromatographic separations using crude enzyme fractions. Finally, a point of unknown significance is the report of Rosen *et al.*[196] that the protein kinase from heart muscle undergoes endogenous autophosphorylation by MgATP; the R subunit is phosphorylated in this reaction. No autophosphorylation occurs, i.e. at an appreciable rate, with the skeletal muscle enzyme.

(e) *Phosphorylase phosphatase and other phosphoprotein phosphatases of muscle*—The characterisation of the phosphoprotein phosphatases in muscle has not kept apace with studies on the protein kinases. With respect to the regulation of glycogen metabolism, interest resides primarily in phosphorylase *a* phosphatase[154, 197], phosphorylase *b* kinase phosphatase[169] and glycogen synthase phosphatase[198]. The latter two activities would appear to reside in the same enzyme[199], which for the sake of simplicity will simply be referred to as phosphoprotein phosphatase in this chapter, implying, as suggested by Zieve and Glinsman[199], that this is an enzyme with a broad specificity comparable to that of its counterpart, i.e. the cyclic AMP-dependent protein kinase. Skeletal muscle phosphorylase *a* phosphatase is presumably a distinct enzyme with a more restricted specificity[169, 197, 200]. It should be noted, however, that neither of the phosphatases has been obtained in a completely pure form and final conclusions with regard to their respective specificities should be held in abeyance.

Phosphorylase *a* phosphatase as studied *in vitro* exhibits properties suggesting that it might be subject to physiological regulation and thus play an active part in controlling the relative amounts of phosphorylase *b* and phosphorylase *a*. Depending upon the stage of purification or possibly the type of assay system employed, the enzyme can be activated[201] or inhibited[202] by divalent cations. It is inhibited by glucose 1-P and inorganic phosphate[201] as well as by AMP (or IMP) and fluoride[203]. Activators of muscle phosphorylase *a* phosphatase of potential physiological significance in addition to the divalent cations include glucose 6-P[204, 205] and glucose[206]. The mechanism of action of the various effectors is of interest in that some of them, e.g. AMP[204, 207-210], influence the phosphatase reaction as a result of their interaction with the substrate, phosphorylase *a*. Others, e.g. inorganic phosphate[210], bind to the phosphatase itself. In addition to the influence of the effectors on the dephosphorylation of phosphorylase *per se*, a second factor that would have to be taken into account, in a complete description of their potential influence, would be how they might affect the catalytic activity of phospho–dephospho phosphorylase hybrids[159, 204]. Glucose 6-P, for example, not only accelerates the phosphatase reaction, but prevents phospho–dephospho phosphorylase hybrids from exhibiting phosphorylase *a* activity. Remaining open as a possible mechanism for the regulation of muscle phosphorylase phosphatase is the interesting question of interconversion of the enzyme between inactive and active forms[211], a problem that has been difficult to resolve since the original suggestion that such a system may exist in the adrenal cortex[212]. Another unexplained observation of great potential significance is the inhibition of muscle phosphorylase *a* phosphatase which occurs in synchrony with phosphorylase *b* kinase activation in the glycogen particle[28]; it is believed that this may be an example of enzyme regulation involving protein–protein interactions.

The phosphoprotein phosphatase from skeletal muscle that is responsible for the dephosphorylation of glycogen synthase D and activated phosphorylase *b* kinase has not been fully characterised, but several properties of possible significance from a regulatory standpoint have been described. With glycogen synthase D as the substrate, the enzyme is inhibited by glycogen at levels greater than 2 mg $ml^{-1}$, but at lower levels of the polysaccharide it is stimulated[200, 213]. Glycogen was also found to be inhibitory when the phosphatase acts on activated phosphorylase *b* kinase[169]. This latter observation would not necessarily imply that the effect of glycogen is through interaction with the phosphatase itself, however, since phosphorylase *b* kinase binds to glycogen even though the latter substance is not a substrate of the kinase[16]. The phosphoprotein phosphatase is stimulated by divalent cations when acting either on glycogen synthase D or activated phosphorylase *b* kinase[169, 213]. With respect to the metal ion effect when the latter substrate is used, Cohen *et al.*[173] have shown that this is strongly influenced by the extent of phosphorylation of the kinase. Phosphorylation of the A subunit, which follows complete activation due to phosphorylation of the B subunit, renders the enzyme susceptible to dephosphorylation by the phosphatase in the absence of metal ions. Glucose 6-P has been shown to stimulate the glycogen synthase D to I reaction in skeletal and cardiac muscle[214] as measured by a change in activity of the synthase, but the phosphorylated sugar did not cause an increase in the rate of dephosphorylation of the enzyme.[213] The mechanism of the glucose 6-P effect has not been elucidated, but it was postulated that it might be due to the presence of different types of bound phosphate in the synthase, or perhaps to the behaviour of phospho–dephospho hybrids that might exist for the enzyme in a manner analogous to that found for phosphorylase (see above).

### 7.4.3 Muscle contraction

The original study of Meyerhof[215], showing that the loss of glycogen from muscle during contraction is paralleled by the appearance of lactic acid, set the stage for investigation of the control of glycogen metabolism as it relates to muscle function. In the nearly 50 years that have passed since this pioneering work, much progress has been made, and it is now possible to present a reasonably convincing story as to how the respective energy-requiring and energy-yielding processes of contraction and glycogenolysis are linked.

#### *7.4.3.1 Activation of phosphorylase* b

The concentrations of AMP and inorganic phosphate in resting muscle are sufficiently high so that if these were the only factors operating, phosphorylase *b* would be active and glycogenolysis would take place even in the absence of contraction[216]. This is not the situation, however, because of the opposing actions of ATP and glucose 6-P, which counteract the effect of these activators[217]. When muscle contracts, the balance between the various effectors is shifted and phosphorylase *b* becomes active. This constitutes a classical example of the regulation of an energy-yielding pathway through a change in adenylate or energy charge[218]. It should be noted that this type of regulation requires that the metabolic products resulting from contraction accumulate before phosphorylase becomes more active, as distinguished from glycogenolysis triggered by the conversion of phosphorylase *b* into phosphorylase *a* (see below). Evidence that the rate of glycogenolysis can in fact be increased as a result of the stimulation of phosphorylase *b* activity is provided by the fact that glycogen depletion occurs in the muscle of I-strain mice subjected to exercise, or when muscle from the mice is stimulated electrically[219, 220].

#### *7.4.3.2 Conversion of phosphorylase* b *into phosphorylase* a

(a) *Site of control*—Shortly after it had been established that phosphorylase in resting muscle exists primarily in the *b* form[221], Cori showed that electrical stimulation of muscle causes conversion of phosphorylase *b* into phosphorylase *a*[222]. This basic observation has been confirmed and refined in numerous studies carried out since that time[146, 223-227]. As it became accepted that contraction is linked to glycogenolysis through a process involving the net conversion of phosphorylase *b* into phosphorylase *a*, attention was turned to the phosphorylase *b* kinase and phosphorylase *a* phosphatase reactions [equations (7.5) and (7.6)] to see which one (or whether both) might be involved in controlling this conversion. Using the isolated frog sartorius muscle system, Danforth *et al.*[146] concluded from kinetic studies that a change in the rate of the kinase reaction was primarily responsible for shifting the balance to phosphorylase *a*. This conclusion was in keeping with the properties of phosphorylase *b* kinase, showing that this enzyme can be regulated *in vitro* by several mechanisms[11]. In more recent work involving the isolated glycogen particle as a model for regulation *in vivo*, Haschke *et al.*[28] point to a possible dual mechanism for regulation of phosphorylase *a* formation brought about by a $Ca^{2+}$-dependent inhibition of phosphorylase phosphatase coincident with the enhancement of phosphorylase *b* kinase activity.

(b) *Stimulation of non-activated phosphorylase* b *kinase by* $Ca^{2+}$—Most investigators now agree that the release of $Ca^{2+}$ from the sarcoplasmic reticulum[227] provides the link through which muscle excitation is coupled to contraction and glycogenolysis (Figure 7.4). The calcium effect on the latter

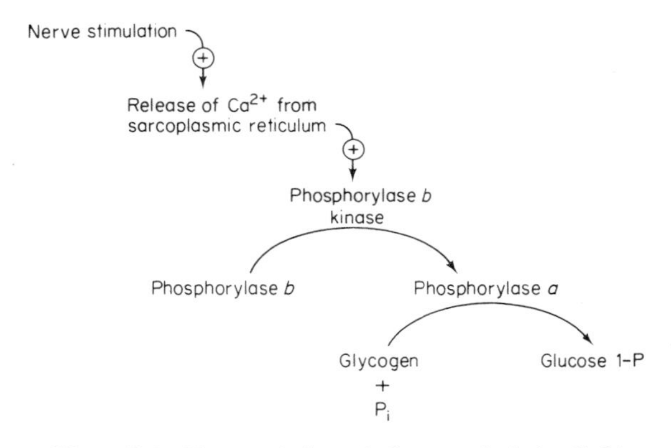


**Figure 7.4** The regulation of glycogenolysis by $Ca^{2+}$

process is due to the requirement of phosphorylase *b* kinase for the metal ion[165] and perhaps to its role in the inhibition of phosphorylase phosphatase (see above). Early experiments involving the role of $Ca^{2+}$ in the regulation of phosphorylase *b* kinase were difficult to interpret, because of two separate calcium effects. One of these was an irreversible conversion of non-activated phosphorylase *b* kinase into its activated form in the presence of a factor originally designated as the kinase activating factor or KAF[165], but later shown to be a $Ca^{2+}$-requiring protease[228]. The other $Ca^{2+}$ effect, also noted by Meyer *et al.*[165], was reversible and consisted of the direct requirement by the enzyme for the metal ion. Ozawa *et al.*[166] examined the latter effect more closely and determined that the sensitivity of the enzyme to $Ca^{2+}$ is very high, with half-maximal activation occurring at *ca.* $2 \times 10^{-7}$ mol $l^{-1}$. Heilmeyer *et al.*[27], however, reported that the kinase requires $2 \times 10^{-6}$ mol $l^{-1}$ free $Ca^{2+}$ for half-maximal activation when it is present in the 'glycogen particle'. These workers attributed the higher $Ca^{2+}$ requirement to macromolecular interactions that might be occurring in the particle, but a more likely explanation of their finding would be that the molar concentration of phosphorylase *b* kinase employed in their experiments was greater than $1 \times 10^{-6}$ mol $l^{-1}$.

Thus it was necessary to add a relatively large amount of $Ca^{2+}$ to reach a concentration equivalent to the amount of enzyme present. Their experimental condition would approximate the intracellular environment more closely, however, than those employed when highly dilute phosphorylase *b* kinase ($10^{-9}$–$10^{-10}$ mol $l^{-1}$) is used in the conventional activity test, since the intracellular phosphorylase *b* kinase concentration is of the order of $10^{-7}$ mol $l^{-1}$ in rabbit skeletal muscle. The concentration of $Ca^{2+}$ that affects the development of tension in muscle[229] is in the range of $0.3 \times 10^{-6}$ to $20 \times 10^{-6}$ mol $l^{-1}$.

No conversion of non-activated phosphorylase *b* kinase to its activated form, i.e. no phosphorylation of the enzyme, accompanies muscle contraction[225,226]. This is consistent with the finding[224-226] that cyclic AMP levels are not altered by electrical stimulation of muscle and that phosphorylase *b* kinase kinase (cyclic AMP-dependent protein kinase) is stimulated by cyclic AMP but not by $Ca^{2+}$. The possibility that phosphorylase *b* kinase might undergo activation by an autocatalytic mechanism[16] would still remain, but there is no evidence that this occurs *in vivo*. Non-activated and activated phosphorylase *b* kinase both require $Ca^{2+}$, but the latter form is more sensitive to the metal ion[170]; this allows for possible interplay between control of the enzyme during muscle contraction and under conditions of epinephrine stimulation.[230]

(c) *Other factors affecting phosphorylase* b *kinase activity*—Danforth[231] called attention to the fact that an increase in muscle pH brought about by the breakdown of phosphocreatine during contraction might well be a factor in regulating the activity of phosphorylase *b* kinase; it will be noted that the pH optimum for the nonactivated form of the kinase (Figure 7.2, curve A) rises sharply between pH 7 and 8. Villar-Palasi and Wei[232] have proposed that a decrease in the ATP concentration of muscle accompanying contraction may increase phosphorylase *b* kinase activity through the liberation of free $Mg^{2+}$ which can increase the activity of the kinase (see earlier section). Several investigators have failed, however, to find changes in ATP levels with moderate electrical stimulation of muscle[233,234].

### *7.4.3.3 Muscle contraction and glycogen synthesis*

The rate of glycogen synthesis, as well as the rate of glycogenolysis, is regulated in response to muscle contraction. Staneloni and Piras[235] showed that a brief period of electrical stimulation causes the conversion of synthase I into synthase D, particularly when the initial levels of the I form are high. The mechanism responsible for this change has not been elucidated. During a period of rest following contraction, synthase D is converted into synthase I[235,236], an effect that is partially blocked by prior treatment of the tissue with epinephrine[236]. These conversions may be regulated in part through the glycogen effect that is described below. Calcium ions are not known to participate in the interconversion reactions involving the synthase.

Danforth[236] found that an inverse relationship exists between the glycogen concentration of muscle and the percentage of synthase found in the active I form. The basis for this effect is thought to be an inhibition of synthase phosphatase by glycogen[200]. By this mechanism glycogen helps to regulate its own level. Thus, when the glycogen concentration is reduced, as in muscle contraction, its resynthesis is favoured by activation of the synthase. When glycogen concentrations have been restored, synthesis of the polysaccharide is slowed owing to inhibition of synthase phosphatase. It has been noted[5] that the upper level of glycogen concentration is 'tightly controlled' at *ca.* 1%. In addition to changes in the relative amounts of glycogen synthetase I and D accompanying muscle contraction, it would seem probable that an increase in glucose 6-P concentration may serve as a feedback mechanism that favours the rebuilding of glycogen stores following contraction[233, 234].

The regulation of glycogen levels by glycogen *per se* may involve an effect on glycogenolysis as well as that on glycogenesis. Glycogen is a strong activator of phosphorylase *b* kinase[16], so that high levels of glycogen would favour phosphorylase *a* formation and glycogenolysis. The possible dual aspect of regulation by glycogen is illustrated in Figure 7.5.

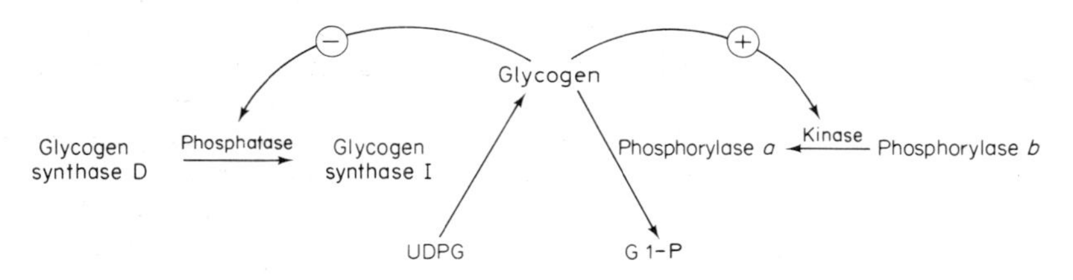


**Figure 7.5** Feedback regulation of glycogen synthesis and glycogen breakdown at the level of interconverting enzymes

In contrast, glycogen does not appear to be a regulatory effector molecule of bacterial glycogen synthesis. Glycogen at 1% concentration does not inhibit *E coli* B or *Arthrobacter viscosus* ADPglucose pyrophosphorylases[135] or the bacterial glycogen synthase[151]. Under certain conditions many bacteria are known to accumulate from 20 to 40% of their dry weight as glycogen[34, 58, 67, 108]. This would suggest that glycogen does not effectively cause feedback inhibition of the glycogen biosynthetic enzymes *in vivo*.

### 7.4.4 Endocrine regulation

#### *7.4.4.1 Epinephrine*

The basic concept that epinephrine causes glycogenolysis in skeletal muscle and other tissues was established in 1928 by Cori and Cori[237]. In 1951, Sutherland showed that the hormone acts by promoting the formation of phosphorylase *a*[203], an observation that was confirmed and extended by Cori and Illingworth[238]. Shortly thereafter, Sutherland and co-workers discovered cyclic AMP and formulated the general concept that this 'second messenger' mediates hormonal responses[239]. In particular, they demonstrated that cyclic AMP produced in response to epinephrine stimulation serves as the agent responsible for phosphorylase activation. All of the work leading to the discovery of cyclic AMP was carried out using the liver phosphorylase system, but subsequent studies concerned with its mechanism of action were carried out in skeletal muscle. Investigators still do not fully understand how cyclic AMP acts in the activation of liver phosphorylase.

Cyclic AMP was shown to promote the activation of muscle phosphorylase *b* kinase *in vitro*[11] and epinephrine to have this effect *in vivo*[224]. The activation reaction was found to involve the phosphorylation of phosphorylase *b* kinase, and subsequently a cyclic AMP-dependent protein kinase[10], which catalysed this reaction, was demonstrated. The mechanism of action of cyclic AMP in the activation of the kinase was elucidated (see above). From these observations it was possible to construct the glycogenolytic 'cascade' that portrays the action of epinephrine in promoting glycogenolysis (Figure 7.6). The increased

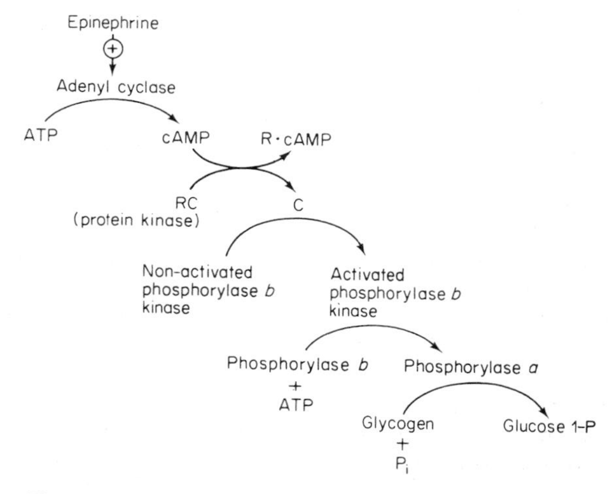


**Figure 7.6** Cascade system for the regulation of glycogenolysis by epinephrine

flow of glucosyl units from glycogen into the glycolytic pathway with epinephrine stimulation is also enhanced by a decreased rate of glycogen synthesis brought about through the conversion of glycogen synthase I to D[12, 236, 240, 241] catalysed by the same cyclic AMP-dependent protein kinase (see Figure 7.3).

The scheme shown in Figure 7.6 has gained wide acceptance, but results at variance with this concept have been reported by Stull and Mayer, who found that very low doses of catecholamines cause conversion of phosphorylase *b* into phosphorylase *a* in skeletal muscle without a detectable increase in cyclic AMP concentrations and without phosphorylase *b* kinase activation[226]. These workers raise the possibility that the catecholamines may be causing a translocation of $Ca^{2+}$ and that this stimulates non-activated phosphorylase *b* kinase. Chelala and Torres[242] have noted changes in the activity of muscle phosphorylase *a* phosphatase in the presence of cyclic AMP which could conceivably also play a part in the stimulation of phosphorylase *a* formation by epinephrine.

### 7.4.4.2 *Insulin*

In addition to the classical effect of insulin in promoting glucose uptake in muscle, it has long been recognised that this hormone has a specific action in facilitating glycogen formation in this tissue. The explanation for this effect was provided by the classical work of Larner and his associates, who showed that insulin causes activation of glycogen synthase by promoting the formation of synthase I through an unknown mechanism[14, 78, 148, 243, 244]. As is evident from equations (7.9) and (7.10), a shift from the phosphorylated to the non-phosphorylated form of the synthase, i.e. synthase I formation, could be brought about either through an effect on the kinase or the phosphatase reaction. Shen *et al.*[245] have provided evidence that insulin acts by decreasing the activity of the kinase.

The most obvious way that the activity of glycogen synthase kinase (cyclic AMP-dependent protein kinase) might be reduced would be through a lowering of cyclic AMP levels, but such a mechanism has not been supported by a study of insulin effects in muscle. On the contrary, insulin was even found to cause a slight elevation of the cyclic nucleotide in this tissue[246]. The possibility that insulin might act by decreasing the sensitivity of the protein kinase to a fixed level of cyclic AMP has been given careful consideration by the Larner group[245], and they have recently isolated a substance from muscle stimulated by insulin which may have such an effect on the kinase[247].

The activity of the cyclic AMP-dependent protein kinase at a given level of cyclic AMP can be influenced by a variety of factors, many of which are not well understood at this time (see earlier section). Any factor that could cause an overall shift in the reactions of equations (7.12a) and (7.12b) to the left would decrease protein kinase activity, and any factor shifting the reactions to the right would increase enzyme activity. It is interesting to speculate that the effect of insulin on the lowering of cyclic AMP levels in certain tissues (not muscle) might be due to the same basic mechanism that decreases the sensitivity of the protein kinase to cyclic AMP. Thus, if insulin promoted the formulation of $R_2C_2(MgATP)_2$, more of the cyclic AMP would be free and subject to attack by the diesterase that catalyses the hydrolysis of cyclic AMP[248]. Without insulin a shift to the right would occur and cyclic AMP would be tied up in the form $R_2$(cyclic AMP)$_2$, which is not subject to attack by the diesterase[249, 250]. The reactions depicted in equations (7.12a) and (7.12b) might be altered by covalent modification of the protein kinase itself; in this connection the report of Rosen *et al.*[196], that the regulatory subunit of the bovine heart cyclic AMP-dependent protein kinase can be phosphorylated by ATP, is of interest.

Another factor to be considered with respect to the action of insulin is the possible role of a heat-stable protein inhibitor of the protein kinase[251-254]. This protein, which combines with the free catalytic subunit of the enzyme[255], is of uncertain physiological significance, but one idea is that it may serve to inactivate any free C that is present at basal levels of cyclic AMP[256]. Under these conditions alterations in the concentration of the inhibitor could have a marked influence on protein kinase activity. Furthermore, Walsh and Ashby have reported that the concentration of the inhibitor increases with insulin administration[257].

#### 7.4.4.3 *Adrenal cortical hormones*

The administration of adrenal cortical hormones to adrenalectomised or normal animals causes an increase in the level of muscle glycogen, the extent of which varies with the nutritional state of the animal and the particular muscle that is examined[258-260]. Evidence has been obtained that this effect is lost in diabetic animals and is thus probably secondary to the release of insulin that normally accompanies hyperglycemia caused by the action of glucocorticoids on the liver[261]. There are some effects of glucocorticoids on muscle glycogen metabolism, however, that are not readily explained on this basis.

Kerpola's study[259] on the effect of cortisone on the relative amounts of phosphorylase *b* and phosphorylase *a* in skeletal muscle were carried out prior to knowledge of the care that is needed to prevent changes in the *a* to *b* ratio *in vitro*. Nevertheless, he showed that muscle from rabbits receiving cortisone was different from that of control animals in that it contained less phosphorylase *a* than the latter, as measured after several fractionation steps. It is probable that *in vitro* changes in the phosphorylase *a* to phosphorylase *b* ratio occurred in this study, but the nature and/or extent of these changes differed, depending upon prior treatment of the animal with cortisone.

It has been shown[262] that adrenalectomy impairs the response of muscle glycogen phosphorylase to epinephrine injection and also the conversion of glycogen synthase I into glycogen synthase D caused by this stimulus[263]. The conversion of non-activated phosphorylase kinase into its activated form with epinephrine injection was not impaired in adrenalectomised animals, nor was their suppression of the conversion of phosphorylase *b* into phosphorylase *a* produced by electrical stimulation. It is difficult to account for these changes on the basis of a change in the concentration of one of the converting enzymes unless the regulation of that enzyme depends upon the nature of the substrate it is acting on. This possibility was raised by Vilchez *et al.*[263] and would not be unusual in a system in which an enzyme serves as the substrate.

## 7.5 THE REGULATION OF BACTERIAL AND PLANT LEAF ADPGLUCOSE PYROPHOSPHORYLASE

### 7.5.1 Allosteric activators and inhibitors

A consistent pattern is usually observed between the source of the ADPglucose pyrophosphorylase, the metabolites that are most effective as activators of the enzyme, and the pathway of carbon utilisation occurring in the micro-organism or tissue. This can be observed in Table 7.4.

The Enterobacteria that accumulate glycogen (*E. coli*, *A. aerogenes*, *A. cloacae*, *S. typhimurium*, *C. freundii*, *E. aurescens*) contain an ADPglucose pyrophosphorylase activated by fructose 1,6-diP, NADPH and pyridoxal 5-P and inhibited by 5′-AMP [116, 130]. These organisms utilise glycolysis as their main route for glucose catabolism. Another class of ADPglucose pyrophosphorylase is found in *A. formicans*[20] or in *M. lysodeikticus*[135], which is activated by either fructose 6-P or fructose 1,6-diP. In contrast, *S. marcescens*, an enteric organism, has an ADPglucose pyrophosphorylase not activated by any metabolite tested[20, 116]. *M. lysodeikticus*, *A. formicans* and *S. marcescens* also degrade glucose via the Embden–Myerhof pathway.

**Table 7.4 Activators and inhibitors of ADPglucose pyrophosphorylases from many sources**

| *Source of ADPglucose pyrophosphorylase* | *Major activators* | *Inhibitors* | *Carbon utilisation pathways* |
|---|---|---|---|
| *E. coli*, *A. aerogenes*<br>*A. cloacae*, *S. typhimurium*<br>*C. freundii*, *E. aurescens*<br>*S. dysenteria* | Fructose 1,6-diP<br>NADPH<br>Pyridoxal 5-P | AMP | Glycolysis |
| *A. viscosus*, *A. tumefaciens*, *Rps. capsulata* | Fructose 6-P<br>Pyruvate | $P_i$, AMP, ADP | Entner–Doudoroff |
| *R. rubrum* | Pyruvate | None | Does not grow on glucose; grows well on Krebs'-cycle intermediates |
| *S. marcescens* | None | AMP | Glycolysis |
| *M. smegmatis*, *A. formicans*, *M. lysodeiktikus* | Fructose 6-P<br>Fructose 1,6-diP | ADP | Glycolysis |
| Leaves of plants, *Chlorella vulgaris*, *Chlorella pyrenoidosa*, *Scenedesmus obliqius*, *Chlamydomonas reinhardii* | 3-P-Glycerate | $P_i$ | Calvin–Bassham cycle, Hatch–Slack pathway |

A group of organisms, *A. tumefaciens*, *A. viscosus* and *Rps. capsulata*, that catabolise glucose via the Entner–Doudoroff pathway contains an ADP-glucose pyrophosphorylase activated both by fructose 6-P and pyruvate[120, 125]. *Rhodospirillum rubrum*, a photosynthetic organism that cannot metabolise glucose, but can grow either as a heterotroph in the light or dark on various tricarboxylic acid intermediates and associated metabolites, or as an auto-troph on $CO_2$ and $H_2$, contains an ADPglucose pyrophosphorylase that is activated specifically by pyruvate[139, 140]. The ADPglucose pyrophosphorylase found in leaves of plants and in green algae is activated by 3-phospho-glycerate[264-268]. The plant and algal ADPglucose pyrophosphorylases are very sensitive to inhibition by inorganic phosphate.

The activators for the various groups of ADPglucose pyrophosphorylase are significant metabolites in the metabolic pathways utilised by the plant tissues or micro-organisms. The significance of these metabolites in their respective pathways and their activation of ADPglucose and 1,4-α-glucan synthesis has been discussed previously[19, 116, 139, 140].

Table 7.4 shows an overlapping of specificity of the activators of the different classes of ADPglucose pyrophosphorylases. For example, fructose 1,6-diP is an effective activator of the enterobacterial enzymes as well as the *A. formicans* and *M. lysodeikticus* enzymes. However, fructose 6-P, an activator for the *A. formicans* and *M. lysodeikticus* ADPglucose pyrophosphorylases, is not an activator for the *E. coli* enzyme. 3-P-Glycerate, a very potent activator for the leaf ADPglucose pyrophosphorylases, can also stimulate the *E. coli* enzyme[130], although poorly; conversely, fructose 1,6-diP is able to activate the spinach-leaf enzyme[265]. However, higher concentrations of fructose 1,6-diP (28 μmol $l^{-1}$) than of 3-P-glycerate (10 μmol $l^{-1}$) are needed for 50% of maximal activation of the spinach-leaf enzyme and the maximum stimulation of $V_{max}$ effected by fructose 1,6-diP (16-fold) is much less than that observed for 3-P-glycerate (58-fold) at pH 8.5. This overlapping of specificity for the activators in the various ADPglucose pyrophosphorylase groups suggests that the activator sites for the different classes are similar to each other. Thus one may hypothesise that mutation of the gene specifying the activator site of the ADPglucose pyrophosphorylase has occurred via evolutionary processes to enable the specificity of the activator site to be compatible or coordinated with the metabolic activities of the organism.

Recently, Robson *et al.*[41,312] have briefly reported the presence of an ADP-glucose pyrophosphorylase in extracts of *C. pasteurianum* that does not appear to be activated by either glycolytic intermediates, pyruvate, NADPH or pyridoxal 5′-P. Furthermore, AMP and $P_i$ did not inhibit the enzyme and only high non-physiological concentrations of ADP inhibited the enzyme. Thus, this ADPglucose pyrophosphorylase does not appear to have any allosteric effectors, as has been noted for the enzyme in other bacterial and plant systems. The absence of allosteric effectors for ADPglucose pyrophosphorylase in *Clostridia* suggests that glycogen synthesis in this organism may represent a less evolved regulatory system. It would be of interest to determine the extent of relatedness between the *Clostridial* and the other bacterial ADPglucose pyrophosphorylases at the polypeptide level.

### 7.5.2 Function of the allosteric activator

#### *7.5.2.1 Kinetic studies*

Studies of the various ADPglucose pyrophosphorylases listed in Table 7.4 show a number of kinetic properties that are modified by the presence of the allosteric activator and inhibitor. The concentrations of the substrates, ATP, glucose 1-P, pyrophosphate and ADPglucose, and the cationic activator $Mg^{2+}$, required to give 50% of maximal velocity ($S_{0.5}$) is significantly lowered in the presence of the allosteric activator. For most ADPglucose pyrophosphorylases the $S_{0.5}$ value of ATP and ADPglucose may be lowered *ca.* 5- to 15-fold by the presence of the activator. In addition, the allosteric activator may stimulate maximal velocity from 2.5- to 60-fold. The magnitude of stimulation of velocity is dependent on the enzyme system and pH. Probably the most important function of the allosteric activator is to modulate the sensitivity of the enzyme to inhibition by the allosteric inhibitors. Inhibition studies have shown that the allosteric inhibitor is non-competitive with respect to the substrates[132]. However, for a number of ADPglucose pyrophosphorylases, inhibition can be completely reversed by increasing concentrations of the activator. The activator fructose 1,6-diP modulates the sensitivity of the *E. coli* B enzyme to inhibition by 5′-AMP[131-133]. The $K_i$ for 5′-AMP at pH 8.5 is 70 μmol $l^{-1}$ at 1.7 mmol $l^{-1}$ fructose 1,6-diP concentration. However, at a lower concentration of activator lesser concentrations of 5′-AMP are required for 50% inhibition. At 0.06 mmol $l^{-1}$ fructose 1,6-diP, only 3.4 μmol $l^{-1}$ 5′-AMP is required for 50% inhibition.

On the basis of these *in vitro* kinetic studies it is believed that the relative concentrations of allosteric inhibitors and activator would modulate the rate of ADPglucose synthesis and therefore the rate of glycogen synthesis. Under conditions where both ADPglucose and glycogen synthesis are minimal, the concentrations of inhibitor and activator are such that the ADPglucose pyrophosphorylase is mainly in the inhibited state; during conditions where there is active glycogen accumulation, either a decrease of inhibitor or an increase of activator has occurred, enabling ADPglucose pyrophosphorylase activity to increase.

### 7.5.3 Evidence in support of the ADPglucose pyrophosphorylase activator–inhibitor interaction being functional *in vivo* in the regulation of glycogen (starch) synthesis

#### *7.5.3.1 Causal relationships*

(a) *Chloroplasts*—Some correlation between the changes in activator and inhibitor concentrations have been observed under conditions where the organism or tissue is either increasing or decreasing the rate of glycogen or starch synthesis.

The results obtained by a number of laboratories on the concentration of glycolytic intermediates[269], phosphate[270, 271] and ATP[271] in the leaf chloroplast in the light and in the dark qualitatively support the hypothesis of regulation of starch synthesis by 3-P-glycerate and inorganic phosphate levels. Starch synthesis rapidly occurs in the leaf chloroplast in the light and is greatly reduced in the dark. Heber and Santarius[270, 271] have shown that the concentration of $P_i$ in the dark is 5–10 mmol $l^{-1}$, and decreases by 30–50% in the light. At these concentrations of $P_i$, *in vitro* kinetic studies with ADPglucose pyrophosphorylase show that at 5 mmol $l^{-1}$ 3-P-glycerate there is an increase of 5-fold in the rate of ADPglucose synthesis when the $P_i$ concentration is decreased from 10 to 7.5 mmol $l^{-1}$ and a 23-fold increase in rate is observed[20, 268] when phosphate is decreased to 5 mmol $l^{-1}$. Thus, under these conditions a decrease of phosphate concentration of only 30–50% in the chloroplast may cause a rapid acceleration of ADPglucose synthesis and therefore of starch synthesis.

(b) *Plant leaves*—MacDonald and Strobel[272] reported that wheat leaves infected with the fungus *Puccinia striiformis* accumulated more starch than non-infected leaves. They could correlate starch accumulation with the decrease observed in $P_i$ levels in diseased leaves during the infection process. This decrease in $P_i$ concentration is most probably due to synthesis of nucleic acids required for fungal growth. Their data suggested that in diseased leaves the variations in the level of $P_i$ and, to a lesser extent, variations in the level of activators of the wheat leaf ADPglucose pyrophosphorylase (3-P-glycerate, fructose 1,6-diP etc.) regulated the rate of starch synthesis via control of ADPglucose pyrophosphorylase activity.

(c) *Algae*—Recently, Kanazawa *et al.*[273] have shown in *Chlorella pyrenoidosa* cells that starch and ADPglucose synthesis occur in the light. Starch synthesis abruptly ceases and the ADPglucose drops to below detectable limits when the light is turned off. UDPglucose levels do not change perceptibly in the light to dark transition. ADPglucose is not detectable at any time later in the dark, despite the high steady state level of ATP and hexose phosphate, the substrates for ADPglucose pyrophosphorylase. Kanazawa *et al.*[273] indicate that this observation provides strong support for the importance of the regulatory role of ADPglucose pyrophosphorylase in starch synthesis *in vivo*. Thus the allosteric effects exerted by 3-P-glycerate and $P_i$ appear to be physiologically important. Since the level of 3-P-glycerate does not appreciably change in the dark to light transition[269, 273], while the phosphate levels appear to increase in the dark and decrease in the light[270, 271], it is suggested that the variation of $P_i$ (the negative effector) is the more important control element.

(d) E. coli—Dietzler *et al.*[274, 275] have shown that when the growth of *E. coli* W4597 (K) ceased because of $NH_4^+$ depletion, the rate of glycogen accumulation increased about 3.3- to 4.2-fold. Concomitant with this, a decrease[274] of fructose 1,6-diP of *ca.* 76% (from 3.1 to 0.71 mmol $l^{-1}$) and an increase in the energy charge[143, 276, 277] from 0.74 in exponential phase to 0.87 in stationary phase occurred[275]. The total concentration of the adenylate pool (ATP + AMP + ADP) in exponential and stationary phase was *ca.* 3 mmol $l^{-1}$. Although the activator (fructose 1,6-diP) concentration decreased 75% from exponential to stationary phase, Dietzler *et al.* have concluded that this decrease in the concentration of allosteric activator was more than offset by the increase in energy charge[275]. Their conclusion was based on the data of Govons *et al.*[114], which showed a 3-fold increase in *E. coli* B ADPglucose pyrophosphorylase activity when the energy charge (as defined by Atkinson[218, 276, 277]) increased from 0.74 to 0.87, with a concomitant decrease of fructose 1,6-diP concentration from 1.5 to 0.5 mmol $l^{-1}$. The total adenosine nucleotide pool was 2 mmol $l^{-1}$ in these *in vitro* experiments. Figure 7.7 shows the effect of fructose 1,6-diP concentration on the response of *E. coli* B ADPglucose pyrophosphorylase to energy charge under the physiological conditions found by Dietzler *et al.* in *E. coli* W4597 (K)[274, 275]. At an adenosine nucleotide pool concentration of 3 mmol $l^{-1}$ with energy charge of 0.75 and 3.0 mmol $l^{-1}$ fructose 1,6–diP, ADPglucose pyrophosphorylase activity is increased to 21% of its maximum velocity when the energy charge is increased to 0.87 and the fructose 1,6-diP concentration is lowered to 0.75 mmol $l^{-1}$. Thus, under these conditions there is an increase of *ca.* 2- to 2.2-fold in the rate of ADPglucose synthesis. Assuming that the concentrations of fructose 1,6-diP, ATP and the energy charge values that are obtained in *E. coli* W4597 (K) are also those for *E. coli* B, the 2.2-fold increase in the rate of ADPglucose synthesis would be sufficient to account for the increase in glycogen accumulation rates seen in *E. coli* B on going from exponential phase to stationary phase in minimal media.

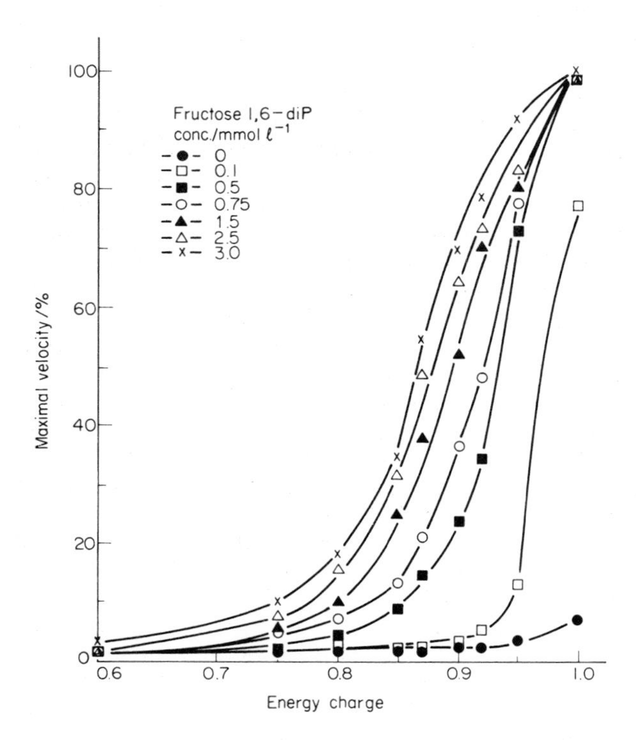


**Figure 7.7** Effect of fructose 1,6-diP concentration on the response of *E. coli* ADPglucose pyrophosphorylase to variation in the adenylate energy charge. ADPglucose synthesis was measured as described previously[265]. The reaction mixture, which was incubated for 10 min at 37 °C, contained 0.5 mM [$^{14}$C]glucose 1-P (specific activity $9.6 \times 10^5$ c.p.m. $\mu mol^{-1}$), a 3 mM adenine nucleotide mixture composed of ATP, AMP and ADP of specified energy charge[276, 277] $\left(\text{energy charge} = 1/2 \dfrac{2\,\text{ATP} + \text{ADP}}{\text{ATP} + \text{ADP} + \text{AMP}}\right)$ as indicated, and calculated on the basis of an equilibrium constant of 0.45 for the adenylate kinase reaction, 7.5 mM $MgCl_2$, 100 mM *N*-2-hydroxyethylpiperazine-*N*′-2-ethanesulphonic acid–NaOH buffer, pH 7.0, 100 μg of bovine plasma albumin, 0.9 μg of crystalline yeast inorganic pyrophosphatase, fructose 1,6-diP as indicated, and enzyme in a total volume of 0.2 ml. The maximal activity which is obtained at energy charge 1.0 is arbitrarily set at 100% and is equal to the activity found when the enzyme is assayed at saturating concentrations of substrates and activator fructose 1,6-diP

Table 7.5 shows the rate of glycogen accumulation found in minimal media and the activity of ADPglucose pyrophosphorylase activity found in exponential and stationary phase. The maximal ADPglucose pyrophosphorylase activity in the exponential phase is 8.4-fold greater than the rate of glycogen accumulation and in stationary phase the rate is about 13-fold greater. However, if the fructose 1,6-diP values and energy charge values observed in *E. coli* W4597 (K) in stationary and exponential phase are taken into consideration, then the ADPglucose pyrophosphorylase activity calculated from the data of Figure 7.7 would be roughly equivalent to the rate of glycogen accumulation in exponential phase, and about 2.5-fold greater in stationary phase. The experiment of Figure 7.7 was done with saturating glucose 1-P values (0.5 mmol $l^{-1}$) and this may not be the physiological concentration of

**Table 7.5 ADPglucose pyrophosphorylase activity and glycogen accumulation rates in *E. coli* B during exponential and stationary growth phases***

| *Growth phase* | *Rate of glycogen accumulation* | *ADPglucose pyrophosphorylase activity* | |
|---|---|---|---|
| | | *Maximum activity* /μmol $g^{-1}$ $h^{-1}$ | *Activity based on energy charge and fructose* 1,6-*diP concentration* |
| Exponential | 15.8 | 132 | 12.5 |
| Stationary | 32 | 360 | 76 |

* Growth of *E. coli* B was on minimal medium[111]. The glycogen accumulation rates were obtained from data reported previously[114,278]. All data are expressed either as μmol of anhydroglucose units (glycogen) or as μmol of ADPglucose formed per gram wet weight *E. coli* cells per hour. The values for maximum activity of ADP-glucose pyrophosphorylase are obtained by assaying crude extracts of cells with saturating concentrations of substrates and the activator fructose 1,6-diP, while the ADPglucose pyrophosphorylase activity, based on the existent energy charges and fructose 1,6-diP concentration found by Dietzler *et al.*[274,275], is obtained from the data of Figure 7.7

glucose 1-P in the cell. No published values for the glucose 1-P concentration are known to the authors. However, if one assumes that the phosphoglucomutase (EC 2.7.5.1) reaction is at equilibrium *in vivo* in *E. coli*, and that the concentration of glucose 6-P in *E. coli* B is equivalent to those found in exponentially growing *E. coli* Hfr 139[279] and *E. coli* W4597 (K)[309], then the glucose 1-P concentration may be estimated as 0.039 mmol $l^{-1}$, a value greater than the $K_m$ for glucose 1-P ($K_m = 0.02–0.033$ mmol $l^{-1}$). No data for glucose 6-P or for glucose 1-P concentrations in $N_2$-limited *E. coli* cells in stationary phase are available.

With these uncertainties, the correlation of the ADPglucose pyrophosphorylase activity with the published fructose 1,6-diP concentrations and energy charge values with the rates of glycogen accumulation in the two growth phases appears to be quite good. However, it should be pointed out that a number of assumptions which may not be valid have been made in determining these correlations. First, the values found for energy charge and fructose 1,6-diP concentrations in *E. coli* W4597 (K) may be different in the *E. coli* B system. Indeed, Lowry *et al.*[279] have shown that fructose 1,6-diP concentrations increase slightly, from 2.6 to 3.2 mmol $l^{-1}$, in *E. coli* Hfr 139 cells when they reach stationary phase in the presence of excess glucose and limiting nitrogen. The variations in fructose 1,6-diP concentrations and in energy charge during growth of *E. coli* B have not been determined. Secondly, the above considerations have assumed no compartmentation of the metabolites in relation to the possible compartmentation of glycogen biosynthetic enzymes. Third, the effects of other cations, anions and metabolites on ADP-glucose pyrophosphorylase activity have not been considered. Nevertheless, the above published data[274, 275] are consistent with the notion of ADPglucose synthesis, and, therefore, bacterial glycogen synthesis, being regulated by energy charge and the various glycolytic intermediates.

#### 7.5.3.2 *Supporting data obtained with mutants of* E. coli *affected at the regulatory site of ADPglucose pyrophosphorylase*

Mutants of *E. coli* B[19, 20, 111, 113-115, 121, 278] and *E. coli* K-12[110, 112, 134], altered in their ability to accumulate glycogen, have been isolated. Some contain ADPglucose pyrophosphorylases modified with respect to their regulatory properties. Table 7.6 compares the rates of accumulation of glycogen, and the maximum amounts of glycogen accumulated, by two glycogen-excess mutants, SG5 and CL1136, with the parent strain *E. coli* B. The maximum amount of glycogen accumulated in SG5 and CL136 is *ca.* three times that accumulated

**Table 7.6 Glycogen accumulation in *E. coli* B and mutants SG5 and CL1136***

| *Strain* | *Generation time at 37 °C* /min | *Medium* | *Glycogen* *Rate of accumulation* /µmol g$^{-1}$ h$^{-1}$ | *Maximal accumulation* /mg g$^{-1}$ |
|---|---|---|---|---|
| *E. coli* B | 24 | Enriched | 23 | 20 |
| | 42 | Minimal | 32 | 20 |
| SG5 | 27 | Enriched | 48 | 53 |
| | 42 | Minimal | 59 | 35 |
| CL 1136 | 24 | Enriched | 77 | 57 |
| | 47 | Minimal | 114 | 74 |

* The enriched and minimal media with glucose as a carbon source and the assays for glycogen accumulation are described in Ref. 111. Accumulation of glycogen is expressed as milligrams of anhydroglucose per gram (wet weight) of cells and the value given is the maximal amount accumulated in stationary phase. The rate of glycogen accumulation is expressed as the change of µmoles of anhydroglucose per gram of cell (wet weight) per hour

by the parent strain in enriched media containing 1% glucose. In minimal media containing 0.6% glucose, the amount of glycogen accumulated in SG5 and CL1136 is two- and four-fold greater, respectively, than that found in *E. coli* B. The rates of glycogen synthesis in both media are about 2- and 3.5-fold greater for the mutants SG5 and CL1136, respectively, The levels of activity of the glycogen biosynthetic enzymes (ADPglucose pyrophosphorylase, glycogen synthase and branching enzyme) in the mutants and in the parent strain are roughly equivalent and, therefore, cannot account for the increased rate of accumulation of glycogen present in the mutants. Furthermore, a study of the kinetic properties of the mutant ADPglucose pyrophosphorylases showed that they had approximately the same apparent affinities as the parent enzyme for the substrates ATP and glucose 1-P, and for the divalent cation activator $Mg^{2+}$ [20, 114]. The modified properties were the apparent affinities of the mutant enzymes for the allosteric effectors. As seen in Table 7.7, the concentration of fructose 1,6-diP required for 50% of maximal activation (defined as $A_{0.5}$) is about 3-fold less for the SG5 ADPglucose pyrophosphorylase and 12-fold less for the CL1136 enzyme. The enzyme from the parent strain is also more dependent on the allosteric activator for maximal activity than are the enzymes obtained from the mutants. At saturating fructose 1,6-diP concentrations there is a 34-fold stimulation of ADPglucose synthesis catalysed by the *E. coli* B enzyme but only a 14-fold and 1.5-fold stimulation for the SG5 and CL1136 ADPglucose pyrophosphorylases, respectively. Table 7.7 also shows that the apparent affinities for the activators NADPH and pyridoxal 5′-P are also greater with the mutant ADPglucose pyrophosphorylases than with the *E. coli* B enzyme.

**Table 7.7 Kinetic constants for allosteric effectors of *E. coli* B, SG5 and CL1136 ADP-glucose pyrophosphorylases**

| *Enzyme* | $A_{0.5}$/μmol $l^{-1}$ *Fructose 1,6-diP* | *NADPH* | *Pyridoxal 5-P* | $I_{0.5}AMP$/μmol $l^{-1}$ *Fructose* 1,6-*diP conc.*/mmol $l^{-1}$ 1.5 | 0.5 | 0.15 |
|---|---|---|---|---|---|---|
| *E. coli* B | 68 | 64 | 16 | 87 | 41 | 16 |
| SG5 | 22 | 31 | 7 | 170 | 74 | 29 |
| CL 1136 | 5.2 | 5.0 | 0.9 | 680 | 380 | 142 |

In addition to having greater apparent affinities for the activators, the mutant ADPglucose pyrophosphorylases are also less sensitive to the allosteric inhibitor 5′-AMP[20, 114, 178]. Table 7.7 shows that higher concentrations of 5′-AMP are required to give 50% inhibition of the mutant ADPglucose pyrophosphorylases than are required for inhibition of the *E. coli* B enzyme. Both mutant and *E. coli* B enzymes become more sensitive to 5′-AMP inhibition in the presence of lower concentrations of fructose 1,6-diP, i.e. lower concentrations of the inhibitors are needed to achieve 50% inhibition. At all concentrations of fructose 1,6-diP the mutant enzymes are generally still less sensitive than the parent enzyme to inhibition. Thus fructose 1,6-diP antagonises or partially reverses the inhibition by 5′-AMP of the mutant ADPglucose pyrophosphorylases in the same manner as has been shown previously for the *E. coli* B enzyme[131, 132].

These studies suggest that the increased accumulation of glycogen in the mutants SG5 and CL1136 is due to an alteration of the ADPglucose pyrophosphorylases, which cause a greater affinity for the activators and a lower affinity for the inhibitor. Correlation of the relative sensitivities of *E. coli* B, SG5 and CL1136 ADPglucose pyrophosphorylases to AMP inhibition and fructose 1,6-diP activation with the increased rates of glycogen accumulation are in agreement with the expressed view that the cellular levels of the allosteric activators and inhibitors of ADPglucose pyrophosphorylase modulate the rate of ADPglucose synthesis and thereby regulate the rate of synthesis and accumulation of glycogen in the cell[19, 20, 114, 178].

### 7.5.4 The physiological activator of ADPglucose pyrophosphorylase in *E. coli*

Although the foregoing discussion provided convincing data for the importance of the allosteric phenomena associated with the *E. coli* ADPglucose pyrophosphorylase in the *in vivo* regulation of glycogen synthesis, the question of which is the more important physiological activator, NADPH fructose 1,6-diP or pyridoxal 5-P, has not been answered in the studies with the mutants, SG5 or CL1136. All three activators are very effective at low concentrations, as indicated in Table 7.7. The stimulation of the rate of ADPglucose synthesis at 1.5 mmol $l^{-1}$ ATP, 0.5 mmol $l^{-1}$ glucose 1-P and 5.0 mmol $l^{-1}$ $MgCl_2$ by saturating concentrations of fructose 1,6-diP, NADPH and pyridoxal 5-P is *ca.* 34.5-, 27- and 38-fold respectively. However, at the physiological energy charge value of 0.85 [275, 280], only fructose 1,6-diP and pyridoxal 5-P are capable of giving significant activation of the *E. coli* ADPglucose pyrophosphorylase[114, 143]. NADPH gives only slight activation below 0.5 mmol $l^{-1}$. This inability of NADPH to activate ADPglucose pyrophosphorylase at a physiological energy charge was first observed by Shen and Atkinson[143]. 'Poor activators', i.e. metabolites such as P-enolpyruvate, 3-P-glycerate and 2-P-glycerate, that only activate at higher concentrations and do not elicit as great a stimulation as observed with NADPH, fructose 1,6-diP or pyridoxal 5-P, give negligible activation at energy charge of 0.85 [114, 143]. Thus, it appears that only pyridoxal 5-P and fructose 1,6-diP would qualify as potential activators under physiological conditions.

The concentration of pyridoxal 5-P required[114] to give 50% maximal activation at energy charge of 0.85 is 57 µmol $l^{-1}$. The concentration of pyridoxal 5-P found[281] in *E. coli* is *ca.* 24–48 µmol $l^{-1}$ and most of this is protein-bound[282] and therefore would not be available for activation of ADPglucose pyrophosphorylase. Although pyridoxal 5-P is capable of activating the *E. coli* B ADPglucose pyrophosphorylase to a great extent *in vitro*, the concentration required for this activation may be considered non-physiological. Thus it is probable that fructose 1,6-diP may be the only important allosteric activator *in vivo*. The concentrations of fructose 1,6-diP found in *E. coli* grown on glucose range from 0.6 to 3.0 mmol $l^{-1}$ and these concentrations are effective in activating the ADPglucose pyrophosphorylase at the physiological range of energy charge of 0.85 [274, 280].

Consistent with this view are the studies by Dietzler *et al.*, where *E. coli* W4597 (K) or G34 were grown under five different nutrient conditions that gave a 10-fold range in the rate of glycogen accumulation in stationary phase[307, 308]. This was achieved by varying the nature of the carbon and/or the nitrogen source of the media. The different rates of glycogen accumulation found in the various nutrient conditions could be linearly related to the square of the fructose 1,6-diP concentration found in the bacteria. The ATP concentrations in the bacteria were the same, and independent of the composition of the media. The results of Dietzler *et al.*[307,308] strongly suggest that of the *in vitro* primary activators, only fructose 1,6-diP is functional *in vivo*.

### 7.5.4.1 E. coli *B mutant SG14*

The view that fructose 1,6-diP may be the only important physiological activator is supported by studies with the *E. coli* B mutant SG14[20,115,310]. SG14 accumulates glycogen at *ca.* 55–65% the rate of *E. coli* B and contains *ca.* 16% of the ADPglucose synthesising activity as *E. coli* B[115]. Yet the activity present is still 3- to 5-fold greater than that required for the observed rate of glycogen accumulation in SG14. The concentrations of ATP and $Mg^{2+}$ required for 50% of maximal activity ($S_{0.5}$) are 4- to 5-fold higher for the SG14 enzyme than the *E. coli* B enzyme[115]. Whereas the $S_{0.5}$ values for ATP and $Mg^{2+}$ are 0.39 and 2.38 mmol $l^{-1}$ respectively, for *E. coli* B enzyme in the presence of 1.5 mmol $l^{-1}$ fructose 1,6-diP, the $S_{0.5}$ values for ATP and $Mg^{2+}$ are 1.6 and 10.2 mmol $l^{-1}$, respectively, for the SG14 ADPglucose pyrophosphorylase in the presence of saturating fructose 1,6-diP (4.0 mmol $l^{-1}$). Reports in the literature[274,275,279,280,283,284,307] indicate that the ATP level in growing *E. coli* ranges from 2 to 6 mmol $l^{-1}$, and the total magnesium level[285-287] is *ca.* 25–40 mmol $l^{-1}$. Therefore, the SG14 ADPglucose pyrophosphorylase would essentially be saturated with respect to these substrates. The apparent affinities ($S_{0.5}$) for glucose 1-P for the *E. coli* B and SG14 enzymes are about the same (0.036 mmol $l^{-1}$).[115]

The major differences between the SG14 and *E. coli* B ADPglucose pyrophosphorylases appear to be their sensitivities toward activation and inhibition. About 12-fold more fructose 1,6-diP is needed for 50% maximal stimulation of the SG14 ADPglucose pyrophosphorylase ($A_{0.5} = 0.82$ mmol $l^{-1}$) than for half-maximal stimulation of *E. coli* B enzyme, while the $A_{0.5}$ value for pyridoxal 5-P for the SG14 enzyme (0.44 mmol $l^{-1}$) is *ca.* 25-fold higher than for the *E. coli* B enzyme. Both fructose 1,6-diP and pyridoxal 5-P stimulate ADPglucose synthesis catalysed by the *E. coli* B enzyme to about the same extent. However, the stimulation of the SG14 ADPglucose pyrophosphorylase seen with pyridoxal 5-P is only one-half that elicited by fructose 1,6-diP. A notable difference is that NADPH, an activator for the *E. coli* B enzyme, does not stimulate the SG14 enzyme[20,115]. Compounds similar to NADPH in structure, such as PRPP and 2′-PADPR, that are capable of activating the *E. coli* B ADPglucose pyrophosphorylase, do not activate the SG14 enzyme.

Since the apparent affinity of the SG14 enzyme for its activators is considerably lower than that observed for the *E. coli* B ADPglucose pyrophosphorylase, it was an unexpected finding that SG14 is capable of accumulating glycogen at one-half the rate observed for the parent strain.

This rate is accounted for by the relative insensitivity of the SG14 enzyme to inhibition by 5′-AMP. The SG14 ADPglucose pyrophosphorylase is much less sensitive to 5′-AMP inhibition in the concentration range of 0–0.2 mmol $l^{-1}$ than is the parent strain enzyme. At a saturating concentration of fructose 1,6-diP for the SG14 enzyme (4.0 mmol $l^{-1}$), only 7% inhibition of the SG14 enzyme is observed at 0.2 mmol $l^{-1}$ 5′-AMP; the same concentration of 5′-AMP gives 40% inhibition of the *E. coli* B enzyme. At a concentration of fructose 1,6-diP which gives 80% of maximal velocity (1.5 mmol $l^{-1}$) for the SG14 enzyme, 0.2 mmol $l^{-1}$ 5′-AMP causes inhibitions of 84% and 33% with the *E. coli* B and SG14 enzymes, respectively. A decrease in fructose 1,6-diP concentration to 1.0 or 0.5 mmol $l^{-1}$ further increases the sensitivity of the *E. coli* ADPglucose pyrophosphorylase activity to inhibition. However, at these concentrations of fructose 1,6-diP the sensitivity of the SG14 ADPglucose pyrophosphorylase to 5′-AMP remains the same, or becomes less than that seen at 1.5 mmol $l^{-1}$ fructose 1,6-diP. At concentrations of 0.5–1.0 mmol $l^{-1}$ of fructose 1,6-diP the *E. coli* B enzyme is inhibited 90% or more by 0.2 mmol $l^{-1}$ AMP while the inhibition of the SG14 enzyme ranges from 12 to 30%. Although the modification of the SG14 enzyme causes it to have a lower apparent affinity for its activators, it also renders the enzyme less sensitive to 5′-AMP inhibition. These two effects appear to compensate for each other and allow SG14 to accumulate glycogen at about one-half the rate of the parent strain.

The data obtained from the kinetic studies of the SG14 ADPglucose pyrophosphorylase suggest that fructose 1,6-diP is the most important physiological activator of the *E. coli* ADPglucose pyrophosphorylase. This is based on the observation that NADPH is not an activator of the SG14 enzyme and that the concentration of pyridoxal 5-P needed for activation of the enzyme ($A_{0.5}$ = 0.44 mmol $l^{-1}$) is considerably higher than that reported to be present in *E. coli* B. The concentration[281] of pyridoxal 5-P is 24 to 48 μmol $l^{-1}$, and most of this metabolite is probably protein bound in the cell and unavailable for activation of the ADPglucose pyrophosphorylase[282].

The concentration[274, 275, 288] of fructose 1,6-diP in *E. coli* is *ca.* 0.6–3.2 mmol $l^{-1}$ and the $A_{0.5}$ of SG14 ADPglucose pyrophosphorylase is 0.82 mmol $l^{-1}$. The concentration of fructose 1,6-diP in the *E. coli* cell required for activation of the SG14 enzyme would, therefore, be sufficient to account for the activation of the SG14 ADPglucose pyrophosphorylase required for synthesis of ADPglucose at the rates required for the glycogen accumulation observed in SG14.

## 7.6 GENETIC REGULATION OF BACTERIAL GLYCOGEN SYNTHESIS

### 7.6.1 Levels of glycogen biosynthetic enzymes during the growth phase

As indicated elsewhere, many groups have reported that glycogen accumulation occurs in stationary phase of growth. Table 7.8 shows the levels of ADPglucose pyrophosphorylase and glycogen synthase in the exponential and the stationary phases of growth of *E. coli* B in enriched and in minimal media. The levels of the glycogen biosynthetic enzymes and the accumulation of glycogen begin to increase dramatically at the end of the late exponential growth phase and reach their maximum in the stationary phase[19, 121, 278]. In enriched media, containing 1% glucose and 0.6% yeast extract, there is an 11- to 12-fold increase in the specific activities of the pyrophosphorylase and

**Table 7.8 Glycogen, ADPglucose pyrophosphorylase and glycogen synthase levels in *E. coli* B and mutant SG3***

| *Growth conditions* | *Glycogen* B (/mg $g^{-1}$, wet weight) | *Glycogen* SG3 (/mg $g^{-1}$, wet weight) | *Pyrophosphorylase* B (/µmol $mg^{-1}$ 10 $min^{-1}$) | *Pyrophosphorylase* SG3 (/µmol $mg^{-1}$ 10 $min^{-1}$) | *Glycogen synthase* B (/µmol $mg^{-1}$ 10 $min^{-1}$) | *Glycogen synthase* SG3 (/µmol $mg^{-1}$ 10 $min^{-1}$) |
|---|---|---|---|---|---|---|
| 1% Glucose–yeast extract | | | | | | |
| Exponential phase cells | 1.3 | 2.9 | 0.04 | 0.39 | 0.10 | 0.41 |
| Stationary phase cells | 19 | 26 | 0.43 | 0.94 | 0.84 | 1.23 |
| 0.6% Glucose–minimal media | | | | | | |
| Exponential phase cells | 4.2 | 7.2 | 0.21 | 0.79 | 0.43 | 0.85 |
| Stationary phase cells | 20 | 37 | 0.43 | 1.29 | 0.93 | 1.38 |

* The growth of the organisms, composition of the media and glycogen and enzyme assays are described elsewhere[111]

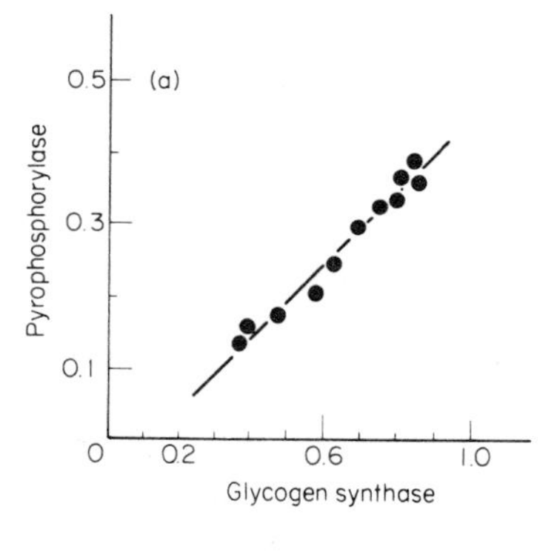


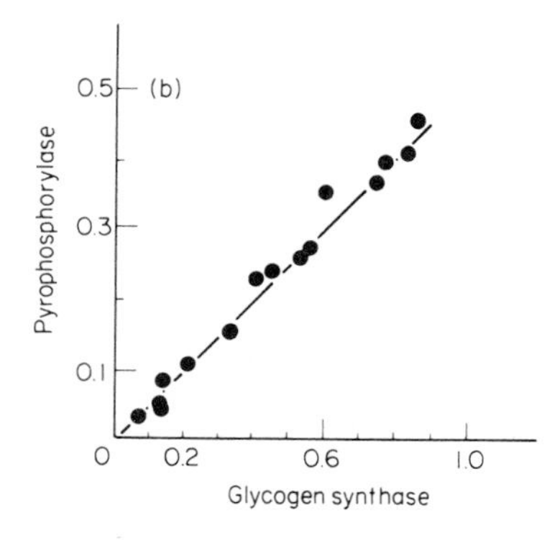


**Figure 7.8** The specific activities of ADP-glucose pyrophosphorylase and ADP-glucose glycogen synthase in crude extracts of *E. coli* B grown on 0.6% minimal glucose medium (a) and yeast extract-1% glucose medium (b). The specific activity values are in µmoles product formed in 10 min per mg of protein. The composition of the media, the enzyme assays and preparation of the crude extracts have been previously reported[111]. The data in the figure have been obtained from previously published growth curves[19, 121, 278]

synthase in the stationary phase of growth (Table 7.8). In minimal media containing 0.6% glucose and limiting $NH_4Cl$, the pyrophosphorylase and glycogen synthase levels in the stationary phase of the growth curve are about the same as those found in the enriched media. In minimal media, however, the levels of these enzymes are less repressed in the exponential phase of growth. Thus only a two- to three-fold increase in the levels of the enzymes is seen during the transition from the exponential phase to the stationary phase. It appears that the synthesis of the pyrophosphorylase and synthase is derepressed as soon as the micro-organism ceases to grow. This derepression occurs in a coordinate manner. Figure 7.8 shows that the levels of the ADP-glucose pyrophosphorylase and glycogen synthase increase in constant proportion to each other. The increase may involve protein synthesis, as is suggested by the studies of Cattaneo *et al.*[289]. They showed that glycogen accumulation in *E. coli* at the beginning of the stationary phase was inhibited by the addition of chloramphenicol to the medium.

Strains of *E. coli* K-12 also show the same type of derepression of the activity of glycogen biosynthetic enzymes upon going from logarithmic phase to stationary phase in enriched media[134]. In *Agrobacterium tumefaciens* it has been shown that derepression of the glycogen biosynthetic enzymes also occurs when the growth of the organism becomes nitrogen-limited[121]. Greater repression is seen when the organism is grown on enriched media. However, this phenomenon does not occur in the photosynthetic organism *Rhodopseudomonas capsulata*.[49] Glycogen accumulation, glycogen synthase activity and ADPglucose pyrophosphorylase activity appear to be at their maximum during the logarithmic phase of growth when the cells are grown on malate as a carbon source either in the light or in the dark. Maximum glycogen accumulation in exponential phase cultures of *Streptococcus mitis* has been reported. It is quite possible that in this organism the levels of the glycogen biosynthetic enzymes are already present at optimal levels[52].

### 7.6.2 Isolation of an *E. coli* B mutant derepressed in the glycogen biosynthetic enzymes

A mutant of *E. coli* B, SG3, has been isolated that is derepressed in both ADPglucose pyrophosphorylase and glycogen synthase activities[19,111]. Table 7.8 shows the levels of glycogen and the glycogen synthetic enzymes during the growth of the mutant on glycose-yeast extract medium and on minimal medium. In enriched medium and in the early exponential phase, SG3 has twice as much glycogen per gram of wet weight cells and contains eight- to eleven-times more ADPglucose pyrophosphorylase activity and three- to four-times more glycogen synthase activity than the parent strain. Thus, both enzymes seem to be appreciably elevated in the exponential phase of growth. There is a further elevation of the pyrophosphorylase and synthase activities of about two- to three-fold when SG3 enters the stationary phase. Table 7.8 also shows that in minimal medium SG3 contains derepressed levels of the glycogen biosynthetic enzymes and about twice as much glycogen as the parent strain in the exponential phase.

Kinetic studies on partially purified preparations of SG3 ADPglucose pyrophosphorylase and glycogen synthase indicate that they are identical to the parent *E. coli* B enzymes with respect to the kinetic constants and parameters of activator specificity, substrate requirement and inhibition characteristics[290]. Thus more of the normal enzymes seem to be synthesised in SG3, rather than more reactive enzyme molecules. However, it should be noted that the relative proportionate increase in the two enzymes in growing cultures of *E. coli* SG3 is different from that observed for *E. coli* B. Whereas the ratio of enzyme activity of ADPglucose pyrophosphorylase to glycogen synthase is about 0.5, the ratio is found to be about 0.94 for SG3.

No information is presently available on the mechanism or factors involved in the derepression of the glycogen biosynthetic enzymes in the stationary phase of growth. A few experiments have been done to determine whether cyclic-AMP affected the levels of the glycogen biosynthetic enzymes. Unpublished experiments have shown that SG3 behaves normally with respect to induction of β-galactosidase by the inducers isopropyl thiogalactoside and methyl thiogalactoside. The induced β-galactosidase activity in SG3 is sensitive to catabolite repression[291] when glucose is present in the medium in the same way as the parent strain and inclusion of 1 mmol $l^{-1}$ cyclic AMP in the medium overcomes the glucose repression of β-galactosidase synthesis in *E. coli* B[292]. Cyclic AMP, however, had no effect on the exponential phase levels of the glycogen biosynthetic enzymes in *E. coli* B. Thus cyclic AMP may not be directly involved in regulation of the synthesis of glycogen biosynthetic enzymes.

### 7.6.3 Location of the glycogen biosynthetic enzyme genes on the *E. coli* genetic linkage map

Damotte and colleagues have isolated three types of glycogen biosynthetic mutants in *E. coli* K-12[110,112,134]. These are (1) ADPglucose:1,4-α-glucan 4-α-glucosyl transferase (glycogen synthase)-negative mutants, (2) branching enzyme-negative mutants, and (3) ADPglucose pyrophosphorylase mutants. Sigal and Puig[293] reported that the glycogen genes glg A, B and C were all cotransducible with the maltose A genes, and subsequently the same group reported that the glg genes are cotransducible with the aspartate semialdehyde gene asd[112]. The catabolic L-glycerol 3-phosphate dehydrogenase gene, glp D, is located very close to the maltose A and asd genes on the *E. coli* genetic linkage map[294]; and with the availability of an L-glycerol phosphate dehydrogenase negative mutant of *E. coli* B, the results of the French workers on the positions of the glycogen genes on the linkage map were confirmed[113]. Furthermore, the SG3 mutation also cotransduces about 80% with glp D, which indicates that it is clustered with the other glg genes[113]. It is therefore quite possible that the glg genes are constituted in an operon. Various observations support this concept: the genes are clustered on the *E. coli* chromosome, the activities of two of the enzymes ADPglucose pyrophosphorylase and glycogen synthase are coordinately derepressed at the end of the exponential growth phase, and the glycogen-excess mutant SG3 appears to be derepressed for the synthesis of both glycogen synthase and ADPglucose pyrophosphorylase under all conditions. This would imply that SG3 has received a mutation in one of the control genes for the operon. The signal for the increase of the glycogen synthetic enzymes in *E. coli* appears to be linked with a decrease in the growth rate. As cell division ceases, the enzymes are synthesised to store any available excess carbon as a reserve. It would be of interest to determine the mechanism of the genetic regulation of the biosynthesis of these enzymes, as it may give us some insight into the processes by which the micro-organism is able to modulate various processes in conjunction with the modulation of its growth rate.

Although the genetic regulation of the biosynthesis of the glycogen biosynthetic enzymes is an important factor in controlling the accumulation of glycogen in *E. coli*, it should be pointed out that in glucose-minimal media there is already an 8- to 20-fold excess of the ADPglucose pyrophosphorylase and glycogen synthase levels in the exponential phase of growth to account for the maximum rate of glycogen accumulation seen in the stationary phase. The activities of these enzymes must therefore be controlled by the intracellular concentrations of their substrates and effectors. This is consistent with the evidence indicating that the allosteric regulation of ADPglucose pyrophosphorylase is physiologically significant in the control of the rate of glycogen synthesis in *E. coli*.

## 7.7 CATABOLISM OF GLYCOGEN IN BACTERIA

### 7.7.1 Polysaccharide phosphorylases

Accumulation of glycogen occurs in the presence of excess carbon source in the medium. When the exogenous carbon source has been completely utilised, or considerably diminished, disappearance of the polysaccharide can be observed. However, there are very few reports on the enzyme systems involved in the catabolism of the intracellular polysaccharide. A polyglucose phosphorylase was identified in maltose-grown cells of *E. coli* by Doudoroff *et al.* in 1949[99]. However, the phosphorylases in *E. coli* and other bacteria were not purified and extensively characterised until after 1967.

Two polyglucose phosphorylases have been identified in *E. coli* K-12[96,97]. The two enzymes may be separated by ammonium sulphate fractionation or by DEAE-cellulose column chromatography. One of the enzymes is a maltodextrin phosphorylase which is present in glucose-grown cells but is induced to a 10-fold higher level in maltose-grown cells. Maltodextrin phosphorylase shows a preference for short-chain oligosaccharides. Maltopentaose, maltohexaose and maltoheptaose were effective substrates[100], and low activity was observed with glycogen and maltotetraose[96,97,100]. This maltodextrin phosphorylase has been purified to homogeneity[100]. It has a molecular weight of 140 000 and contains one mole of pyridoxal 5-P per 143 000 grams. Thus it appears to be monomeric in structure.

Chen and Segel have identified the second polyglucose phosphorylase in *E. coli* K-12 as a glycogen phosphorylase[96,97]. This enzyme is constitutive and its level of activity is independent of the carbon source used for growth. This enzyme showed three times as much activity with glycogen as the primer than with maltodextrins. Molecular weight estimates from measurements of electrophoretic mobilities on gels of various concentrations gave a value of *ca.* 250 000. Spectrophotometric evidence suggested the presence of pyridoxal 5-P.

The regulatory properties of both phosphorylases are not as impressive as those found for the mammalian counterparts. No evidence has been obtained for the existence of, or interconversion of, inactive and active forms of bacterial phosphorylase. Both enzymes are stimulated only slightly by 5'-AMP[96,97]. The glycogen phosphorylase is activated by NaF four-fold and by $Na_2SO_4$ 10-fold. The maltodextrin phosphorylase was not appreciably activated by salt. Glycogen phosphorylase was competitively inhibited by ADPglucose ($K_i$ = 0.2 mmol $l^{-1}$), TDPglucose ($K_i$ = 1 mmol $l^{-1}$) and UDPglucose ($K_i$ = 1 mmol $l^{-1}$). Chen and Segel[96,97] have suggested that high intracellular concentrations of ADPglucose would stimulate glycogen synthesis and inhibit glycogen degradation. ADPglucose levels would be expected to be high with the availability of a carbon source in the medium, but absent or minimal with the depletion from carbon in the medium. Thus ADPglucose levels may regulate the glycogen phosphorylase activity. Nevertheless, it should be pointed out that in most bacterial systems the glycogen biosynthetic enzyme activities in glucose-grown cells appear to be 10- to 400-fold greater than the corresponding phosphorylase activity. This is consistent with the observation that the rate of glycogen biosynthesis (accumulation) is much greater than the rate of glycogen degradation. Although the activities of the 1,4-poly-α-glucose phosphorylases are low, they are still compatible with the *in vivo* rates of glycogen degradation[97].

The glycogen phosphorylase from glucose-grown cells of *Streptococcus salivarius* has been purified[98]. In contrast to the *E. coli* K-12 glycogen phosphorylase, this enzyme was more active with low molecular weight maltodextrins from corn than with oyster or *S. salivarius* glycogen. The enzyme is activated three- to four-fold in the direction of phosphorolysis by the simultaneous presence of 5'-AMP and NaF[295]. This activated form of the enzyme can be further activated two- to three-fold by 3.0 mmol $l^{-1}$ phosphoenolpyruvate and inhibited 40% by 1.5 mmol $l^{-1}$ ADPglucose[295]. The physiological significance of the phosphoenolpyruvate activation at present is obscure.

The granulose phosphorylase of *Clostridium pasteuranium* ATCC 6013 was inhibited by sugar nucleotides; ADPglucose ($K_i$ = 20 μmol $l^{-1}$) was a potent competitive inhibitor[311].

### 7.7.2 Degradation of glycogen by the enzymes of the maltose operon and isoamylase

The complete degradation of glycogen necessitates the involvement of an enzyme(s) that is able to hydrolyse the 1,6-α-glucosidic branch linkages. In yeast and in mammals this is effected by the 4-α-glucanotransferase–amylo-1,6-glucosidase system (EC 3.2.1.3). In bacteria, however, two different enzymes capable of hydrolysing the α-1,6-linkages of amylopectin or glycogen have been described.

Pullulanase is an enzyme that was originally found in *Aerobacter aerogenes*[256] and is capable of hydrolysing the 1,6-α-glucosidic linkages of pullulan (poly-6³-α-maltotriosylmaltotriose), and of amylopectin, to yield the respective unbranched oligosaccharide chains. The enzyme is specifically induced by maltose, higher maltodextrins and pullulan, but not by glucose[17, 297]. Similar activities have been reported in *E. coli* strain ML[17], *Streptococcus mutans*[17], *Escherichia intermedia*[298], *Streptococcus mitis*[299] and *Pseudomonas stutzeri*[94]. For *A. aerogenes* and *E. coli* strain ML the pullulanase activity exists in a lipopolysaccharide–phospholipoprotein particulate complex in association with the outer membrane system[300]. It thus appears that the function of the pullulanase is to debranch exogenous branched maltodextrins in the media. Pullulanase, however, cannot appreciably degrade glycogen[17, 301] and thus is not the enzyme involved in the debranching of the intracellular glycogen formed on various carbon sources.

Another enzyme capable of debranching glycogen or amylopectin (but not pullulan) has been found in a number of bacterial species: *Cytophaga*[302], a pseudomonad[303] and in *E. coli* NCTC 5928 [17]. The enzyme appears to be intracellular and is found in the soluble portion of the cell-free extracts of the organisms[17]. Palmer *et al.*[17] have postulated that this enzyme may participate in the degradation of glycogen in the stationary phase and have invoked the series of reactions depicted in Figure 7.9 for the degradation of bacterial

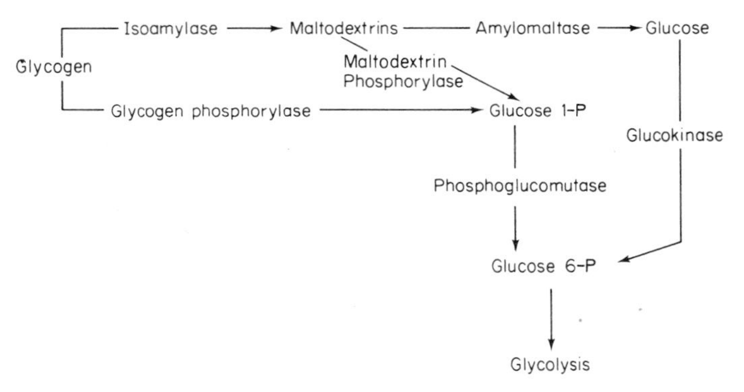


**Figure 7.9** A scheme for the catabolism of glycogen in bacteria as proposed by Palmer *et al.*[17]

glycogen. The combined actions of isoamylase, glycogen phosphorylase, maltodextrin phosphorylase and amylomaltase would lead to formation of glucose 1-P and glucose, which could then be utilised for energy via glycolysis. The levels of enzyme activities of the above enzymes appear to be low in glucose-grown *E. coli* NCTC 5928 cells. However, they increase about two-fold in the stationary phase and are compatible with rates of glycogen degradation observed in many of the *Enterobacteriacea*. It would be of interest to utilise the available mutants of the maltose operon enzymes of *E. coli* K-12[102, 103] and determine whether glycogen breakdown is affected. Similarly, a search for mutants which may have a defective isoamylase may provide some enlightening studies on the subject. A mutant strain of *Clostridium pasteuranium* defective in granulose degradation is deficient in granulose phosphorylase activity[313]. The physiological consequences of this deficiency have not been reported.

Palmer *et al.*[17] also pointed out that the metabolism of branched 1,4-α-glucans in bacteria and plants are in many respects similar. In both pathways ADPglucose is the specific precursor in biosynthesis and its formation, catalysed by ADPglucose pyrophosphorylase, is rate limiting and subject to regulatory control. In catabolism of starch in plants it is believed that degradation occurs via reactions catalysed by R-enzyme, D-enzyme and starch phosphorylase[304]. R-Enzyme, D-enzyme and starch phosphorylase catalyse almost similar reactions as isoamylase, amylomaltose and maltodextrin (glycogen) phosphorylase, respectively. In addition, the plant systems also contain β-amylase and α-amylase. β-Amylase has not been reported to be present in bacteria, but there are many reports of α-amylase being present in the extracellular fluids of many bacterial cultures[305]. Presumably these extracellular enzymes are not involved in the metabolism of the intracellular reserve polysaccharide. In many cases the α-amylase is produced in organisms not noted for accumulating glycogen. However, there is one report of the presence of an intracellular α-amylase in *E. coli* K-12 [306]. The enzyme required chloride ions for optimal activity in contrast to other bacterial amylases and its pH optimum was in the neutral range. The enzyme was quite active with linear substrates such as amylose and soluble starch. However, the rate of reducing sugar release from either *E. coli* or rabbit liver glycogen was only 4–5% of the rate observed for amylose.

The quantitative roles of all the above degradative enzymes, both in plants as well as in bacteria, in the catabolism of glycogen and starch remain to be determined. Similarly, the possible modes of regulation of these enzymes during the period of active glycogen or starch accumulation remain to be uncovered.

**Acknowledgements**

The authors would like to thank Dr. W. J. Whelan and Dr. D. N. Dietzler for copies of references 17 and 275 respectively, prior to their publication.

**Note added in proof**

Since this work was submitted, new findings of interest with respect to the phosphorylation and dephosphorylation of glycogen synthase have appeared. Nimmo and Cohen[314] and Huang *et al.*[315] described the existence in muscle of an enzyme other than the cyclic AMP-dependent protein kinase that catalyses the phosphorylation of glycogen synthase D. This enzyme also phosphorylates phosvitin and will utilise GTP as well as ATP. The reports from the two laboratories differ somewhat with respect to the possible role of the enzyme in the regulation of synthase activity, but it seems likely that it may prove to have a physiological role. Huang *et al.*[315] presented a model showing the possible existence of three interacting phosphorylation sites and two kinases, i.e. the new kinase and the cyclic AMP-dependent protein kinase, in order to explain their findings with respect to the role of phosphorylation in the synthase I to D conversion. Meanwhile, Schlender and Reimann[316] found what would appear to be still another cyclic AMP-independent protein kinase in the kidney medulla that catalyses a clear-cut glycogen synthase I to D conversion by itself. Soderling[317] has reported that the complete glycogen synthase I to synthase D reaction catalysed by the cyclic AMP-dependent protein kinase requires incorporation of 2 moles of phosphate per enzyme subunit. The major activity change occurs as a result of the introduction of the second mole, which requires a very high protein kinase concentration, and causes a further decrease in the ratio of activity minus glycose 6-P to activity plus glucose 6-P. Nakai and Thomas[318] have studied phosphoprotein phosphatases in heart muscle and have obtained good evidence that glycogen synthase phosphatase and phosphorylase phosphatase activities may reside in the same enzyme. Highly purified rabbit liver phosphorylase phosphatase[319] has also been shown to dephosphorylate liver glycogen synthase[320].

## References


1. Newsholme, E. A. and Start, C. (1973). *Regulation in Metabolism*, 146 (London: Wiley)
2. Fischer, E. H., Pocker, A. and Saari, J. C. (1970). In *Essays in Biochemistry*, Vol. 6, 23 (P. N. Campbell and F. Dickens, editors) (New York and London: Academic Press)
3. Fischer, E. H., Heilmeyer, L. M. G., Jr. and Haschke, R. H. (1971). In *Current Topics in Cellular Regulation*, Vol. 4, 211 (B. L. Horecker and E. R. Stadtman, editors) (New York and London: Academic Press)
4. Graves, D. J. and Wang, J. H. (1972). In *The Enzymes*, 3rd ed., Vol. 7, 435 (P. D. Boyer, editor) (New York and London: Academic Press)
5. Larner, J. and Villar-Palasi, C. (1971). In *Current Topics in Cellular Regulation*, Vol. 3, 195 (B. L. Horecker and E. R. Stadtman, editors) (New York and London: Academic Press)
6. Stalmans, W. and Hers, H. G. (1973). In *The Enzymes*, 3rd ed., Vol. 9, 309 (P. D. Boyer, editor) (New York and London: Academic Press)
7. Krebs, E. G. (1972), In *Current Topics in Cellular Regulation*, Vol. 5, 99 (B. L. Horecker and E. R. Stadtman, editors) (New York and London: Academic Press)
8. Walsh, D. A. and Krebs, E. G. (1973). In *The Enzymes*, 3rd ed., Vol. 8, 55 (P. D. Boyer, editor) (New York and London: Academic Press)
9. Robison, C. A., Butcher, R. W. and Sutherland, E. W. (1971). *Cyclic AMP*, 1 (New York and London: Academic Press)
10. Walsh, D. A., Perkins, J. P. and Krebs, E. G. (1968). *J. Biol. Chem.*, **243,** 3763
11. Krebs, E. G., Graves, D. J. and Fischer, E. H. (1959), *J. Biol. Chem.*, **234,** 2867
12. Belocopitow, E. (1961). *Arch. Biochem. Biophys.*, **93,** 457
13. Krebs, E. G., Love, D. S., Bratvold, G. E., Trayser, K. A., Meyer, W. L. and Fischer, E. H. (1964). *Biochemistry*, **3,** 1022
14. Rossel-Perez, M. and Larner, J. (1964). *Biochemistry*, **3,** 773
15. Appleman, M. M., Belocopitow, E. and Torres, H. N. (1964). *Biochem. Biophys. Res. Commun.*, **14,** 550
16. Delange, R. J., Kemp, R. G., Riley, W. D., Cooper, R. A. and Krebs, E. G. (1968). *J. Biol. Chem.*, **243,** 2200
17. Palmer, T. N., Wöber, G. and Whelan, W. J. (1973). *Eur. J. Biochem.*, **39,** 601
18. Brown, D. H. and Illingworth, B. (1962). *Proc. Nat. Acad. Sci. USA*, **48,** 1783
19. Preiss, J. (1969). In *Current Topics in Cellular Regulation*, Vol. 1, 125 (B. L. Horecker and E. R. Stadtman, editors) (New York: Academic Press)
20. Preiss, J. (1973). In *The Enzymes*, 3rd ed., Vol. 8, 73 (P. D. Boyer, editor) (New York and London: Academic Press)

21. Dawes, E. A. and Senior, P. J. (1973). In *Advances in Microbiol Physiology*, Vol. 10, 135 (A. H. Rose and D. W. Tempest, editors) (New York and London: Academic Press)
22. Dawes, E. A. and Ribbons, D. (1962). *Ann. Rev. Microbiol.*, **16**, 241
23. Dawes, E. A. and Ribbons, D. (1964). *Bacteriol Rev.*, **28**, 126
24. Wilkinson, J. F. (1959). *Exp. Cell Res. Suppl.*, **7**, 111
25. Leavitt, C. A. and Krebs, E. G. (1974). Unpublished results
26. Villar-Palasi, C. (1969). *Ann. N.Y. Acad. Sci.*, **166**, 719
27. Heilmeyer, L. M. G., Jr., Meyer, F., Haschke, R. H. and Fischer, E. H. (1970). *J. Biol. Chem.*, **245**, 6649
28. Haschke, R. H., Heilmeyer, L. M. G., Jr., Meyer, F. and Fischer, E. H. (1970). *J. Biol. Chem.*, **245**, 6657
29. Strange, R. E., Dark, F. A. and Ness, A. G. (1961). *J. Gen. Microbiol.*, **25**, 61
30. Madsen, N. B. (1961). *Biochim. Biophys. Acta*, **50**, 194
31. Madsen, N. B. (1963). *Can. J. Biochem. Physiol.*, **41**, 561
32. Boylen, C. W. and Ensign, J. C. (1970). *J. Bacteriol.*, **103**, 578
33. Zevenhuizen, L. P. T. M. (1964). *Biochem. Biophys. Acta*, **81**, 608
34. Zevenhuizen, L. P. T. M. (1966). *J. Microbiol. Serol.*, **32**, 356
35. Mulder, E. G. and Zevenhuizen, L. P. T. M. (1967). *Arch. Mikrobiol.*, **59**, 345
36. Mulder, E. G., Deinema, M. H., Van Veen, W. L. and Zevenhuizen, L. P. T. M. (1962). *Rec. Trav. Chim. Pays-Bas*, **81**, 797
37. Ghosh, H. P., and Preiss, J. (1965). *Biochim. Biophys. Acta*, **104**, 274
38. Barry, C., Gavard, R., Milhaud, G. and Aubert, J. P. (1952). *C. R. Acad. Sci.*, **235**, 1062
39. Goldemberg, S. H. (1972). In *Biochemistry of the Glycosidic Linkage*, 621 (R. Piras and H. G. Pontis, editors) (New York and London: Academic Press)
40. Garvard, R. and Milhaud, G. (1952). *Ann. Inst. Pasteur, Paris*, **82**, 471
41. Robson, R. L., Robson, R. M. and Morris, J. G. (1972). *Biochem. J.*, **130**, 4P
42. Levine, S., Stevenson, H. J. R., Tabor, E. C., Bordner, R. H. and Chambers, L. A. (1953). *J. Bacteriol.*, **66**, 664
43. Holme, T. and Palmstierna, H. (1956). *Acta Chem. Scand.*, **10**, 578
44. Sigal, N., Catteneo, J. and Segel, I. H. (1964). *Arch. Biochem. Biophys.*, **108**, 440
45. Antoine, A. D. and Tepper, B. S. (1969). *Arch. Biochem. Biophys.*, **134**, 207
46. Elbein, A. D. and Mitchell, M. (1973). *J. Bacteriol.*, **113**, 863
47. Chargaff, E. and Moore, D. H. (1944). *J. Biol. Chem.*, **155**, 483
48. Antoine, A. D. and Tepper, B. S. (1969). *J. Bacteriol.*, **100**, 538
49. Eidels, L. and Preiss, J. (1970). *Arch. Biochem. Biophys.*, **140**, 75
50. Stanier, R. Y., Doudoroff, M., Kunisawa, R. and Contopoulou, R. (1959). *Proc. Nat. Acad. Sci. USA*, **45**, 1246
51. Steiner, K. and Preiss, J. (1973). Unpublished results
52. Gibbons, R. J. and Kapsimalis, B. (1963). *Arch. Oral. Biol.*, **8**, 319
53. Houte, J. V. and Jansen, H. M. (1970). *J. Bacteriol.*, **101**, 1083
54. Houte, J. V. and Saxton, C. A. (1971). *Caries. Res.*, **5**, 30
55. Houte, J. V. and Jansen, H. M. (1968). *Caries Res.*, **2**, 47
56. Khandelwal, R. L., Spearman, T. N. and Hamilton, I. R. (1972). *Can. J. Biochem.*, **50**, 440
57. Eisenberg, R. J., Elchisak, M. and Lai, C. (1974). *Biochem. Biophys. Res. Commun.*, **57**, 959
58. Segel, I. H., Cattaneo, J. and Sigal, N. (1965). *Proc. Int. CNRS Symp., Mech. Regulation Cellular Activities Microorganisms*, **1963**, 337
59. Mallette, M. F. (1963). *Ann. N.Y. Acad. Sci.*, **102**, 521
60. Marr, A. G., Nilson, E. H. and Clark, D. J. (1963). *Ann. N.Y. Acad. Sci.*, **102**, 536
61. McGrew, S. B. and Mallette, M. F. (1965). *Nature*, **208**, 1096
62. Pirt, S. J. (1965). *Proc. Roy. Soc. B*, **163**, 224
63. McGrew, S. B. and Mallette, M. F. (1962). *J. Bacteriol.*, **83**, 844
64. Ribbons, D. W. and Dawes, E. A. (1963). *Ann. N.Y. Acad. Sci.*, **102**, 464
65. Strange, R. E. (1968). *Nature*, **220**, 606
66. Tempest, D. W. and Strange, R. E. (1966). *J. Gen. Microbiol.*, **44**, 273
67. Mackey, B. M. and Morris, J. G. (1971). *J. Gen. Microbiol.*, **19**, 210
68. Strasdine, G. A. (1972). *Can. J. Microbiol.*, **18**, 211

69. Strasdine, G. A. (1968). *Can. J. Microbiol.*, **14**, 1059
70. Burleigh, I. G. and Dawes, E. A. (1967). *Biochem. J.*, **102**, 236
71. Brown, D. H., Gordon, R. B. and Brown, B. I. (1973). *Ann. N.Y. Acad. Sci.*, **210**, 238
72. Sutherland, E. W. and Cori, C. F. (1948). *J. Biol. Chem.*, **172**, 737
73. Sutherland, E. W. and Cori, C. F. (1951). *J. Biol. Chem.*, **188**, 531
74. Cori, G. T. (1945). *J. Biol. Chem.*, **158**, 333
75. Leloir, L. F. and Cardini, C. E. (1957). *J. Amer. Chem. Soc.*, **79**, 6340
76. Leloir, L. F., Olavarria, J. M., Goldemberg, S. H. and Carminatti, H. (1959). *Arch. Biochem. Biophys.*, **81**, 508
77. Traut, R. R. (1962). *Thesis* (New York: Rockefeller Institute)
78. Rosell-Perez, M., Villar-Palasi, C. and Larner, J. (1962). *Biochemistry*, **1**, 763
79. Sutherland, E. W. (1949). *J. Biol. Chem.*, **180**, 1279
80. Milstein, C. (1961). *Biochem. J.*, **79**, 584
81. Ray, W. J., Jr. and Peck, E. J., Jr. (1972). In *The Enzymes*, 3rd ed., Vol. 6, 407 (Paul D. Boyer, editor) (New York and London: Academic Press)
82. Peck, E. J., Jr. and Ray, W. J., Jr. (1971). *J. Biol. Chem.*, **246**, 1160
83. Turnquist, R. L. and Hansen, R. G. (1973). In *The Enzymes*, 3rd ed., Vol. 8, 51 (Paul D. Boyer, editor) (New York and London: Academic Press)
84. Hestrin, S. (1960). In *The Bacteria*, Vol. 3, 373 (I. C. Gunsalus and R. Y. Stanier, editors) (New York: Academic Press)
85. Hehre, E. J. and Hamilton, D. M. (1946). *J. Biol. Chem.*, **166**, 77
86. Hehre, E. J. and Hamilton, D. M. (1948). *J. Bacteriol.*, **55**, 197
87. Hehre, E. J., Hamilton, D. M. and Carlson, A. S. (1949). *J. Biol. Chem.*, **177**, 267
88. Hehre, E. J. (1951). *Advan. Enzymol.*, **11**, 297
89. Okada, G. and Hehre, J. (1974). *J. Biol. Chem.*, **249**, 126
90. Carlson, A. S. and Hehre, E. J. (1949). *J. Biol. Chem.*, **177**, 281
91. Monod, J. and Torriani, A. M. (1948). *C. R. Acad. Sci., Paris*, **227**, 240
92. Walker, G. J. (1966). *Biochem. J.*, **101**, 861
93. Lacks, S. (1968). *Genetics*, **60**, 685
94. Wöber, F. (1973). *Z. Physiol. Chem.*, **354**, 75
95. Chao, J. and Weathersbee, C. J. (1974). *J. Bacteriol.*, **117**, 181
96. Chen, G. S. and Segel, I. H. (1968). *Arch. Biochem. Biophys.*, **127**, 164
97. Chen, G. S. and Segel, I. H. (1968). *Arch. Biochem. Biophys.*, **127**, 175
98. Khandelwal, R. L., Spearman, T. N. and Hamilton, I. R. (1973). *Arch, Biochem. Biophys.*, **154**, 295
99. Doudoroff, M., Hassid, W. Z., Putman, E. W., Porter, A. L. and Lederberg, J. (1949). *J. Biol. Chem.*, **179**, 921
100. Schwartz, M. and Hofnung, M. (1967). *Eur. J. Biochem.*, **2**, 132
101. Walker, G. J., Lukas, M. C. and Lavrova, A. (1969). *Carbohyd. Res.*, **9**, 381
102. Schwartz, M. (1965). *C. R. Acad. Sci., Paris*, **260**, 2613
103. Schwartz, M. (1966). *J. Bacteriol.*, **92**, 1083
104. Madsen, N. B. (1961). *Biochim. Biophys. Acta*, **50**, 194
105. Greenberg, E. and Preiss, J. (1964). *J. Biol. Chem.*, **239**, 4314
106. Shen, L., Ghosh, H. P., Greenberg, E. and Preiss, J. (1964). *Biochim. Biophys. Acta*, **89**, 370
107. Zevenhuizen, L. P. T. M. (1964). *Biochim. Biophys. Acta*, **81**, 608
108. Sigal, N., Cattaneo, J., Chambost, J. P. and Favard, A. (1965). *Biochem. Biophys. Res. Commun.*, **20**, 616
109. Fox, J., Kennedy, L. D., Hawker, J. S., Ozbun, J. L., Greenberg, E., Lammel, C. and Preiss, J. (1973). *Ann. N.Y. Acad. Sci.*, **210**, 265
110. Damotte, M., Cattaneo, J., Sigal, N. and Puig, J. (1968). *Biochem. Biophys. Res. Commun.*, **32**, 916
111. Govons, S., Vinopal, R., Ingraham, J. and Preiss, J. (1969). *J. Bacteriol.*, **97**, 970
112. Cattaneo, J., Damotte, M., Sigal, N., Sanchez-Medina, G. and Puig, J. (1969). *Biochem. Biophys. Res. Commun.*, **34**, 694
113. Preiss, J., Ozbun, J. L., Hawker, J. S., Greenberg, E. and Lammel, C. (1973). *Ann. N.Y. Acad. Sci.*, **210**, 265
114. Govons, S., Gentner, N., Greenberg, E. and Preiss, J. (1973). *J. Biol. Chem.*, **248**, 1731
115. Preiss, J., Sabraw, A. and Greenberg, E. (1971). *Biochem. Biophys. Res. Commun.*, **42**, 180

116. Ribereau-Gayon, G., Sabraw, A., Lammel, C. and Preiss, J. (1971). *Arch. Biochem. Biophys.*, **142**, 675
117. Kindt, T. J. and Conrad, H. E. (1967). *Biochemistry*, **6**, 3718
118. Gahan, L. C. and Conrad, H. E. (1968). *Biochemistry*, **7**, 3979
119. Crawford, K. and Preiss, J. (1974). Unpublished results
120. Eidels, L., Edelmann, P. and Preiss, J. (1970). *Arch. Biochem. Biophys.*, **140**, 60
121. Preiss, J., Govons, S., Eidels, L., Lammel, C., Greenberg, E., Edelmann, P. and Sabraw, A. (1970). In *Miami Winter Symp.*, Vol. 1, 122 (W. J. Whelan and J. Schultz, editors) (Amsterdam: North-Holland)
122. Greenberg, E. and Preiss, J. (1965). *J. Biol. Chem.*, **240**, 2341
123. Shen, L. and Preiss, J. (1964). *Biochem. Biophys. Res. Commun.*, **17**, 424
124. Shen, L. and Preiss, J. (1965). *J. Biol. Chem.*, **240**, 2334
125. Shen, L. and Preiss, J. (1966). *Arch. Biochem. Biophys.*, **116**, 375
126. Goldemberg, S. H. and Algranati, I. D. (1969). *Biochim. Biophys. Acta*, **177**, 166
127. Parsons, T. and Preiss, J. (1974). Unpublished results
128. Preiss, J. and Greenberg, E. (1965). *Biochemistry*, **4**, 3228
129. Preiss, J., Shen, L. and Partridge, M. (1965). *Biochem. Biophys. Res. Commun.*, **18**, 180
130. Preiss, J., Shen, L., Greenberg, E. and Gentner, N. (1966). *Biochemistry*, **5**, 1833
131. Gentner, N. and Preiss, J. (1967). *Biochem. Biophys. Res. Commun.*, **27**, 417
132. Gentner, N. and Preiss, J. (1963). *J. Biol. Chem.*, **243**, 5882
133. Gentner, N., Greenberg, E. and Preiss, J. (1969). *Biochem. Biophys. Res. Commun.*, **36**, 373
134. Creuzet-Sigal, N., Latil-Damotte, M., Cattaneo, J. and Puig, J. (1972). *Biochemistry of the Glycosidic Linkage*, 647 (R. Pites and H. G. Pontis, editors) (New York: Academic Press)
135. Shen, L. and Preiss, J. (1974). Unpublished results
136. Elbein, A. D. and Mitchell, M. (1973). *J. Bacteriol.*, **113**, 863
137. Lapp, D. and Elbein, A. D. (1972). *J. Bacteriol.*, **112**, 327
138. Dietzler, D. N. and Strominger, J. L. (1973). *J. Bacteriol.*, **113**, 946
139. Furlong, C. E. and Preiss, J. (1969). In *Progress in Photosynthesis Research*, Vol. 3, 1604 (H. Metzner, editor) (Germany: Tubingen Press)
140. Furlong, C. E. and Preiss, J. (1969). *J. Biol. Chem.*, **244**, 2539
141. Builder, J. E. and Walker, G. J. (1970). *Carbohyd. Res.*, **14**, 35
142. Morris, B. and Preiss, J. (1974). Unpublished results
143. Shen, L. and Atkinson, D. E. (1970). *J. Biol. Chem.*, **245**, 3996
144. Cahill, G. F., Jr. (1970). *New Engl. J. Med.*, **282**, 668
145. Peter, J. P., Barnard, R. J., Edgerton, V. R., Gillespie, C. A. and Stenpel, K. E. (1972). *Biochemistry*, **11**, 2627
146. Danforth, W. H., Helmreich, E. and Cori, C. F. (1962). *Proc. Nat. Acad. Sci. USA*, **48**, 1191
147. Cori, C. F. and Cori, G. T. (1936). *Proc. Soc. Exp. Biol. Med.*, **34**, 702
148. Parnas, J. K. and Mochnacko, I. (1936). *Compt. Rend. Soc. Biol.*, **123**, 1173
149. Parmeggiani, A. and Morgan, H. E. (1962). *Biochem. Biophys. Res. Commun.*, **9**, 252
150. Madsen, N. B. (1964). *Biochem. Biophys. Res. Commun.*, **15**, 390
151. Fox, J. and Preiss, J. (1974). Unpublished results
152. Krebs, E. G. (1954). *Biochim. Biophys. Acta*, **15**, 508
153. Heilmeyer, L., Jr. and Haschke, R. H. (1972). In *Protein–Protein Interaction*, 299 (R. Jaenicke and E. Helmreich, editors) (New York, Heidelberg, Berlin: Springer-Verlag)
154. Cori, G. T. and Green, A. A. (1943). *J. Biol. Chem.*, **151**, 31
155. Fischer, E. H. and Krebs, E. G. (1955). *J. Biol. Chem.*, **216**, 121
156. Sutherland, E. W. and Woselait, W. D. (1955). *Nature*, **175**, 169
157. Krebs, E. G. and Fischer, E. H. (1956). *Biochim. Biophys. Acta*, **20**, 150
158. Wang, J. H. and Graves, D. J. (1964). *Biochemistry*, **3**, 1437
159. Fischer, E. H., Hurd, S. S., Kob, P., Seery, V. L. and Teller, D. (1968). In *Control of Glycogen Metabolism*, 19 (W. J. Whelan, editor) (New York and London: Academic Press)
160. Hayakawa, T., Perkins, J. P., Walsh, D. A. and Krebs, E. G. (1973). *Biochemistry*, **12**, 567

161. Cohen, P. (1973). *Eur. J. Biochem.*, **34,** 1
162. Graves, D. J., Hayakawa, T., Horvitz, R. A., Beckman, E. and Krebs, E. G. (1973). *Biochemistry*, **12,** 580
163. Fischer, E. H., Graves, D. J., Crittenden, E. R. S. and Krebs, E. G. (1959). *J. Biol. Chem.*, **234,** 1698
164. Stull, J. T., Brostrom, C. O. and Krebs, E. G. (1972). *J. Biol. Chem.*, **247,** 5272
165. Meyer, W. L., Fischer, E. H. and Krebs, E. G. (1964). *Biochemistry*, **3,** 1033
166. Ozawa, E., Hosoi, K. and Ebashi, S. (1967). *J. Biochem.*, **61,** 531
167. Huijing, F. and Larner, J. (1966). *Proc. Nat. Acad. Sci. USA*, **56,** 647
168. Parmeggiani, A. and Krebs, E. G. (1974). Unpublished results
169. Riley, W. D., DeLange, R. J., Bratvold, G. E. and Krebs, E. G. (1968). *J. Biol. Chem.*, **243,** 2209
170. Brostrom, C. O., Hunkeler, F. L. and Krebs, E. G. (1971). *J. Biol. Chem.*, **246,** 1961
171. Walsh, D. A., Perkins, J. P., Brostrom, C. O., Ho, E. S. and Krebs, E. G. (1971). *J. Biol. Chem.*, **246,** 1968
172. Hayakawa, T., Perkins, J. P. and Krebs, E. G. (1973). *Biochemistry*, **12,** 574
173. Cohen, P. and Antoniw, J. F. (1973). *FEBS Lett.*, **34,** 43
174. Soderling, T. R., Hickenbottom, J. P., Reimann, E. M., Hunkeler, F. L., Walsh, D. A. and Krebs, E. G. (1970). *J. Biol. Chem.*, **245,** 6317
175. Brown, N. E. and Larner, J. (1971). *Biochim. Biophys. Acta*, **242,** 81
176. Larner, J. and Sanger, F. (1965). *J. Mol. Biol.*, **11,** 491
177. Schlender, K. K., Wei, S. H. and Villar-Palasi, C. (1970). *Biochim. Biophys. Acta*, **191,** 272
178. Rall, T. W., Sutherland, E. W. and Berthet, J. (1957). *J. Biol. Chem.*, **224,** 463
179. Krebs, E. G., DeLange, R. J., Kemp, R. G. and Riley, W. D. (1966). *Pharmacol. Rev.*, **18,** 163
180. Langan, R. A. (1968). *Science*, **162,** 579
181. Corbin, J. D., Reimann, E. M., Walsh, D. A. and Krebs, E. G. (1970). *J. Biol. Chem.*, **245,** 4849
182. Huttunen, J. K., Steinberg, D. and Mayer, S. E. (1970). *Proc. Nat. Acad. Sci. USA*, **67,** 290
183. Huijing, F. and Larner, J. (1966). *Proc. Nat. Acad. Sci. USA*, **56,** 647
184. Krebs, E. G. and Walsh, D. A. (1969). *FEBS Symp.*, **19,** 121
185. Kuo, J. F. and Greengard, P. (1969). *J. Biol. Chem.*, **244,** 3417
186. Brostrom, M. A., Reimann, E. M., Walsh, D. A. and Krebs, E. G. (1970). *Advan. Enzyme Regul.*, **8,** 191
187. Gill, G. N. and Garren, L. D. (1970). *Biochem. Biophys. Res. Commun.*, **39,** 335
188. Tao, M., Salas, M. L. and Lipmann, F. (1970). *Proc. Nat. Acad. Sci. USA*, **67,** 408
189. Kumon, A., Yamamura, H. and Nishizuka, Y. (1970). *Biochem. Biophys. Res. Commun.*, **41,** 1290
190. Reimann, E. M., Brostrom, C. O., Corbin, J. P., King, C. A. and Krebs, E. G. (1971). *Biochem. Biophys. Res. Commun.*, **42,** 187
191. Beavo, J. A., Bechtel, P. J. and Krebs, E. G. (1974). Unpublished results
192. Erlichman, J., Rubin, C. S. and Rosen, O. M. (1973). *J. Biol. Chem.*, **248,** 7607
193. Haddox, M. K., Newton, N. E., Hartle, D. K. and Goldberg, N. D. (1973). *Biochem Biophys. Res. Commun.*, **47,** 653
194. Bechtel, P. J. and Beavo, J. A. (1974). *Fed. Proc.*, **33,** 1362
195. Hofmann, F. and Krebs, E. G. (1974). *Fed. Proc.*, **33,** 1324
196. Rosen, O. M., Erlichman, J. and Rubin, C. S. (1974). In *Metabolic Interconversion of Enzymes*, 1973, 143 (E. H. Fischer, E. G. Krebs and E. R. Stadtman, editors) (Berlin: Springer-Verlag)
197. Graves, D. J., Fischer, E. H. and Krebs, E. G. (1960). *J. Biol. Chem.*, **235,** 805
198. Friedman, D. L. and Larner, J. (1963). *Biochemistry*, **2,** 669
199. Zieve, F. J. and Glinsman, W. H. (1973). *Biochem. Biophys. Res. Commun.*, **50,** 872
200. Villar-Palasi, C. (1969). *Ann. N.Y. Acad. Sci.*, **166,** 719
201. Cori, G. T. and Cori, C. F. (1945). *J. Biol. Chem.*, **158,** 321
202. Keller, P. J. and Cori, C. F. (1955). *J. Biol. Chem.*, **214,** 127
203. Sutherland, E. W. (1951). In *Phosphorus Metabolism*, Vol. 1, 53 (W. D. McElroy and B. Glass, editors) (Baltimore: The Johns Hopkins Press)
204. Hurd, S. S. (1967). *Ph.D. Thesis*, University of Washington

205. Bot, G. and Dósa, I. (1967). *Acta Biochim. Biophys. Acad. Sci. Hung.*, **2,** 335
206. Holmes, P. A. and Mansour, T. E. (1968). *Biochim. Biophys. Acta*, **156,** 275
207. Krebs, E. G., Gonzalez, C., Posner, J. B., Love, D. S., Bratvold, G. E. and Fischer, E. H. (1964). In *Control of Glycogen Metabolism*, 200 (W. J. Whelan and M. P. Cameron, editors) (London: Churchill)
208. Nolan, C., Novoa, W. B., Krebs, E. G. and Fischer, E. H. (1964). *Biochemistry*, **3,** 542
209. Bot, G. and Dosa, I. (1971). *Acta Biochim. Biophys. Acad. Sci. Hung.*, **6,** 73
210. Martensen, T. M., Brotherton, J. E. and Graves, D. J. (1973). *J. Biol. Chem.*, **248,** 8323
211. Chelala, C. A. and Torres, H. N. (1970). *Biochim. Biophys. Acta*, **198,** 504
212. Riley, G. A. and Haynes, R. C., Jr. (1963). *J. Biol. Chem.*, **238,** 1563
213. Kato, K. and Bishop, J. S. (1972). *J. Biol. Chem.*, **247,** 7420
214. Thomas, J. A. and Nakai, C. (1973). *J. Biol. Chem.*, **248,** 2208
215. Meyerhof, A. (1930). *Die Chemischen Vorgänge im Muskel und ihr Zusammeuhang mit Arbeitsleistung und Wärmebildung* (Berlin: Springer-Verlag)
216. Helmreich, E. and Cori, C. F. (1964). *Proc. Nat. Acad. Sci. USA*, **51,** 131
217. Morgan, H. E. and Parmeggiani, A. (1964). *J. Biol. Chem.*, **239,** 2440
218. Atkinson, D. E. (1971). *In Metabolic Pathways*, Vol. 5, 1 (H. J. Vogel, editor) (New York and London: Academic Press)
219. Lyon, J. B. and Poerter, J. (1963). *J. Biol. Chem.*, **238,** 1
220. Danforth, W. H. and Lyon, J. B. (1964). *J. Biol. Chem.*, **239,** 4047
221. Krebs, E. G. and Fischer, E. H. (1955). *J. Biol. Chem.*, **216,** 113
222. Cori, C. F. (1956). In *Enzymes: Units of Biological Structure and Function*, 573 (O. H. Gaebler, editor) (New York: Academic Press)
223. Rulon, R. R., Schottelius, D. D. and Schottelius, B. A. (1962). *Ann. J. Physiol.*, **202,** 821
224. Posner, J. B., Stern, R. and Krebs, E. G. (1965). *J. Biol. Chem.*, **240,** 982
225. Drummond, G. I., Harwood, J. P. and Powell, C. A. (1969). *J. Biol. Chem.*, **244,** 4235
226. Stull, J. T. and Mayer, S. E. (1971). *J. Biol. Chem.*, **246,** 5716
227. Ebashi, S. (1965). In *Molecular Basis of Muscular Contraction* (BBA Library, Vol. 9), 197 (S. Ebashi *et al.*, editors) (Tokyo: Igaku Shoin; Amsterdam: Elsevier)
228. Huston, R. B. and Krebs, E. G. (1968). *Biochemistry*, **7,** 2116
229. Ebashi, S. and Endo, M. (1968). In *Progress in Biophysics and Molecular Biology*, 123 (I. A. V. Butler and D. Noble, editors) (New York: Pergamon Press)
230. Gross, S. R. and Mayer, S. E. (1974). *Life Sci.*, **14,** 401
231. Danforth, W. H. (1965). In *Control of Energy Metabolism*, 287 (B. Chance, R. W. Estabrook and J. R. Williamson, editors) (New York and London: Academic Press)
232. Villar-Palasi, C. and Wei, S. H. (1970). *Proc. Nat. Acad. Sci. USA*, **67,** 345
233. Wilson, J. E., Sacktor, B. and Tiekert, C. G. (1967). *Arch. Biochem. Biophys.*, **120,** 542
234. Piras, R. and Staneloni, R. (1969). *Biochemistry*, **8.** 2153
235. Staneloni, R. and Piras, R. (1969). *Biochem. Biophys. Res. Commun.*, **36,** 1032
236. Danforth, W. H. (1965). *J. Biol. Chem.*, **240,** 588
237. Cori, C. F. and Cori, G. T. (1928). *J. Biol. Chem.*, **79,** 309
238. Cori, G. T. and Illingworth, B. (1955). *Biochim. Biophys. Acta*, **21,** 105
239. Sutherland, E. W. and Rall, T. W. (1960). *Pharmacol. Rev.*, **12,** 265
240. Craig, J. W. and Larner, J. (1964). *Nature*, **202,** 971
241. Rosell-Perez, M. and Larner, J. (1964). *Biochemistry*, **3,** 773
242. Chelala, C. A. and Torres, H. N. (1969). *Biochim. Biophys. Acta*, **178,** 423
243. Larner, J., Villar-Palasi, C. and Richman, D. J. (1960). *Arch. Biochem. Biophys.*, **86,** 56
244. Villar-Palasi, C. and Larner, J. (1961). *Arch. Biochem. Biophys.*, **94,** 436
245. Shen, L. C., Villar-Palasi, C. and Larner, J. (1970). *Physiol. Chem. Phys.*, **2,** 536
246. Goldberg, N. D., Villar-Palasi, C., Sarko, H. and Larner, J. (1967). *Biochem. Biophys. Acta*, **148,** 665
247. Larner, J., Huang, L. C., Brooker, G., Murad, F. and Miller, T. B. (1974). *Fed. Proc.*, **33,** 261
248. Sutherland, E. W. and Rall, T. W. (1958). *J. Biol. Chem.*, **232,** 1077.
249. Brostrom, C. O., Corbin, J. P., King, C. A. and Krebs, E. G. (1971). *Proc. Nat. Acad. Sci. USA*, **68,** 2444

250. O'Dea, R. F., Haddox, M. K. and Goldberg, N. D. (1971). *J. Biol. Chem.*, **246,** 6183
251. Gonzalez, C. (1962). *M.S. Thesis*, University of Washington
252. Posner, J. B., Stern, R. and Krebs, E. G. (1965). *J. Biol. Chem.*, **240,** 982
253. Appleman, M. M., Birnbaumer, L. and Torres, H. N. (1966). *Arch. Biochem. Biophys.*, **116,** 39
254. Walsh, D. A., Ashby, C. D., Gonzalez, C., Calkins, D., Fischer, E. H. and Krebs, E. G. (1971). *J. Biol. Chem.*, **246,** 1977
255. Ashby, C. D. and Walsh, D. A. (1972). *J. Biol. Chem.*, **247,** 6637
256. Beavo, J. A., Bechtel, P. J. and Krebs, E. G. (1974). *Proc. Nat. Acad. Sci. USA*, in press
257. Walsh, D. A. and Ashby, C. D. (1973). *Recent Progress in Hormone Research*, Vol. 29, 239 (New York and London: Academic Press)
258. Long, C. N. H., Katzin, B. and Fry, E. G. (1940). *Endocrinology*, **26,** 309
259. Kerppola, W. (1952). *Endocrinology*, **51,** 192
260. Leonard, S. L. (1953). *Endocrinology*, **53,** 226
261. Bergamini, E., Gagliardi, C. and Pelligrino, C. (1969). *FEBS Lett.*, **4,** 1
262. Vilchez, C., Piras, N. and Piras, R. (1972). *Mol. Pharmacol.*, **8,** 780
263. Schaeffer, L. D., Chenoweth, N. and Dunn, A. (1969). *Biochim. Biophys. Acta*, **192,** 304
264. Ghosh, H. P. and Preiss, J. (1965). *J. Biol. Chem.*, **240,** 960
265. Ghosh, H. P. and Preiss, J. (1966). *J. Biol. Chem.*, **241,** 4491
266. Sanwal, G. G. and Preiss, J. (1967). *Arch. Biochem. Biophys.*, **119,** 454
267. Preiss, J., Ghosh, H. P. and Wittkop, J. (1967). In *The Biochemistry of Chloroplasts*, Vol. 2, 131 (T. W. Goodwin, editor) (New York: Academic Press)
268. Sanwal, G. G., Greenberg, E., Hardie, J., Cameron, E. C. and Preiss, J. (1968). *Plant Physiol.*, **43,** 417
269. Heber, U. W. (1967). In *The Biochemistry of Chloroplasts*, Vol. 2, 71 (T. W. Goodwin, editor) (New York: Academic Press)
270. Heber, U. W. and Santarius, K. A. (1965). *Biochim. Biophys. Acta*, **109,** 390
271. Santarius, K. A. and Heber, U. W. (1965). *Biochim. Biophys. Acta*, **102,** 39
272. MacDonald, P. W. and Strobel, G. A. (1970). *Plant Physiol.*, **46,** 126
273. Kanazawa, T., Kanazawa, K., Kirk, M. R. and Bassham, J. A. (1972). *Biochim. Biophys. Acta*, **256,** 656
274. Dietzler, D. N., Leckie, M. P. and Lais, C. J. (1973). *Arch. Biochem. Biophys.*, **156,** 684
275. Dietzler, D. N., Lais, C. J. and Leckie, M. P. (1974). *Arch. Biochem. Biophys.*, **160,** 14
276. Atkinson, D. E. and Walton, G. M. (1967). *J. Biol. Chem.*, **242,** 3239
277. Atkinson, D. E. (1968). *Biochemistry*, **7,** 4030
278. Preiss, J. (1972). *Intra-Sci. Chem. Rep.*, **6,** 13
279. Lowry, O. H., Carter, J., Ward, J. B. and Glaser, L. (1971). *J. Biol. Chem.*, **246,** 6511
280. Chapman, A. G., Fall, L. and Atkinson, D. E. (1971). *J. Bacteriol.*, **108,** 1072
281. Dempsey, W. B. (1972). *Biochim. Biophys. Acta*, **264,** 344
282. Dempsey, W. B. and Arcement, L. J. (1971). *J. Bacteriol.*, **264,** 344
283. Cole, H. A., Wimpenny, J. W. T. and Hughes, D. E. (1967). *Biochim. Biophys. Acta*, **143,** 445
284. Bagnara, A. S. and Finch, L. R. (1968). *Biochem. Biophys. Res. Commun.*, **33,** 15
285. Lubin, M. and Ennis, H. L. (1964). *Biochim. Biophys. Acta*, **80,** 614
286. Webb, B. M. (1968). *J. Gen. Microbiol.*, **43,** 401
287. Silver, S. (1969). *Proc. Nat. Acad. Sci. USA*, **62,** 764
288. Hempfling, W. P., Hofer, M., Harris, E. J. and Pressman, B. C. (1967). *Biochim. Biophys. Acta*, **141,** 391
289. Cattaneo, J., Sigal, N., Favard, A. and Segel, I. H. (1966). *Bull. Soc. Chim. Biol.*, **48,** 441
290. Govons, S. and Preiss, J. (1974). Unpublished experiments
291. Magasanik, B. (1961). *Cold Spring Harbor Symp. Quant. Biol.*, **26,** 193
292. Perlman, R. L., DeCrombrugghe, B. and Pastan, I. (1969). *Nature*, **223,** 810
293. Sigal, N. and Piug, J. (1968). *C. R. Acad. Sci., Paris*, **267,** 1223
294. Cozzarelli, N. R., Freedberg, W. E. and Lin, E. C. C. (1968). *J. Mol. Biol.*, **31,** 371

295. Spearman, T. N., Khandelwal, R. L. and Hamilton, I. R. (1973). *Arch. Biochem. Biophys.*, **154,** 306
296. Bender, H. and Wallenfels, K. (1961). *Biochem. Z.*, **334,** 79
297. Bender, H. and Wallenfels, K. (1966). *Methods Enzymol.*, **8,** 555
298. Ueda, S. and Nauri, N. (1967). *Appl. Microbiol.*, **15,** 492
299. Walker, E. J. (1966). *Biochem. J.*, **101,** 861
300. Bender, H. (1970). *Arch. Mikrobiol.*, **71,** 331
301. Mercier, C. and Whelan, W. J. (1970). *Eur. J. Biochem.*, **16,** 579
302. Gunja-Smith, Z., Marshall, J. J., Smith, E. E. and Whelan, W. J. (1970). *FEBS Lett.*, **12,** 96
303. Yokobayaski, K., Misaki, A. and Harada, T. (1970). *Biochim. Biophys. Acta*, **212,** 458
304. Lee, E. Y. C., Smith, E. E. and Whelan, W. J. (1970). In *Miami Winter Symp.*, Vol. 1, 139 (W. J. Whelan and J. Schultz, editors) (Amsterdam: North Holland)
305. Breed, R. S., Murry, E. G. D. and Smith, N. R. (1957). *Bergey's Manual of Determinative Bacteriology*, 7th ed., (Baltimore: Williams and Wilkins)
306. Chambost, J. P., Favard, A. and Cattaneo, J. (1967). *Bull. Soc. Chim. Biol.*, **49,** 1231
307. Dietzler, D. N., Leckie, M. P., Lais, C. J. and Magnani, J. L. (1974). *Arch. Biochem. Biophys.*, **162,** 602
308. Dietzler, D. N., Leckie, M. P., Lais, C. J. and Magnani, J. L. (1975). *J. Biol. Chem.*, **250,** 2383
309. Dietzler, D. N., Lais, C. J., Magnani, J. L. and Leckie, M. P. (1974). *Biochem. Biophys. Res. Commun.*, **60,** 875
310. Preiss, J., Greenberg, E. and Sabraw, A. (1975). *J. Biol. Chem.*, **250,** in press
311. Robson, R. L. and Morris, G. J. (1974). *Biochem. J.*, **144,** 513
312. Robson, R. L., Robson, R. M. and Morris, J. G. (1974). *Biochem. J.*, **144,** 503
313. Mackey, B. H. and Morris, J. G. (1974). *FEBS Lett.*, **48,** 64
314. Nimmo, H. G. and Cohen, P. (1974). *FEBS Lett.*, **47,** 162
315. Huang, K.-P., Huang, F. L., Glinsmann, W. H. and Robinson, J. C. (1975). *Biochem. Biophys. Res. Commun.*, **65,** 1163
316. Schlender, K. K. and Reimann, E. M. (1975). *Proc. Nat. Acad. Sci. USA*, **72,** 2197
317. Soderling, T. R. (1975). *J. Biol. Chem.*, **250,** 5407
318. Nakai, C. and Thomas, J. A. (1974). *J. Biol. Chem.*, **249,** 6459
319. Brandt, H., Capulong, Z. L. and Lee, E. Y. C. (1975). *J. Biol. Chem.*, in press
320. Killilea, S. D., Lee, E. Y. C., Brandt, H. and Whelan, W. J. (1975). *Metabolic Interconversions of Enzymes* (New York, Heidelberg, Berlin: Springer Verlag), in press

# 8
# Some Inborn Errors of Carbohydrate Metabolism

**D. H. BROWN and B. I. BROWN**
Washington University School of Medicine

---



---

## 8.1 INTRODUCTION

This chapter contains a discussion of some of the genetically determined diseases of carbohydrate metabolism from the point of view of the biochemical and clinical consequences of the enzymatic lesions. Emphasis has been placed on more recent studies, and, where appropriate, unresolved questions are discussed at some length. It is recognised that many other diseases of equal importance, such as galactosemia, have not been mentioned. Instead, the authors have chosen to treat the material which has been selected in greater depth.

## 8.2 INBORN ERRORS OF GLYCOGEN METABOLISM

### 8.2.1 Introduction

Glycogen is the only reserve polysaccharide in mammalian tissues and, as such, it serves as a source of glucose which can be mobilised for many metabolic processes. Ordinarily, the glycogen content of mammalian cells does not exceed an upper level which is somewhat characteristic for each tissue. However, in man a number of metabolic diseases are known which are distinguished by excessive storage of glycogen due to deficiencies of various enzymes which normally are active in the degradative pathways of metabolism of this polysaccharide. By studying patients with these metabolic defects, much has been learned about the normal metabolism of glycogen. The history of investigation of the glycogenoses, as they are called, is illustrative of the stages involved in studying many other heritable polysaccharide storage diseases. After initial clinical description of each syndrome and chemical characterisation of the structure of the accumulated polysaccharide, research has been directed towards discovery of an enzymatic defect. After the biochemical lesion has been found, heterozygous carriers of the disease have often been identifiable and it has then been possible to make accurate prenatal diagnoses, at least in some of these syndromes. Thus a basis has been laid for intelligent genetic counseling. However, there has not always been rapid progress from one stage of such investigations to the next. Thus glycogen storage diseases Types I and II were described clinically first in 1929[1] and 1932[2], respectively. Not until 1952 was Type I disease, then known as von Gierke's disease, shown by Cori and Cori to be due to a deficiency of glucose 6-phosphatase (EC 3.1.3.9)[3]. The later discovery that this hydrolytic enzyme can also act as an inorganic pyrophosphatase and a phosphotransferase under suitable conditions[4,5] was confirmed in human tissue by the demonstration that these additional activities, as well, of course, as hydrolytic activity toward glucose 6-phosphate, were not demonstrable in extracts prepared from the livers of patients with Type I disease[6].

The enzyme deficiency which characterises Type II glycogen storage disease ('Pompe's disease') was reported by Hers in 1963 to be the lack of an enzyme, presumably located in lysosomes, which acts optimally at acid pH to hydrolyse 1,4-α-glucosidic bonds in maltose and glycogen[7]. Subsequent work revealed that an α-glucosidase (EC 3.2.1.3) purified from rat liver lysosomes hydrolysed both the 1,6- and the 1,4-α-glucosidic linkages in glycogen at acid pH[8], and that this 1,6-α-glucosidase debranching activity also was not present in the tissues of children affected with Type II disease[9].

In Types III and IV glycogen storage diseases, the polysaccharides which are present in affected tissues are known to have structures different from that of glycogen. The polysaccharide from Type III tissues was found to have a greater branch point content and shorter outer chains than glycogen[10]. The characteristic polysaccharide of what was later referred to as Type IV glycogen storage disease was shown to be more sparsely branched than glycogen and to have longer than normal outer and inner chains[10]. In this latter disease, excessive polysaccharide storage usually does not occur, since, in the original case and in most cases reported subsequently, the content of the abnormal polysaccharide present in affected tissues is less than that of the glycogen which would be present in the same tissues of normal individuals. On the basis of the structures of the isolated polysaccharides which were established by stepwise enzymatic degradation *in vitro* using phosphorylase (EC 2.4.1.1) and amylo-1,6-glucosidase (EC 3.2.1.33) as a debranching enzyme, the probable enzyme deficiency in Type III disease was predicted to be that of the debranching enzyme, while that in the Type IV syndrome was suggested to be a deficiency of the glycogen branching enzyme (EC 2.4.1.18)[10]. Later, in 1956, it was confirmed that Type III patients do lack the ability to debranch the phosphorylase limit dextrin which accumulates in their affected tissues[11], and in 1966 it was shown that there is a demonstrable branching enzyme deficiency in tissues of patients with Type IV disease[12]. Both of these discoveries depended on the development of specific enzyme assays applicable to small amounts of human tissue, as well, of course, as on the availability of tissue samples from numerous affected and normal individuals. Review of the clinical literature reveals that van Creveld had described Type III disease in two different children as early as 1928 and 1932 without, of course, being aware of its cause. It is interesting that later, after the work of Illingworth *et al.* had demonstrated the enzymatic basis for this syndrome, van Creveld and Huijing were able to establish by enzyme assay in leukocytes that the original patients, who by then had reached adulthood, actually had Type III disease[13].

Since a recent comprehensive review on glycogen and its metabolism has been published[14], as well as several reviews on the glycogen storage diseases[15-18], the following discussion of four of these syndromes will emphasise recent contributions in the field, and some consideration will be given also to the relationship between the biochemical lesion and the physiological consequences in each of these enzyme deficiency diseases.

### 8.2.2 Type I glycogen storage disease

Since the action of glucose 6-phosphatase, a microsomal enzyme present in liver, kidney and intestinal mucosa, is required for the maintenance of blood glucose, its absence in Type I disease accounts for the hypoglucemia and lack of glucemic response which follows the administration of galactose, epinephrine or glucagon to affected patients. However, the marked glycogen storage in liver and kidney and the consequent hepatorenomegaly are not as easily explained on the basis of this enzyme deficiency. In addition to the clinical features already mentioned, patients with this disease usually are of relatively short stature and have hyperlacticacidemia, hyperuricemia, hyperlipemia, xanthomas, retinal changes and a tendency to bleeding. Long-term follow-up studies of four patients with Type I disease, who were from 13 to 21 years of age, showed that although their birth weights had been normal for their gestational ages, their rates of growth were so slow that both their weights and heights were below the third percentile at and after their first year of life[19]. Although puberty was delayed, Fine *et al.*[19] found no evidence for abnormal endocrine status, as judged by tests for thyroid and adrenal function and by measurements of the circulating level of growth hormone and the urinary excretion of gonadotropin. Other patients with this disease have been reported to have subnormal plasma insulin levels, due probably to chronic and variable hypoglucemia. Thus, Lockwood *et al.*[20] studied five Type I patients between the ages of 16 and 47 years and found that their blood insulin levels were about one-half of normal and that the increase in insulin elicited by ingestion of glucose likewise was significantly less than in normal subjects. Other investigators have found that after an eight hour fast the plasma insulin level in Type I patients is less than that in control subjects, while the concentrations of growth hormone and of cortisol in blood do not differ significantly from normal[21]. These observations suggest that in Type I patients the manifestations of a chronic low level of insulin may be especially prominent because of concurrent chronic hypoglucemia. Starzl *et al.*[22] have recently suggested that the beneficial effects which result from portal diversion by surgical means in some patients with glycogen storage disease is due to the increased supply of insulin to peripheral tissues which may result. Three of seven patients treated surgically by Starzl had Type I disease and the two children who have survived now for more than one year have shown accelerated growth accompanied by a lessening of their hyperlipemia and improvement in their tendency to hypoglucemia. Two brothers prepared for surgery by a protracted period of hyperalimentation, by Folkman *et al.*[23], prior to the establishment of a porta-caval shunt, have also shown marked clinical improvement but have remained unable to withstand fasting. Similar results had been obtained by Boley *et al.*[24] with a 5-year-old child with Type I disease. Deglycogenation of the liver did not occur in any case. It is clear that the surgical procedure of portal diversion can be of considerable value in alleviating many of the secondary effects of glucose 6-phosphatase deficiency if the patients to be subjected to this procedure are carefully selected. The alternative management of Type I disease by dietary manipulations has shown that frequent feedings of glucose may result in lessened hyperlacticacidemia and hyperlipemia, but that the child then may have more frequent and severe hypoglucemic episodes when these do occur[25]. Experience in this respect has not been uniform. Cuttino *et al.*[26] have reported that the level of serum lipids was reduced, as was the liver size in an infant whose triglyceride intake was restricted predominantly to neutral fats containing medium chain length fatty acids. It has been reported by Hulsmann *et al.*[27] that the rate of synthesis of long-chain fatty acids from [$^{14}$C]citrate as the added carbon source was 3–4 times higher than normal in suitably fortified homogenates of liver tissue removed by biopsy from two patients with Type I disease. These same homogenates showed a normal rate of fatty acid synthesis when [$^{14}$C]malonyl coenzyme A was substituted for [$^{14}$C]citrate. Since no specific assays were carried out for any of the several enzymes involved in fatty acid synthesis from citrate, it is not possible to decide to what the marked elevation in rate of fatty acid synthesis in this case is to be attributed. The finding made by Hulsmann *et al.* may have general physiological significance in view of the known high blood lactate and pyruvate levels in this disease and the marked hyperlipemia which occurs. That the glycolytic pathway in the liver of Type I patients is functionally intact and responsive to stimulation by glucagon *in vivo* has been shown by Sarcione *et al.*[28], who administered [$^{14}$C]glucose orally to a 5 year old patient and then, after several hours, obtained one liver biopsy specimen at laparotomy before, and another specimen 30 minutes after, beginning the intravenous infusion of glucagon. Extraction and analysis of the [$^{14}$C]glycogen from the liver showed that, in the 30 minute interval, 20% of the total glycogen had been metabolised and that 60% of the [$^{14}$C]-glucose units which were in the polysaccharide before glucagon administration were no longer present there. As expected, the blood glucose concentration did not increase, but the concentration of lactate in the patient's blood increased about four-fold[28]. This direct demonstration of unimpaired glycogenolysis provides support for the possibility of an increased rate of fatty acid synthesis from pyruvate *in vivo* in the liver of the Type I patient, and it also suggests that the high liver glycogen content is more likely to be due to an increased rate of glycogen synthesis than it is to a decrease in the rate of its catabolism. Since the possibility of glucose 6-phosphate being converted into glucose is excluded from consideration in Type I liver, the steady state concentration of this phosphate ester might be elevated above the normal level, and especially so if phosphofructokinase (EC 2.7.1.11) activity were to be somewhat inhibited by a metabolite such as citrate. An increased intracellular level of glucose 6-phosphate might lead to activation of that form of glycogen synthase (EC 2.4.1.11) (the D-form) which is now considered to be physiologically inactive under normal circumstances. While this possibility has been mentioned by several investigators, it has been discounted by Hers on the basis of some results involving direct measurement of the concentration of glucose 6-phosphate in Type I liver tissue[29]. Later, Ockerman made a more detailed study of this question in liver samples from nine affected individuals and found that in at least six cases the glucose 6-phosphate content of the tissue was significantly elevated[30]. Freshly removed samples taken by needle biopsy had the highest concentrations and these were from 6 to 9 times the level of control tissues. Although these findings suggest that serious consideration should be given to the hypothesis that there is an increased rate of glycogen synthesis in affected tissue, the question cannot be settled finally by assays of metabolites in whole liver since the possibility of intracellular compartmentalisation of the hexose phosphates cannot be satisfactorily investigated experimentally. In addition to the effects of glucose 6-phosphate on glycogen synthase, both forms of this enzyme, as well as the protein kinase (EC 2.7.1.37) and glycogen synthase D phosphatase which effect their interconversion, are now known to be subject to allosteric regulation by other cellular metabolites. Thus, an increased rate of glycogen deposition in Type I disease remains as a possible explanation for the high content of this polysaccharide in the liver. Any such effect would have to oppose the decreased rate of glycogen synthesis which would be expected to result from the lower than normal levels of circulating insulin which have been reported in Type I patients[20]. Only direct measurement by isotopic means of the rate of glycogen synthesis in the liver of Type I patients *in vivo* compared with that in normal subjects and others who have different types of glycogenosis can give an unambiguous answer to the problem of the etiology of glycogen accumulation in Type I glycogen storage disease.

Recently, Garland and Cori have shown clearly that endogenous microsomal phospholipid is required for the activity of the glucose 6-phosphatase of rat liver[31]. They resolved a partially purified preparation of this enzyme and then showed that it could be reactivated by the addition under suitable conditions of certain unsaturated phospholipids of exogenous origin. Monooleylphosphatidylcholine was especially effective in this respect. The possible relationship of the phospholipid requirement of this enzyme to the existence and expression of the enzymatic defect in human tissue has not yet been studied adequately in patients with Type I glycogen storage disease.

### 8.2.3 Type II glycogen storage disease

This syndrome was described clinically in 1932 by several groups of investigators and is often referred to as Pompe's disease. The pathological involvement of heart muscle is striking and led to the use of the term 'cardiomegalia glycogenica' in early clinical reports. However, it is now known that when the syndrome occurs in its most severe form in infants, glycogen storage in the tissues is so widespread throughout the body that reference to Type II glycogen storage disease as 'generalised glycogenosis' is more appropriately descriptive. Evidence that a glucosidase which is presumably lysosomal is absent from such affected tissues was obtained first by Hers[7], and this conclusion was supported by the finding that a large amount of the glycogen found in the liver of young children with this disease is segregated within vacuoles which are enclosed by a single membrane and which presumably are secondary lysosomes[32]. Several other investigators have confirmed this finding, and Type II glycogen storage disease is now recognised to be the only glycogenosis in which glycogen accumulation within lysosomal-like structures occurs. However, even in Type II a considerable portion of the glycogen which is present in the liver of such patients is in the cytosol of the hepatocytes, and so it is understandable that the Type II child has a normal hyperglucemic response to administered epinephrine or glucagon and that hypoglucemia is not a clinical feature of this disease. It has been shown also that the enzyme which is missing is a glucosidase which has dual specificity, in that it can hydrolyse both the 1,4- and the 1,6-$\alpha$-glucosidic bonds of glycogen, and which, because of this property, has the ability to convert glycogen totally into glucose at acid pH[8,9]. As far as is known, this enzyme is absent from all tissues of children who have the typical severe form of Type II disease. In view of this fact, it might be expected that glycogen accumulation within lysosomes would also be demonstrable in all affected tissues. This does not seem to be the case, since several reports have appeared in which it has been shown that large deposits of glycogen can be present in heart and skeletal muscle of Type II patients without this polysaccharide being enclosed within the membrane of any microscopically visible organelle. It should also be mentioned that, although such glycogen-containing organelles are invariably present in affected liver, centrifugal studies designed to separate intact sub-cellular particles from this tissue have not yet been successful in showing that there is a population of isolable, glycogen-containing organelles having the expected content of known lysosomal marker enzymes. A reason for this could be that the membrane of human liver lysosomes, and especially of those which contain glycogen, is less stable than the membrane of the lysosomes which are isolable from rat liver. It may also be possible that the degree of heterogeneity of lysosomes with respect to their content of various enzymes is greater in human than in rat liver. Since the hydrolytic degradation of glycogen at an acidic pH seems to be a physiologically important process in normal human tissues, it is conceivable that lysosomes containing α-glucosidase are a distinct population having very few other hydrolytic enzymes and so not recognisable under circumstances where the α-glucosidase is absent. The techniques which have been used to obtain liver lysosomes from Type II tissue have not taken account of the experimental difficulty which unusual fragility or unique enzyme content could make. Since histological studies have led to the impression that there is poor correlation between the presence of an abnormal quantity of glycogen in Type II muscle and its sequestration within structures having a unit membrane, the possibility might be considered that the α-glucosidase which is active at acid pH is not exclusively a lysosomal enzyme in normal human muscle. Recently, Angelini and Engel[33] have reported that this α-glucosidase is present at its highest relative specific activity in the particulate fraction, prepared centrifugally from human skeletal muscle, which contains several other acid hydrolases usually considered to be lysosomal markers. However, it is possible to calculate from the data published by these authors that in their experiments only 5% of the total recovered α-glucosidase activity (and 8% of the β-glucuronidase activity) was in the lysosomal fraction, while 50% of the α-glucosidase (and 28% of the β-glucuronidase) was wholly soluble. Thus, it is uncertain whether these data indicate unusual fragility and some heterogeneity of the lysosomes, or whether they might also provide evidence for a bimodal distribution of the α-glucosidase. There is difficulty in accounting for all of the clinical, morphological and biochemical findings in Type II glycogenosis by considering it to be simply a lysosomal storage disease, especially now that it is known that older children and even adults may have a nearly total functional deficiency of α-glucosidase active at acid pH without having the severe symptomatology shown by the infant with the classic infantile form of Type II disease. These clinically less affected older patients have a myopathy which manifests itself either in dystrophic changes or only as muscular weakness. There is usually no cardiac involvement which is noticeable clinically. This situation, which has become known as the adult form of acid maltase deficiency, was first described clinically in three patients in 1965[33,34], and since that time numerous other patients have been studied[15,35-38]. In most of these cases where liver tissue was also analysed, the deficiency of the α-glucosidase was found to extend to that organ also, although hepatomegaly does not occur typically in the adult cases. A peculiarity of the adult form of this disease is the variable pathological involvement of different muscles in the same patient and their different glycogen contents which have been found to vary from the level characteristic of normal muscle to as high as 20% of the wet weight. It should be emphasised that in all muscles analysed the glucosidase active at pH 4 has been found to have virtually no activity. Although the impression has been gained that the functional disability of a muscle tends to parallel its degree of glycogen storage, it seems equally clear that glycogen accumulation does not result inevitably from α-glucosidase deficiency.

As a result of findings in the uniformly fatal, infantile form of Type II disease compared with those in the less severe adult form of α-glucosidase deficiency at acid pH, the question of the significance of this enzyme and of its presumed lysosomal localisation in the catabolism of glycogen in normal tissues becomes particularly important. Recently, Brown and Brown have found that the glycogen content of cultured human fibroblasts derived from patients with the fatal form of Type II disease is often approximately twice as high as that of normal fibroblasts, and they have also shown that the half-life of [$^{14}$C]-labelled glycogen in affected cells is three times longer than that in normal cells[39,40]. It is known that cultured fibroblasts from Type II patients lack α-glucosidase activity at acid pH[39-41]. These observations with fibroblast cultures seem to indicate that the α-glucosidase normally plays an important role in glycogenolysis in these cells. Glycogen storage in cultured cells from two infants with Type II disease has been confirmed by DiMauro *et al.*[42]. In this latter work it was shown that the mutant cells, when deprived of glucose, utilised their glycogen reserves at approximately the same rate as normal cells. On the basis of this finding the authors questioned whether the stored glycogen could have been sequestered within lysosomes. The occurrence of such organelles which contain glycogen in fibroblasts cultured from some patients with the adult form of Type II disease has been reported by Angelini *et al.*[43]. The biochemical results obtained by DiMauro *et al.* are not necessarily at variance with those of Brown and Brown[40], since the latter were obtained with cultures in which both glycogen synthesis and glycogen degradation were occurring at normal rates, and the cells were not exposed to the metabolic stress of being deprived of extracellular glucose. Angelini *et al.*[43] reported that while fibroblasts from five patients with the adult form of acid α-glucosidase deficiency had no more than 10% of the normal level of activity of this enzyme, these cells showed no significant increase in their total glycogen content compared with normal cells. If a difference in the steady state rates of glycogen synthesis and degradation exists in the fibroblasts of children with the infantile form of the disease compared with patients who have the adult form in such a way that glycogen accumulates in the fibroblasts of one group of patients to higher than normal levels but not in those of the other group, it may eventually be possible to find the biochemical basis for such a difference. It has been reported[44] but not confirmed[45] that an electrophoretic variant of α-glucosidase active at pH 4 is present in cultured fibroblasts from the adult form of α-glucosidase deficiency. Whether a small residual activity of a structurally modified form of this hydrolytic enzyme could have major metabolic significance is uncertain. However, in view of the findings made in some of the lysosomal lipid storage diseases, this possibility deserves careful investigation.

In many heritable diseases which are due to enzyme deficiencies, individual patients within a group show marked variation in the severity of the symptomatology associated with the defect. As has been discussed above, this certainly appears to be true of Type II glycogen storage disease, in which, however, the range of phenotypic expression is extreme, extending from certain death in infancy to only a moderate degree of muscular incapacity in middle age. Future biochemical studies may show that these remarkable differences in severity are explicable on an enzymatic basis and that a new classification of subtypes of this syndrome can be made, based on clear enzymatic findings.

### 8.2.4 Type III glycogen storage disease

Glycogen phosphorylase is able to catalyse the phosphorolytic cleavage of a large number of the 1,4-α-linked glucose units which are present in the outer chains of mammalian glycogen. The action of this enzyme can be very rapid in both liver and muscle, and the α-glucose 1-phosphate which is formed is an important substrate for subsequent cellular metabolism. Phosphorylase action stops after each outer chain has been shortened to the point where four glucose units remain attached to the triply bonded branch point glucose unit[46]. Further action of phosphorylase on this 'limit dextrin' cannot occur until the 1,6-bonded α-glucose unit has been removed from the branch point. This occurs by action of the glycogen 'debranching enzyme'. Two steps are involved in the overall reaction catalysed by this enzyme. The first is a rearrangement of the remaining outer chain glucose units brought about because the debranching enzyme is also a transglycosylase which is able to move maltotriosyl or maltosyl segments from the terminus of a donor chain and to attach these in a new 4-position on the non-reducing end of a suitable acceptor chain[46-48]. The specificity of the transferase is such that maltotriosyl segment transfer occurs to the greatest extent[47,48]. This fact has physiological relevance, since the 1,6-linked α-glucose unit of each outer side chain of the limit dextrin is covered by three 1,4-linked α-glucose units[46]. After these three units are removed as a maltotriosyl residue in a single transfer act, the 1,6-bonded unit becomes exposed, and it is then susceptible to hydrolytic cleavage by the amylo-1,6-glucosidase activity of the debranching enzyme[46-48]. In this way one molecule of glucose is obtained from each branch point in glycogen when phosphorylase and the debranching enzyme act consecutively to degrade the polysaccharide completely. This fact allows the degree of branching in glycogen to be determined exactly, and use of this method made possible the structure analysis which showed that the polysaccharide which accumulates in the tissues of patients with Type III glycogen storage disease has relatively short outer chains and, hence, approaches the structure of a limit dextrin[10]. Illingworth *et al.* found in 1956 that such patients lack the debranching enzyme in affected tissues[11]. Since the pathway for glycogen synthesis is unimpaired in this disease, the outer structure of the polysaccharide present reflects the nutritional history of the patient during the hours immediately preceding the biopsy procedure. If tissue samples are taken during a period of glycogen deposition, the polysaccharide will have a nearly normal structure and will not resemble a limit dextrin. About 35% of the molecule of the glycogen of normal human tissues is present as outer chain glucose units susceptible to cleavage by phosphorylase. Hence, it is understandable that many patients with Type III disease, and especially those who are older, do not usually have severe hypoglycemic episodes when a moderate degree of fasting occurs after a period of active glycogen synthesis. In the fasting state such patients do fail to show a glucemic response to epinephrine or glucagon, and this finding is expected also in view of the presence of a near limit dextrin in their liver under these circumstances. In the Type III patient all of the metabolic pathways for the disposition of α-glucose 1-phosphate and for the maintenance of blood glucose level are functional, and the rates of glycolysis, pyruvate oxidation and carboxylation, fatty acid synthesis, and fatty acid oxidation appear to be essentially normal in the liver, as judged from the fact that persistently high blood levels of lactate, ketone bodies and lipids do not occur typically.

In view of the fact that the reaction catalysed by the debranching enzyme is complex and involves two different activities, it is of importance to consider whether more than one protein is involved. All efforts to show that more than one protein is present in highly purified preparations of the debranching enzyme from rabbit skeletal muscle have failed[48-50], and, similarly, the enzyme purified from rabbit liver has been shown to be a single protein with the two kinds of activity[51]. If, as seems most probable, the debranching enzymes of human liver and muscle are also single proteins having both transferase and glucosidase activities, it might be expected that the patient with Type III glycogen storage disease would lack both of these activities in every affected tissue. As has been discussed previously, there is good evidence that in both Types I and II glycogen storage diseases the characteristic enzyme deficiency is generalised and extends to all tissues which contain the implicated enzyme. Such appears not to be the case in Type III disease. For example, in 7 out of 29 patients from whom biopsy samples of both liver and skeletal muscle were obtained, Brown and Brown[16] reported in 1968 that the debranching enzyme was inactive in the liver but fully active in the muscle, and, in contrast to the liver, that the muscle contained a normal amount of glycogen which had a normal structure. In the remaining 22 cases the debranching enzyme was absent from both muscle and liver.

In 1962, Illingworth and Brown discovered an oligosaccharide, $6^3$-α-glucosylmaltotetraose, which can serve as a substrate with which the glucosidase activity of the debranching enzyme can be assayed specifically, and its usefulness in measuring this enzymatic activity in human tissues was shown[52]. Another substrate, 1,6-α-glucosyl cyclomaltohexaose, which is equally suitable for this purpose, was discovered by Taylor and Whelan in 1966[53]. Assays which are apparently specific for the oligosaccharyltransferase activity of the debranching enzyme have also been described. One such assay involves the chromatographic separation of various 1,4-linked [$^{14}$C]oligosaccharides formed by α-maltosyl and α-maltotriosyl unit transfer from the outer chains of glycogen to [$^{14}$C]maltose acting as an acceptor[48]. Another assay involves a measurement of the decrease in [$^{14}$C]formic acid produced when a limit dextrin, which has been labelled previously in its outermost 1,4-linked α-glucose units by pre-incubation with [$^{14}$C]glucose 1-phosphate and phosphorylase, is oxidised by periodate before and after the labelled polysaccharide is incubated with the debranching enzyme which has transglycosylase activity[54]. Both of these methods are cumbersome and present problems in the interpretation of the results owing to potential interference from the action of 1,4-α-glucosidases, and even from the disproportionating action of glycogen phosphorylase in tissue extracts.

A commonly used assay for the debranching enzyme is the incorporation of [$^{14}$C]glucose into glycogen which was described first by Hers[55]. The simplest interpretation of the significance of this assay is that it depends upon the very slight reversibility of the amylo-1,6-glucosidase reaction[56], and, as such, it might be considered specific for the glucosidase activity itself. However, it has been clearly demonstrated that many of the 1,6-α-glucose units introduced in this way are covered subsequently by unlabelled glucose units through the action of the oligosaccharyltransferase activity of the debranching enzyme[57,58]. For this reason it is possible that the action of the transglycosylase has an indirect influence on the rate of glucose incorporation by the glucosidase. Since the site(s) of incorporation of 1,6-bonded α-[$^{14}$C]-glucose units on the outer chains of glycogen is not known, it cannot be excluded that during the incorporation assay the transglycosylase may play an important role in rearranging the outer chain 1,4-linked α-glucose units of glycogen in such a way that a structure more favourable for glucose incorporation results. That at least a critical minimum chain length exists for this reaction to occur at a significant rate is shown by the fact that Nelson *et al.* found that a limit dextrin of glycogen formed by phosphorylase action, which has a uniform outer chain length of four glucose units, is a very much poorer acceptor than glycogen in the incorporation assay[59,60], even though hydrolysis of many of its outer branch points occurs simultaneously, leading to the appearance of outer chains which may be from six to ten glucose units in length. Because of these considerations and the unexpected relative ineffectiveness of the limit dextrin as an acceptor, it should be emphasised that this assay for the debranching enzyme is understood only imperfectly. This is highly relevant to a discussion of Type III glycogen storage disease since van Hoof and Hers have published the results of a study of the tissues of 45 patients with this disease and have concluded that while 34 of them lacked the debranching enzyme in both liver and muscle, the remaining 11 seemed to have some glucosidase or transglycosylase activity in one or the other of these tissues[61]. The latter conclusion was based on the results of assay by the $^{14}$C incorporation method as well as by the assays which have been discussed in the preceding paragraph and by others not described here. van Hoof and Hers proposed that variable degrees of loss of the two catalytic functions of the debranching enzyme can occur by mutational events. The apparent incomplete expression of debranching enzyme deficiency has been seen more often in the muscle than in the liver. Whether this signifies that the liver and muscle enzymes are under separate genetic control cannot be decided from present information.

The question of the extent to which Type III glycogen storage disease is expressed in fibroblasts grown in tissue culture has been studied by Brown and Brown, who found that in three cell lines the quantity of polysaccharide present was not greater than normal but that its structure resembled that of a limit dextrin[39,40]. In this work, assay of the debranching enzyme—which even in normal cells is present at a rather low level compared with other enzymes of glycogen metabolism—showed that the mutant cells had from zero to less than 15% of the activity of normal cells when the ability to form glucose specifically from a limit dextrin was measured. Because of the nature of this assay it is not possible to assign qualitative significance to an apparent, small residual activity. That the enzyme deficiency has a physiological significance in fibroblasts was shown by the finding that the half life of [$^{14}$C]-glycogen in such cells is intermediate in value between normal cells and those cultured from patients with Type II glycogen storage disease[40]. That the heterozygote state can also be detected in fibroblasts using the limit dextrin assay has been shown[39]. On the other hand, the [$^{14}$C]glucose incorporation method has been reported to indicate that fibroblasts from patients with Type III disease have from 10% to 70% of normal activity in this respect[42,62], and it was also reported that heterozygote detection by this method was not successful[62]. DiMauro *et al.*[42] found higher than normal contents of polysaccharide in the cultured fibroblasts from one such patient. It appears that differences in expression of the genetic defect may be evident in cultured fibroblasts just as they are in the liver and muscle of some individuals with Type III glycogen storage disease. Resolution of some of these problems could come from a detailed structural and functional study of the human muscle and liver debranching enzymes.

### 8.2.5 Type IV glycogen storage disease

This is a rare congenital disease which is due to a deficiency of the glycogen branching enzyme[12]. Infants with the disorder fail to thrive, are hypotonic, exhibit progressive hepatosplenomegaly, and usually die from liver failure in their second year of life. Progressive cirrhosis of the liver with ascites has been a common finding at necropsy. Although the liver has been the organ usually examined, histological studies and electron microscopy have sometimes been carried out on the heart, kidney, smooth and striated muscles and on the reticuloendothelial and nervous systems as well. The first case of this disease was described clinically by Andersen[63], who isolated a polysaccharide from the liver by digestion of the tissue with strong alkali followed by ethanol precipitation. After several reprecipitations the material was found by Illingworth and Cori to be sparingly soluble in cold but readily soluble in hot water and to show a tendency to precipitate from aqueous solution[10]. These properties were different from those of normal human liver glycogen isolated similarly. Furthermore, when this abnormal homopolymer of glucose was mixed with $I_2$–KI, the complex formed had an absorption maximum at 530 nm, and it was as chromogenic on a weight basis as was corn amylopectin (absorption maximum, 550 nm). Stepwise enzymatic degradation established the fact that this polysaccharide had outer chains which, on average, were much longer than those of human liver glycogen, and, further, that an abnormality of structure in this respect extended into the inner portion of the molecule as well. This conclusion was reached because the limit dextrin which remained after 82% of the mass of the polysaccharide had been removed by three consecutive, alternating treatments with phosphorylase and debranching enzyme gave a complex with iodine whose absorption maximum[10] was at 550 nm. A limit dextrin of normal glycogen prepared by a similar series of enzymatic treatments is virtually achroic when tested with iodine. In recognition of all of these properties, the abnormal polysaccharide was referred to as 'amylopectin-like' and this general conclusion has been reached also by every investigator subsequently who has isolated the total alkali-resistant polysaccharide fraction from livers of other patients having this disease. In the authors' laboratory, polysaccharides have been isolated subsequently from the livers of eight additional children with Type IV disease and all of these products have had amylopectin-like character to a greater or lesser degree. Variation has been observed in the total branch point content which has varied from 4.7 to 6.7%. In every case, however, the polysaccharides have been highly chromogenic when tested with iodine.

Recently, Mercier and Whelan have reported on their studies of the structure of a polysaccharide isolated by alkali digestion from the liver of one child who died with this disease at 11 months of age[64,65]. Fernandes and Huijing had reported earlier on the clinical and biochemical findings in this case, and in the course of their study they treated the child for a period of 6 days, a few weeks before his death, with a preparation of fungal α-glucosidase[66]. This treatment reduced the polysaccharide content of the liver substantially. The value for the content reported before treatment (10.7%) is much higher than that found in the liver of any other known patient with this disease and it must, therefore, be considered as atypical. The residual polysaccharide isolated at autopsy was subjected by Mercier and Whelan to direct debranching by pullulanase[64] and by a bacterial isoamylase[65,67] in order to study the unit chains of which the substance was composed. The extent of debranching by pullulanase was 43%, and the major fraction of the substance which was degraded consisted of chains with an average length of 25 glucose units. A smaller fraction contained short chains with an average length of 5 glucose units. Maltose, maltotriaose and maltotetraose were found in the latter fraction. In contrast, amylopectin was completely debranched by pullulanase to give chains of an average length of 20 glucose units and no very short chains. These observations provided direct evidence that this sample of Type IV polysaccharide did not have the same structure as amylopectin, although it did have 4.6% total branch points and inner and outer chains which on the average were much longer than those of glycogen[64]. The bacterial isoamylase was found to degrade 78% of the polysaccharide to give a major fraction of chains with an average length of about 22 glucose units, and, as in the work with pullulanase, there were many short chains[65]. These results are important in indicating that the polysaccharide from Type IV liver does not always resemble amylopectin in all structural respects. The fact that only one sample was studied by the authors does not permit a general conclusion to be drawn on this point. However, it is probably not likely that any Type IV human liver will contain a polysaccharide with exactly the same average structure as plant amylopectin.

In view of the clear demonstration of the absence of the normal glycogen branching enzyme in the liver and leukocytes of Type IV patients[12,66,68], the existence in the liver of a glucose polymer which has 1,6-branch points in it certainly requires explanation. Brown and Brown suggested that such tissues may contain a second branching enzyme with a different substrate specificity whose activity is either absent from, masked, or reduced in amount in normal human liver, and that this enzyme may require different assay conditions for detection[12]. In this connection it is interesting that cultures of human skin fibroblasts can be used to detect the branching enzyme deficiency in Type IV disease[68], but that such cells retain 5–10% of the branching activity found in control fibroblasts[39]. The polysaccharide present in the mutant cells has a structure more closely resembling that of normal glycogen than the polysaccharide which is present in the liver of the same individual[39]. Further investigation is under way in the authors' laboratory to compare the catalytic properties of the residual branching enzyme which is present in Type IV fibroblasts with those of the normal fibroblast branching enzyme. It has been found that the detection of the heterozygote state in cultured fibroblasts from parents of Type IV patients is possible[68], and, hence, it appears that there can be full phenotypic expression in this cell type of the genetic lesion. This is important in indicating that studies of the residual branching activity in Type IV fibroblasts are likely to be relevant to the other affected tissues of these patients.

There have been two other interesting suggestions made with regard to the enzymatic origin of branch points in the polysaccharide present in Type IV tissues. Manners has discussed the possibility that Type IV liver may contain a branching enzyme during foetal development and that this activity decreases rather than increases at birth[69]. It could even be possible that such a foetal enzyme, which might have different substrate specificity, could persist in the mutant tissue. In unpublished work the authors of this chapter have examined one saline-aborted foetus from a pregnancy which was found to be at risk by ante-natal diagnosis based on branching enzyme assay in foetal cells cultured from amniotic fluid. The family was already known to have had one child who had died with Type IV disease. The foetal liver did not appear to contain any branching enzyme when assayed as described[12]. However, the time of death of the foetus *in utero* was unknown and post-mortem changes in enzyme activity could not be ruled out. Further work with this difficultly obtainable tissue is in progress. From a general point of view the suggestion of Manners[69] deserves careful consideration, since there are examples of congenital diseases, such as hereditary fructose intolerance, which is discussed later in this chapter, where persistence of a foetal liver enzyme in the affected adult organ has been demonstrated.

A different possible explanation for the existence of a branched polysaccharide in Type IV liver has been offered by Huijing *et al.*[70]. These investigators reported that, under appropriate experimental conditions *in vitro*, the glycogen debranching enzyme can act in reverse (in the presence of 5 mM glucose) to synthesise branch points and, thus, indirectly to stimulate the synthetic action of phosphorylase. The assay system was essentially that described by Brown and Brown for the normal glycogen branching enzyme[12]. The incorporation of [$^{14}$C]glucose into glycogen by the debranching enzyme acting in reverse, and the covering of the introduced 1,6-linked units by maltotriosyl chains through transglycosylase action, has been discussed at length above in the section on Type III glycogen storage disease. Whether this is a possible route of synthesis of branch points in Type IV tissues depends upon many factors such as (i) the intracellular concentration of glucose in contact with the debranching enzyme in such tissue, and (ii) the ability of UDPglucose–glycogen transglucosylase (glycogen synthase) to elongate the new short outer side chains which might be produced in this way. There is no direct information available on either of these points, except that Parodi *et al.* have shown that liver glycogen synthase acts by a multirepetitive chain elongation mechanism and that this enzyme does not utilise as acceptors more than *ca.* 50% of all of the outer chains of liver glycogen[71]. It should be noted that Mercier and Whelan have found a substantial number of short outer side chains on the sample of Type IV polysaccharide which they have analysed, and this fact could be interpreted as supporting the mode of branch point biosynthesis discussed here, although alternative explanations are also possible[64,65]. It is clear that additional experimental work is required to explain the origin of the branch points in Type IV polysaccharides.

It should be mentioned finally that Type IV disease does not appear to be restricted to the liver in most cases, although in at least one instance glycogen of normal structure was isolated in normal amount from a skeletal muscle biopsy[12]. Very often there is involvement of the heart[72,73,74] as well as sometimes of the central nervous system[72,75]. It is a peculiarity of Type IV disease that the polysaccharide present in liver, heart and, sometimes, skeletal muscle seems to be heterogeneous in composition with respect to its solubility and branch point content[12,73,74]. The biochemical reason for this is unknown. Histological studies have usually agreed in indicating that the polysaccharide in tissue sections is relatively inert to α-amylase digestion. This is most probably due to its physical surroundings in the cell, since the authors have found in unpublished work that it is necessary to use strong alkali extraction to recover all of the material in a form which is totally degradable *in vitro* by glycogen phosphorylase plus debranching enzyme. It should be emphasised, however, that in a physiological sense the Type IV polysaccharide can be mobilised in the liver since patients usually have a normal response to a glucagon tolerance test. The relative insolubility of the polysaccharide in the liver appears to trigger the onset of cirrhosis. Type IV disease presently can be considered to be universally fatal in early childhood inasmuch as no patient with branching enzyme deficiency is known to have survived to adulthood.

## 8.3 INBORN ERRORS OF FRUCTOSE METABOLISM

### 8.3.1 Introduction

Fructose, either as the free monosaccharide or as part of the disaccharide sucrose, is an important source of dietary carbohydrate in most areas of the world. The sugar alcohol glucitol (sorbitol), which is present in many commercially prepared food products, can be converted in some tissues into fructose and metabolised there via the fructose pathway. It has been estimated that the average human intake of fructose is of the order of 50–70 g $d^{-1}$. The utilisation of dietary sucrose requires that this disaccharide first be converted into glucose and fructose by the action of invertase in the intestine. Dahlqvist has reported recently that human intestinal invertase develops very early in foetal life and that the level of activity of this enzyme in various human population groups does not depend upon the average content of sucrose in their diets[76]. Some clinical aspects of the metabolism of fructose were reviewed in 1972 in a series of papers presented at a symposium in Helsinki and these contain recent information as well as useful literature references[77]. Two of the hereditary disorders of fructose metabolism which are discussed rather briefly below have been reviewed in detail by Froesch[78]. These are essential fructosuria and hereditary fructose intolerance.

### 8.3.2 Essential fructosuria

In human liver, fructose is phosphorylated at C-1 by fructokinase (EC 2.7.1.3), a relatively specific kinase with a $K_m$ for fructose of *ca.* 0.5 mmol $l^{-1}$. Mammalian liver also contains three different 'hexokinases' (EC 2.7.1.1) which are able to phosphorylate fructose at C-6, all of which have $K_m$ values for this ketose of *ca.* 3 mmol $l^{-1}$. These three hexokinases also phosphorylate glucose, and their $K_m$ values for this aldose are from 0.01 to 0.2 mmol $l^{-1}$. Because of the large difference in their apparent affinities for glucose and fructose, and the much higher concentration of glucose than of fructose in plasma under normal circumstances, it is likely that glucose is an extremely effective inhibitor of the direct conversion of fructose into fructose 6-phosphate in the liver through the action of these enzymes. Even more significant in considering the relative importance of C-1 phosphorylation by fructokinase, and C-6 phosphorylation by hexokinase, may be the finding that rat liver hepatocytes are virtually devoid of the three low $K_m$ hexokinase isoenzymes, although they do contain fructokinase[79]. The parenchymal cells of rat liver contain the high $K_m$ hexokinase isoenzyme which has been called 'glucokinase' (EC 2.7.1.2) and this enzyme has virtually no affinity for fructose. If human liver is similar to rat liver in these respects, fructose metabolism in the hepatocyte can be considered to be wholly dependent on the formation initially of fructose 1-phosphate. For many years a benign condition, essential fructosuria, has been recognised clinically. Individuals with essential fructosuria lack hepatic fructokinase, as was demonstrated by Schapira *et al.*[80], and since the renal threshold for fructose is low, fructose is excreted by these patients in variable amounts depending upon its content in the diet. Since the condition is benign and asymptomatic, it is often discovered only by those diabetic screening programmes which are based on measurement of the urinary level of total reducing sugars rather than on specific enzymatic assay for glucose. In these cases the identity of the urinary sugar must be ascertained to avoid a diagnosis of diabetes in individuals with essential fructosuria. Affected individuals should also be made aware of the problem of an adequate caloric intake if diets are consumed which contain predominantly fruits. Essential fructosuria is very rare and appears to be inherited as an autosomal recessive trait.

### 8.3.3 Hereditary fructose intolerance

A more serious disturbance of fructose metabolism, hereditary fructose intolerance, was first described in 1956 by Chambers and Pratt[81] in a single patient, and in the following year Froesch *et al.*[82] reported the typical symptomatology in two siblings and two relatives, thus establishing the hereditary nature of the disease. Ingestion of fructose by individuals with hereditary fructose intolerance results in severe hypoglucemia and vomiting. Adults develop a strong aversion to fruits and thus avoid the symptoms, whereas infants fed fructose fail to thrive, develop hepatomegaly and jaundice and frequently have albuminuria and aminoaciduria. In the first decade after recognition of this syndrome, 25 families with a total of 45 patients were described[78] and many more cases have been recorded since. Perheentupa *et al.* reported 18 patients in 14 Finnish families[83]. Froesch *et al.* have published the pedigree of a family in which there were numerous consanguineous marriages which resulted in eight cases of hereditary fructose intolerance[78,84]. Although most family histories suggest an autosomal recessive mode of inheritance, several families of affected children with affected parents have been reported.

In hereditary fructose intolerance the primary enzyme defect is the virtual absence of the liver aldolase (EC 4.1.2.13) which splits fructose 1-phosphate to D-glyceraldehyde and dihydroxyacetone phosphate. Activity measurements on whole liver homogenates with fructose 1-phosphate as substrate have given results from 0 to 12% of normal, while the activity toward fructose 1,6-diphosphate generally has been much less depressed. The ratio of activities in normal liver toward these two substrates is usually found to be *ca.* 1, whereas homogenates of livers obtained from fructose intolerant patients give ratios of activity toward the diphosphate compared to fructose 1-phosphate of 6 to 10.

Most of the early work on the action of aldolase was carried out on the rabbit muscle enzyme. In 1953, Leuthardt *et al.*[85] and Hers and Kusaka[86] reported the presence of aldolase in mammalian liver. Aldolases from liver and muscle differ in their relative activity toward fructose 1,6-diphosphate and fructose 1-phosphate, the activity of muscle being 50 times greater toward the diphosphate whereas the liver enzyme cleaves both substrates at approximately equal rates (see above). In 1958, Peanasky and Lardy[87] succeeded in crystallising the enzyme from liver and proved that a single protein catalyses the cleavage of both substrates. Three distinct isoenzymes with aldolase activity, which have been designated as A, B and C, have been identified in various tissues and characterised on the basis of their activity ratios toward the two substrates, inhibition by specific antibodies, and by electrophoretic mobility differences. One of these, aldolase C, occurs in brain. Muscle contains aldolase A, while the predominant aldolase of liver has been designated as aldolase B. Hybrids can also occur, as shown by specific staining procedures which depend upon enzymatic activity, following electrophoresis of tissue extracts. Horecker *et al.*[88] have reviewed the chemistry, kinetic properties and reaction mechanisms proposed for the aldolases. Although much of the early work was carried out on enzymes purified from rabbit tissues, Lebherz and Rutter[89] have shown that aldolases A, B and C are widely distributed among vertebrates, and that human aldolases A and B can be distinguished by selective inhibition using specific antibodies prepared against aldolases A and B of the rabbit. Human foetal liver has a higher ratio of activity toward fructose 1,6-diphosphate compared with fructose 1-phosphate than does adult liver and this ratio decreases with age[89]. This latter finding suggests that an A to B transition normally occurs early in foetal life.

Schapira and co-workers[90,91] have studied the question of what kinds of aldolase are present in the livers of patients with hereditary fructose intolerance by using antibodies prepared in chickens against either crystalline rabbit muscle aldolase (anti-aldolase A) or liver aldolase (chiefly anti-aldolase B). In liver extracts from three patients with the disease, cross-reacting, precipitating protein was present when tested with the anti-B sera, and from the extent of the reaction observed it was calculated that from 18 to 30% as much of such protein was present as in normal human liver. Since the enzymatic activity of the patients' liver aldolase in splitting fructose 1-phosphate was only *ca.* 3% of the normal level, the conclusion was drawn that a structurally similar although enzymically inactive aldolase B is indeed synthesised in the mutant tissue[91]. In this work there was also a suggestion of the persistence of a foetal aldolase pattern in the liver of fructose intolerant patients, as judged by the large degree of inhibition of activity toward fructose 1,6-diphosphate given by anti-aldolase A sera. However, it was also found by these authors that not all affected livers contain a significant quantity of the protein which cross reacts with anti-aldolase B. Schapira *et al.* have studied the electrophoretic pattern of the enzymatically active aldolase which is still present in the mutant tissue and have concluded that hybrids of aldolase A and aldolase C are present[91], such as are found normally in adult human brain and embryonic human heart[89]. Schapira *et al.* have also reported some kinetic experiments in which the apparent $K_m$ for fructose 1-phosphate of the residual aldolase B activity of normal and affected livers was measured[91]. In normal foetal or adult livers the value found was 2–3 mmol $l^{-1}$, but in ten affected livers homogenised in water the values found ranged from 10 to 60 mmol $l^{-1}$. In addition to this evidence for alteration in structure of the small amount of enzymatically active aldolase which is still present in fructose intolerant patients, these authors found that homogenisation of the affected livers in a buffer (composition unspecified) containing β-mercaptoethanol and EDTA resulted in an extract in which the $K_m$ of the aldolase for fructose 1-phosphate had the same value as that characteristic of normal livers. The role of SH groups in the activity of various aldolases has been reviewed by Horecker *et al.*[88]. In the light of present information, derived mostly from studies of the rabbit enzymes, it is possible to consider that among the structural changes which characterise the mutant aldolase B are those which render some of this protein's SH groups readily susceptible to oxidation during tissue homogenisation, or, alternatively, that abnormal intrachain disulphide bridges exist in the native mutant aldolase such that its binding sites for fructose 1-phosphate are markedly affected.

Fructose administration to intolerant individuals results in profound hypoglucemia and a depletion of serum phosphorus. Fructose 1-phosphate accumulates intracellularly in liver and, as a consequence, there may be a depletion of intracellular ATP. Bode *et al.*[92] have shown that intravenous infusion of fructose or glucitol into normal human subjects (about 700 mg per kg of body weight) reduces the total hepatic ATP level to 50% of its initial value within 30 minutes and leads to a 2- to 3-fold accumulation of fructose 1,6-diphosphate, glucose 6-phosphate, α-glycerophosphate and dihydroxyacetone phosphate, as well as to the accumulation of a high level of fructose 1-phosphate. The latter observation particularly indicates that in man, as in the rat[93], the fructose 1-phosphate aldolase reaction is the limiting step in fructose metabolism. The hypoglucemia which occurs in the fructose intolerant individual as a result of fructose ingestion seems to be explicable in general terms on the basis of an inhibition of glucose release from the liver rather than being due to stimulation of insulin release[78]. Patients show a lack of responsiveness to glucagon during fructose-induced hypoglucemia but they can effectively convert galactose into glucose and, in so doing, their hypoglucemia can be partially relieved. This seems to indicate that neither phosphoglucomutase nor glucose 6-phosphatase can be the site of a primary inhibitory action of intracellular fructose 1-phosphate. On the other hand, the failure to respond to glucagon suggests that the phosphorolysis of glycogen is inhibited in such patients, although, in itself, this would not account for hypoglucemia in the face of active gluconeogenesis. Recently, Kaufmann and Froesch have reported that dog liver phosphorylase is competitively inhibited by fructose 1-phosphate ($K_i = 4$ mmol $l^{-1}$) and that the extent of inhibition is dependent upon the concentration of inorganic phosphate[94]. These authors have suggested that inhibition of phosphorylase by fructose 1-phosphate may have an especially important physiological significance in the liver of the fructose intolerant patient where fructose 1-phosphate accumulation may be accompanied by low intracellular levels of inorganic phosphate. As yet, assays for active liver phosphorylase have not been made in the liver of such patients. Since none of these observations seems to offer a wholly adequate explanation for fructose-induced hypoglucemia in the patient with hereditary fructose intolerance, it might be considered whether the native form of fructose 1,6-diphosphatase (EC 3.1.3.11), which is maximally active at neutral pH in mammalian liver[95], can be partially inhibited by the high concentration of intracellular fructose 1-phosphate which may exist in the liver of affected individuals. Such an inhibition could result in the formation of less glucose by gluconeogenesis. Adequate studies of gluconeogenesis in the liver of patients with hereditary fructose intolerance have not been made, and, accordingly, it is not possible to decide whether this possible explanation has validity.

### 8.3.4 Hepatic fructose 1,6-diphosphatase deficiency

In 1970, Baker and Winegrad reported the first case of hepatic fructose 1,6-diphosphatase deficiency[96]. The patient was a 5-year-old female child who had fasting hypoglucemia and metabolic acidosis. A male sibling had died at 6 months of age with similar symptoms. Two additional cases were reported in 1971[97,98] and two others have been diagnosed since and studied[99,100]. The case described by Baerlocher *et al.*[97] was the third affected child of healthy, consanguineous parents. Examination of tissue samples has revealed essentially no fructose 1,6-diphosphatase in the livers from all cases or in the leukocytes which have been examined in one case[100]. Fructose 1,6-diphosphatase activity has been normal in the skeletal muscles of two cases studied[99,100]; this finding might have been predicted from the known differences in structure and enzymatic properties of the corresponding diphosphatases of rabbit muscle and rabbit liver[101,102]. Since the activity of hepatic fructose 1,6-diphosphatase is required for gluconeogenesis, its absence accounts for the fasting hypoglucemia often exhibited by affected children and for the elevation in blood level of the glucogenic amino acids and particularly of alanine which several patients have had[97,99,100]. The common finding of hyperlacticacidemia can also be explained by a block in gluconeogenesis. The most important clinical features from the viewpoint of patient management are the hypoglucemia and the episodic metabolic acidosis which can lead to serious hyperventilation. Administration of fructose by feeding or by infusion has led to severe hypoglucemia in four of the five patients. The cause of this is not apparent. It is possible that fructose 1,6-diphosphate accumulates in the liver under these circumstances and acts in some way to block glucose release from whatever glycogen can be mobilised in the glycolytic pathway. It should be mentioned that the content of glycogen in the livers of two patients has been reported to be high (9.5% of the wet weight in one case[97] and 7.7% in another[100]), as it is in children with Type I glycogen storage disease where hyperlacticacidemia and hypoglucemia also occur owing to a different block in the gluconeogenic pathway. The reason for the inability of the fructose 1,6-diphosphatase deficient child to mobilise liver glycogen to combat episodes of hypoglucemia is not known. Dietary restriction of fructose, sucrose and glucitol has resulted in some improvement in one patient[99], particularly in controlling the tendency to metabolic acidosis. There was also a significant decrease in the extent of fatty infiltration of the liver in another patient as a result of fructose restriction[100]. Since fructose 1,6-diphosphatase deficiency is the heritable disorder of fructose metabolism to be described most recently, the number of reported cases of this disease is small and its rarity as a metabolic disease cannot yet be determined. It is potentially fatal, but it is possible that very early detection and careful patient management can prolong the life of the affected child significantly.

## 8.4 INBORN ERRORS OF GLYCOSAMINOGLYCAN METABOLISM

### 8.4.1 Introduction

Historically, the recognition of the enzymatic basis for each of the heritable disorders of glycogen metabolism, as well as for those involving monosaccharide utilisation for which biochemical explanations are known, came only after the principal pathways for the intermediary metabolism of these substances had been elucidated. The principal exception to this is the pathway for the total conversion of glycogen into glucose which, at least in some tissues, seemingly occurs within lysosomes, and which is now known to be due to the action of a specific α-glucosidase capable of acting on both 1,4- and 1,6-α-glucosidic bonds. As has been discussed above, the existence of this lysosomal pathway for glycogen catabolism came to light only as a result of the discovery by Hers[7] of the enzymatic basis of Type II glycogen storage disease. A somewhat similar situation exists with respect to those heritable disorders of connective tissue which involve primarily its glycosaminoglycan (acid mucopolysaccharide) components. These diseases will be discussed below. Biochemical knowledge of the pathways of anabolism and catabolism of these heteropolymers of complex structure has lagged far behind that of glycogen. A principal reason for this is that the necessary chemical studies on composition and structure of the acid mucopolysaccharides have progressed very slowly, owing first to the difficulty of separating them in a native state from each other and from the other constituents of connective tissue, and, second, to the unexpected complexity of their structures*. Even now, not one of the members of this class of substances can be regarded as having its structure fully described. Much of the earlier chemical work was done with preparations which are now recognised to have been partially degraded because of the methods used in their isolation. This is particularly true with respect to the covalent bonds by which most of these polysaccharides in their native state are now known to be linked to protein in the extracellular ground substance of connective tissue. A second major problem in the study of the metabolism of these substances is the fact that there is no mammalian organ which is composed mostly of connective tissue cells having a sufficient metabolic activity to permit the kind of productive studies of relevant metabolic pathways which have characterised, e.g. the history of research on glycogen metabolism in liver and muscle during the past 40 years. In fact, not until the very recent development of adequate tissue culture methods have the metabolic characteristics of fibroblasts begun to be described in a comprehensive way. This work, too, is still in its early stages, but the fact that the living cell type which is responsible for the biosynthesis of the acid mucopolysaccharides (as well as of collagen) can be studied in pure culture gives a very powerful tool to the investigator of metabolic diseases. The important results which have been obtained by the use of fibroblast cultures derived from patients with mucopolysaccharide storage diseases will be discussed below, and it will be seen that these studies have opened windows through which the outlines of the pathway of catabolism of these compounds may now begin to be seen.

### 8.4.2 Early clinical and biochemical studies

The first detailed clinical description of what is now recognised to have been one of the mucopolysaccharide storage diseases was published by Hunter in 1917, who described[103] two young brothers of dwarfed stature who had protuberant abdomens due to hepatosplenomegaly. These children also had

* See Chapters by Rodén and by Muir and Hardingham.

rather grotesque facial features as well as hands that were deformed, with fingers held so as to suggest the shape of a claw. Two years later Hurler described[104] two infants with deformed features who had clouding of the cornea as well as distinct mental retardation and conspicuous gibbus. In these three latter respects the affected children differed from Hunter's patients, who had none of these characteristics. These important clinical differences between the two sets of patients were not accorded much attention during the next 30 years, and the many additional children who were described during this time as having similar clinical features were all classified as patients with Hurler–Pfaundler disease or 'gargoylism'. Brante reported[105] in 1952 that a sulphated mucopolysaccharide of seemingly simple constitution, similar to that of chondroitin sulphate, was present in large amount in a liver specimen removed at autopsy from one patient with this disease. In this same report he also applied the term 'mucopolysaccharidosis' to the syndrome of gargoylism which until that time had been regarded as a lipochondrodystrophy. The belief that the Hurler syndrome was a disorder of lipid metabolism had been widespread. Stacey and Barker in 1956 reported[106] the presence of a sulphated polysaccharide (as well as of non-sulphated fractions) in the liver of a patient with Hurler's disease. In 1957, Brown studied[107] liver samples removed at autopsy from two cases of Hurler's disease in girls, as well as a liver sample removed by biopsy from a third, living male patient. In all instances the tissues had been preserved in a frozen condition without contact with organic solvents until they could be studied in the laboratory. In this work mucopolysaccharide fractions of quite different properties from those of any previously described type of chondroitin sulphate were purified from each of the liver samples. These polysaccharide fractions were found to contain glucosamine and glucuronic acid, and it was shown that both *N*-acetyl and *N*-sulphate groups were present in them in a ratio of approximately 2:1. In addition, *O*-sulphate groups were present in varying amount. The high positive optical rotation of the substances indicated that the molecules contained many glycosidic bonds of $\alpha$-configuration. The surprising observation was also made that the substances were of low molecular weight. It was concluded that the principal storage substance characteristic of the liver of patients with Hurler's disease was heparan sulphate (formerly called 'heparitin sulphate'). It was found to be present to the extent of 0.5–1.5% of the wet weight of the organ in these first three cases studied, whereas none of this substance could be isolated from a control liver sample. In 1957, Dorfman and Lorincz reported[108] on the finding in the urines of two patients with Hurler's disease of a considerable amount of dermatan sulphate (formerly called chondroitin sulphate B) and of a small amount of a mucopolysaccharide fraction which may have been related in structure to the heparan sulphate found by Brown in the livers of affected children. With the report one year later by Meyer *et al.*[109] of the urinary excretion of large quantities of these two types of sulphated polysaccharides by patients with Hurler's syndrome, the occurrence of mucopolysacchariduria was firmly established as a prominent feature of this disease or group of diseases. Meyer *et al.*[110] in 1959 and Meyer[111] in 1961 reported the results of analysis of the polysaccharides present in various organs removed at autopsies of seven patients with this syndrome. In the livers of five out of six cases, from

70% to over 90% of the total mucopolysaccharide fraction was heparan sulphate, the rest being dermatan sulphate. In one of these cases the liver contained approximately equal quantities of the two polysaccharides. In all of Meyer's cases the spleen contained almost entirely dermatan sulphate with only a minor amount of heparan sulphate. In brain, kidney and other organs the distribution of the stored mucopolysaccharides varied, but in most instances dermatan sulphate predominated. Numerous additional results of the analyses of the storage mucopolysaccharide fraction from the organs of other patients have been published during the 12 years subsequent to these first studies, and all have confirmed that storage and excretion of heparan sulphate and dermatan sulphate are characteristic of the Hurler syndrome. However, some other mucopolysaccharidoses may also involve the accumulation and excretion of these polysaccharides as well as of certain others, as will be discussed below.

The term 'Hurler syndrome', as used in the preceding discussion of the early clinical and biochemical investigations, is meant in a generic sense only, because it is now recognised that individuals who have had any one of five or more mucopolysaccharidoses which are, in fact, genetically distinct, often have been classified as having Hurler's disease in one or more variant forms. Comparison of such patients on the basis of clinical studies alone, or even with the availability of reliable information about their individual mucopolysaccharide excretion patterns, led to the impression that an unusually wide range of phenotypic expression characterises this group of disorders. However, a distinct advance in the more precise classification of these diseases was made, even before the definitive biochemical work of Neufeld and her co-workers (discussed below), when it became clear that some male patients have a milder form of Hurler's disease in which corneal opacity does not occur. These patients really have what is now recognised to be a distinct clinical entity known as the 'Hunter syndrome', which is unique among all other known mucopolysaccharidoses in the fact that it is inherited as an X-linked recessive trait. A detailed and comprehensive review of the literature concerning these patients, as well as of descriptions of individuals affected by other mucopolysaccharidoses, has been published recently by McKusick[112]. It is not the purpose of the present chapter to include detailed clinical information about all of these diseases. Rather, in what follows, emphasis will be placed on the contribution of biochemical investigations to the differentiation and more rational classification of these syndromes.

### 8.4.3 Lysosomal nature of these diseases

Shortly after the discovery that the congenital absence of a presumably lysosomal α-glucosidase is the underlying enzymatic defect in Type II glycogen storage disease (see above), Van Hoof and Hers[113] published in 1964 a report of an electron micrographic study of liver samples from four children who were diagnosed as having had Hurler's disease. In retrospect, from the reported pattern of the mucopolysaccharide excretion by these patients, it seems possible that two of them actually had the Sanfilippo syndrome (Harris[114], Sanfilippo *et al.*[115]) which will be discussed below. In any event, these four children's livers were all found to contain large, single membrane-delimited vacuoles, many of which were from one-third to one-half as large as the nuclei of the hepatocytes. These vacuoles were rather uniformly filled with finely granular, amorphous masses of a substance which was not well-defined by the fixation and staining methods used. From this result, Van Hoof and Hers concluded that it was probable that these vacuoles contained the mucopolysaccharides which were already known to accumulate in the liver of such patients. Furthermore, by comparison with electron micrographs of liver from animals which had been pretreated by injection of dextrans or of Triton WR 1339, as well as of liver tissue from patients with Type II glycogen storage disease, Van Hoof and Hers proposed the hypothesis that Hurler's syndrome is a lysosomal disease which is due either to a genetic defect of a lysosomal hydrolase required for mucopolysaccharide degradation, or to the abnormal accumulation of mucopolysaccharides within lysosomes because of abnormalities in the structure, or in the quantity of these substances, due to genetically determined defects in the pathways of their biosyntheses. It was suggested by these investigators that the polysaccharides might accumulate intrahepatically either by autophagic uptake of molecules synthesised within the hepatocyte itself or by endocytosis of molecules brought to the lives by way of the blood. These important suggestions by Hers and his colleagues, and especially those pertaining to the lysosomal nature of the disease, soon received support from similar studies of patients with related syndromes (Loeb *et al.*[116], Wallace *et al.*[117]). The work of Hers and his colleagues set the stage for tissue culture studies of fibroblasts derived from the skin of patients with various mucopolysaccharide storage diseases.

### 8.4.4 Studies utilising fibroblasts in culture

The first indication that fibroblasts from patients with mucopolysaccharidoses would be useful in biochemical studies of the genetic lesions in these diseases came from the finding by Danes and Bearn[118,119] that fibroblasts cultured from patients with Hurler's syndrome, as well as from heterozygote parents, contain metachromatic granules. Shortly afterwards, Matalon and Dorfman reported[120] the results of chemical analyses which showed that fibroblasts from affected patients contain 5–10 times as much of a total acid mucopolysaccharide fraction as cells derived from normal children. Approximately 70% of the polysaccharide fraction from one cell line was found to be dermatan sulphate and the remaining 30% appeared to be hyaluronic acid. No heparan sulphate was detected. These authors also showed that growth of the cells in tee presence of [$^{35}$S]sulphate or of [$^{3}$H]acetate resulted in the labelling of the intracellular mucopolysaccharide fraction and also of a similar fraction which could be isolated from the culture medium. The incorporation of these radioactive precursors into both normal and Hurler fibroblasts could be inhibited by puromycin, and this finding suggested that a protein–polysaccharide complex is formed as a biosynthetic product in both kinds of cells. This latter observation by Dorfman and co-workers was especially important, because otherwise it might have been postulated that the basic enzyme defect in the Hurler cell is that sulphated mucopolysaccharide chains are synthesised independently of a protein acceptor. Such an explanation, which would imply a defect in biosynthesis, had seemed plausible on the basis of the early observation by Brown[107] that the heparan sulphate of Hurler liver had a molecular weight of less than 5000, as well as from the more detailed studies of Knecht, Cifonelli and Dorfman, who reported[121] in 1967 that the heparan sulphate which they obtained from the livers of two patients with Hurler's disease and from one child with Hunter's syndrome was, in these three instances, a mixture of two classes of low molecular weight substances with different properties. One class (molecular weight 3600–5500) contained serine which was covalently linked in only *ca.* 20% of all the molecules. In these oligosaccharides, more of the glucosamine units were *N*-acetylated than were *N*-sulphated. In the other class (molecular weight 2700–4000) there was no significant amount of serine, and *ca.* 75% of all of the glucosamine was *N*-sulphated. In the light of the puromycin inhibition data cited above, Dorfman and his colleagues concluded that the two classes of Hurler heparan sulphate oligosaccharides probably had been formed by partial enzymatic hydrolysis of a parent heparan sulphate molecule which had originally been linked to protein in the same manner as the heparan sulphate (molecular weight 24 000) which they isolated from normal human aorta. Their work gave no information about the site of this presumed partial degradation—whether within the hepatocyte itself, or, possibly, within whatever cell type is responsible for the major synthesis of heparan sulphate in the body. In this connection it may be recalled that the earlier work of Matalon and Dorfman[120] had failed to give evidence of heparan sulphate biosynthesis by fibroblasts in tissue culture. In other work, Dorfman had shown[122] that the storage form of dermatan sulphate from Hurler tissues, like heparan sulphate, contains less than one serine residue per polysaccharide chain. In contrast, Matalon and Dorfman found[123] that the dermatan sulphate present in Hurler fibroblasts grown in tissue culture has the relatively high molecular weight (20 000–40 000) and the serine content (one residue per polysaccharide chain) which is characteristic of that in normal human tissues.

The fact that heparan sulphate is one of the two principal storage polysaccharides in several of the mucopolysaccharidoses, and that it is also a prominent component of the urinary polysaccharides excreted by patients with these connective tissue diseases, makes the question of whether it is synthesised by the fibroblast particularly important. In 1968, Kraemer isolated[124] heparan sulphate from Chinese hamster cells grown in suspension culture, and in 1971 he showed[125] that six established mammalian cell lines adapted to growth in suspension culture, as well as a diploid mouse embryo lung strain grown as a monolayer on glass, all synthesised heparan sulphate during exponential growth. The cell lines studied included human HeLa cells of epithelial origin, a mouse fibroblast strain which had lost some differentiated functions, and a strain of hamster fibroblasts which had retained the ability to synthesise both hyaluronic acid and chondroitin sulphate. In other work, Bates and Levene[126] showed that heparan sulphate was formed when a mouse fibroblast cell line (3T6) was grown in monolayer culture. In Kraemer's work, as well as in that of Bates and Levene, the observation was made that part of the intracellular heparan sulphate was soluble in trichloroacetic acid, but that a significant amount was acid-insoluble before proteolytic digestion, which suggested that this substance could occur covalently bound to some cellular components. Although not yet demonstrated conclusively, it seems probable, from the work cited here, that normal human fibroblasts growing in primary culture are also able to synthesise heparan sulphate (as well as dermatan sulphate) as a proteoglycan. Strong support for this view has come recently from the work of Kresse and Neufeld[127], who have isolated a polysaccharide structurally similar to heparan sulphate from fibroblasts grown from skin biopsies of patients with the Sanfilippo syndrome (see below).

Of fundamental importance in understanding the biochemical basis for any mucopolysaccharide storage disease is a clear demonstration of whether such storage is the result of excessive synthesis, incomplete degradation, or impaired secretion of either the polysaccharide in the form in which it is synthesised or of products of related structure formed from it by partial enzymatic degradation. Fratantoni, Hall and Neufeld showed[128], by kinetic studies involving $^{35}S$ labelling of intracellular and extracellular polysaccharides synthesised by human fibroblasts growing in tissue culture, that two metabolic pools of such polysaccharides seem to exist within the cell: a small secretory pool with rapid turnover, and a separate pool of larger size and slower turnover which serves as a reservoir of molecules awaiting degradation. It was shown that cells from Hurler and Hunter patients steadily accumulate intracellular sulphated polysaccharides even though neither the rate of their synthesis ([$^{35}S$]sulphate incorporation) nor of their secretion is abnormal. This finding clearly indicated that the primary defect is one of intracellular degradation, as had been suggested earlier by Van Hoof and Hers[113].

The experimental approach which has given the clearest picture of the true genetic and biochemical relationships among various mucopolysaccharidoses has been that of growing fibroblasts in mixed culture to demonstrate genotypic differences. Neufeld and her associates pioneered this work, and they showed[129] in 1968 that the excessive and continuing intracellular accumulation of sulphated polysaccharides in cultured fibroblasts from Hurler or Hunter patients could be corrected, in that these cells became like normal human fibroblasts in their attainment of a steady state with respect to sulphated mucopolysaccharide synthesis and secretion, when the abnormal cells were grown in mixed culture either with normal cells or with mutant cells of a different genotype. In contrast, a mixed culture of cells from two different Hurler patients showed no such correction effect. In this important paper it was also shown that cells from a patient with the Sanfilippo syndrome could correct the metabolic defect in Hurler cells. This observation emphasised the likelihood that this clinically distinct syndrome[112,114,115] has a different biochemical basis from that of Hurler's disease. The fact that Sanfilippo patients excrete almost exclusively heparan sulphate in their urine[112] had been only suggestive of a genotypic difference between this syndrome and Hurler's disease. Fratantoni *et al.* also showed[129,130] that the corrective effect which they had found by the use of mixed cell culture could be demonstrated when mutant cells were grown in the presence of cell-free medium which previously had been exposed to growing cells of a different genotype. This discovery led to the preparation[130] of concentrates containing corrective factor activity from a serum-free medium in which cells of various types had been grown. Preliminary characterisation of the Hunter and Hurler factors suggested that they were heat-labile macromolecules which were neither nucleic acids nor polysaccharides. Although Neufeld's group carried out further investigations of the properties of the fibroblast corrective factors, the small quantity of each which could be obtained from the tissue culture medium in which cells had grown prevented definitive characterisation. However, as a result of their finding that normal human urine contains a protein (0.01% of the total urinary protein) which has Hurler corrective factor activity, Barton and Neufeld purified[131] this substance 1000-fold and showed that the preparation was distinct from any of ten known lysosomal hydrolases for which assay was made, and, in particular, that it also was not a sulphatase capable of acting to degrade the [$^{35}$S]mucopolysaccharide fraction which had been isolated from Hurler fibroblasts.

### 8.4.5 Hurler syndrome: MPS I H

Chemical studies of the structure of the two known storage mucopolysaccharides found in tissues of patients with the Hurler and the Hunter syndrome—dermatan sulphate and heparan sulphate—had shown that L-iduronic acid is a constituent of both substances. Accordingly, Matalon, Cifonelli and Dorfman[132] studied the question of whether an L-iduronidase (EC 3.2.1.76) activity might be detectable in human liver or human fibroblasts. They found that chemically desulphated heparan sulphate and dermatan sulphate served as substrates from which L-iduronic acid could be formed by enzyme systems present either in the 600 × *g* supernatant fraction of whole liver or in a whole homogenate of human fibroblasts when these extracts were incubated at pH 4.5 for a prolonged period of time. When Hurler fibroblasts were used, a much diminished yield of iduronic acid was found. Preliminary evidence that α-L-iduronidase is, in fact, a lysosomal enzyme in rat liver has been presented recently by Weissmann and Santiago[133], who used a chemically synthesised substrate, phenyl α-L-iduronide, to show that this activity was concentrated to about the same extent as some marker enzymes in a centrifugally prepared lysosomal pellet. With the availability of the synthetic substrate it became possible for Matalon and Dorfman[134] and, simultaneously, Bach *et al.*[135] to show that the enzymatic defect in the Hurler syndrome (Mucopolysaccharidosis I H) is the lack of α-L-iduronidase. This demonstration was made in the first instance by incubating whole homogenates of cultured human fibroblasts or the 10 000 × *g* supernatant fraction from human liver at pH 4.0 for 24 hours, following which the expected product of the enzyme's action (phenol) could be detected. Extracts prepared similarly from cells derived from patients with the Hunter or the Sanfilippo syndromes had normal levels of this enzymatic activity. Neufeld's group[135] carried out similar experiments, and, in addition, found that the active fractions of Hurler corrective factor which had been isolated from urine[131], and which contained two chromatographically separable 'isofactors', also contained two isozymes of α-L-iduronidase. These enzymes correspond exactly in their chromatographic behaviour to the 'isofactors'. What the physiological significance of the existence of these isoenzymes might be has not yet been determined.

### 8.4.6 Scheie syndrome: MPS I S

The Scheie syndrome [Mucopolysaccharidosis I S (formerly called Mucopolysaccharidosis V)][112] was first described in 1962 as a 'forme fruste' of Hurler's disease[136] in which prominent corneal clouding and aortic valve involvement occur accompanied by some degree of deformity of the hands and feet. However, the patients with this rare disease appear to have a normal lifespan and, usually, they have either normal or superior intelligence. Wiesmann and Neufeld[137] found that fibroblasts cultured from the skin of Scheie and Hurler patients are deficient in the same specific factor required for a normal rate of mucopolysaccharide degradation. It also has been shown by Bach *et al.*[135] that cultured fibroblasts from a Scheie patient have no α-L-iduronidase activity. Although both the Hurler and Scheie syndromes are inherited as autosomal recessive disorders, it remains to be discovered whether or not they are the result of homozygosity for allelic genes with resulting differences in the kind of structural changes in the gene product (α-L-iduronidase) which make this enzyme inactive, but which are not the same in the two diseases. Nevertheless, both syndromes are properly classified as α-L-iduronidase deficiency diseases on the basis of the present experimental evidence.

### 8.4.7 Sanfilippo syndrome: MPS III A and III B

The Sanfilippo syndrome (Mucopolysaccharidosis III)[112] has been mentioned above. Patients with this disease show relatively mild skeletal abnormalities, but they have severe disturbance of central nervous system function. Affected children may be of either sex and there may be more than one affected child in a family. Severe mental deterioration of these patients and urinary excretion of a large amount of heparan sulphate as the only mucopolysaccharide are characteristic of the clinical findings in all patients. As has been stated above, Kresse and Neufeld[127] have found that fibroblasts cultured from the skin of several Sanfilippo patients accumulated an excessive amount of what seemed to be heparan sulphate which had an especially high content of *N*-sulphate (sulphamido) groups. Kresse *et al.* had reported earlier[138] that the Sanfilippo syndrome is biochemically heterogeneous in the sense that cultured fibroblasts from such patients fall into two subgroups, as judged by their having distinguishable cross-correction activity with respect to decreasing the pool size and shortening the half-life of their stored mucopolysaccharide (heparan sulphate). The chemical observations on the nature of this stored, sulphated material were made with cultures derived from patients who had been classified as having the Type A form of the Sanfilippo syndrome[127]. Kresse and Neufeld partially purified a protein fraction from normal human urine which was active as a Sanfilippo A corrective factor and which had demonstrable sulphatase activity at pH 4.5 toward the heparan sulphate fraction which such 'uncorrected' fibroblasts accumulated[127]. These authors suggested that the specificity of the enzyme is probably directed toward the *N*-sulphate groups on the glucosamine residues in heparan sulphate, although this point remains to be shown conclusively. With this reservation, it seems appropriate to consider provisionally that the Sanfilippo A syndrome (Mucopolysaccharidosis III A) is an *N*-sulpho-D-glucosamine sulphatase deficiency disease.

The enzymatic defect which is characteristic of Sanfilippo Type B (Mucopolysaccharidosis III B)[112] patients was found by O'Brien in 1972 to be the lack of an α-*N*-acetylglucosaminidase (EC 3.2.1.50) which has its maximal activity at pH 4.5 when assayed using *p*-nitrophenyl *N*-acetyl-α-D-glucosaminide as the substrate[139]. This finding was made by direct assay for the enzyme in cultures of fibroblasts from Sanfilippo patients who had previously been shown to be Type B on the basis of lack of response of their cultured cells to Type A corrective factor (sulphamido sulphate sulphatase, *vide supra*) in the intracellular heparan sulphate accumulation assay. O'Brien also found that the frozen liver and kidney removed at autopsy from another patient lacked this enzyme activity, as did autopsy liver tissue from still another individual. Confirmation of O'Brien's identification of the enzyme defect in Sanfilippo Type B disease came almost simultaneously when von Figura and Kresse[140] partially purified the Type B corrective factor from normal human urine and showed that the final preparation had α-N-acetylglucosaminidase activity which remained in constant ratio to factor activity through the various steps of the procedure. By using uridine diphosphate *N*-acetylglucosamine as a substrate at pH 4.4, these investigators also showed that skin fibroblasts from Type B patients were deficient specifically in α-*N*-acetylglucosaminidase. In view of these findings, and O'Brien's observation that the activity of this enzyme in skin fibroblasts cultured from both parents of one Type B patient was markedly reduced, suggesting the existence of the heterozygous state, it seems to be established now that the Sanfilippo Type B syndrome is an α-*N*-acetylglucosaminidase deficiency disease.

McKusick has stated that he regards the Type A and Type B Sanfilippo syndromes as being 'completely indistinguishable phenotypically'[112]. Their biochemical differentiation suggests, however, that the chemical structure of the 'heparan sulphate' which is excreted by children who have these two diseases, as well as of that which may be stored in their various organs, may not be the same in the two types. Since a different degradative enzyme is missing in each type, it would be expected that especially the degree of *N*-sulphation and perhaps the average molecular weight of the stored polysaccharides as well might be different. The normal occurrence of α-glycosidic (including α-*N*-acetylglucosaminidic) bonds and of *N*-sulpho-D-glucosamine in heparan sulphate has been discussed above. However, at the present time, knowledge of the detailed molecular architecture of this polysaccharide is so incomplete that meaningful prediction of the changes in structure which may be characteristic of the heparan sulphate which is accumulated in the tissues of and excreted by Types A and B Sanfilippo patients is not possible. The two known enzyme deficiencies characteristic of these types account satisfactorily, however, for the fact that dermatan sulphate, which contains *N*-acetylgalactosamine, is not stored or excreted by these patients.

### 8.4.8 Hunter syndrome: MPS II

The Hunter syndrome (Mucopolysaccharidosis II)[112] is the only known mucopolysaccharidosis which is transmitted as an X-linked recessive trait. In keeping with this fact it was shown by Neufeld and her co-workers, early in their investigation of factors which are active in effecting normal catabolism of mucopolysaccharides in cultured skin fibroblasts, that the 'Hunter factor' is distinct from the 'Hurler factor'[129]. Although patients with the Hunter syndrome may show some of the skeletal changes which are characteristic of Hurler patients, others, such as lumbar gibbus, do not occur, and, most distinctive of all, clinically significant clouding of the cornea does not occur in the Hunter syndrome. Cantz *et al.*[141] have partially purified the Hunter corrective factor from normal human urine and have shown that it acts by restoring the genetically impaired ability of Hunter fibroblasts to degrade dermatan [$^{35}$S]sulphate when it is added exogenously to the medium in which the cells are growing. During the incubation, inorganic [$^{35}$S]sulphate appears in the culture medium. Recently, Bach *et al.* have reported[142] that the purified Hunter factor also acts as a sulphatase when it is incubated *in vitro* at pH 4.0 with the total [$^{35}$S]mucopolysaccharide fraction which can be isolated from Hunter fibroblasts grown in the presence of inorganic [$^{35}$S]sulphate. In this work it was found that only a few per cent of all of the sulphate residues in the polymer appeared to be susceptible to cleavage by the Hunter factor. Identification of the location of these residues was suggested by the finding that the factor could also be shown to act together with added α-L-iduronidase at pH 4.0 on a disaccharide, 4-*O*-α-L-sulpho-iduronosyl-D-sulphoanhydromannose, which had been derived from heparin by deaminative degradation[143]. The product of the combined action of the two enzymes on this substrate was L-iduronic acid. Inasmuch as many of the iduronic acid residues in heparin are known to be *O*-sulphated at C-2[143], the chemically derived disaccharide used in this work, presumably also had a sulphate group on this position of its terminal, non-reducing iduronic acid residue. The experimental evidence presented by Bach *et al.* is not decisive with respect to whether sulphatase action preceded or followed the α-L-iduronidase cleavage of the glycosidic bond in the disaccharide. This point is of considerable importance in extrapolating from this result to any prediction about the structure of the partially degraded heparan sulphate and dermatan sulphate which may be accumulated in Hunter cells which lack the seemingly specific 2-sulpho-L-iduronate sulphatase. It may be pointed out that iduronic acid residues which are *O*-sulphated at C-2 have not been shown conclusively to be present in heparan sulphate. However, sulphated iduronic acid residues are known to occur in dermatan sulphate[144], and both heparan sulphate and dermatan sulphate are known to be stored and excreted by patients with the Hunter syndrome. However, on the whole there is considerable evidence to support the proposal that the Hunter syndrome is a 2-sulpho-L-iduronate sulphatase deficiency disease.

### 8.4.9 Other mucopolysaccharidoses

Several other types of mucopolysaccharidosis have been recognised and described clinically. These include the Morquio syndrome (Mucopolysaccharidosis IV)[112] and the Maroteaux–Lamy syndrome (Mucopolysaccharidosis VI)[112]. Patients with the former disease are strikingly dwarfed and develop a characteristic stature, severe neurological symptoms and corneal clouding. They may or may not excrete excessive amounts of keratan sulphate in their urine, especially at a young age. The detailed chemical structure of this heterogeneous and perhaps chemically most complex of all glycosaminoglycans has not been fully elucidated. It occurs as a proteoglycan and, as such, seems to be linked to the same protein backbone which has some chondroitin sulphate chains also attached. However, keratan sulphate itself is regarded as having no uronic acid in its structure. In addition, it is distinguished from the other mucopolysaccharides by the large amount of D-galactose which it contains, much of which is *O*-sulphated at C-6. The biochemical basis for the Morquio syndrome is unknown. It may be expected, by analogy with the other mucopolysaccharidoses discussed above, that one of the specific enzymes necessary for the normal catabolism of keratan sulphate will eventually be found to be missing in at least that subgroup of Morquio patients in whom keratan sulphaturia is prominent.

Patients with the Maroteaux–Lamy syndrome somewhat resemble those with Hurler's disease in skeletal abnormalities and presence of corneal opacity, but they usually have normal intellect[112]. They excrete typically large amounts of dermatan sulphate in the urine. Barton and Neufeld found that the Hurler corrective factor (α-L-iduronidase) has no effect on the mucopolysaccharide accumulation which occurs in cultured fibroblasts from patients with the Maroteaux–Lamy syndrome[131]. This clearly showed that these two diseases are biochemically distinct. Preliminary observations on the partial purification of a specific Maroteaux–Lamy factor from normal human urine indicate that it is a protein which is also different from the Hunter and Sanfilippo factors discussed above[145]. What step in dermatan sulphate degradation is catalysed by this presumably specific hydrolytic enzyme remains to be discovered.

Recently, another distinct type of mucopolysaccharidosis has been described by Sly *et al.* in a two-year-old male child who was found to have a mild degree of mucopolysacchariduria and profound hepatosplenomegaly[146]. The skeletal involvement in this patient was moderately severe and his relative shortness of stature became more evident as he approached his third year of life. The urine contained approximately equal quantities of what were provisionally identified as chondroitin 4-sulphate and chondroitin 6-sulphate. These are constituents of urine from normal individuals but in markedly lower concentrations. The leukocytes of the patient studied by Sly *et al.* were found to be deficient in β-glucuronidase (EC 3.2.1.31) activity. Hall *et al.*[147] found that cultured skin fibroblasts from this child likewise had no more than 2% of the β-glucuronidase activity of normal cells when tested with *p*-nitrophenyl β-D-glucuronide as the substrate. Hall *et al.* also showed that the patient's fibroblasts accumulated [$^{35}$S]mucopolysaccharide when grown in the presence of [$^{35}$S]sulphate and that addition of β-glucuronidase prepared from bovine liver to the culture medium corrected the biochemical defect in these cultured cells. The discovery of this disease, which McKusick has classified[112] as Mucopolysaccharidosis VII, is so recent that it is not possible to compare its frequency of occurrence with that of the other syndromes which have been discussed above.

## 8.5 CONCLUDING REMARKS

The development in recent years of techniques for the successful growth of human fibroblasts in tissue culture has made possible many of the advances in biochemical knowledge of the inborn errors of carbohydrate metabolism which have been discussed in this chapter. This is particularly true of the mucopolysaccharidoses about which more has been learned in the past five years than was known as the result of a decade of work before that time. There are some physiologically essential enzymatic reactions which fibroblasts are unable to catalyse, e.g. the formation of glucose by glucose 6-phosphatase action. However, the large number and variety of enzymatic conversions for which these cells are competent is impressive. The fact that a single cell line can be studied over a period of months has permitted detailed investigation of metabolic disorders which are so rare that their study would have been impractical if it had been necessary to rely wholly on the availability of other affected tissues from these patients. It is to be expected that further significant advances in the understanding of normal human cell function will result from continued study of normal and abnormal fibroblast cultures.

## References


1. von Gierke, E. (1929). *Beitr. z. path. Anat. u.z. allg. Path.*, **82,** 497
2. Pompe, J. C. (1932). *Nederl. T. Geneesk.*, **76,** 304
3. Cori, G. T. and Cori, C. F. (1952). *J. Biol. Chem.*, **199,** 661
4. Nordlie, R. C. and Arion, W. J. (1964). *J. Biol. Chem.*, **239,** 1680
5. Stetten, M. R. and Taft, H. L. (1964). *J. Biol. Chem.*, **239,** 4041
6. Illingworth, B. and Cori, C. F. (1965). *Biochem. Biophys. Res. Commun.*, **19,** 10
7. Hers, H. G. (1963). *Biochem. J.*, **86,** 11
8. Jeffrey, P. L., Brown, D. H. and Brown, B. I. (1970). *Biochemistry*, **9,** 1403
9. Brown, B. I., Brown, D. H. and Jeffrey, P. L. (1970). *Biochemistry*, **9,** 1423
10. Illingworth, B. and Cori, G. T. (1952). *J. Biol. Chem.*, **199,** 653
11. Illingworth, B., Cori, G. T. and Cori, C. F. (1956). *J. Biol. Chem.*, **218,** 123
12. Brown, B. I. and Brown, D. H. (1966). *Proc. Nat. Acad. Sci. USA*, **56,** 725
13. van Creveld, S. and Huijing, F. (1964). *Metab.*, **13,** 191
14. Ryman, B. E. and Whelan, W. J. (1971). In *Advances in Enzymology*, Vol. 34, 285 (F. F. Nord, editor) (New York: Interscience)
15. Hers, H. G. and van Hoof, F. (1968). In *Carbohydrate Metabolism and Its Disorders*, Vol. 2, 151 (F. Dickens, P. J. Randle and W. J. Whelan, editors) (London: Academic Press)
16. Brown, B. I. and Brown, D. H. (1968). In *Carbohydrate Metabolism and Its Disorders*, Vol. 2, 123 (F. Dickens, P. J. Randle, and W. J. Whelan, editors) (London: Academic Press)
17. Howell, R. R. (1972). In *The Metabolic Basis of Inherited Disease*, 3rd ed., 149 (J. B.

Stanbury, J. B. Wyngaarden and D. S. Fredrickson, editors) (New York: McGraw-Hill)
18. Brown, B. I. and Brown, D. H. (1974). In *Metabolic, Endocrine and Genetic Disorders of Children*, Vol. 2, Chapter 39, 733 (V. C. Kelley, editor) (Hagerstown, Md.: Harper and Row)
19. Fine, R. N., Frasier, S. D. and Donnell, G. N. (1969). *Amer. J. Dis. Child.*, **117**, 169
20. Lockwood, D. H., Merimee, T. J., Edgar, P. J., Greene, M. L., Fujimoto, W. Y., Seegmiller, J. E. and Howell, R. R. (1969). *Diabetes*, **18**, 755
21. Sadeghi-Nyad, A., Loridan, L. and Senior, B. (1970). *J. Pediat.*, **76**, 561
22. Starzl, T. E., Putnam, C. W., Porter, K. A., Halgrimson, C. G., Corman, J., Brown, B. I., Gotlin, R. W., Rodgerson, D. O. and Greene, H. L. (1973). *Ann. Surgery*, **178**, 525
23. Folkman, J., Philippart, A., Tze, W.-J. and Crigler, J. (1972). *Surgery*, **72**, 306
24. Boley, S. J., Cohen, M. I. and Gliedman, M. L. (1970). *Pediatrics*, **46**, 929
25. Kelsch, R. C. and Oliver, W. J. (1969). *Pediat. Res.*, **3**, 160
26. Cuttino, J. T., Summer, G. K., Hill, H. D. and Mitchell, B. J. (1970). *Pediatrics*, **46**, 925
27. Hulsmann, W. C., Eijkenboom, W. H. M., Koster, J. F. and Fernandes, J. (1970). *Clin. Chim. Acta*, **30**, 775
28. Sarcione, E. J., Sokal, J. E. and Lowe, C. U. (1970). *Biochem. Med.*, **3**, 337
29. Hers, H. G. (1964). In *Advances in Metabolic Disorders*, Vol. 1, 1 (R. Levine and R. Luft, editors) (New York: Academic Press)
30. Ockerman, P. A. (1965). *Clin. Chim. Acta*, **12**, 445
31. Garland, R. C. and Cori, C. F. (1972). *Biochemistry*, **11**, 4712
32. Baudhuin, P., Hers, H. G. and Loeb, H. (1964). *Lab. Invest.*, **13**, 1139
33. Angelini, C. and Engel, A. G. (1973). *Arch. Biochem. Biophys.*, **156**, 350
34. Courtecuisse, V., Royer, P., Habib, R., Monnier, C. and Demos, J. (1965). *Arch. Franc. Pediat.*, **22**, 1153
35. Zellweger, H., Illingworth Brown, B., McCormick, W. F. and Tu, J. (1965). *Ann. Paediat.*, **205**, 413
36. Hudgson, P., Gardner-Medwin, D., Worsfold, M., Pennington, R. J. T. and Walton, J. N. (1968). *Brain*, **91**, 435
37. Engel, A. G. (1970). *Brain*, **93**, 599
38. Angelini, C. and Engel, A. G. (1972). *Arch. Neurol.*, **26**, 344
39. Brown, D. H. and Brown, B. I. (1971). In *Biochemistry of the Glycosidic Linkage*, 687 (R. Piras and H. G. Pontis, editors) (New York: Academic Press)
40. Brown, B. I. and Brown, D. H. (1972). *Biochem. Biophys. Res. Commun.*, **46**, 1292
41. Nitowsky, H. M. and Grunfeld, A. (1967). *J. Lab. Clin. Med.*, **69**, 472
42. DiMauro, S., Rowland, L. P. and Mellman, W. J. (1973). *Pediat. Res.*, **7**, 739
43. Angelini, C., Engel, A. G. and Titus, J. L. (1972). *New Engl. J. Med.*, **287**, 948
44. Dreyfus, J. C. and Alexandre, Y. (1972). *Biochem. Biophys. Res. Commun.*, **48**, 914
45. Koster, J. F., Slee, R. G., Hulsmann, W. C. and Niermeijer, M. F. (1972). *Clin. Chim. Acta*, **40**, 294
46. Walker, G. J. and Whelan, W. J. (1960). *Biochem. J.*, **76**, 264
47. Brown, D. H. and Illingworth, B. (1962). *Proc. Nat. Acad. Sci. USA*, **48**, 1783
48. Brown, D. H. and Brown, B. I. (1966). In *Methods in Enzymology*, Vol. 8, 415 (E. F. Neufeld and V. Ginsburg, editors) (New York: Academic Press)
49. Nelson, T. E., Kolb, E. and Larner, J. (1969). *Biochemistry*, **8**, 1419
50. Brown, D. H., Gordon, R. B. and Brown, B. I. (1973). *Ann. N.Y. Acad. Sci.*, **210**, 238
51. Gordon, R. B., Brown, D. H. and Brown, B. I. (1972). *Biochim. Biophys. Acta*, **289**, 97
52. Illingworth, B. and Brown, D. H. (1962). *Proc. Nat. Acad. Sci. USA*, **48**, 1619
53. Taylor, P. M. and Whelan, W. J. (1966). *Arch. Biochem. Biophys.*, **113**, 500
54. Hers, H. G., Verhue, W. and Van Hoof, F. (1967). *Eur. J. Biochem.*, **2**, 257
55. Hers, H. G. (1959). *Rev. Int. Hepatol.*, **9**, 35
56. Larner, J. and Schliselfeld, L. H. (1956). *Biochim. Biophys. Acta*, **20**, 53
57. Brown, D. H. and Illingworth, B. (1963). In *Control of Glycogen Metabolism*, 139 (W. J. Whelan and M. P. Cameron, editors) (London: Churchill)
58. Hers, H. G., Verhue, W. and Mathieu, M. (1963). *Control of Glycogen Metabolism*, 157 (W. J. Whelan and M. P. Cameron, editors) (London: Churchill)
59. Nelson, T. E., Kolb, E. and Larner, J. (1968). *Biochim. Biophys. Acta*, **167**, 212
60. Nelson, T. E. and Larner, J. (1970). *Anal. Biochem.*, **33**, 87

61. van Hoof, F. and Hers, H. G. (1967). *Eur. J. Biochem.*, **2,** 265
62. Justice, P., Ryan, C. and Hsia, D. Y.-Y. (1970). *Biochem. Biophys. Res. Commun.*, **39,** 301
63. Andersen, D. H. (1956). *Lab. Invest.*, **5,** 11
64. Mercier, C. and Whelan, W. J. (1970). *Eur. J. Biochem.*, **16,** 579
65. Mercier, C. and Whelan, W. J. (1973). *Eur. J. Biochem.*, **40,** 221
66. Fernandes, J. and Huijing, F. (1968). *Arch. Dis. Child.*, **43,** 347
67. Gunja-Smith, Z., Marshall, J. J., Smith, E. E. and Whelan, W. J. (1970). *FEBS Lett.*, **12,** 96
68. Howell, R. R., Kaback, M. M. and Brown, B. I. (1971). *J. Pediat.*, **78,** 638
69. Manners, D. J. (1967). In *Control of Glycogen Metabolism*, 83 (W. J. Whelan, editor) (Oslo: Universitetsforlaget; London: Academic Press)
70. Huijing, F., Lee, E. Y. C., Carter, J. H. and Whelan, W. J. (1970). *FEBS Lett.*, **7,** 251
71. Parodi, A. J., Morodi, J., Krisman, C. R. and Leloir, L. F. (1970). *Eur. J. Biochem.*, **16,** 499
72. Sidbury, J. B., Jr., Mason, J., Burns, W. B., Jr. and Ruebner, B. H. (1962). *Bull. J. Hopkins Hosp.*, **111,** 157
73. Holleman, L. W. J., van der Haar, J. A. and de Vaan, G. A. M. (1966). *Lab. Invest.*, **15,** 357
74. Reed, G. B., Jr., Dixon, J. F. P., Neustein, H. B., Donnell, G. N. and Landing, B. H. (1968). *Lab. Invest.*, **19,** 546
75. Schochet, S. S., McCormick, W. F. and Zellweger, H. (1970). *Arch. Path.*, **90,** 354
76. Dahlqvist, A. (1972). *Acta Med. Scand. Suppl.*, **542,** 13
77. Nikkila, E. A. and Huttunen, J. K. (editors) (1972). *Acta Med. Scand. Suppl.*, **542,** 1
78. Froesch, E. R. (1972). In *The Metabolic Basis of Inherited Disease*, 3rd ed., 131 (J. B. Stanbury, J. B. Wyngaarden and D. S. Fredrickson, editors) (New York: McGraw-Hill)
79. Sapag-Hagar, M., Marco, R. and Sols, A. (1969). *FEBS Lett.*, **3,** 68
80. Schapira, F., Schapira, G. and Dreyfus, J. C. (1961). *Enzymol. Biol. Clin.*, **1,** 170
81. Chambers, R. A. and Pratt, R. T. C. (1956). *Lancet*, **2,** 240
82. Froesch, E. R., Prader, A., Labhart, A., Stuber, H. W. and Wolf, H. P. (1957). *Schweiz. Med. Wschr.*, **87,** 1168
83. Perheentupa, J., Raivo, K. O. and Nikkila, E. A. (1972). *Acta Med. Scand. Suppl.*, **542,** 65
84. Froesch, E. R., Wolf, H. P., Baitsch, H., Prader, A. and Labhart, A. (1963). *Amer. J. Med.*, **34,** 151
85. Leuthardt, F., Testa, E. and Wolf, H. P. (1953). *Helv. Chim. Acta*, **36,** 227
86. Hers, H. G. and Kusaka, T. (1953). *Biochim. Biophys. Acta*, **11,** 427
87. Peanasky, R. J. and Lardy, H. A. (1958). *J. Biol. Chem.*, **233,** 365
88. Horecker, B. L., Tsolas, O. and Lai, C. Y. (1972). In *The Enzymes*, Vol. 7, 213 (P. D. Boyer, editor) (New York: Academic Press)
89. Lebherz, H. G. and Rutter, W. J. (1969). *Biochemistry*, **8,** 109
90. Nordmann, Y. and Schapira, F. (1972). *Biochimie*, **54,** 741
91. Schapira, F., Nordmann, Y. and Gregori, C. (1972). *Acta Med. Scand. Suppl.*, **542,** 77
92. Bode, J. C., Zelder, O., Rumpelt, H. J. and Wittkamp, U. (1973). *Eur. J. Clin. Invest.*, **3,** 436
93. Burch, H. B., Lowry, O. H., Meinhardt, L., Max, P., Jr. and Chyu, K. (1970). *J. Biol. Chem.*, **245,** 2092
94. Kaufmann, U. and Froesch, E. R. (1973). *Eur. J. Clin. Invest.*, **3,** 407
95. Traniello, S., Pontremoli, S., Tashima, Y. and Horecker, B. L. (1971). *Arch. Biochem. Biophys.*, **146,** 161
96. Baker, L. and Winegrad, A. I. (1970). *Lancet*, **2,** 13
97. Baerlocher, K., Gitzelmann, R., Nussli, R. and Dumermuth, G. (1971). *Helv. Paediat. Acta*, **26,** 489
98. Hulsmann, W. C. and Fernandes, J. (1971). *Pediat. Res.*, **5,** 633
99. Pagliara, A. S., Karl, I. E., Keating, J. P., Brown, B. I. and Kipnis, D. M. (1972). *J. Clin. Invest.*, **51,** 2115
100. Melancon, S. B., Khachadurian, A. K., Nadler, H. L. and Brown, B. I. (1973). *J. Pediat.*, **82,** 650

101. Traniello, S., Pontremoli, S., Tashima, Y. and Horecker, B. L. (1971). *Arch. Biochem. Biophys.*, **146**, 161
102. Black, W. J., Van Tol, A., Fernando, J. and Horecker, B. L. (1972). *Arch. Biochem., Biophys.*, **151**, 576
103. Hunter, C. (1917). *Proc. Roy. Soc., Med.*, **10**, 104
104. Hurler, G. (1919). *Z. Kinderheilk*, **24**, 220
105. Brante, G. (1952). *Scand. J. Clin. Lab. Invest.*, **4**, 43
106. Stacey, M. and Barker, S. A. (1956). *J. Clin. Path.*, **9**, 314
107. Brown, D. H. (1957). *Proc. Nat. Acad. Sci. USA*, **43**, 783
108. Dorfman, A. and Lorincz, A. E. (1957). *Proc. Nat. Acad. Sci. USA*, **43**, 443
109. Meyer, K., Grumbach, M. M., Linker, A. and Hoffman, P. (1958). *Proc. Soc. Exp. Biol. Med.*, **97**, 275
110. Meyer, K., Hoffman, P., Linker, A., Grumbach, M. M. and Sampson, P. (1959) *Proc. Soc. Exp. Biol. Med.*, **102**, 587
111. Meyer, K. (1961). *Can. Med. Assoc. J.*, **84**, 851
112. McKusick, V. A. (1972). *Heritable Disorders of Connective Tissue*, 4th ed. (Saint Louis: C. V. Mosby)
113. Van Hoof, F. and Hers, H.-G. (1964). *C. R. Acad. Sci. (Paris)*, **259**, 1281
114. Harris, R. C. (1961). *Amer. J. Dis. Child.*, **102**, 741
115. Sanfilippo, S. J., Podosin, R., Langer, L. O., Jr. and Good, R. A. (1963). *J. Pediat.*, **63**, 837
116. Loeb, H., Jonniaux, G., Resibois, A., Cremer, N., Dodion, J., Tondeur, M., Gregoire, P. E., Richard, J. and Cieters, P. (1968). *J. Pediat.*, **73**, 860
117. Wallace, B. J., Kaplan, D., Adachi, M., Schneck, L. and Volk, B. W. (1966). *Arch. Pathol.*, **82**, 462
118. Danes, B. S. and Bearn, A. G. (1965). *Science*, **149**, 987
119. Danes, B. S. and Bearn, A. G. (1966). *J. Exp. Med.*, **123**, 1
120. Matalon, R. and Dorfman, A. (1966). *Proc. Nat. Acad. Sci. USA*, **56**, 1310
121. Knecht, J., Cifonelli, J. A. and Dorfman, A. (1967). *J. Biol. Chem.*, **242**, 4652
122. Dorfman, A. (1964). *Biophys. J.*, **4**, 155
123. Matalon, R. and Dorfman, A. (1968). *Proc. Nat. Acad. Sci. USA*, **60**, 179
124. Kraemer, P. M. (1968). *J. Cell Physiol.*, **71**, 109
125. Kraemer, P. M. (1971). *Biochemistry*, **10**, 1445
126. Bates, C. J. and Levene, C. I. (1971). *Biochim. Biophys. Acta*, **237**, 214
127. Kresse, H. and Neufeld, E. F. (1972). *J. Biol. Chem.*, **247**, 2164
128. Fratantoni, J. C., Hall, C. W. and Neufeld, E. F. (1968). *Proc. Nat. Acad. Sci. USA*, **60**, 699
129. Fratantoni, J. C., Hall, C. W. and Neufeld, E. F. (1968). *Science*, **162**, 570
130. Fratantoni, J. C., Hall, C. W. and Neufeld, E. F. (1969). *Proc. Nat. Acad. Sci. USA*, **64**, 360
131. Barton, R. W. and Neufeld, E. F. (1971). *J. Biol. Chem.*, **246**, 7773
132. Matalon, R., Cifonelli, J. A. and Dorfman, A. (1971). *Biochem. Biophys. Res. Commun*, **42**, 340
133. Weissmann, B. and Santiago, R. (1972). *Biochem. Biophys. Res. Commun.*, **46**, 1430
134. Matalon, R. and Dorfman, A. (1972). *Biochem. Biophys. Res. Commun.*, **47**, 959
135. Bach, G., Friedman, R., Weissmann, B. and Neufeld, E. F. (1972). *Proc. Nat. Acad. Sci. USA*, **69**, 2048
136. Scheie, H. G., Hambrick, G. W., Jr. and Barness, L. A. (1962). *Amer. J. Ophthal.*, **53**, 753
137. Wiesmann, U. and Neufeld, E. F. (1970). *Science*, **169**, 72
138. Kresse, H., Wiesmann, U., Cantz, M., Hall, C. W. and Neufeld, E. F. (1971). *Biochem. Biophys. Res. Commun.*, **42**, 892
139. O'Brien, J. S. (1972). *Proc. Nat. Acad. Sci. USA*, **69**, 1720
140. von Figura, K. and Kresse, H. (1972). *Biochem. Biophys. Res. Commun.*, **48**, 262
141. Cantz, M., Chrambach, A., Bach, G. and Neufeld, E. F. (1972). *J. Biol. Chem.*, **247**, 5456
142. Bach, G., Eisenberg, F., Jr., Cantz, M. and Neufeld, E. F. (1973). *Proc. Nat. Acad. Sci. USA*, **70**, 2134
143. Lindahl, U. and Axelsson, O. (1971). *J. Biol. Chem.*, **246**, 74
144. Malmstrom, A. and Fransson, L.-A. (1971). *Eur. J. Biochem.*, **18**, 431

145. Barton, R. W. and Neufeld, E. F. (1972). *J. Pediat.*, **80,** 114
146. Sly, W. S., Quinton, B. A., McAlister, W. H. and Rimoin, D. L. (1973). *J. Pediat.*, **82,** 249
147. Hall, C. W., Cantz, M. and Neufeld, E. F. (1973). *Arch. Biochem., Biophys.*, **155,** 32

# Index